Invitation to Linear Programming and Game Theory

Written in a conversational tone, this classroom-tested text introduces the fundamentals of linear programming and game theory, showing readers how to apply serious mathematics to practical real-life questions by modeling linear optimization problems and strategic games. The treatment of linear programming includes two distinct graphical methods. The game theory chapters include a novel proof of the minimax theorem for 2×2 zero-sum games. In addition to zero-sum games, the text presents variable-sum games, ordinal games, and n-player games as the natural result of relaxing or modifying the assumptions of zero-sum games. All concepts and techniques are derived from motivating examples, building in complexity, which encourages students to think creatively and leads them to understand how the mathematics is applied. With no prerequisite besides high school algebra, the text will be useful to motivated high school students and undergraduates studying business, economics, mathematics, and the social sciences.

David C. Vella is Professor of Mathematics at Skidmore College. He has taught college mathematics for more than 35 years and has been an organizer of the Hudson River Undergraduate Mathematics Conference (HRUMC) since its inception in 1993. He is a member of the American Mathematical Society (AMS), the Mathematical Association of America (MAA), and a charter member of the New York Alpha Theta chapter of Pi Mu Epsilon. His Erdős number is four.

Invitation to Linear Programming and Game Theory

DAVID C. VELLA
Skidmore College

CAMBRIDGE
UNIVERSITY PRESS

CAMBRIDGE
UNIVERSITY PRESS

University Printing House, Cambridge CB2 8BS, United Kingdom

One Liberty Plaza, 20th Floor, New York, NY 10006, USA

477 Williamstown Road, Port Melbourne, VIC 3207, Australia

314–321, 3rd Floor, Plot 3, Splendor Forum, Jasola District Centre, New Delhi – 110025, India

79 Anson Road, #06–04/06, Singapore 079906

Cambridge University Press is part of the University of Cambridge.

It furthers the University's mission by disseminating knowledge in the pursuit of education, learning, and research at the highest international levels of excellence.

www.cambridge.org
Information on this title: www.cambridge.org/9781108476256
DOI: 10.1017/9781108568166

© David C. Vella 2021

First published 2021

A catalogue record for this publication is available from the British Library.

ISBN 978-1-108-47625-6 Hardback
ISBN 978-1-108-70002-3 Paperback

Contents

Preface   *page* vii
Acknowledgments   xi

1   Preliminaries   1
1.1   Mathematical Models   1
1.2   Systems of Linear Equations   3
1.3   Elimination and Matrices   10
1.4   Vectors, Linear Combinations, and Bases   30
1.5   Three Points of View   44
1.6   Some Nonlinear Models   49

2   Matrix Algebra   63
2.1   Matrices   63
2.2   Operations on Matrices I   65
2.3   Operations on Matrices II: Matrix Inversion   78
2.4   Solving Linear Systems by Matrix Inversion   84
2.5   Some Applications to Cryptography and Economics   90

3   Graphical Linear Programming   102
3.1   Introduction and Graphical Solutions   102
3.2   The Decision Space: $m \times 2$ Problems   115
3.3   Convex Sets and the Corner Point Theorem   121
3.4   Problems with No Solution or Infinitely Many Solutions   124
3.5   The Constraint Space: $2 \times n$ Problems   132

4   Sensitivity Analysis and Duality   148
4.1   Introduction to Sensitivity Analysis   148
4.2   Changes in the Objective Coefficients   150
4.3   Marginal Values Associated with Constraints   156
4.4   Other Changes: Drawbacks of the Graphical Methods   166
4.5   Duality   173

5   The Simplex Algorithm   185
5.1   Standard Form Maximization Problems   186

5.2   Phase II Pivoting   194
5.3   Standard Form Minimization Problems   212
5.4   Solving Linear Programming Problems with a Computer   220

6   Game Theory   242
6.1   Introduction   242
6.2   Dominant Strategies and Nash Equilibrium Points   252
6.3   Mixed-Strategy Constant-Sum Games   269
6.4   Solving Mixed-Strategy Games: The Minimax Theorem in the 2×2 Case   281

7   More Game Theory   301
7.1   Solving Larger Constant-Sum Games   301
7.2   Solving Constant Sum-Games with Linear Programming   313
7.3   Using a Computer to Solve Constant-Sum Games   326
7.4   Variable-Sum Games   335

8   Sensitivity Analysis, Ordinal Games, and *n*-Person Games   352
8.1   Sensitivity Analysis in Game Theory   352
8.2   Ordinal Games   373
8.3   Sequential Games: The Theory of Moves   388
8.4   A Brief Introduction to *n*-Player Games   398
8.5   Legislative Voting and Political Power   409

9   More Linear Programming   423
9.1   Phase I Pivoting   423
9.2   Alternate Approach to Minimization Problems   435
9.3   Sensitivity Analysis and the Simplex Algorithm   441

Appendix: A Rapid Review of Sets and Probability   480
A.1   A Review of Sets   480
A.2   Enumeration   488
A.3   Experiments, Sample Spaces, and Probability Models   497
A.4   Uniform Sample Spaces   502
A.5   Conditional Probability and Independent Events   505
A.6   Bayes's Theorem   512

References   518
Index   519

Preface

This is a text about linear programming and strategic games, mathematical models which are each a type of optimization problem. Linear programming may be viewed as a technique to determine the best way to allocate scarce resources. Game theory is concerned with the best course of action in a situation of strategic conflict. The similarity in these descriptions goes beyond merely that they both give an optimal solution to some problem – it turns out that linear programming can be used to solve certain games. Thus, the two topics are complementary, and it is natural to cover them together. Linear programming and game theory provide two extremely useful lenses through which to regard the world.

The treatment of this material is aimed at first- or second-year undergraduate students, as well as motivated high school students. With these audiences in mind as I wrote, I constrained myself to keep the formal mathematics at a level that would be inviting but not overwhelming. So the tone of the writing is more conversational than some textbooks, and some theorems and propositions are stated without proof, if including the proofs would have been too technical or detract too much from the flow of ideas. Furthermore, the field of mathematical modeling is quite broad, and so another constraint that guided the writing was to make a coherent selection of topics rather than a brief sampling of many different topics.

In the end, a solution that seemed to satisfy these constraints (and one that is hopefully close to optimal) was to focus on deterministic models rather than stochastic ones and to avoid models that required calculus and/or differential equations. Instead, the topics chosen are two examples of *linear models* – those that use the solution of systems of linear equations and matrix algebra. So one perspective for viewing the text is that it presents linear algebra and matrix algebra in action. However, we do not assume that the reader has taken a formal course in linear algebra. Part of the purpose of the text is to introduce topics that might make the further study of linear algebra appealing to the reader.

We can identify at least three distinct groups within the target audience for this text. For the first- or second-year undergraduate student (probably the largest group), a course based around this text would be an excellent *introductory mathematics course for non-science majors*. The applications covered in the text cover a wide range of disciplines such as management and business, economics, political science, and international relations. The mathematics that the students will learn from this text goes beyond minimal quantitative reasoning skills. As such, a course based on this text would probably be better preparation for a future major in management and business, economics, etc., than many "mathematics for liberal arts" courses.

I do not mean to imply that the text would not be useful or appreciated by science majors, but most of these students begin their college-level training in mathematics with the standard calculus track or even with linear algebra. Yet, this text could also be of interest to the potential mathematics major. If the reader enjoys the topics covered here, perhaps the text could serve as a broader invitation than just one to linear programming and game theory; perhaps it could serve as an *invitation to the mathematics major* and provide an entrance into the study of mathematics that is an alternative to the previously mentioned standard calculus track taken by most science majors and mathematics majors.

A second group of students for whom this text was written is the mathematics majors who have already taken a course in linear algebra. For such students, reading this text (either independently or as part of a course) may help fill in a few holes in their training or answer some lingering questions they may have about linear algebra by providing them the opportunity to see linear algebra used in the context of other disciplines.

To cite a specific example of this, linear algebra students learn that vector spaces have many different bases, and they learn how to convert from one basis to another, but it may be lost on those students *why* one would want to convert to a different basis. One answer, perhaps, is given by the study of eigenvectors and diagonalization, for those students who learn those topics. Another answer, which is somewhat less technical than the study of diagonalization, occurs in Chapter 3 of this text, when the reader encounters the solution to linear programming problems via the technique of *graphing in the constraint space*. Each "basic solution" of the problem is obtained precisely by choosing a suitable basis for the constraint space. Readers will find this idea reinforced in Chapter 5, when they realize that every time a pivot is performed in the simplex algorithm, what is really happening is a change from one basis to another. While this text does not cover enough theory to be appropriate as a main text in such a course, it is suitable for consideration as a *supplementary text for linear algebra courses*.

Finally, a third target audience is motivated high school students who are anxious to enrich their education by learning some mathematics that does not appear in the usual K–12 curriculum. The technical requirements for reading this book are minimal, and I believe the style is friendly to the reader who wants to study independently. All that is really assumed in terms of prerequisites is basic high school algebra: how to plot lines and curves and some experience in solving linear systems. It would be helpful in Chapters 8 and Appendix for the reader to also know some very basic set theory – mainly the language of sets and how to form unions, intersections, and complements of subsets. Overall, the prerequisites are very modest.

The material in this text has been classroom tested – I have used the material on matrix algebra and game theory for more than 10 years as the topic for a first-year seminar course in game theory and voting theory at Skidmore College. Prior to that, for several years, I used the linear programming material for an elementary modeling course in mathematics at Skidmore College (part of a series of courses entitled Mathematics in Context), which was designed for liberal arts students with an interest in business, economics, or social sciences. Additionally, I have used nearly all of the material for 18 years as the basis for a course in probability and game theory for the Johns Hopkins University Center for Talented Youth (CTY) program, an intensive summer program for gifted high school students aged 12–16. The material has been successful with both high school and college audiences.

While the focus of the text is on models that are deterministic and linear, nothing in mathematics is truly self-contained. There are so many interconnections between topics that I could not completely exclude stochastic and nonlinear concepts. Stochastic ideas appear in game theory, when probability is used to solve mixed strategy games in Chapters 6–8. However, we keep the discussion of probability to a minimum. In a few pages in Chapter 6, we outline all the probability theory needed for the rest of the book. This approach seemed better than interrupting the flow of ideas for a formal treatment of probability theory. In practice, I have found that most students in my classes at both levels – the college student and the motivated high school student – have either had some exposure to probability in high school or else learn the needed material very quickly; they often have a good intuitive understanding of the topic. Nevertheless, I have placed a more systematic treatment of probability in the Appendix – A Rapid Review of Sets and Probability – for those readers seeking a more complete treatment.

Furthermore, some nonlinear ideas make two brief appearances. The first is in Chapter 1, where there are some exercises in quadratic curves (putting them into standard form) and also some exercises in quadric surfaces (particularly "saddle surfaces" – again putting them into a standard form). These appearances are meant to lend support to the other, linear models that are the main focus of the text. The quadratic curves arise in simple optimization problems that can be compared and contrasted with the later optimization problems that arise in linear programming in Chapter 3. The quadric surfaces are used in Section 6.4, where a novel approach to the minimax theorem is presented for 2×2 constant-sum games. (This novel proof does not extend to games of larger sizes than 2×2, and for the general case, we must rely on linear programming.)

Finally, Sections 8.4 and 8.5 also contain material that lies outside of the linear model focus. The mathematics needed for analyzing n-player games (where $n > 2$) is different from the mathematics of two-player game theory, where matrix algebra is the staple. When $n > 2$, instead of matrix algebra and linear programming, the mathematics used is closer to set theory than it is to linear algebra, another topic that many students have seen or can learn quickly. Even though the topics in this chapter diverge from the linear algebra focus of the rest of the text, including them makes for a somewhat more complete treatment of game theory. Although we only scratch the surface of the theory of n-player games, I felt that I would be remiss to try to introduce the reader to game theory and to pass over n-player games in complete silence.

In short, even with these brief transgressions away from linear models, the focus of our text is narrower than books that attempt to survey many different types of mathematics, such as a typical text in math modeling or finite mathematics. Unlike many of these texts, we do not include such standard topics as financial mathematics, statistics, dynamical systems, graph theory and networks, or logic.

What is gained from omitting these topics is the room to go into greater depth. The treatment of linear programming, for example, includes two different graphical approaches. One is the standard method of graphing in the decision space (often referred to as simply "the graphical method" in most texts), which can handle problems with just two decision variables but an arbitrary number of constraints (problems of size $m \times 2$). But, as mentioned before, we also present a method of graphing in the "constraint space," which can handle problems with just two constraints but an arbitrary number of decision variables (problems of size $2 \times n$.) This is a unique feature of this text. I have never encountered a text at this level that covered this method in detail; indeed, I have only seen it mentioned in one other text (see Loomba, 1976), where it was only briefly covered, and for 2×2 problems only.

Similarly, our treatment of game theory is more complete than what is found in finite mathematics texts, which typically restrict their attention to two-player zero-sum games. We discuss variable-sum games, ordinal games, and also n-player games (Chapter 8); and also go into more detail for zero-sum games than the typical finite mathematics text does, by providing a proof of the minimax theorem.

On the other hand, the treatment of linear programming or game theory here could not compete with more advanced texts that only focus on just one of these topics and/or are targeted at the mathematics major. Linear Programming texts such as Karloff (1991) or Nash and Sofer (1996) are bound to be more complete than the treatment here. Indeed, these texts cover topics such as numerical stability of the simplex algorithm, complexity theory and more sophisticated methods for solving these problems such as the "ellipsoid" algorithm and Karmarkar's algorithm, which we do not even mention. Likewise, game theory texts such as Luce and Raiffa (1957) or the first great treatise on game theory (von Neumann and Morgenstern, 1944), as well as more modern texts such as Taylor and Zwicker (1999), are more comprehensive treatments of game theory than what appears in this text and are aimed at a more mathematically sophisticated reader. Also, there are texts at a higher level that cover both topics, such as Brickman (1989).

This text does not try to compete with these more specialized texts, which address a more mathematically mature reader. For a book at this level, I am comfortable with sacrificing the completeness of those texts in order to keep the readability at a level commensurate with first-year college students, plus or minus a year or two. Additional resources are available online at cambridge.org/9781108476256.

In terms of other texts that are more akin to the present text in style and level, the comparison is closer with the texts by Taylor and Pacelli (2008) or Straffin (1993) for game theory (neither of which, however, covers matrix algebra or linear programming and their applications to games); and to Calvert and Voxman (1989) for linear programming (which omits other linear models such as game theory). Indeed, these three texts were used in early versions of my courses and influenced my understanding of linear programming and game theory to a great extent.

In addition to Calvert and Voxman (1989), I also mention the text of Loomba (1976), which influenced my early thinking on linear programming more than 30 years ago. I am indebted to Alan Taylor and William Zwicker for a number of conversations about game theory and voting theory, for several inspiring lectures I have had the pleasure to hear by each of them, and for recommendations for further reading, especially for pointing me to the work of political scientist Steven Brams in the publications Brams (1994, 1985a, and 1985b). In addition to mathematical influences, music of various genres has always been important to me, and like-minded readers may enjoy the fact that this book is riddled with references in homage to songs and artists.

Acknowledgments

I want to thank Skidmore College for its support in this writing project and both Skidmore College and the Johns Hopkins University CTY program for the opportunity to teach various incarnations of this course over the last 22 years. I also thank the students who have taken these classes and who have helped me hone both the lecture material and the exercises. Finally, I am grateful to Joanne for her patience and support while I was engaged in the writing process, and to our daughters, Angelina and Chiara, for their support as well. In particular, the little roadside lemonade stand they ran for a few days one summer when they were young children provided me with more food for thought than they could have possibly imagined.

1 Preliminaries

1.1 Mathematical Models

As a young child, I once strayed into an older cousin's workroom in the basement of his home, where he was assembling model airplanes. I was fascinated by the detail of the models, the sheets of decals, the plastic pieces of the fuselage and wings attached to plastic rods when the box was first opened, and what appeared to me to be very complex directions for assembly. However, I was disappointed when I discovered that the little models did not fly. I did not quite understand that the models were not supposed to be an *exact* duplicate of a real airplane. The model was intended to capture some aspects of a real plane – the design, the shape and proportions, the working wheels, and perhaps movable propellers or movable ailerons and rudder. Other aspects of a real plane are deliberately ignored. No jet engines, no rows of over-cramped seats, no oxygen masks dropping down from the ceiling, and no flight attendants serving beverages and snacks.

Mathematical models work the same way. If you have a physical problem to solve that is quite complicated, you might build a mathematical model of the problem. Models are designed to capture key aspects of the problem you solve but deliberately ignore other aspects of the problem. They are therefore simpler than the real-life problem with which you started. Otherwise, why build a model at all, unless it is easier to solve than the original problem?

By a mathematical model, we mean any construction using mathematics. For example, an equation or a system of equations is an example of a model. So is a function. A network of points (called vertices) and lines (called edges) connecting them is a type of mathematical model called a *graph*. A model might be a mathematical system defined by a collection of *axioms*, such as a finite projective geometry, a Euclidean plane, or an algebraic system such as a group or a ring. If you have not heard of projective geometries, groups, or rings before, don't worry – we will not be using those more advanced models in this book. Our models will consist of systems of linear equations, systems of linear inequalities, and some easy generalizations such as matrices, linear programming models, and game theory models. Presumably, linear programming and game theory will be new to the reader, although you are probably already familiar with systems of equations.

Before we start working with any specific models, we outline and summarize the modeling process in general. The schematic diagram in Figure 1.1 is helpful.

We begin with a physical problem (upper-left corner of the diagram), and we seek a physical solution. It may be difficult to get there directly (the dashed line in the diagram), so instead we take the more circuitous route following the solid arrows. The first step is to build a mathematical model. For a simple example, suppose our model is a system of equations. In arriving at the system, we had to decide which aspects of reality to ignore. That is, we try to retain just the relevant bits of information to formulate the model and strip away the inessential extraneous information. This

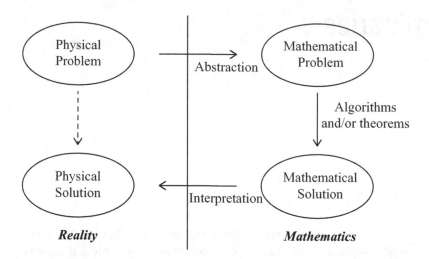

Figure 1.1 The modeling process.

step of the modeling process is called *abstraction* – one abstracts or takes away only the parts that are needed.

Abstraction serves several purposes. First, as noted before, it could make the mathematical problem simpler than the physical one (for example, if you are trying to understand the trajectory of a baseball, you might decide to ignore air resistance and just concentrate on understanding how gravity affects the motion). The abstraction process helps by focusing your attention on just what is essential in the problem. Finally, by throwing away irrelevant information, your model becomes more *general* than the physical problem that inspired it. Solving a system of equations requires the same skills whether the equations came from a study of traffic patterns in a city neighborhood, a problem in deciding in investment options, or a problem in designing a diet.

The second step is to solve the mathematical problem and arrive at a mathematical solution. This depends on what the model is, but it often is the case that there is a standard way to do this. In the preceding example where the model is a system of equations, there are standard methods for solving these systems (some of which we review in this chapter). On the other hand, if your problem translates into a question about a graph or a finite geometry or a group, there may be a theorem or algorithm within these fields of mathematics that already exists and is ready-made to solve your problem.

For example, suppose your company is planning to hire a large number of new workers to fill many different positions, where different skills are needed for various positions. You receive many applications from people with different sets of skills. How do you match the applicants with the jobs efficiently? It turns out you can model this question by using a graph, and there are theorems in graph theory that tell you when a solution is possible and algorithms that address exactly how to find these types of matchings. (See chapter 4 of Clark and Holton, 1991, for example.) Even if your problem translates to a question that has not yet been answered, for example, to a new open question in graph theory, you can pass the question on to graph theorists who can then try to answer it without worrying about what physical problem inspired it.

The third step is *interpretation*, which basically means looking at the mathematical solution and seeing if it makes sense in the context of your physical problem. For example, if your model is a system of equations, the mathematical solution may have solutions that are real or complex numbers, but in your original problem, perhaps negative or fractional or imaginary solutions may not make sense.

After having completed this cycle, you now have the physical solution you sought, with the caveat that your solution depends on the assumptions you built into the model and on what aspects of reality you chose to ignore in the abstraction step. Ideally, now you can revisit those assumptions, adjust your model, and repeat the cycle, aiming for a solution that is better or more realistic. (Indeed, there are model airplanes that do fly!) For example, for the baseball trajectory problem, after solving the problem ignoring air resistance, you may then go back and try to adjust the model by including air resistance.

Another way to study the model further is *sensitivity analysis*. Suppose you solved a problem telling you how to make an investment portfolio – how much to invest in various stocks, how much to invest in bonds, etc. You may wonder what would happen if you made a small change – perhaps to invest just a little more in a particular stock. How would that affect your expected earnings? How would the change affect your risk? You may be able to answer such questions without solving the entire problem over. With sensitivity analysis, you study how sensitive your solution is to small changes in the original problem or to the assumptions you built into the model.

In this chapter, we review systems of linear equations, and in the next chapter, we study matrix algebra, which is closely related to systems of linear equations. These topics are interesting and useful mathematical models but are only a "warm-up" to the two main topics in this book: linear programming and game theory. Rather than introducing a plethora of other important and useful models, this text stays with linear programming and game theory to go into enough depth to illustrate the full modeling process, including sensitivity analysis and varying the assumptions that go into building these models.

1.2 Systems of Linear Equations

Consider the following scenario:

Example 1.1 Peter, Paul, and Mary are comparing salaries. Peter works on a lemon tree orchard and hasn't gotten a raise in eight years. Paul works in a tool factory producing hammers and received a raise last year for the first time in three years. Mary is a mechanic for jet planes and gets a raise every year. This year, Paul made $22,000 more than Peter, and Mary made $16,000 more than Paul. If the average of their three salaries is $72,000 for this year, determine each person's individual salary for the year.

A problem like this is familiar to most students from high school algebra. They often are referred to as story problems or word problems. When first learning about problems like this, students know the goal is to set up an equation or a system of equations and often lament that setting up the equations is more difficult than solving them. They are correct, because setting up the equations is the first step, abstraction, in the modeling process. Until one has had some practice, it can be tricky to sift through all the information and decide what to keep and what to discard when setting up the model. The fact that the solving part (step 2 of the modeling process) is easier than step 1 is exactly the point of modeling. The model is supposed to clarify and simplify the original problem.

Recall that you should begin by naming variables for the important unknowns, so let x be Peter's salary, y be Paul's salary, and z be Mary's salary. It should be clear that the type of jobs they have and how long it has been since they have gotten a raise are not important, so we ignore that information in the construction of our model. What is important is the relationships between the salaries, which are given in the last two sentences of the problem. The reader should easily be able to obtain the following equations:

$$x + 22{,}000 = y$$
$$y + 16{,}000 = z$$
$$\frac{x + y + z}{3} = 72{,}000$$

This is a system of three linear equations in three unknowns. They are linear because the unknowns only appear to the first power. Not all story problems will reduce to linear equations, but in this book, linear systems will play a large role. There is a standard form to write such systems – we usually put all the variables on the left, each with its own numerical coefficient, and constants on the right. Thus, the system can be written as

$$-x + y = 22{,}000$$
$$-y + z = 16{,}000$$
$$\frac{1}{3}x + \frac{1}{3}y + \frac{1}{3}z = 72{,}000.$$

Of course, there are other equivalent ways to write the same system. You could replace any equation with an equivalent equation such as multiplying the top equation through by a factor of -1 to obtain $x - y = -22{,}000$, or multiplying through the third equation by a factor of 3 to clear the denominators. Or you could write the equations in a different order. All these variations produce an *equivalent system*, which means the solution set is the same.

At this point, the abstraction process is done, and now we are ready for step 2 – find the mathematical solution to this system. The reader is probably familiar with more than one way to solve this system. Common methods of solution include *substitution* and *elimination*. In the method of substitution, one solves one of the equations for one variable in terms of the other variables and then substitutes this expression for that variable in the other equations. This is especially easy in the preceding example because the first two equations each involve only two of the variables instead of all three. Thus, the first equation yields

$$y = x + 22{,}000,$$

and the second equation yields

$$z = y + 16{,}000.$$

Combining these by substituting for y in this last equation yields

$$z = (x + 22{,}000) + 16{,}000 = x + 38{,}000.$$

Now we have expressions for each of y and z in terms of x, which we can substitute into the third equation and simplify:

$$\frac{x + (x + 22{,}000) + (x + 38{,}000)}{3} = 72{,}000$$

$$\frac{3x + 60{,}000}{3} = 72{,}000$$

$$x + 20{,}000 = 72{,}000$$

$$x = 52{,}000.$$

Thus, we have the value of x, and combined with the preceding expressions for y and z in terms of x, we obtain

$$x = 52{,}000$$

$$y = x + 22{,}000 = 74{,}000$$

$$z = x + 38{,}000 = 90{,}000.$$

Finally, we proceed to step 3 of the modeling process – the interpretation of our solution. Since the mathematical solution consists of positive real numbers, which are valid salaries, and because the solution is unique, we have solved the original problem:

Peter made $52,000 this year

Paul made $74,000 this year, and

Mary made $90,000 this year.

You can verify that the solution is correct by checking all the original conditions set forth in the problem. Furthermore, since the only information we neglected in the model was irrelevant information such as what their actual jobs were, this solution is an accurate reflection of reality. There are no improvements to the model that we can make.

Another method that the reader may have seen seen for step 2, obtaining the mathematical solution, is elimination. Elimination is based on manipulating the equations to force some of the coefficients to be 0, which means that corresponding variable has been "eliminated" from the equation. The manipulations must produce equivalent systems, of course. For example, if we multiply the third equation by 3, we obtain

$$x + y + z = 216{,}000.$$

If we add the first equation $-x + y = 22{,}000$ to this, we obtain

$$2y + z = 238{,}000.$$

The x coefficient has been forced to be 0, and so x has been eliminated. What is left is an equation with just the two unknowns, y and z. But the second equation, $-y + z = 16{,}000$, also just involves those two variables. Subtracting the second equation from the previous equation will eliminate the z:

$$3y + 0z = 222{,}000$$
$$3y = 222{,}000$$
$$y = 74{,}000.$$

We combine this information with the first two given equations again to obtain the complete solution:

$$y = x + 22{,}000$$
$$74{,}000 = x + 22{,}000$$
$$x = 52{,}000$$

and

$$z = y + 16{,}000$$
$$z = 74{,}000 + 16{,}000 = 90{,}000,$$

and we are done.

In case the reader is unsure of what sort of manipulations are allowed while performing elimination, we will clarify that in the next section. For the rest of this section, we practice more examples of the abstraction part of the modeling process.

Example 1.2 Walter, a chemist at Royal Charlemagne Labs has two containers of nitric acid. One is a 40% concentration mix and one is a 10% concentration mix. How many liters of each should Walter mix together if he needs 6 liters of a 30% concentration for an experiment?

In this example, the only irrelevant information is the name of the laboratory – the rest of the given information is important. So it should be a straightforward process to set up the equations in the model. Begin by declaring the unknown variables, so let x be the number of liters of the 40% solution and y be the number of liters of the 10% solution. Since we want 6 liters of the mixture altogether, one equation is obvious:

$$x + y = 6.$$

But how do we extract a second equation? The key to this is to notice that this problem is about concentrations, which are ratios of volumes. Specifically, the concentration is the volume of pure acid divided by the total volume. The previous equation keeps track of the total volume – what we need is a separate equation which keeps track of the volume of (pure) acid involved. If we have x liters of a 40% solution, how much pure acid is that? The answer is 40% of x, or $0.4x$ liters. Similarly, y liters of a 10% solution contains $0.1y$ liters of pure acid. When we mix them together, we get

$$0.4x + 0.1y$$

liters of pure acid. Since we want the resulting mixture to be 30% acid, the total amount of pure acid should be $0.3(6) = 1.8$ liters. Thus, our second equation is

$$0.4x + 0.1y = 1.8.$$

In summary, the system is

$$x + y = 6 \text{ (total liters of mixture)}$$
$$0.4x + 0.1y = 1.8 \text{ (total liters of pure acid)}$$

If desired, you can multiply through the second equation by a factor of 10 to get rid of the decimal terms and use the equation $4x + y = 18$ instead. Since this system is just two equations and two unknowns, it is easy to solve by either substitution or elimination. We leave the solution as an exercise.

Example 1.3 Paradise Vacations is having a "buy one, get one free" sale on their three-day ocean cruises. Eddie Moolah wants to buy two tickets from Paradise, but the only money he can spare is a large jar of coins in his closet. The jar contains dollar coins, quarters, and dimes only. There are 3,260 coins altogether. The number of quarters plus twice the number of dollar coins is equal to the number of dimes. The number of quarters plus the number of dimes is 100 less than seven times the number of dollar coins. If a ticket costs $850, does Eddie have enough?

What Eddie's money is being used for is irrelevant. The only question is whether or not he has at least $850 in his jar. We need to know the number of each coin. So let x be the number of dollar coins, let y be the number of quarters, and z be the number of dimes. Converting the stated conditions into equations and putting them into standard form the reader should obtain

$$x + y + z = 3,260$$
$$2x + y - z = 0$$
$$7x - y - z = 100.$$

We leave the solution as an exercise. What about step 3, the interpretation step? After all, it is not just the values of x, y, and z we needed to see whether or not Eddie goes on vacation. Once the system is solved, you can add up the values of the coins to see if it makes the $850 threshold or not.

Example 1.4 We can use a linear system of equations to model traffic flow. Consider the map in Figure 1.2 of one-way streets in a city neighborhood.

The arrows indicate the direction of traffic flow on each street. The number labels indicate the number of vehicles per hour that traverse the streets entering or leaving the area. The unknowns v, w, x, y, and z are the number of vehicles per hour traversing the streets within the indicated grid. If we stipulate that at each intersection A, B, C, and D, the number vehicles entering that intersection per hour equals the number of vehicles leaving the intersection per hour, we can derive an equation for each intersection.

The reader should be able to derive the following four equations:

$$v + w = 50 \text{ (intersection A)}$$
$$v + x + 80 = y + 170 \text{ (intersection B)}$$
$$w + 150 = x + z \text{ (intersection C)}$$
$$y + z = 110 \text{ (intersection D)}.$$

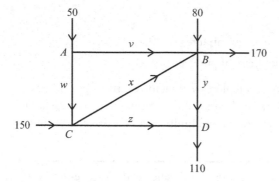

Figure 1.2 A traffic network.

In standard form, these equations are

$$v + w = 50$$
$$v + x - y = 90$$
$$-w + x + z = 150$$
$$y + z = 110.$$

This system is more difficult to solve because there are five variables and four equations. The reader may enjoy attempting to find a solution now, but in the next section, we will give a general algorithm that can solve any system. For now, let us conclude by saying that with more variables than equations, you should expect that the solution is not unique. Indeed, if you look at the map and think about it, you can probably see various different values for the variables that will work. For example, try $v = 20$, $w = 30$, $x = 150$, $y = 80$, and $z = 30$. Try to write down a few more that work. Our goal in the next section will be to find all of the solutions of a system.

Exercises

1. Solve the system of equations we derived for the Royal Charlemagne Labs problem (Example 1.2).
2. Solve the system of equations we derived for the Paradise Vacations problem (Example 1.3), and interpret the solution in terms of Eddie's goals.
3. Johnny Vishnu leads a yoga class at Industrial Nirvana Studio and Gym. Five times the number of men in the class is four more than three times the number of women. If there are 36 students altogether, how many more women than men are there?

   a. Set up a system of linear equations for this problem.
   b. Solve the system by substitution, and answer the question.
   c. Solve the system by elimination.

4. Mr. Gilmour invested a total of $10,000 in two mutual funds. The Windpillow fund earns 6% interest, and the Crazy Diamond fund earns 8% interest. If he earned $730 interest in one year, how much did he invest in each fund?

a.   Set up a system of linear equations for this problem.

b.   Solve the system by substitution, and answer the question.

c.   Solve the system by elimination.

5.   The Astral Moon Dance Troupe is taking its show on the road. The troupe has 26 members and 1300 cubic feet of props, supplies, and luggage to transport. The troupe calls Morrison Vans and Trucks, a company that rents panel trucks and 10-foot moving vans. Each panel truck can seat four people and carry 80 cubic feet of luggage. Each moving van can seat two people and carry 300 cubic feet of luggage. How many of each type of truck should the dancers rent?

a.   Set up a system of linear equations for this problem.

b.   Solve the system by substitution, and answer the question.

c.   Solve the system by elimination.

6.   The painter Peter Black has made 14 paintings for an exhibit in the Gabriel Gallery in Solsbury. The paintings are all exactly the same size and have the same size border around the paintings. If the length of his paintings (without the border) is 8 inches less than twice the width, and if the 2-inch border adds exactly 224 square inches to the area of the painting, what are the dimensions of the paintings without the border? Set up the system, and solve by whatever method you prefer.

7.   Leo, a mechanic at Freeway Star Automotive, needs to winterize a fleet of race cars. He has a vat containing 80% concentration antifreeze and a vat containing 30% concentration antifreeze. How many liters of each should he mix together in order to obtain 50 liters of antifreeze at 60% concentration? Set up the system, and solve by whatever method you prefer.

8.   Leo's friend Ritchie has a vehicle with an 8-liter radiator, which is currently filled with antifreeze at 50% concentration. How many liters should Leo drain from the radiator and replace with 80% concentration in order to leave the radiator full with a 60% concentration? Set up the system and solve by whatever method you prefer.

9.   Mr. Nelson travels a lot. In one pocket, he has US dollar coins; in another, he has euro coins; and in another, he has Hong Kong dollar coins. He has five more Hong Kong dollars than euros, and two-thirds as many Hong Kong dollars as US dollars. If he has 58 coins altogether, how many of each type of coin does he have? Set up the system of equations and solve by whatever method you prefer.

10.   Squeeze from a Stranger Coffee Shop sells hot coffee for $3.00 per cup, iced coffee for $4.00 per cup, and lattes for $5.00 per cup. One day they sold 495 cups altogether and made $2,035 in revenue. If the coffee shop sold 25 more iced coffees than lattes, how many cups of each type of beverage did it sell? Set up the system, and solve by whatever method you prefer.

11.   The music department at Tippett College sponsored a lecture/concert featuring free jazz legend Julie Keith. Tickets for students were $15, tickets for faculty were $25, and tickets for the general public were $40. A total of 970 tickets were sold, for a total of $31,100. If the total number of general public tickets sold was 30 more than twice the number of student tickets plus the number of faculty tickets, how many of each type of ticket were sold? Set up the system of equations, and solve by whatever method you prefer.

12.   Consider the network of one-way streets shown in Figure 1.3.

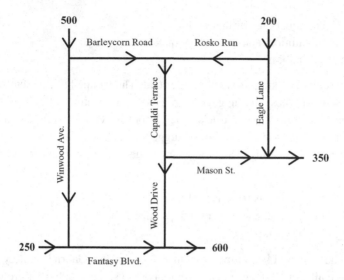

Figure 1.3 Another traffic network.

The Berkshire city planners want to study the traffic patterns through this network. The numbers are the number of vehicles per day that enter or leave the network. Set up the system of linear equations for this network. Clearly indicate to which street each variable is associated. Do not solve the system of equations!

1.3   Elimination and Matrices

In this section, we clarify and streamline the process of solving a linear system by elimination. First, we establish something about the nature of the solutions, which we do by a graphical approach. So consider the following system of equations in two variables:

Example 1.5

$$6x + 3y = 30$$
$$2x + 10y = 28.$$

The reader will recognize that each equation has a graph that is a straight line. Furthermore, by definition, the graph is the set of all (x, y) that make the equation true, which is to say that the graph is a picture of the solution set of the equation. The solution to the system is the set of all common solutions to each equation, which means the solution set for the system is the set of points that are on both graphs – that is, the intersection of the two lines. It is easy to see by solving this system that the solution is $(x, y) = (4, 2)$. Figure 1.4 provides a graphical depiction of the situation.

Solving graphically is not practical in general, especially if there are more than two variables, since in that case, the graph has more dimensions than two. However, the two-dimensional case can give us some insight into what happens in general, an approach we exploit a number of times

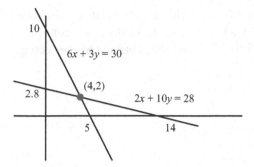

Figure 1.4   The point of intersection of the two lines is the solution.

in this book. In this case, the key observation is that the solution is unique, because the two lines intersect at only one point.

What other possibilities are there? Well, if the lines are parallel, there is no intersection point. Thus, there is no common solution to the two equations. Finally, what if there are (at least) two distinct solution points P and Q? In that case, since two distinct points determine a unique line, it follows that the line containing PQ must be the same as each of the lines in the system of equations, so that the two lines are coincident. In that case, every point on the line PQ is a common solution, so there are infinitely many solutions. The reader can check the following systems for themselves.

$$6x + 3y = 30$$
$$2x + y = 20$$

The reader should plot the lines and see in this case that the lines are parallel, so there is no common solution.

$$6x + 3y = 30$$
$$-4x - 2y = -20$$

In this case, the reader should see that the two lines are coincident, so there are infinitely many solutions.

These exhaust the possibilities, because two lines in Euclidean space can intersect at a single point (unique solution) or not at all if they are parallel (no solution) or at every point of coincident lines (infinitely many solutions).

Now, for larger systems, we want to draw the same conclusions. We are not going to prove anything mathematically here, but rather give a heuristic argument. A single equation in three variables has a graph that is a flat plane in three-dimensional Euclidean space. A plane is determined by three non-collinear points, such as the intercepts with the three coordinates axes of Euclidean three-space. For example, consider the equation:

$$x + 2y + 5z = 20.$$

The three intercepts are $(20, 0, 0)$ on the x-axis, $(0, 10, 0)$ on the y-axis and $(0, 0, 4)$ on the z-axis, which is drawn perpendicular to the xy-plane. The convention is that the positive direction on the x-axis comes forward, the positive direction on the y-axis is to the right, and the positive direction on the z-axis is upwards. (Such a convention yields what is called a right-handed coordinate system.) Figure 1.5 shows the coordinate system and the three intercepts of the plane. Since the plane passes through all three points, it passes through the drawn triangle obtained by connecting the intercepts with straight lines, as shown in Figure 1.5.

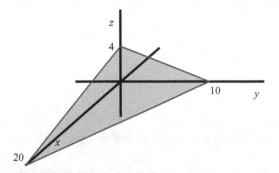

Figure 1.5 A plane in three-dimensional space.

The reader will not have to master graphing in three-dimensional space for this book. However, once you know that a linear equation in three variables represents a plane, you can visualize what happens in a system. If each equation has a solution set that is a plane, the entire system has a solution set that is the common intersection of all the planes in the system. Again, if any two of the planes are parallel, there is no common intersection so no solution to the system. If two of the planes are neither parallel nor coincident, then their intersection will be a line. This line may be the entire solution set (for example, if this line lies on all the other planes in the system). So it is possible that there are infinitely many solutions. Another way to obtain infinitely many solutions is if all the planes are coincident, so the solution set is an entire plane. If there are three equations, suppose two of them intersect in a line, and that line is not contained in the third plane. Then either that line is parallel to the third plane (so there are no solutions to the system) or the line crosses this plane at a single point (so there is a unique solution). The reader can imagine systems with more than three equations, but again, the common intersection is either empty, a single point, or infinite (a line or a plane).

We won't try to illustrate what happens in higher dimensions, when there are even more variables, but in all cases, the solution set to a system of linear equations has just one of three possibilities:

- The solution is unique (called a *non-degenerate* system).
- There is no solution (called an *inconsistent* system).
- There are infinitely many solutions (called a *redundant* system).

In the latter two cases, we also say the system is *degenerate*.

We now focus on the process of elimination. First, let's recall and formalize a definition from the last section.

DEFINITION 1.6   *Two systems of linear equations are called* equivalent *if they have the same solution set.*

We mentioned earlier that elimination proceeds by performing certain manipulations on the system designed to eliminate variables from some of the equations. But we emphasize that these manipulations also must replace one system with an equivalent system in order to be helpful – otherwise, the solutions set could change. So the first thing to establish is a list of what manipulations will replace a system with an equivalent one. The answer is the following meager list of three so-called *elementary operations*:

1. Interchange the position of any two equations in the list of equations.
2. Replace any equation by a nonzero multiple of itself.
3. Replace any equation with itself plus any multiple of one of the other equations in the system.

The first thing one must realize is that elementary operations are something that is done to the entire system, not just to one equation. In the salary example from the last section, we briefly illustrated elimination, but when we manipulated one of the equations to obtain the equation $2y + z = 238,000$, we were sloppy and did not write down the entire system. In future examples, *we will explicitly write down the entire system for each step.*

Secondly, why do the elementary operations produce an equivalent system? Well, the first operation just results in the exact same equations in the system, but just written in a different order. Since the order does not matter when finding the intersection of the solution sets of the individual equations, this means the solution set for the system is unchanged. The second operation is even more obvious. It replaces a single equation with another equation that has the exact same solution set, because the graph of the line is unchanged. Thus, the entire graph of the system is also unchanged. (Notice for this to be true, it must be a nonzero multiple. If we multiply both sides of an equation by 0, we just get the equation $0 = 0$, which is always true, so we will have lost information – namely the solution set of the equation before we multiplied by 0. Thus, the solution set does change when the multiplier is zero.)

The third operation is the least intuitive one, yet that is the most important one for forcing coefficients to vanish. First of all, we can justify it algebraically. If the two equations in question are abbreviated by

$$a = b$$
$$c = d$$

and r is any real number, then the following equation follows by adding the same thing to both sides of the first equation – in this case, adding rc (which may be 0 if r is):

$$a + rc = b + rc.$$

However, we know $rc = rd$ by multiplying the second equation through by r, so by substitution on the right side, we have

$$a + rc = b + rd.$$

While this is a true statement, don't forget we need to write the entire system. Thus, we replace the original system with

$$a + rc = b + rd$$
$$c = d.$$

This illustrates that we have replaced the first equation by itself plus a multiple of the second equation. But this is an equivalent system because we are retaining the fact that $c = d$, so we can recover $a = b$ from the first line. Note that had we retained $a = b$ and the new equation instead, we would not necessarily be able to recover $c = d$ because r might be 0. This is why in the third elementary operation, it is the equation to which we add the multiple that is replaced.

While this explains the third operation algebraically, the reader may still want to know what is happening geometrically. After all, when this operation is used, the equation itself, and its graph, is changed. To illustrate that the new system is equivalent, regardless of that change, it helps to look at an example.

So consider Example 1.1. The system is

$$6x + 3y = 30$$
$$2x + 10y = 28.$$

Suppose we use the third elementary operation to replace the second equation with itself plus (-2) times the first equation. The new system is

$$6x + 3y = 30$$
$$-10x + 4y = -32.$$

Although this looks like a different system, it is equivalent to the original. The graphs of both systems together are shown in Figure 1.6.

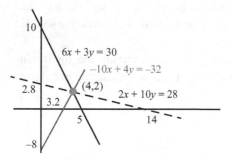

Figure 1.6 The dashed line is replaced by a new line passing through the same intersection point.

Here the new line replaces the dashed line. The new system still has the unique solution $(4, 2)$. Clearly, the two systems are equivalent. However, the new system doesn't seem any easier to solve than the original. That's because neither variable has been eliminated. Instead of replacing the second equation by itself minus 2 times the first, we should replace it by itself minus $\frac{1}{3}$ times the first. Algebraically, the reason is obvious – we eliminate the x coefficient that way:

$$2x + 10y = 28$$
$$\underline{-\tfrac{1}{3}(6x + 3y = 30)}$$
$$0x + 9y = 18.$$

The resulting equation, $9y = 18$ (which is equivalent to $y = 2$ by the second elementary operation), has a very simple graph. The new system is

$$6x + 3y = 30$$
$$y = 2$$

with graphs and equivalent solution set shown in Figure 1.7.

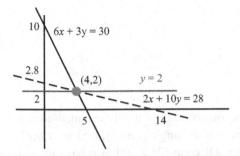

Figure 1.7 The dashed line is replaced by the horizontal line.

To continue the solution by elimination, now we eliminate y from the first equation, by replacing it with itself minus 3 times the second equation:

$$6x + 3y = 30$$
$$\underline{-3(y = 2)}$$
$$6x + 0y = 24.$$

The resulting equation $6x = 24$ can be written as $x = 4$ by the second elementary operation, and its graph is also very simple. The new system is

$$x = 4$$
$$y = 2.$$

Its graph is shown in Figure 1.8.

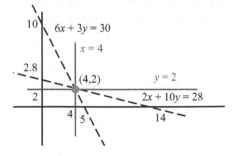

Figure 1.8 Both dashed lines have been replaced by the solid lines.

In short, the operations serve to redraw the graphs keeping the same point as the solution, while the specific operations chosen to eliminate variables result in lines that are parallel to the coordinate axes, making the solution easy to read off.

Presumably, if one had a system of three equations and three unknowns – x, y, and z – one would apply a series of elementary operations to the system, eliminating two variables from each equation until, hopefully, the end result is a system of this form:

$$x + 0y + 0z = a$$
$$0x + y + 0z = b$$
$$0x + 0y + z = c$$

or, more simply,

$$x = a$$
$$y = b$$
$$z = c.$$

Then the (unique) solution can be just read off. Geometrically, each of the these equations is a plane parallel to the coordinate planes, and their intersection is the single point $(x, y, z) = (a, b, c)$. The same ideas would generalize to even larger systems. Of course, we still also have to deal with the degenerate cases. Rather than try to deal with degenerate cases with an ad hoc geometric approach, we promised in this section to streamline the entire process. Once we do this, it will be easy to see what happens in the degenerate cases.

The key observation in the preceding example is that the operation of replacing the second equation with itself minus 2 times the first equation didn't help solve the system, but replacing it with itself minus $\frac{1}{3}$ times the first equation did help because it eliminated the x. In fact, the multiple $r = \frac{1}{3}$ is the only multiple r for which the second equation minus r times the first has no x coefficient. This is obviously because the coefficient of x in the second equation, 2, is exactly $\frac{1}{3}$ of the coefficient of x in the first equation, 6. In other words, deciding which multiple r of one equation to subtract from another is determined by the values of the *coefficients* in the system. The actual names of the variables are irrelevant – it's the coefficients that matter. So to streamline the entire process of elimination, we'd like to write just the coefficients and not bother writing the variables.

This leads to the notion of an *(augmented) coefficient matrix* for a system of equations. As we shall see in Chapter 2, a *matrix* is simply a rectangular array of numbers. In this case, each row (horizontal) will have the coefficients for each equation, and each column (vertical) corresponds to a variable, except the rightmost column, which represents the numbers on the right sides of the equations. Thus, the system

$$6x + 3y = 30$$
$$2x + 10y = 28$$

has augmented coefficient matrix

$$\begin{bmatrix} 6 & 3 & 30 \\ 2 & 20 & 28 \end{bmatrix}.$$

It is understood that the first column corresponds to the variable x and the second to the variable y, while the last column is the right-hand side of the equations. But we do not write the variables in because the algorithm for elimination can be expressed entirely in terms of the coefficients. Incidentally, the smaller matrix without the last column

$$\begin{bmatrix} 6 & 3 \\ 2 & 20 \end{bmatrix}$$

is called the *coefficient matrix of the system*. The only difference between this and the augmented version is the augmented version has an extra column for the right-hand side of the equations. In this chapter, we only use the augmented version. In Chapter 2, we'll see the unaugmented version also has some uses.

The first thing to do is to translate the elementary operations on the preceding into operations to be performed on these augmented matrices. Since the equations correspond to rows of the matrix, they are called *elementary row operations*:

1.   You may interchange any two rows of the matrix.
2.   You may replace any row with a nonzero multiple of itself.
3.   You may replace any row with itself plus any multiple of another row.

Clearly, when a sequence of elementary row operations is used on an augmented matrix, the result is the augmented matrix of some equivalent system.

1.3.1 Non-degenerate Examples

We illustrate the process with the following system:

Example 1.7

$$10x + 7y = 5$$
$$3x + 2y = 1.$$

We begin by writing the augmented coefficient matrix:

$$\begin{bmatrix} 10 & 7 & 5 \\ 3 & 2 & 1 \end{bmatrix}.$$

We would like to put a coefficient of 1 where the 10 is. You can do this in a number of ways. For example, you use use the second row operation to replace row 1 with $\frac{1}{10}$ times row 1, which we symbolize as $R_1 \to \frac{1}{10}R_1$. Another option is to use the third row operation to replace row 1 with itself minus 3 times row 2, which we symbolize as $R_1 \to R_1 - 3R_2$. Row 2, of course, is unchanged in either case. Using the second option, we can avoid fractions, at least in this first step, so we opt for it. We obtain the new matrix:

$$\begin{bmatrix} 1 & 1 & 2 \\ 3 & 2 & 1 \end{bmatrix}.$$

The reader should, for practice, write down the associated (equivalent) linear system for this matrix. Now that the x variable has a coefficient of 1 in the first equation, the next step is to eliminate x from all the rest of the equations (in this case, just the second equation.) This means making all the coefficients in the first column (except the first row) into a 0, or "clearing out the first column." So we must convert the 3 to a 0. What row operation would accomplish that? Clearly, switching rows will not help, and neither will replacing the second row with a nonzero multiple of itself. Only the third row operation can do that – as noted before, this is the operation designed to eliminate variables. Here we must replace row 2 by itself minus 3 times row 1, or $R_2 \rightarrow R_2 - 3R_1$. The new matrix is

$$\begin{bmatrix} 1 & 1 & 2 \\ 0 & -1 & -5 \end{bmatrix}.$$

Now that the first column is finished, we move to the next column and eliminate y from all equations except the second, where we put a coefficient of 1. The operation $R_2 \rightarrow (-1) \cdot R_2$ takes care of the second row (putting the coefficient of y equal to 1), which should be done first, just as we did for the first column. We obtain

$$\begin{bmatrix} 1 & 1 & 2 \\ 0 & 1 & 5 \end{bmatrix}.$$

Finally, we use the operation $R_1 \rightarrow R_1 - R_2$ to eliminate the y from the first equation:

$$\begin{bmatrix} 1 & 0 & -3 \\ 0 & 1 & 5 \end{bmatrix}.$$

We have succeeded. Converting this back to a linear system, we obtain

$$1x + 0y = -3$$
$$0x + 1y = 5$$

or simply

$$x = -3$$
$$y = 5,$$

with the obvious (unique) solution $(x, y) = (-3, 5)$. The reader can easily check the solution in the original equations. Observe that there is more than one way to obtain the answer – you can be creative in your use of the elementary row operations. For example, here is another approach to solve the system. We indicate each step as follows:

$$\begin{bmatrix} 10 & 7 & 5 \\ 3 & 2 & 1 \end{bmatrix} \quad R_1 \rightarrow \tfrac{1}{10}R_1$$

$$\begin{bmatrix} 1 & \tfrac{7}{10} & \tfrac{5}{10} \\ 3 & 2 & 1 \end{bmatrix} \quad R_2 \rightarrow R_2 - 3R_1$$

$$\begin{bmatrix} 1 & \tfrac{7}{10} & \tfrac{5}{10} \\ 0 & -\tfrac{1}{10} & -\tfrac{5}{10} \end{bmatrix} \quad R_2 \rightarrow (-10)R_2$$

$$\begin{bmatrix} 1 & \tfrac{7}{10} & \tfrac{5}{10} \\ 0 & 1 & 5 \end{bmatrix} \quad R_1 \rightarrow R_1 - \tfrac{7}{10}R_2$$

$$\begin{bmatrix} 1 & 0 & -3 \\ 0 & 1 & 5 \end{bmatrix}.$$

We obtain the same solution, of course. It is really up to the reader what operations to use. But notice that in either solution we worked column by column, first placing a coefficient of 1 in the appropriate spot and then clearing the rest of the column. This may seem like more work than other methods of solution, but if the system is much larger than two equations and two unknowns, elimination turns out to be the most efficient method of solution. Here is a larger example.

Example 1.8 Solve the following system by elimination:

$$5x - y + 2z = 21$$
$$-4x + 3y + z = 8$$
$$x + y + 3z = 26.$$

The augmented coefficient matrix is

$$\begin{bmatrix} 5 & -1 & 2 & 21 \\ -4 & 3 & 1 & 8 \\ 1 & 1 & 3 & 26 \end{bmatrix}.$$

Again, we focus on the first column and begin by putting a coefficient of 1 into the first entry. There are various row operations that will do this. For example, $R_1 \to R_1 + R_2$ will do. Another option is to use the first row operation to switch the first and third rows $R_1 \leftrightarrow R_3$. We illustrate with the latter choice, leaving the former for an exercise. We obtain the following matrix, and the next two operations are indicated, which will clear out the rest of the first column:

$$\begin{bmatrix} 1 & 1 & 3 & 26 \\ -4 & 3 & 1 & 8 \\ 5 & -1 & 2 & 21 \end{bmatrix} \begin{array}{l} \\ R_2 \to R_2 + 4R_1 \\ R_3 \to R_3 - 5R_1. \end{array}$$

Performing these operations, we obtain

$$\begin{bmatrix} 1 & 1 & 3 & 26 \\ 0 & 7 & 13 & 112 \\ 0 & -6 & -13 & -109 \end{bmatrix}.$$

Each time one uses a set of row operations to put a 1 in some position and then clear out the rest of the column, we refer to that process as a *pivot*. The position in the matrix where the 1 has been placed is called the *pivot position*. When the pivot has been completed so that the column is all 0s except for the 1 in the pivot position, the resulting column is called a *unit column*.

At this point, we move to the next column and convert it to a unit column by a pivot on the second row and second column, where the entry is 7. This means convert the 7 to a 1 and clear out the rest of the column. To convert the 7 to a 1, we could use $R_2 \to \frac{1}{7}R_2$. But instead we will use $R_2 \to R_2 + R_3$ in order to put off dealing with fractions for as long as possible. The result is as follows, with the rest of the pivot operations indicated:

$$\begin{bmatrix} 1 & 1 & 3 & 26 \\ 0 & 1 & 0 & 3 \\ 0 & -6 & -13 & -109 \end{bmatrix} \begin{array}{l} R_1 \to R_1 - R_2 \\ \\ R_3 \to R_3 + 6R_2. \end{array}$$

Completing the pivot yields

$$\begin{bmatrix} 1 & 0 & 3 & 23 \\ 0 & 1 & 0 & 3 \\ 0 & 0 & -13 & -91 \end{bmatrix}.$$

Finally, we move to the third column and pivot on the position where the -13 is located, in the third row and third column. We begin by $R_3 \to -\frac{1}{13}R_3$:

$$\begin{bmatrix} 1 & 0 & 3 & 23 \\ 0 & 1 & 0 & 3 \\ 0 & 0 & 1 & 7 \end{bmatrix} \quad R_1 \to R_1 - 3R_3.$$

Performing the final row operation needed to complete the pivot we arrive at

$$\begin{bmatrix} 1 & 0 & 0 & 2 \\ 0 & 1 & 0 & 3 \\ 0 & 0 & 1 & 7 \end{bmatrix},$$

which corresponds to the unique solution $(x, y, z) = (2, 3, 7)$.

To summarize the procedure, if we have a non-degenerate system, there are the same number of equations as variables, and so the augmented matrix has the form

$$[A|B],$$

where the coefficient matrix A has the same number of rows and columns (called a *square* matrix), and B is a single column representing the constants on the right side of the equations in the system. One performs a pivot on each column except the last, moving from left to right (using elementary row operations) to end up with a matrix of the form

$$[I|C],$$

where I is a square matrix consisting of unit columns (with the nonzero entries running down the diagonal of the matrix from the upper-left corner to the lower-right corner. That is, the matrix I has the form

$$I = \begin{bmatrix} 1 & 0 & 0 & \cdots & 0 \\ 0 & 1 & 0 & \cdots & 0 \\ 0 & 0 & 1 & \cdots & 0 \\ \vdots & \vdots & \vdots & \ddots & \vdots \\ 0 & 0 & 0 & \cdots & 1 \end{bmatrix}.$$

In this case, the (unique) solution to the system can be read off from the column C.

The matrix I is called an *identity matrix* (explained in Chapter 2) and has some special properties that we use in later chapters.

Exercises

1–11. Solve Exercises 1–11 from Section 1.2 by pivoting the augmented coefficient matrix.

1.3.2   Degenerate Examples and the General Gauss-Jordan Elimination Algorithm

The difficulty with the simple procedure in the last subsection is that if the system is degenerate, one will not be able to perform the pivots to obtain an identity matrix. To handle the more general cases, the idea is to modify the procedure so that the goal is to obtain something that is as close to looking like the identity matrix as is possible. For our purposes, being close to the identity matrix means being in *reduced row echelon form.*

DEFINITION 1.9   *A matrix A is in* reduced row echelon form *if the following is true:*

1.   *Any row consisting of all zeros appears below any row containing a nonzero entry.*
2.   *In any row containing a nonzero entry, the first nonzero entry in the row, reading from left to right, is a 1. It is called a* leading 1 *for the matrix.*
3.   *Any column containing a leading 1 is a unit column (so all other entries in the column aside from the leading 1 are 0).*
4.   *Any leading 1 appears to the right of any leading 1 in the rows above it. (So the leading 1s in the matrix go down to the right.)*

An identity matrix is always in reduced echelon form:

$$I = \begin{bmatrix} 1 & 0 & 0 & 0 \\ 0 & 1 & 0 & 0 \\ 0 & 0 & 1 & 0 \\ 0 & 0 & 0 & 1 \end{bmatrix}.$$

As is evident, every row has a leading 1, and they go down to the right, and every column is a unit column because every row has a leading 1. The following matrices are also in reduced echelon form:

$$A = \begin{bmatrix} 1 & 0 & 0 & 3 \\ 0 & 1 & 0 & 5 \\ 0 & 0 & 1 & 7 \end{bmatrix}$$

$$B = \begin{bmatrix} 1 & 0 & 0 & 3 \\ 0 & 1 & 0 & 0 \\ 0 & 0 & 1 & 7 \\ 0 & 0 & 0 & 0 \end{bmatrix}$$

$$C = \begin{bmatrix} 1 & 0 & 7 & 0 & 4 \\ 0 & 1 & 1 & 0 & 4 \\ 0 & 0 & 0 & 1 & 4 \\ 0 & 0 & 0 & 0 & 0 \\ 0 & 0 & 0 & 0 & 0 \end{bmatrix}$$

$$D = \begin{bmatrix} 1 & 0 & 0 & 1 & 4 & 7 \\ 0 & 1 & 0 & 2 & 5 & 8 \\ 0 & 0 & 1 & 3 & 6 & 9 \\ 0 & 0 & 0 & 0 & 0 & 0 \end{bmatrix}.$$

The reader should observe that every column containing a leading 1 is a unit column in these examples, but not all columns contain a leading 1. In particular, the second 1 in the second row of matrix C is not a leading 1 nor is the second 1 in the first row of matrix D. The number of leading 1s in a matrix in reduced echelon form is called the *rank* of the matrix. All four of the examples $A-D$ have rank 3.

The following matrices are not in reduced echelon form:

$$M = \begin{bmatrix} 1 & 0 & 0 & 2 \\ 0 & 0 & 0 & 0 \\ 0 & 1 & 0 & -1 \\ 0 & 0 & 1 & 5 \end{bmatrix}$$

$$N = \begin{bmatrix} 0 & 1 & 0 & 3 \\ 1 & 0 & 0 & 6 \\ 0 & 0 & 1 & 9 \end{bmatrix}$$

$$P = \begin{bmatrix} 1 & 0 & 8 & 2 & 7 \\ 0 & 1 & -3 & 4 & 7 \\ 0 & 0 & 1 & 6 & 2 \\ 0 & 0 & 0 & 0 & 0 \end{bmatrix}$$

$$Q = \begin{bmatrix} 1 & 0 & 0 & 0 \\ 0 & 3 & 0 & 9 \\ 0 & 0 & 1 & 10 \end{bmatrix}.$$

Indeed, in matrix M, there is a row of all zeros that does not appear at the bottom of the matrix, violating condition 1. In matrix N, condition 4 is violated; in matrix P, condition 3 is violated as the third column is not a unit column. Finally, in matrix Q, condition 2 is violated, as the leading nonzero entry in the second row is a 3, not a 1. With practice, it is easy to recognize when a matrix is in reduced echelon form.

The astute reader will observe, however, that each of the matrices $M, N, P,$ and Q can be put into reduced echelon form by performing elementary row operations. Indeed, in matrix M, we perform the two switches $R_2 \leftrightarrow R_3$ followed by $R_3 \leftrightarrow R_4$ to put the row of all zeros at the bottom of the matrix, and the result is then in reduced row echelon form. Similarly, in N, the switch $R_1 \leftrightarrow R_2$ will result in a matrix in reduced echelon form. In matrix Q, the operation $R_2 \to \frac{1}{3}R_2$ will replace the 3 with a leading 1. Finally, in matrix P, all that is needed is to finish the pivot on the third column to make it into a unit column, so the operations needed are $R_1 \to R_1 - 8R_3,$ and $R_2 \to R_2 + 3R_3$. The reader should verify all this.

In fact, although we will not prove this, any matrix can be put into reduced echelon form using elementary row operations. A proof can be found in any linear algebra text, such as Lay (2012). After trying several examples, the reader will no doubt be convinced of the truth of this claim anyway. Now, the question is, how does this help when solving systems of linear equations?

Consider a very simple example:

$$3x + 7y = 23$$
$$9x + 21y = 69.$$

The associated augmented coefficient matrix is

$$\begin{bmatrix} 3 & 7 & 23 \\ 9 & 21 & 69 \end{bmatrix}.$$

Pivoting on the 3 to make the first column a unit column, first by using the operation $R_1 \to \frac{1}{3}R_1$ to put a leading 1 in the pivot position, and then using the operation $R_2 \to R_2 - 9R_1$ to clear the 9 and make the first column a unit column, we obtain

$$\begin{bmatrix} 1 & \frac{7}{3} & \frac{23}{3} \\ 0 & 0 & 0 \end{bmatrix}.$$

This matrix is already in reduced row echelon form. Because of the row of all zeros, there are no further operations that will produce any more leading 1s. But now if we consider the equations that correspond to this augmented matrix (which we know is equivalent to the original system), we obtain

$$x + \frac{7}{3}y = \frac{23}{3}$$
$$0x + 0y = 0.$$

Of course, the second equation, $0 = 0$, is satisfied for all values of x and y, so the set of common solutions is the same as the solution set to just the first equation. In other words, this is a redundant system – the two lines are coincident, so all points on the line $3x + 7y = 23$ are solutions to the system. But conveniently, this line is written as

$$x + \frac{7}{3}y = \frac{23}{3},$$

which means we can solve for x in terms of y. That is, we can write

$$x = \frac{23}{3} - \frac{7}{3}y,$$

where y is an arbitrary real number.

Since there are no restrictions on y, we say y is a *free variable*. Since we often like to write solutions as ordered pairs, we can write this solution in an alternate form, called *parametric form*, as follows. We assign a new letter (called a *parameter*) for any free variable and solve for all the variables in terms of the parameters. So, if $y = t$, then $x = \frac{23}{3} - \frac{7}{3}t$, and we can write the solution as all the ordered pairs of the form:

$$(x, y) = \left(\frac{23}{3} - \frac{7}{3}t, t \right).$$

The reader can check that each value of the parameter t gives a different point on the line $3x + 7y = 23$.

So the presence of a free variable (or more than one free variable) will always lead to a solution set that is infinite, one of the two types of degenerate systems. Furthermore, we know there will be a free variable by looking at the matrix in reduced row echelon form. Indeed, each free variable corresponds to a column (except for the rightmost column) of the augmented coefficient matrix that is not a unit column (that is, a column that does not contain a leading 1). The same approach will work for a larger system with free variables. For example, suppose we are solving a system and end up with a matrix in reduced row echelon form, as follows:

$$\begin{bmatrix} 1 & 0 & -1 & 0 & 2 & 0 & 5 \\ 0 & 1 & 2 & 0 & 4 & -2 & 0 \\ 0 & 0 & 0 & 1 & -7 & -3 & 10 \\ 0 & 0 & 0 & 0 & 0 & 0 & 0 \\ 0 & 0 & 0 & 0 & 0 & 0 & 0 \end{bmatrix}.$$

Because this matrix has five rows, the original system had five equations, and because it has seven columns, there are six variables (remember the last column does not correspond to a variable.) Suppose the names of the variables are u, v, w, x, y, z, in that order. While this matrix is in reduced form, there are only three leading $1s$, corresponding to the variables u, v, x. The columns for w, y, z are not unit columns, so they are free variables. So to write down the general solution, we solve these equations for u, v, x in terms of w, y, z, and replace free variables with parameters, say r, s, t. The equations read

$$u - w + 2y = 5$$
$$v + 2w + 4y - 2z = 0$$
$$x - 7y - 3z = 10.$$

Moving the free variables to the right side:

$$u = 5 + w - 2y$$
$$v = -2w - 4y + 2z$$
$$x = 10 + 7y + 3z.$$

Replacing the free variables with the parameters, the solutions look like

$$u = 5 + r - 2s$$
$$v = -2r - 4s + 2t$$
$$w = r$$
$$x = 10 + 7s + 3t$$
$$y = s$$
$$z = t.$$

Or, writing them as ordered 6-tuples, the solutions are

$$(u, v, w, x, y, z) = (5 + r - 2s, \, -2r - 4s + 2t, r, 10 + 7s + 3t, s, t)$$

It is always understood that the parameters can take on any real number values, so there are always infinitely many solutions when free variables are present. (In fact, if you study linear algebra, you will find that the number of parameters is the dimension of the solution set. So one parameter is a line, two parameters are a plane, etc.)

What about the other type of degenerate systems – ones with no solution? Consider this simple example:

$$3x + 7y = 23$$
$$9x + 21y = 70.$$

You will notice this is almost exactly the same system as the preceding one, but by changing the 69 to 70 on the right side of the second equation, the system geometrically becomes two parallel lines rather than coincident lines, so there is no solution. What happens algebraically? We again put the matrix in reduced echelon form, using in fact the exact same row operations we used before. Dividing the first row by 3 to put a leading 1 in the first row and column, and then replacing the second row by itself minus 9 times the (revised) first row, we end up with the following matrix:

$$\begin{bmatrix} 1 & \frac{7}{3} & \frac{23}{3} \\ 0 & 0 & 1 \end{bmatrix}.$$

We know this corresponds to a system equivalent to the original. But the second equation reads:

$$0x + 0y = 1$$

which clearly has no solution since $0 = 1$ is never true! Thus, the entire system's lack of a solution is revealed. It turns out that this always happens. If the system has no solution, in the course of performing the elementary row operations, you will always end up with at least one row of the form

$$\begin{bmatrix} 0 & 0 & 0 & \cdots & 0 & a \end{bmatrix},$$

with $a \neq 0$. This represents an equation with no solution. So if such a row is obtained in the process, even if the matrix is not completely row-reduced, you can stop – the system has no solution.

Summarizing, we now have an efficient algorithm for solving any system of liner equations using augmented coefficient matrices:

ALGORITHM 1.10   Gauss-Jordan Elimination.

1.   Convert the system to matrix form (that is, write down the augmented coefficient matrix.)
2.   Use elementary row operations to put the matrix in reduced row echelon form. This is done as outlined in the previous subsection – work column by column, performing pivots on any column with a leading 1 to convert them to unit columns. Use row switch operations as necessary to make sure the leading 1s go down to the right, and any rows of all zeros are moved to the bottom of the matrix.

   a.   Along the way, if you obtain any row which is all zeros except for the rightmost entry, which is nonzero, then stop. This system has no solution.
   b.   Otherwise, continue until the matrix is in reduced row echelon form.

3.   If there are no free variables, then the system has a unique solution that can be read off from the last column. (In this case, the echelon form of the matrix will be $[I|C]$, where I is an identity matrix and C is a column listing the solution to the system.)
4.   If there are free variables, there are infinitely many solutions. A convenient way to write down the solutions is in parametric form. To obtain parametric form, convert the final matrix back to equation form, and move all the free variables to the right side (that is, solve for the other variables in terms of the free variables). Then replace the free variables with parameters in order to express all the variables in terms of the parameters.

We conclude this section by solving the traffic flow problem given in Example 1.4, Section 1.2. Recall that the system had the form:

$$v + w = 50$$
$$v + x - y = 90$$
$$-w + x + z = 150$$
$$y + z = 110.$$

We convert to the augmented coefficient matrix and perform the indicated row operations:

$$
\begin{bmatrix}
1 & 1 & 0 & 0 & 0 & 50 \\
1 & 0 & 1 & -1 & 0 & 90 \\
0 & -1 & 1 & 0 & 1 & 150 \\
0 & 0 & 0 & 1 & 1 & 110
\end{bmatrix}
\quad R_2 \to R_2 - R_1
$$

$$
\begin{bmatrix}
1 & 1 & 0 & 0 & 0 & 50 \\
0 & -1 & 1 & -1 & 0 & 40 \\
0 & -1 & 1 & 0 & 1 & 150 \\
0 & 0 & 0 & 1 & 1 & 110
\end{bmatrix}
\quad R_2 \to (-1) \cdot R_2.
$$

This puts a leading 1 in the second column, which we then clear out:

$$
\begin{bmatrix}
1 & 1 & 0 & 0 & 0 & 50 \\
0 & 1 & -1 & 1 & 0 & -40 \\
0 & -1 & 1 & 0 & 1 & 150 \\
0 & 0 & 0 & 1 & 1 & 110
\end{bmatrix}
\quad
\begin{array}{l}
R_1 \to R_1 - R_2 \\[4pt]
R_3 \to R_3 + R_2
\end{array}
$$

$$
\begin{bmatrix}
1 & 0 & 1 & -1 & 0 & 90 \\
0 & 1 & -1 & 1 & 0 & -40 \\
0 & 0 & 0 & 1 & 1 & 110 \\
0 & 0 & 0 & 1 & 1 & 110
\end{bmatrix}.
$$

Notice that we have two identical rows, so the operation $R_4 \to R_4 - R_3$ can now be used to put a row of all zeros at the bottom:

$$
\begin{bmatrix}
1 & 0 & 1 & -1 & 0 & 90 \\
0 & 1 & -1 & 1 & 0 & -40 \\
0 & 0 & 0 & 1 & 1 & 110 \\
0 & 0 & 0 & 0 & 0 & 0
\end{bmatrix}
\quad
\begin{array}{l}
R_1 \to R_1 + R_3 \\[4pt]
R_2 \to R_2 - R_3.
\end{array}
$$

The form of this matrix shows there cannot be a leading 1 in the third column, so that variable, x, is a free variable. We move on to the fourth column and perform the row operations indicated earlier to obtain

$$
\begin{bmatrix}
1 & 0 & 1 & 0 & 1 & 200 \\
0 & 1 & -1 & 0 & -1 & -150 \\
0 & 0 & 0 & 1 & 1 & 110 \\
0 & 0 & 0 & 0 & 0 & 0
\end{bmatrix}.
$$

We are done as the matrix is in reduced row echelon form. There are two free variables, x and z. We now write the solution in parametric form. First, convert back to equations:

$$
v + x + z = 200
$$
$$
w - x - z = -150
$$
$$
y + z = 110.
$$

Solving for v, w, and y in terms of the free variables, x and z, we obtain

$$v = 200 - x - z$$
$$w = -150 + x + z$$
$$y = 110 - z.$$

Assigning parameters, the full solution is

$$v = 200 - s - t$$
$$w = -150 + s + t$$
$$x = s$$
$$y = 110 - t$$
$$z = t$$

or, as an ordered quintuple, we can write

$$(v, w, x, y, z) = (200 - s - t, -150 + s + t, s, 110 - t, t),$$

which illustrates the complete solution of a degenerate system with free variables.

However, we should now go back to the interpretation step of the modeling process. If you were studying the traffic patterns, you would want to know how to draw real conclusions from your mathematical solution. The first thing you should realize is that because the streets are all one-way, and the variables, v, w, x, y, z indicate the number of vehicles per hour that traverse these streets in the prescribed direction, none of the variables should be negative, which would indicate a traffic snarl since "reverse flow" is not really possible. Thus, s and t are not completely unrestricted. Since $v \geq 0$, this means $s + t$ (which are nonnegative since x and z are nonnegative) must satisfy

$$s + t \leq 200.$$

Similarly, since $w \geq 0$, we see

$$s + t \geq 150.$$

Combining these, we see

$$150 \leq s + t \leq 200.$$

Finally, since $y \geq 0$, we must have

$$t \leq 110.$$

Furthermore, since these variables represent numbers of vehicles, fractional answers do not make sense. Thus, the real-life solution to the traffic system is determined by the set of all ordered pairs of *integers* (s, t) such that

$$0 \leq s$$
$$0 \leq t \leq 110$$
$$150 \leq s + t \leq 200,$$

in addition to the preceding parametric equations. In particular, while the mathematical solution to the system was infinite, there are only a finite number of integer solutions that satisfy the preceding inequalities. We will not pursue this any further here, but the example does serve to remind us that the mathematical solution to the model is not the end of the story.

Exercises

1. In Example 1.7, we suggested that there were several ways to start the elimination process to put a leading 1 in the first row and column. One of the suggestions, which we did not illustrate, was the operation $R_1 \rightarrow R_1 + R_2$. Finish the example using this suggestion and verify that your answer is the same as the one we obtained in the text.

   In Exercises 2–9, solve the system using the Gauss-Jordan algorithm. If the solution is infinite, write the solution in parametric form.

2.
$$10x + 3y = 20$$
$$2x + y = 2$$

3.
$$u + 2v + 3w = 11$$
$$2u + 4v - 4w = 2$$
$$3u + v + 5w = 25$$

4. a.
$$x + 2y - z = 70$$
$$2x - y + 8z = 150$$

   b. Observe that, thinking geometrically, each of these equations is a plane in three-dimensional space, and since there are only two planes, their intersection cannot be a single point. The intersection of two planes is empty (if they are parallel), or a line, or a plane (if they are coincident). Thus, even before starting the elimination process, we know there cannot be a unique solution. More generally, whenever there are more variables than there are equations, the system must be degenerate, with infinitely many (or no) solutions. Explain why, but do not use a geometric argument. Instead, appeal to the augmented coefficient matrix. [**Hint**: Think about free variables.]

5.
$$6x + 3y + 4z = 41$$
$$x - y + 2z = 17$$
$$2x + 5y + z = 22$$
$$7x + 7y + 3z = 46$$

6.

$$3w + 2x - y - 2z = 12$$
$$2w - x + 5y + 4z = 10$$
$$8w + 3x + 3y = 34$$
$$2w + 6x - 12y - 12z = 4$$

7.

$$2x + 3y + 5z = 15$$
$$6x - 3y + 10z = 18$$
$$20x - 5z = 23$$

8.

$$5w + 8x - y + z = 100$$
$$w + 2x + y + 2z = 60$$
$$2w + 4y - z = 50$$
$$4w + 5x + 2y + z = 100$$

9.

$$-2p + 3q + 6r + s - 2t + 6u = 0$$
$$p + 4q + 4s + 2t - u = 60$$
$$q + 3r + 2s - 3u = 30$$
$$2p - 3q + 9r + 2t + 3u = 50$$
$$4p + 6q + 10r - 12s + 8t - 6u = 20$$

10. Consider the linear system:

$$p + 20q + r - 2s + t = 10$$
$$q + 10r - 6s + t = 216$$
$$r + 8s - t = 18$$
$$s + 2t = 257$$
$$t = 121$$

a. Notice that the last equation tells us that $t = 121$. Substitute this into the previous equation, which involves only s and t, and solve it for s. Then substitute the values for s and t into the third equation and solve for r. Keep going, substituting in the values of the solved variables into the previous equation at each step until you obtain the complete solution. This method of solution is called *back substitution*, and it works because the augmented coefficient matrix has all 0 entries below each leading 1.

b. Solve the system as usual using Gauss-Jordan elimination. You should obtain the same solution as in Part a.

REMARK 1.11   *If we relax the definition of reduced row echelon form so that condition 3 reads that in every column with a leading 1, every entry below the leading 1 is a 0, the resulting matrix is said to be in* row echelon form *(as opposed to reduced row echelon form.) A variation of the Gauss-Jordan elimination method to solve a system is then to use row operations to put the augmented coefficient matrix into row echelon form, and then use back substitution as in Part a.*

11.   a.   Solve the following system via Gauss Jordan elimination:

$$x + 2y + 5z = 30$$
$$3x + 6y - 10z = 65$$
$$5x + 10y = 125$$

   b.   Instead, solve it using back substitution as indicated in Remark 1.11.

12.   Solve the system arising from the traffic flow problem in Exercise 12 of Section 1.2. What other conditions on the solution are there arising from the restriction that all the variables must be nonnegative integers?

1.4   Vectors, Linear Combinations, and Bases

In this section, we take a brief break from the study of linear systems to introduce vectors. A vector is a quantity that has both a magnitude and a direction associated with it. By contrast, a real number, also known as a *scalar*, simply has a magnitude. Real numbers are excellent models for the measurement of quantities where direction is irrelevant or unimportant, such as the length of a piece of string, the age of a tree, or the temperature of a heated piece of metal. Vectors would be useful in modeling quantities where direction is important. If you are a meteorologist studying wind, it would be incomplete information to say that the wind is blowing at 30 miles per hour. You also want to know from what direction the wind is blowing. The same ideas apply for the flow of any fluid – whether it is a gas such as wind or a liquid such as the flow of a water current (useful for sailors at sea) or the flow of blood in a human heart (people who design artificial heart valves need that kind of detailed information, for example). The same is true for more abstract flows such as the flow of heat ("heat flux") through a solid material, or electromagnetic flux. It is also important in physics and engineering for the study of any force whatsoever – not only do you want to know the magnitude of a force (how hard it is pushing?) but also in what direction it is pushing. Velocities and accelerations are also vector quantities when describing the motion of an object through space.

A vector quantity is depicted geometrically by drawing an arrow. The arrow obviously tells you the direction, while the magnitude can be encoded as the length of the arrow. We illustrate in the case of vectors in a plane in Figure 1.9.

Each vector has an *initial point* from where the arrow emanates and a *terminal point* where the arrowhead is drawn. We can designate the vector from P to Q by the symbol $\overrightarrow{PQ}$ (write the initial point on the left and the terminal point on the right.) The other vectors pictured in Figure 1.9 can be written as $\overrightarrow{RS}$ and $\overrightarrow{TU}$. However, the alert reader may notice that the three vectors drawn all look like they are parallel and have the same length. Indeed, they do. Since each vector is determined by the two points, one can easily compute the slopes of each line segment. We leave it to the reader

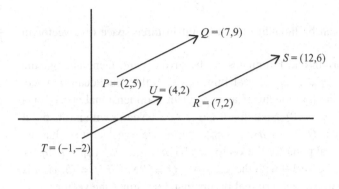

Figure 1.9 Vectors in a plane.

to verify that all three vectors shown have slope $\frac{4}{5}$ so they are parallel. Also, by using the distance formula (or the Pythagorean theorem) for points in the plane, we can easily verify that all three vectors shown have length $\sqrt{4^2 + 5^2} = \sqrt{41}$.

Since magnitude and direction, and not the specific initial point, are all that matter for a vector, any two vectors with the same magnitude and direction should be considered the same vector. Thus, in Figure 1.9, there is really only one vector drawn (but it is drawn with three different initial points.) One may think of moving or translating the vector around the plane, and as long as we keep the vector "parallel to itself" and preserve the length as we move it around, we are not changing the vector. So in Figure 1.9, we have $\overrightarrow{PQ} = \overrightarrow{RS} = \overrightarrow{TU}$.

Here is why this is helpful. If the initial point does not matter, why not move it so that it emanates from the origin $O = (0,0)$? If the origin O is considered the initial point, then the entire vector is determined by just the terminal point. In this way, every vector in the plane can be written down as a single ordered pair. Thus, when you see an ordered pair – say, $(3, -7)$ – you may interpret it in two equally useful ways. You may regard it as simply a point in the Euclidean plane, or you may think of it as a vector that has initial point $(0,0)$ and that has terminal point $(3, -7)$. Furthermore, we mention in passing here that we often find it useful to write these ordered pairs vertically instead of horizontally. Thus, $(3, -7)$ and $\left(\begin{smallmatrix} 3 \\ -7 \end{smallmatrix}\right)$ both refer to the same vector, shown in Figure 1.10.

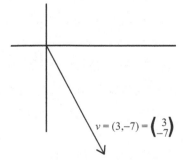

Figure 1.10 An ordered pair regarded as a vector.

Similarly, in three-dimensional Euclidean space, any vector can be translated so that its initial point is the origin and, thus, is determined by its terminal point that is an ordered triple, $v = (a, b, c)$

or $v = \begin{pmatrix} a \\ b \\ c \end{pmatrix}$. Any ordered triple can be thought of as a point in three-space or a vector in three-space that emanates from the origin and terminates at the given point. Generalizing, the collection of all ordered n-tuples $(a_1, a_2, \ldots, a_n)$ of real numbers is called Euclidean n-space, often written $\mathbb{R}^n$. An element of $\mathbb{R}^n$ can be thought of as a point in n-dimensional space, or a vector with initial point at the origin $(0, 0, \ldots, 0)$. Note that if $v = \overrightarrow{PQ}$ is a vector with initial point $P = (p_1, p_2, \ldots, p_n)$ and terminal point $Q = (q_1, q_2, \ldots, q_n)$, then if we translate it so that the initial point is the origin, then the terminal point (or the vector itself) becomes $v = (q_1 - p_1, q_2 - p_2, \ldots, q_n - p_n)$. This is illustrated in Figure 1.9 with the vectors $\overrightarrow{PQ} = \overrightarrow{RS} = \overrightarrow{TU} = (5, 4)$. It is the change in coordinates between the initial and terminal points that determines the vector.

The sides of a triangle, or of any polygon, can be thought of as vectors if desired. So learning how to manipulate vectors can lead to techniques to prove theorems in Euclidean geometry. We briefly illustrate this idea in the exercises at the end of the section (see Exercise 2), but only briefly, as this is not our main interest in vectors in this book. The fact that we can represent vectors as ordered n-tuples allows us to connect the geometric properties of vectors to algebra, and that is our main interest in vectors. In particular, we will show how they are related to systems of linear equations.

By the algebra of vectors, we mean how to combine them with each other (vector addition) and how to combine them with real numbers (scalar multiplication.) We define these operations presently.

DEFINITION 1.12   *Given a vector $v \in \mathbb{R}^n$ and a real number $s \in \mathbb{R}$ (called a scalar), we can form a new vector w, written $w = sv$, called the scalar multiple of v by the formula*

$$w = s \begin{pmatrix} x_1 \\ x_2 \\ \vdots \\ x_n \end{pmatrix} = \begin{pmatrix} sx_1 \\ sx_2 \\ \vdots \\ sx_n \end{pmatrix}. \tag{1.1}$$

That is, to scale v by a factor of s, simply multiply each coordinate by s.

It is important to connect the algebraic definition to what happens geometrically. Figure 1.11 illustrates this connection.

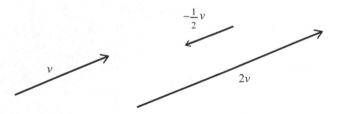

$-\frac{1}{2}v$

v

$2v$

Figure 1.11 Scaling a vector.

The reader should observe that the magnitude of the vector sv is just $|s|$ times the magnitude of v, and the direction of v is unchanged, although if the scalar is negative, the orientation reverses – the arrow points in the opposite direction. Remember that the initial point does not matter. If these vectors were all drawn so that they emanate from the origin, they all would lie on the line through v. In particular, all the multiples of the form sv where $s \geq 0$ lie on the ray (or half-line) through v.

DEFINITION 1.13 *Given two vectors* $v = \begin{pmatrix} x_1 \\ x_2 \\ \vdots \\ x_n \end{pmatrix}, w = \begin{pmatrix} y_1 \\ y_2 \\ \vdots \\ y_n \end{pmatrix} \in \mathbb{R}^n$, *their sum is defined as the vector given by adding the corresponding coordinates:*

$$v + w = \begin{pmatrix} x_1 \\ x_2 \\ \vdots \\ x_n \end{pmatrix} + \begin{pmatrix} y_1 \\ y_2 \\ \vdots \\ y_n \end{pmatrix} = \begin{pmatrix} x_1 + y_1 \\ x_2 + y_2 \\ \vdots \\ x_n + y_n \end{pmatrix}. \tag{1.2}$$

A similar definition works for subtraction $v - w = (\ldots, x_i - y_{i,\ldots})$ – subtract the corresponding coordinates. Here is an example in the plane: If $v = (5, 2)$ and $w = (3, 4)$, then $v + w = (5 + 3, 2 + 4) = (8, 6)$. The vectors are drawn in Figure 1.12.

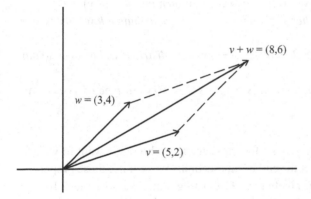

Figure 1.12 Adding vectors.

Again, we would like to understand the connection between the algebra and the geometry. The picture suggests that the vector $v + w$ is the diagonal of the parallelogram with sides v and w. This can be verified as follows. The opposite sides of a parallelogram are parallel and have the same length, whence they are the same vector. Thus, if we translate the vector w so that its initial point is $(5, 2)$, the terminal point P of v, then we know from before that the terminal point $Q = (x, y)$ of w must have the property that

$$w = \overrightarrow{PQ} = (x - 5, y - 2)$$
$$(3, 4) = (x - 5, y - 2).$$

Solving for $Q = (x, y)$, we obtain $Q = (5 + 3, 4 + 2) = (8, 6)$, which is exactly the vector $v + w$. Thus, $v + w$ is the vector emanating from the origin and with terminal point Q, which is the opposite corner of the parallelogram. This works for the plane – what about higher dimensions? Well, any two vectors emanating from the origin will lie in a plane (in fact, a unique plane, unless they are scalar multiples of one another – also called *parallel* vectors). One may then work in that plane, so essentially, the problem reduces to a plane problem. This is not a rigorous argument, but hopefully, the reader can believe that in all dimensions, the connection between the algebra of vector addition and the geometry is the following law.

CONCLUSION 1.14   The Parallelogram Law. *The sum of two vectors in $\mathbb{R}^n$ is the vector which is the diagonal of the parallelogram formed with the sides being v and w.*

Vectors are usually studied in a linear algebra course. In that course, one proves the following properties of these operations. Here, we present these properties without proof.

THEOREM 1.15   *The addition of vectors and scalar multiplication satisfy the following properties:*

a. $v + w = w + v$ *for all vectors* v, w *(commutative law)*

b. $u + (v + w) = (u + v) + w$ *for all vectors* u, v, w *(associative law)*

c. $v + 0 = v$ *for any vector v, where by 0 we mean the vector* $0 = (0, 0, \ldots, 0)$ *representing the origin (0 is called the additive identity),*

d.   *Given any vector v, there is an* additive inverse, *a vector* $-v$ *such that* $v + (-v) = 0$

e.   *If r is a scalar and* v, w *are vectors, then* $r(v + w) = rv + rw$ *(distributive law over vector addition)*

f.   *if r and s are scalars and v is a vector, then* $(r + s)v = rv + sv$ *(distributive law over scalar addition)*

g.   *If r and s are scalars and v is a vector, then* $r(sv) = (rs)v = s(rv)$ *(a type of associative law for scalar multiplication)*

h. $1v = v$ *for any vector v.*

Many of these properties have very easy proofs, for those readers who don't want to wait to take a linear algebra course (see Exercise 3 at the end of this section.) We will use these properties without further discussion. Of course, linear algebra is a fascinating topic that goes much further than merely proving trivial facts about n-tuples. For example, vectors are treated *axiomatically*, meaning that any set of elements that satisfy the properties of Theorem 1.15 (called *axioms*) can be called "vectors," and the entire set is called a "vector space." In this way, one may think of many things aside from ordered n-tuples as vectors. Among the objects that can be considered as vectors under this more general framework are polynomials, functions, matrices, solutions to certain differential equations, and more. The modern concept of a vector is far more general than just n-tuples, which makes linear algebra a very general and powerful tool. In our text, however, we will not pursue this abstract approach to vectors. For us, a vector is always an n-tuple of real numbers.

We now combine the two operations.

DEFINITION 1.16   *Let* $S = \{v_1, v_2, \ldots, v_m\}$ *be a collection of vectors in* $\mathbb{R}^n$. *A* linear combination *of the vectors in S is any expression of the form*

$$w = s_1 v_1 + s_2 v_2 + \cdots + s_m v_m \qquad (1.3)$$

where the s_i *are scalars (real numbers).*

In other words, a linear combination of vectors in S is simply a sum of scalar multiples of the vectors in S. Here are some examples and questions one could ask. Let $v = \begin{pmatrix} 1 \\ 3 \\ 5 \end{pmatrix}$ and $w = \begin{pmatrix} 3 \\ -1 \\ 1 \end{pmatrix}$. Find $3v + 2w$ and $6v - 2w$. The solution is

$$3v + 2w = 3 \begin{pmatrix} 1 \\ 3 \\ 5 \end{pmatrix} + 2 \begin{pmatrix} 3 \\ -1 \\ 1 \end{pmatrix} = \begin{pmatrix} 3 \\ 9 \\ 15 \end{pmatrix} + \begin{pmatrix} 6 \\ -2 \\ 2 \end{pmatrix} = \begin{pmatrix} 9 \\ 7 \\ 17 \end{pmatrix}.$$

Notice we used both the definition of scalar multiplication and the definition of vector addition. Similarly,

$$6v - 2w = 6 \begin{pmatrix} 1 \\ 3 \\ 5 \end{pmatrix} - 2 \begin{pmatrix} 3 \\ -1 \\ 1 \end{pmatrix} = \begin{pmatrix} 0 \\ 20 \\ 28 \end{pmatrix}.$$

Given a set S of vectors, the collection of all linear combinations of them is called their *span*, written Span(S) or $< S >$.

DEFINITION 1.17   *Let* $S = \{v_1, v_2, \ldots, v_m\}$ *be a set of vectors in* $\mathbb{R}^n$. *The* Span *of S is the collection of all possible linear combinations:*

$$Span(S) = < S > = \{z = s_1 v_1 + s_2 v_2 + \cdots + s_m v_m | s_i \in \mathbb{R}, i = 1, 2, \ldots, m\}$$

So the two examples show that $\begin{pmatrix} 9 \\ 7 \\ 17 \end{pmatrix}$ and $\begin{pmatrix} 0 \\ 20 \\ 28 \end{pmatrix}$ are members of Span(v, w). What vectors belong to the span of v, w? Can we describe them geometrically? Observe that any multiple rv is on the line through v, and any multiple sw is on the line through w. That means any linear combination $rv + sw$ is obtained from rv and sw using the parallelogram law, so the sum lies in the plane of the parallelogram containing rv and sw as sides; that is, every linear combination lies in the plane determined by v and w (if v and w happen to be parallel, it is even more restrictive – every linear combination will lie on the line through v). In particular, there must be some vectors in three-dimensional space that are not on that plane and so are not in the span of v and w. Therefore, an important question to ask is, given a set S of vectors and a vector w, is w a member of the span of S or not?

For a very simple example, consider

$$S = \{U_1, U_2\} = \left\{ \begin{pmatrix} 1 \\ 0 \\ 0 \end{pmatrix}, \begin{pmatrix} 0 \\ 1 \\ 0 \end{pmatrix} \right\}$$

(so the vectors are unit vectors on the x- and y-coordinate axes.) It should be clear than any vector in the xy-plane belongs to the span of S. Indeed, if $v = \begin{pmatrix} a \\ b \\ 0 \end{pmatrix}$ is an arbitrary vector in the xy-plane, then clearly,

$$v = \begin{pmatrix} a \\ 0 \\ 0 \end{pmatrix} + \begin{pmatrix} 0 \\ b \\ 0 \end{pmatrix} = aU_1 + bU_2.$$

On the other hand, it is equally clear that any vector not in the xy place has a nonzero z-coordinate, and such a vector cannot be written as a linear combination of U_1 and U_2 because every linear combination of these two vectors has a z-coordinate equal to 0. This shows that Span(S) is precisely the xy-plane in $\mathbb{R}^3$. This question was easy to answer in this example because of the special form of the vectors in S.

In the preceding example, we know Span$\{v, w\}$ is a plane in $\mathbb{R}^3$ (a plane through the origin, in fact, since $O = 0v + 0w$ is always a linear combination), but it is less obvious which plane it is. How would we answer the question if a given vector is in the span or not? If it is not clear how to answer this geometrically, we return to the definition, which will lead to an algebraic solution. For example, let $z = \begin{pmatrix} -9 \\ 13 \\ 11 \end{pmatrix}$. To see if z is an element of the span of $\{v, w\}$, we must find scalars x and y such that

$$xv + yw = z,$$

$$x \begin{pmatrix} 1 \\ 3 \\ 5 \end{pmatrix} + y \begin{pmatrix} 3 \\ -1 \\ 1 \end{pmatrix} = \begin{pmatrix} -9 \\ 13 \\ 11 \end{pmatrix}.$$

By the definition of scaling and adding, this means

$$\begin{pmatrix} x + 3y \\ 3x - y \\ 5x + y \end{pmatrix} = \begin{pmatrix} -9 \\ 13 \\ 11 \end{pmatrix}.$$

But to say that two vectors are equal is to say that their coordinates agree. Therefore, we must find x and y such that

$$x + 3y = -9$$
$$3x - y = 13$$
$$5x + y = 11.$$

This is a linear system. Thus, to answer whether or not a given vector z is in the span of a set of vectors, we must solve a linear system with the given vector z on the right side of the equations and the set of vectors are the columns of the coefficient matrix on the left side. The reader can check (Exercise 5) that the unique solution to this system is $x = 3$ and $y = -4$. This shows the given vector z does belong to the span of $\{v, w\}$:

$$z = 3v - 4w.$$

On the other hand, we claim that $z = \begin{pmatrix} 1 \\ 2 \\ 1 \end{pmatrix}$ is not in the span of $\{v, w\}$. Indeed, if it were, we could find scalars x and y such that

$$xv + yw = z$$

$$x \begin{pmatrix} 1 \\ 3 \\ 5 \end{pmatrix} + y \begin{pmatrix} 3 \\ 2 \\ -1 \end{pmatrix} = \begin{pmatrix} 1 \\ 2 \\ 1 \end{pmatrix}.$$

This would require solving the system:

$$x + 3y = 1$$
$$3x - y = 2$$
$$5x + y = 1.$$

We leave it to the reader to check that there is no solution to this system (Exercise 6). Thus, z is not an element of $\mathrm{Span}\{v, w\}$.

Another type of question we could ask is, if a vector z is in $\mathrm{Span}(S)$, then can we write z as a linear combination of the elements of S in more than one way, or is such an expression unique? Consider the following example: Let

$$S = \{u, v, w\} = \left\{ \begin{pmatrix} 4 \\ -1 \\ 2 \end{pmatrix}, \begin{pmatrix} 1 \\ 3 \\ 3 \end{pmatrix}, \begin{pmatrix} -2 \\ 7 \\ 4 \end{pmatrix} \right\}.$$

Let $z = \begin{pmatrix} 14 \\ 16 \\ 22 \end{pmatrix}$. Show that $z \in Span(S)$, and determine if z can be written as a linear combination of the elements in S in a unique way. To solve this, we try to find scalars $r, s,$ and t such that $ru + sv + tw = z$; that is

$$r \begin{pmatrix} 4 \\ -1 \\ 2 \end{pmatrix} + s \begin{pmatrix} 1 \\ 3 \\ 3 \end{pmatrix} + t \begin{pmatrix} -2 \\ 7 \\ 4 \end{pmatrix} = \begin{pmatrix} 14 \\ 16 \\ 22 \end{pmatrix}.$$

Again, by definition of scaling and adding vectors, this means r, s, t must satisfy

$$\begin{pmatrix} 4r + s - 2t \\ -r + 3s + 7t \\ 2r + 3s + 4t \end{pmatrix} = \begin{pmatrix} 14 \\ 16 \\ 22 \end{pmatrix},$$

which leads to the system:

$$4r + s - 2t = 14$$
$$-r + 3s + 7t = 16$$
$$2r + 3s + 4t = 22.$$

Again, we leave it to the reader to find a complete solution to this system (Exercise 7). Here, we will be content to give two distinct solutions:

$(r, s, t) = (3, 4, 1)$ works. Also, $(r, s, t) = (2, 6, 0)$ works, and so we see that, indeed, z belongs to the span of S, but not in a unique way. Our final example is this: Suppose

$$S = \left\{ \begin{pmatrix} 1 \\ 1 \\ 0 \end{pmatrix}, \begin{pmatrix} 0 \\ 1 \\ 1 \end{pmatrix}, \begin{pmatrix} 1 \\ 1 \\ 1 \end{pmatrix} \right\}.$$

The question is to show that any vector $v = \begin{pmatrix} a \\ b \\ c \end{pmatrix}$ is in the span of S, and the expression writing z as a linear combination of the elements of S is always unique. Here, we must find scalars x, y, z such that

$$x \begin{pmatrix} 1 \\ 1 \\ 0 \end{pmatrix} + y \begin{pmatrix} 0 \\ 1 \\ 1 \end{pmatrix} + z \begin{pmatrix} 1 \\ 1 \\ 1 \end{pmatrix} = \begin{pmatrix} a \\ b \\ c \end{pmatrix}.$$

As in previous examples, this leads to a system of linear equations:

$$x + z = a$$
$$x + y + z = b$$
$$y + z = c.$$

Again, we ask the reader to solve the system (Exercise 8). This time, what should happen is that no matter what $a, b,$ and c are, one obtains a unique solution to the system. In fact, the solution is $(x, y, z) = (b - c, b - a, a - b + c)$. This last example, where every element of Span(S) has a unique expression in terms of S, is important.

DEFINITION 1.18 *A set S of vectors is* linearly independent *if every element of Span(S) can be expressed uniquely as a linear combination of the vectors in S.*

So our preceding examples show that the set

$$S = \left\{ \begin{pmatrix} 1 \\ 1 \\ 0 \end{pmatrix}, \begin{pmatrix} 0 \\ 1 \\ 1 \end{pmatrix}, \begin{pmatrix} 1 \\ 1 \\ 1 \end{pmatrix} \right\}$$

is linearly independent, but the set

$$S = \left\{ \begin{pmatrix} 4 \\ -1 \\ 2 \end{pmatrix}, \begin{pmatrix} 1 \\ 3 \\ 3 \end{pmatrix}, \begin{pmatrix} -2 \\ 7 \\ 4 \end{pmatrix} \right\}$$

is not linearly independent because we found a vector that can be expressed as a linear combination of the vectors in S in more than one way. Sets that are not linearly independent are called *linearly dependent*. We remark that most textbooks in linear algebra give an alternate definition of linear independence, but that definition is equivalent to ours (see Exercise 9).

DEFINITION 1.19 *If a set of vectors S is linearly independent, then we say S is a* basis *for Span(S). In particular, if Span(S) = $\mathbb{R}^n$, and S is linearly independent, we say S is a basis for $\mathbb{R}^n$.*

Example 1.20 The set $S = \{U_1, U_2\} = \left\{ \begin{pmatrix} 1 \\ 0 \end{pmatrix}, \begin{pmatrix} 0 \\ 1 \end{pmatrix} \right\}$ is a basis for the plane $\mathbb{R}^2$. It is called

the *standard basis*. To see it is a basis, observe that any vector $\begin{pmatrix} a \\ b \end{pmatrix} = aU_1 + bU_2$, so any

vector in the plane belongs to Span(S). Furthermore, the set S is linearly independent, since the

only solution to $\begin{pmatrix} a \\ b \end{pmatrix} = xU_1 + yU_2$ is $x = a$ and $y = b$. (The reader should check this by

writing down the corresponding system of equations – note that the augmented coefficient matrix
is already in reduced row echelon form!) This example generalizes. In $\mathbb{R}^n$, let U_i be the unit vector
with a 1 in the ith coordinate and 0 in every other coordinate. Then $S = \{U_1, U_2, \ldots, U_n\}$ is a basis
for $\mathbb{R}^n$ called the *standard basis*.

The past few examples clearly indicate a connection between vectors and linear systems. Apparently, whenever we try to express one vector as a linear combination of other vectors, we are led naturally to a linear system. We elaborate on that connection in the next section.

We remark that in a course in linear algebra, one spends a lot more time exploring the subtle concepts of linear independence, spanning sets, and bases. We have only scratched the surface here. Our purpose was to introduce just enough terminology to be useful in some later sections of the text, particularly in Chapters 3 and 5.

We close this section with one more observation that will be useful in Chapters 3 and 5. Earlier, we pointed out that the collection of all scalar multiples of a single nonzero vector v is simply the line through v. Or, in the language of spanning sets, Span$\{v\}$ is the line through v. But we also observed that the collection of nonnegative scalar multiples of v is the positive ray through v. What happens when S has two (or more) vectors? Consider the case when $S = \{v, w\}$, where S is a linearly independent set. Then Span(S) $= \{xv + yw | x, y \in \mathbb{R}\}$ By the parallelogram law for vector addition, as we noted before, every one of these linear combinations lies in the plane P containing v and w. Although we won't prove it, the converse is true also – any vector in the plane of v and w can be written as a sum of a multiple of v and a multiple of w, whence belongs to Span(S). So just as the span of a single vector is a line through the origin and containing v, the span of a pair of (linearly independent) vectors is a plane through the origin containing both v and w.

But what about the nonnegative linear combinations? Well, again, if $x \geq 0$, then xv lies on the positive ray through v, and similarly, $y \geq 0$ implies yw lies on the positive ray through w. It follows, again, by the parallelogram law, that $xv + yw$ is the diagonal of the parallelogram with sides xv and yw. In particular, not only is $xv + yw$ in the plane of v and w, but it is between these positive rays. Figure 1.13 illustrates the idea. The plane shown is the plane of v and w. The lines through v and w divide the plane into four regions much like the x- and y-axes divide the plane into quadrants. The nonnegative linear combinations are those in the "first" quadrant – they lie between the "positive v-axis" and the "positive w-axis." The region is shaded below in Figure 1.13.

Finally, we note that if $\{v, w\}$ is linearly dependent, then w and v are parallel (Exercise 10). Thus, w is a scalar multiple of v, so the entire span of $\{v, w\}$ collapses to Span$\{v\}$, the line through v.

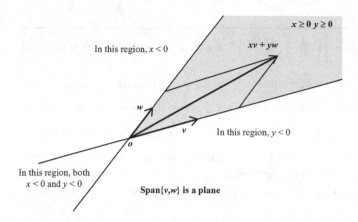

Figure 1.13 Nonnegative linear combinations of v and w.

Exercises

1. Let $v = \begin{pmatrix} -2 \\ 5 \end{pmatrix}$ and $w = \begin{pmatrix} 6 \\ 1 \end{pmatrix}$. Plot v, w and the following vectors in the plane:

    a. $v + w$ (draw the parallelogram illustrating the parallelogram law).
    b. $v - w$ (draw the parallelogram illustrating the parallelogram law. [**Hint**: $v - w = v + (-w)$.]
    c. The vector $\overrightarrow{PQ}$, where $P = (3, -4)$ and $Q = (1, 3)$.
    d. $z = -3v$ and $z = \frac{1}{3}w$.
    e. $z = 3v + w$.
    f. $z = -4v + 2w$.
    g. Shade in the region of the plane that contains all the nonnegative linear combinations of v and w.

2. In this exercise, we illustrate how theorems in Euclidean geometry may be proved using vectors. Let $ABCD$ be an arbitrary quadrilateral in the plane. Suppose the coordinates are $A = (x_1, y_1)$, $B = (x_2, y_2)$, $C = (x_3, y_3)$, and $D = (x_4, y_4)$. Let P be the midpoint of side AB, Q the midpoint of side BC, R the midpoint of side CD, and S the midpoint of side DA.

    a. Find the coordinates of P, Q, R, S (using the midpoint formula for points in the plane you learned in your high school algebra class).
    b. Compute the coordinates of the vector $\overrightarrow{PQ}$ and the coordinates of the vector $\overrightarrow{RS}$. Show that when you add these vectors, you obtain $\overrightarrow{PQ} + \overrightarrow{RS} = 0$.
    c. Similarly, show that $\overrightarrow{QR} + \overrightarrow{SP} = 0$.
    d. Explain why the results of Parts b and c imply that the quadrilateral $PQRS$ is a parallelogram.

    You have just proved the following theorem.

    THEOREM 1.21   *The quadrilateral $PQRS$ obtained by joining the midpoints of an arbitrary quadrilateral $ABCD$ is always a parallelogram.*

Figure 1.14 illustrates the theorem.

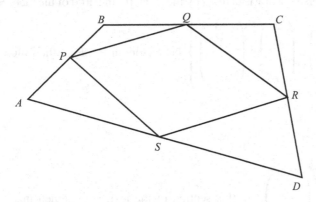

Figure 1.14 $PQRS$ is a parallelogram.

3. We mentioned that some of the properties of Theorem 1.15 are easy to verify. For example, to verify the commutative law for addition of vectors, let $u = \begin{pmatrix} a_1 \\ a_2 \\ \vdots \\ a_n \end{pmatrix}$ and $v = \begin{pmatrix} b_1 \\ b_2 \\ \vdots \\ b_n \end{pmatrix}$.

Then

$$u + v = \begin{pmatrix} a_1 \\ a_2 \\ \vdots \\ a_n \end{pmatrix} + \begin{pmatrix} b_1 \\ b_2 \\ \vdots \\ b_n \end{pmatrix} = \begin{pmatrix} a_1 + b_1 \\ a_2 + b_2 \\ \vdots \\ a_n + b_n \end{pmatrix}$$

by definition of vector addition. But now look at each coordinate – these are real numbers, and we already know the commutative law holds for real numbers. Thus, for each $i = 1, 2, \ldots, n$, we know $a_i + b_i = b_i + a_i$. Thus, the vector on the right is the same vector as

$$\begin{pmatrix} b_1 + a_1 \\ b_2 + a_2 \\ \vdots \\ b_n + a_n \end{pmatrix} = \begin{pmatrix} b_1 \\ b_2 \\ \vdots \\ b_n \end{pmatrix} + \begin{pmatrix} a_1 \\ a_2 \\ \vdots \\ a_n \end{pmatrix} = v + u$$

again by using the definition of vector addition. Thus, the commutative law for vectors follows from the commutative law for real numbers. Use similar ideas, working with coordinates, to prove the following parts of Theorem 1.15:

a. Part b of the theorem
b. Part d of the theorem
c. Part e of the the theorem
d. Part g of the theorem
e. Part h of the theorem

f.   Prove that the additive inverse $-v$ of v is the same vector as $(-1)v$ (scalar multiple.) [**Hint**: the additive inverse $-v$ of v is defined by the property in Part d of the theorem.]

4.   Let $S = \{u, v, w\} = \left\{ \begin{pmatrix} 1 \\ 2 \\ 1 \\ 3 \end{pmatrix}, \begin{pmatrix} 2 \\ 0 \\ 0 \\ 1 \end{pmatrix}, \begin{pmatrix} -4 \\ 3 \\ 2 \\ -1 \end{pmatrix} \right\}$ be vectors in $\mathbb{R}^4$. Find the following elements of Span(S):

a. $2u + v + w$
b. $5u + 2w$
c. $7u - 2v - 3w$

5.   Let $v = \begin{pmatrix} 1 \\ 3 \\ 5 \end{pmatrix}$ and $w = \begin{pmatrix} 3 \\ -1 \\ 1 \end{pmatrix}$. In this section of the text, we claimed that $z = \begin{pmatrix} -9 \\ 13 \\ 11 \end{pmatrix}$ is an element of Span$\{v, w\}$. To check this, we were led to solving the following system:

$$x + 3y = -9$$
$$3x - y = 13$$
$$5x + y = 11$$

Solve the system and verify that the unique solution is $(x, y) = (3, -4)$ as we claimed.

6.   In this section of the text, we claimed that $z = \begin{pmatrix} 1 \\ 2 \\ 1 \end{pmatrix} \notin \text{Span}\{v, w\}$, where v and w are the vectors in Exercise 5. Verify this claim by showing that the following system has no solution:

$$x + 3y = 1$$
$$3x - y = 2$$
$$5x + y = 1$$

7.   In this section of the text, we considered the example

$$S = \{u, v, w\} = \left\{ \begin{pmatrix} 4 \\ -1 \\ 2 \end{pmatrix}, \begin{pmatrix} 1 \\ 3 \\ 3 \end{pmatrix}, \begin{pmatrix} -2 \\ 7 \\ 4 \end{pmatrix} \right\}$$

and we claimed that the vector $z = \begin{pmatrix} 14 \\ 16 \\ 22 \end{pmatrix}$ was an element of Span(S), but that the expression of z as a linear combination of elements of S was not unique. Verify this by finding the complete solution to the system we derived:

$$4r + s - 2t = 14$$
$$-r + 3s + 7t = 16$$
$$2r + 3s + 4t = 22.$$

Write the solution in parametric form, and verify that the two specific solutions we gave in the text are elements of this parametric family.

8. In this section of the text, we considered this example:

$$S = \left\{ \begin{pmatrix} 1 \\ 1 \\ 0 \end{pmatrix}, \begin{pmatrix} 0 \\ 1 \\ 1 \end{pmatrix}, \begin{pmatrix} 1 \\ 1 \\ 1 \end{pmatrix} \right\}.$$

We made two claims about this set:

i. Any vector $z = \begin{pmatrix} a \\ b \\ c \end{pmatrix}$ in $\mathbb{R}^3$ is an element of Span(S) (that is, Span$(S) = \mathbb{R}^3$), and

ii. For any vector z, the expression of z as a linear combination of the elements of S is unique (that is, S is a linearly independent set.) In order to verify these claims, we would have to show that for any vector z, the following system has a unique solution:

$$x + z = a$$
$$x + y + z = b$$
$$y + z = c.$$

Verify these claims by solving this system, and show that the unique solution is the one we gave in the text.

9. a. Consider the following set:

$$S = \{u, v, w\} = \left\{ \begin{pmatrix} 1 \\ 1 \\ 1 \end{pmatrix}, \begin{pmatrix} 2 \\ 0 \\ 1 \end{pmatrix}, \begin{pmatrix} -1 \\ 2 \\ 3 \end{pmatrix} \right\}.$$

We mentioned in this section that Span(S) must always contain the vector 0. This is because we always have the following linear combination, called the trivial linear combination:

$$0u + 0v + 0w = 0.$$

In this exercise, show that the trivial linear combination is the *only* linear combination of the 0 vector. That is, show that the only solution to the following system is the trivial solution $x = y = z = 0$:

$$x + 2y - z = 0$$
$$x + 2z = 0$$
$$x + y + 3z = 0.$$

REMARK 1.22 *A system of linear equations with all 0s on the right-hand side is said to be a* homogeneous *system.*

b.   By contrast, consider the following set:

$$T = \{u, v, w\} = \left\{ \begin{pmatrix} 1 \\ -2 \\ 3 \end{pmatrix}, \begin{pmatrix} 4 \\ 1 \\ 2 \end{pmatrix}, \begin{pmatrix} 11 \\ -4 \\ 13 \end{pmatrix} \right\}.$$

Show that, in addition to the trivial linear combination, there are nontrivial linear combinations of the 0 vector. Do this by solving the system and writing the solution in parametric form:

$$x + 4y + 11z = 0$$
$$-2x + y - 4z = 0$$
$$3x + 2y + 13z = 0$$

c.   An alternate way to show that there are nontrivial linear combinations of the vectors in T giving the zero vector 0 is the following. Show that the third vector w is in fact in the span of the first two. That is, show that you can find scalars s, t such that

$$su + tv = w.$$

Do this by solving the appropriate system of linear equations. Once you have the solution, put all the vectors on one side of the equation, writing

$$su + tv - w = 0,$$

and simply observe that this is a nontrivial (because the coefficient of w is not zero) linear combination of the elements of T equal to 0.

REMARK 1.23   *Generalizing this example, it is not hard to prove the following:*

a.   *A set S of vectors is linearly independent if and only if the only linear combination of the elements of S yielding the zero vector 0 is the trivial one. (This is the alternate definition of linear independence given in most linear algebra textbooks. For example, see Lay, 2012.)*

b.   *A set T of vectors is linearly dependent if and only if one of the vectors in T can be expressed as a linear combination of the other vectors in T.*

10.   Let $S = \{u, v\}$ be a set of two vectors. Show that S is a linearly dependent set if and only if v and u are parallel (that is, if and only if v is a scalar multiple of u.)

1.5   Three Points of View

In the last section, we saw that certain questions about linear combinations, spanning sets, and linear independence reduce to solving a system of linear equations. To summarize, if $S = \{v_1, v_2, \ldots, v_m\}$ is a set of vectors and z is a vector, then the expression

$$x_1 v_1 + x_2 v_2 + \cdots + x_m v_m = z$$

led to a system of equations where the augmented coefficient matrix of the system had columns made from the vectors v_i and z. If the system has no solution, z is not in the span of S. If there is a solution, $z \in \text{Span}(S)$. If the solution is unique, then S is a linearly independent set. In fact, the time has come to reveal that the language of vectors and linear combinations is an entirely different way of viewing a system of equations. Any system of equations can be rephrased as a linear combination equation. This is the main reason why we introduced vectors. For example, the system

$$5x + 3y + 2z = 25$$
$$2x - y + 8z = 18$$
$$x + y - 4z = 10$$
$$9x - 7z = 15$$

can be rewritten as a vector equation by reversing the steps we executed in the last section. Thus, think of each side as a vector in $\mathbb{R}^4$. The system becomes a statement of equality of the two vectors:

$$\begin{pmatrix} 5x + 3y + 2z \\ 2x - y + 8z \\ x + y - 4z \\ 9x - 7z \end{pmatrix} = \begin{pmatrix} 25 \\ 18 \\ 10 \\ 15 \end{pmatrix}.$$

Now use the definition of vector addition and scalar multiplication to rewrite the vector on the left as a sum of three vectors, separating the unknowns into different columns and then factoring out the unknown scalars:

$$\begin{pmatrix} 5x + 3y + 2z \\ 2x - y + 8z \\ x + y - 4z \\ 9x - 7z \end{pmatrix} = \begin{pmatrix} 5x \\ 2x \\ x \\ 9x \end{pmatrix} + \begin{pmatrix} 3y \\ -y \\ y \\ 0 \end{pmatrix} + \begin{pmatrix} 2z \\ 8z \\ -4z \\ -7z \end{pmatrix}$$

$$= x\begin{pmatrix} 5 \\ 2 \\ 1 \\ 9 \end{pmatrix} + y\begin{pmatrix} 3 \\ -1 \\ 1 \\ 0 \end{pmatrix} + z\begin{pmatrix} 2 \\ 8 \\ -4 \\ -7 \end{pmatrix}.$$

Therefore, the system of equations becomes

$$x\begin{pmatrix} 5 \\ 2 \\ 1 \\ 9 \end{pmatrix} + y\begin{pmatrix} 3 \\ -1 \\ 1 \\ 0 \end{pmatrix} + z\begin{pmatrix} 2 \\ 8 \\ -4 \\ -7 \end{pmatrix} = \begin{pmatrix} 25 \\ 18 \\ 10 \\ 15 \end{pmatrix}.$$

So solving the system is tantamount to expressing $\begin{pmatrix} 25 \\ 18 \\ 10 \\ 15 \end{pmatrix}$ as a linear combination of the three column vectors on the left side.

We now have three different points of view for dealing with linear systems. The first is the system itself, the second is the augmented coefficient matrix (and the associated elementary row operations), and the third is this language of linear combinations of vectors. While we will almost always use the point of view of row operations and Gauss-Jordan elimination to actually solve a system, the ability to consider the system from these alternate perspectives will give us a great deal of flexibility. For example, in Chapter 3, we will find it very convenient to express systems as linear combinations in order to develop certain techniques to solve linear programming problems. Once certain information has been gleaned from this point of view, we convert everything back to the coefficient matrix and elimination perspective in Chapter 5 to derive the simplex algorithm for solving these problems.

This illustrates the power of linear algebra. It is the ability to view the same thing from these different perspectives that makes it so useful. In fact, in Chapter 2, once we learn matrix algebra, we will add a fourth point of view for regarding a system of equations that will extend our flexibility even more. We will exploit this additional perspective in Chapter 9 when we study sensitivity analysis using the simplex algorithm.

Even that is not the end of the story. Anyone who has studied linear algebra knows there is yet a fifth point of view for linear systems, using the notions of *linear transformations*. We will not pursue that point of view in this text.

Table 1.1 is a summary of how to translate between the various points of view. So let $S = \{v_1, v_2, \ldots, v_n\}$ be a set of vectors in $\mathbb{R}^m$. Let $z \in \mathbb{R}^m$, and consider the question of writing z as a linear combination of the vectors in S:

$$x_1 v_1 + x_2 v_2 + \cdots + x_n v_n = z.$$

Here, $x_1, \ldots, x_n$ are unknown scalars. The corresponding system of equations has m equations (because the vectors have m coordinates) and n unknowns.

Let A be the $m \times n$ coefficient matrix, and $[A|z]$ the corresponding augmented coefficient matrix of the system.

Table 1.1 Translating between the three points of view

System of linear equations	Augmented coefficient matrix	Vector equation (linear combination)
No solution (inconsistent system)	Reduced row echelon form has a row of the form $(0, 0, \ldots, 0, a)$, $a \neq 0$	$z \notin \mathrm{Span}(S)$
Has a solution (consistent system)	Reduced row echelon form has no rows of the form $(0, 0, \ldots, 0, a)$, $a \neq 0$	$z \in \mathrm{Span}(S)$
Unique solution (non-degenerate system)	Reduced row echelon form is $[I|x]$, where I is an identity matrix, x is the unique solution, and reduced echelon form has rows of all 0 if $m > n$. No free variables.	$z \in \mathrm{Span}(S)$ S is linearly independent. The only linear combination which yields the 0 vector is the trivial one.
Infinitely many solutions (redundant system)	Reduced echelon form has at least one free variable	$z \in \mathrm{Span}(S)$ S is linearly dependent. There are nontrivial linear combinations which yield the 0 vector.

We close this section by defining certain types of solutions to redundant systems that will be important in Chapters 3 and 5.

DEFINITION 1.24   *Consider a system of m linear equations, and write it as a vector equation (linear combination) in $\mathbb{R}^m$:*

$$x_1 v_1 + x_2 v_2 + \cdots + x_n v_n = b.$$

Assume that the system is redundant so that there are infinitely many solutions. Consider a given solution $x_i = a_i$ for some real numbers a_i, and let $S = \{v_i \mid a_i \neq 0\}$. The solution is said to be a basic solution of the system if S is a basis of $\mathbb{R}^m$.

Consider the example:

$$4w + 3x + y + 4z = 11$$
$$2w - x + y = 3.$$

This must be a redundant system if it is consistent, because if you solved it via Gauss-Jordan elimination, it is impossible for each column to have a leading 1, so there are free variables in the solution. Consider the specific solution $(w, x, y, z) = (0, 5, 8, -3)$. In order to determine whether it is basic or not, we rewrite the system as a vector equation:

$$w \begin{pmatrix} 4 \\ 2 \end{pmatrix} + x \begin{pmatrix} 3 \\ -1 \end{pmatrix} + y \begin{pmatrix} 1 \\ 1 \end{pmatrix} + z \begin{pmatrix} 4 \\ 0 \end{pmatrix} = \begin{pmatrix} 11 \\ 3 \end{pmatrix}.$$

In the solution, $w = 0$, so the set $S = \left\{ \begin{pmatrix} 3 \\ -1 \end{pmatrix}, \begin{pmatrix} 1 \\ 1 \end{pmatrix}, \begin{pmatrix} 4 \\ 0 \end{pmatrix} \right\}$. This set is not a basis of $\mathbb{R}^2$. Students who have taken linear algebra will recognize this because they know any set of vectors in $\mathbb{R}^2$ with more than two vectors cannot be linearly independent (because the plane is two-dimensional). For students who have not taken linear algebra, just observe that

$$\begin{pmatrix} 3 \\ -1 \end{pmatrix} + \begin{pmatrix} 1 \\ 1 \end{pmatrix} = \begin{pmatrix} 4 \\ 0 \end{pmatrix},$$

and use Part b of the remark in Exercise 9 from the last section. Thus, the given solution is not basic.

On the other hand, consider the solution $(w, x, y, z) = (2, 1, 0, 0)$. In this case, the set $S = \left\{ \begin{pmatrix} 4 \\ 2 \end{pmatrix}, \begin{pmatrix} 3 \\ -1 \end{pmatrix} \right\}$. This set is a basis of the plane. Indeed, the two vectors are not parallel, so by Exercise 10 from Section 1.4, this set is linearly independent. We still have to check that it spans the plane, which means showing that for any vector $c = \begin{pmatrix} r \\ s \end{pmatrix}$ in the plane, there is a solution to

$$x \begin{pmatrix} 4 \\ 2 \end{pmatrix} + y \begin{pmatrix} 3 \\ -1 \end{pmatrix} = \begin{pmatrix} r \\ s \end{pmatrix}.$$

We do this by converting to the augmented matrix point of view

$$\begin{bmatrix} 4 & 3 & r \\ 2 & -1 & s \end{bmatrix}$$

The reader should check that the reduced row echelon form of this matrix is

$$\begin{bmatrix} 1 & 0 & \frac{r+3s}{10} \\ 0 & 1 & \frac{r-2s}{5} \end{bmatrix}.$$

This shows directly that there is a (unique) solution for every r, s. Thus, the set S is a basis for $\mathbb{R}^2$, and so the given solution is a basic solution.

Exercises

1. Convert the following system of equations to a vector equation – that is, into a linear combination. Do not solve the vector equation, but write down explicitly what the set S of vectors is and what the vector to be expressed in terms of S is.

$$19x + 5y - 12z = 109$$
$$24x + 11y + z = 168$$
$$16x - 2y + 30z = 150$$

2. Follow the same directions as Exercise 1 for this system:

$$2p + q + 5r - 3s + 10t = 800$$
$$p + r + t = 90$$
$$-p + 2q - 3r + 4s - 5t = 0$$
$$10p + 15q + 6r - 7s + 3t = 450$$
$$4p + 10q + r - 12t = 227$$

3. The following matrix is the augmented matrix for a linear system.

$$\begin{pmatrix} 1 & 0 & 4 & -2 & 3 & 99 \\ 0 & 1 & 6 & 2 & 0 & 50 \\ 5 & -2 & 3 & 3 & 3 & 39 \\ 4 & 1 & 4 & 1 & 4 & 41 \\ 6 & 5 & 4 & 3 & 2 & 1 \end{pmatrix}$$

    a. Write down explicitly the linear system it represents. Do not solve it.
    b. Write down explicitly the corresponding vector equation (linear combination).

4. In the Table 1.1, in the case where there is a unique solution, explain why there must be rows of all 0s at the bottom of the augmented matrix in reduced echelon form in the case when $m > n$. [**Hint**: Think about free variables and leading 1s.]

5. In case $m < n$, every row has a leading 1; explain why there cannot be a unique solution. More generally, recall that the *rank* of a matrix is the number of leading 1s when the matrix is put in reduced row echelon form. Explain why, if rank($[A|z]$) $< n$, there cannot be a unique solution.

6. In the third row of Table 1.1, in the case of a unique solution, we are suggesting that if there is a unique solution for this one particular z, then S must be a linearly independent set. However, the alert reader will remember from the definition of linearly independent is that

there must be a unique solution for <u>every</u> z in the span of S, not just for one particular z. Explain why the discrepancy doesn't matter. That is, explain why if the solution is unique for one z in the span, it must be unique for every z in the span.

a.   Use the perspective of vector equations to answer this. [**Hint**: If you have two distinct solutions for a particular z, subtract them to obtain a nontrivial linear combination yielding 0. Once you have that, you can add it to ANY solution for any z to produce a different solution for the same z.]

b.   Instead, answer this from the perspective of the augmented coefficient matrix. [**Hint**: If you have a unique solution for a particular z, can there be a leading 1 in the last column? If not, then what happens when you change the z (i.e., change the last column)? Where do the leading 1s end up?]

7.   Consider the system:

$$12w + 4x + 2y - 3z = 14$$
$$4w + x + y + 3z = 6$$

a.   Verify that $(w, x, y, z) = (2, -3, 1, 0)$ is a solution. Determine whether or not it is a basic solution.

b.   Verify that $(w, x, y, z) = (0, 1, 5, 0)$ is a solution. Deteremine whether or not it is a basic solution.

c.   Given that $S = \left\{ \begin{pmatrix} 2 \\ 1 \end{pmatrix}, \begin{pmatrix} -3 \\ 3 \end{pmatrix} \right\}$ is a basis for $\mathbb{R}^2$, find the corresponding basic solution.

8.   Consider the system:

$$w + 4x + 2y + 2z = 14$$
$$2w + 5x + 2y = 19$$
$$3w + 6x + 2y - 3z = 24$$

Verify that $(w, x, y, z) = (4, 1, 3, 0)$ is a solution, but not a basic solution.

9.   Consider the system:

$$p + q + r + 5s + t = 100$$
$$q + r + 8s - 3t = 225$$
$$r + 10s + 7t = 169$$

Explain why any solution where $s = t = 0$ must be a basic solution.

1.6 Some Nonlinear Models

In this section, we make a brief digression form all things linear to discuss some nonlinear topics. The main item from this section that we need later in the text is the material on a particular type of quadric surface we encounter Section 1.6.3. This material will be used in Chapter 6 when we discuss zero-sum games. We build up to this by discussing quadratic curves first in Section 1.6.1, and related applications of this in Section 1.6.2.

1.6.1   Standard Form for Quadratic Curves

In the other sections of this chapter, we discussed systems of equations without restriction on the number of variables involved. However, we did confine our attention to just linear functions (that is, the degree of the variables was no larger than 1) so things would not be too complicated. In this section, we are interested in some models where the functions involved can be degree 2. Since second-degree functions are more complicated than linear functions, we restrict the number of variables involved instead in order to retain some clarity. So, in this subsection, we consider certain quadratic curves in the plane (so the number of variables is just two).

Since there are just two variables involved, the general quadratic curve is the graph of an equation of this form:

$$Ax^2 + Bxy + Cy^2 + Dx + Ey + F = 0.$$

The expressions x^2 and y^2 are quadratic terms – degree 2. Also, the term xy is the product of two degree 1 terms, so it is also considered to be "total degree" 2. If A, B, and C are all 0, then the equation reduces to just linear (degree 1) and constant (degree 0) terms, and we know the graph is a straight line. In your previous algebra or precalculus classes, you may have studied all the other cases. You might recall that if $B = 0$ and $A = C$, the graph reduces to a circle. If $B = 0$ and $A \neq C$, but they have the same sign, then it becomes an ellipse, and if A and C have opposite signs, it becomes a hyperbola. If either A or C is 0 in addition to $B = 0$, then the graph becomes a parabola, with the axis of symmetry parallel to one of the coordinates axes. Finally, if $B \neq 0$, one still obtains an ellipse, parabola, or hyperbola, but rotated so that its axes of symmetry are not parallel to the coordinate axes.

Our intention here is not to cover all these cases but to focus on one particular case – the case when $B = C = 0$ and $E \neq 0 \neq A$. In this case, you can solve for y in terms of x and obtain a function of the form

$$y = f(x) = ax^2 + bx + c.$$

By comparing the expression $y = f(x) = ax^2 + bc + c$ to the general quadratic curve above with equation $Ax^2 + Bxy + Cy^2 + Dx + Ey + F = 0$, you can determine that $a = -\frac{A}{E}$, $b = -\frac{D}{E}$ and $c = -\frac{F}{E}$, although these relationships will not be used in what follows. So this is the case of a parabola where the axis of symmetry is vertical – the parabola opens upward if $a > 0$ and opens downward if $a < 0$. The point where the parabola intersects its axis of symmetry is called the *vertex* of the parabola. Two cases are sketched in Figure 1.15.

Our only goal in this subsection is to be able to graph these functions quickly, much like we can graph straight lines quickly. The key to doing this is to think about transforming the graph of the simplest parabola – with equation $y = x^2$ – into the more general case by manipulations such as translating and scaling.

The reader is probably familiar with the graph of the parabola $y = x^2$, shown in Figure 1.16.

Note that the vertex is the origin $(0,0)$, and the axis of symmetry is the y-axis with equation $x = 0$. Consider an equation of the form $y = ax^2$. To obtain the graph, think of a point on the graph of $y = x^2$, such as $(1,1)$ or $(2,4)$, and multiply the y-coordinate by a to obtain the

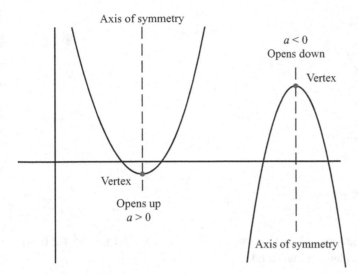

Figure 1.15 Typical parabolas.

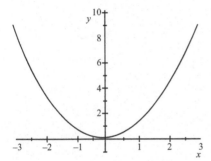

Figure 1.16 The graph of $y = x^2$ has vertex at the origin and axis of symmetry is the y-axis.

corresponding point on the graph of $y = ax^2$. For example, if $a = 2$, we double the y value. Since we do this for every point, the graph of $y = 2x^2$ is obtained by vertically stretching the original graph by a factor of 2. Similarly, if $a = \frac{1}{2}$, we multiply all the y values by $\frac{1}{2}$, which is referred to as a vertical compression. The graphs of all three functions are shown together in Figure 1.17.

The graph of $y = 2x^2$ is the narrowest parabola (the top one), and $y = \frac{1}{2}x^2$ is the widest parabola (the bottom one). You can see in each case how the function has been stretched or compressed vertically. If $a < 0$, the compression or stretching is combined with a reflection across the x-axis, so that the curve now opens downward. For our purposes, the important thing to notice is that no matter what a is, the location of the vertex and the axis of symmetry are unchanged. All parabolas of the form $y = ax^2$ have a vertex at the origin and the y-axis as axis of symmetry.

We can change the location of the vertex by vertical or horizontal *translations*, as opposed to stretches and compressions. If we consider the graph of $y = g(x) = x^2 + k$ and compare it to the graph of $f(x) = x^2$, the difference is that we add k to each y value on $f(x)$ to obtain the corresponding point on $g(x)$. This amounts to a vertical translation (or vertical *shift*) of the entire graph by k units. If $k > 0$, the shift is upward and if $k < 0$ it is downward. Note that vertex moves

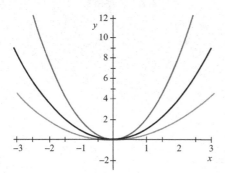

Figure 1.17 Vertical stretches and compressions.

to the point $(0, k)$. To illustrate, the graphs of $y = x^2$, $y = x^2 - 1$ (bottom) and $y = x^2 + 2$ (top) are graphed on the same coordinate plane in Figure 1.18.

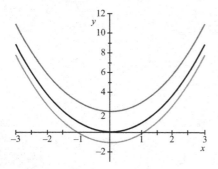

Figure 1.18 Vertical translations.

Finally, a horizontal translation, or shift, is obtained by adjusting the x-coordinate, rather than the y-coordinate. If you replace x with $x - h$, the result is a translation of h units to the right if $h > 0$. Replacing x with $x + h = x - (-h)$ results in a shift of $-h$ units to the right or h units to the left. To illustrate, the graphs of $y = x^2$, $y = (x - 3)^2$ (rightmost), and $y = (x + 1)^2$ (leftmost) are graphed on the same coordinate plane in Figure 1.19.

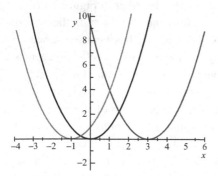

Figure 1.19 Horizontal translations.

Observe that with a horizontal shift, the axis of symmetry moves to the line $x = h$, and in particular, the vertex moves to the point $(h, 0)$.

If the reader would like more practice working with shifts, stretches, and compressions, more examples can be found in a precalculus text such as Kaufmann (1995). Next, we combine all of these transformations. We begin with $y = x^2$ and apply a horizontal translation by h units to the right, obtaining the graph of $y = (x - h)^2$. Then we apply a vertical stretch/compression y multiplying by a, obtaining the graph of $y = a(x - h)^2$. At this point, note that the vertex is still at $(h, 0)$. Finally, we apply a vertical shift of k units to obtain the graph of $y = a(x - h)^2 + k$, with vertex at (h, k).

For example, sketch a graph of $y = \frac{1}{2}(x - 4)^2 - 2$. The solution is that the vertex is at $(h, k) = (4, -2)$, and since $a = \frac{1}{2}$ is positive but less than 1, the graph opens upward and is somewhat wider that the parabola $y = x^2$ because of the vertical compression. The result is easy to plot, just by knowing where the vertex is, as shown in Figure 1.20.

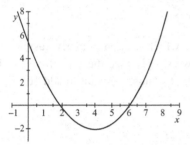

Figure 1.20 Since $a > 0$, the graph opens upward.

For another example, plot $y = -2(x - 1)^2 + 7$. In Figure 1.21, the vertex is $(1, 7)$, the graph opens downward since $a < 0$, and since $|a| > 1$, the graph is narrow because of the vertical stretch, as shown in Figure 1.21.

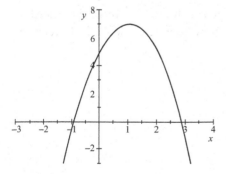

Figure 1.21 Since $a < 0$, the graph opens downward.

Since graphs in this form are so easy to plot, we give them a name.

DEFINITION 1.25 *A quadratic function in the form* $y = a(x - h)^2 + k$ *is said to be in* standard form. *The vertex is located at* (h, k), *and if* $a > 0$, *the parabola opens upward, and if* $a < 0$, *it opens downward.*

Now, what do we do if the equation is not in standard form? We know $y = ax^2 + bx + c$ (which is said to be in *general form*) must have a parabolic graph, but we would like to avoid having to plot many points to see what it looks like. The trick to turn any general form equation into standard form should be familiar to the reader – it is called *completing the square*. For the reader who needs to brush up on the technique, consult a precalculus text such as Kaufmann (1995). We'll do one example here to remind the reader of the method.

Example 1.26 Put the quadratic function $y = 3x^2 + 24x + 21$ into standard form, locate the vertex, and sketch the function.

To solve this, we focus our attention on the terms with x in them, and factor out the $a = 3$, ignoring the constant term:

$$y = 3x^2 + 24x - 10 = (3x^2 + 24x) + 21$$
$$= 3(x^2 + 8x) + 21.$$

Now, the expression in parenthesis $x^2 + 8x$ is a quadratic for which it is particularly easy to complete the square. When the coefficient of the x^2 term is 1, we simply take the x coefficient, divide it by 2, and square it. In this case, half of 8 is 4, and $4^2 = 16$. Thus, we want to add 16 to $x^2 + 8x$ to make it a perfect square:

$$x^2 + 8x + 16 = (x + 4)^2$$

Of course, this entire expression is multiplied by 3 in our function, so we really added $3 \cdot 16 = 48$. However, we do not want to change the value of y, so we must balance this by subtracting 48 outside of the parenthesis. The result is

$$y = 3(x^2 + 8x + 16) + 21 - 48$$
$$y = 3(x + 4)^2 - 27.$$

The square is now complete, and the equation is in standard form. The vertex is located at $(h, k) = (-4, -27)$, and the graph opens upward (and is narrow because of the vertical stretch by a factor of $a = 3$), as shown in Figure 1.22.

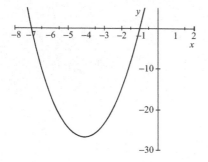

Figure 1.22 Complete the square to put the quadratic function in standard form for easy sketching.

Exercises on this will be at the end of the section.

1.6.2 Optimization

In this subsection, we apply what we learned in the previous subsection. We suppose we have a function whose graph is a curve in the plane. We seek the highest point (that is, the largest y value) on the function (called a *maximum* point), or the lowest point (the smallest y value) on the function (called a *minimum* point). Problems like this are called *optimization* problems.

For a general curve with perhaps many oscillations, we also might need to distinguish between points that are higher than all nearby points but not the highest point overall. These are called *local* maximum points (or *relative* maximum points), and similarly for local minimum points. We are interested in finding the highest local maximum, also called a *global* maximum, and the global minimum. Figure 1.23 illustrates this.

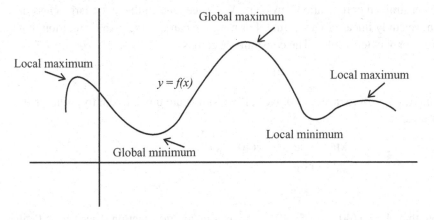

Figure 1.23 Local and global extreme points.

Locating all these local maxima and minima, for a general function, requires calculus, which we do not assume the reader knows. However, if the function is quadratic, then if there is a maximum or minimum point, it must be the vertex of the parabola, and we do know how to locate that. Therefore, in this subsection, we illustrate the process by solving some optimization problems when the function happens to be quadratic.

Example 1.27 With 80 feet of fencing forming the perimeter, what dimensions $x \times y$ form the rectangular garden of maximum area?

Let A represent the area of the garden, which has length x and width y. Thus, $A = xy$. We are seeking the largest-possible value for A, so it is an optimization problem, specifically a maximization problem. However, at the moment there are two obstacles to solving this problem. The first problem is there are two many variables – A, x, and y. We've written A as a function of x and $y - A = f(x, y) = xy$ – but being a function of two variables its graph is a surface, not a curve. The other obstacle is that we have not used all the information – namely that there is 80 feet of fencing available. It's a pleasant surprise that removing the second obstacle also removes the first at the same time! Indeed, by considering the perimeter (where the fence is located) of the garden, we see that

$$2x + 2y = 80.$$

This is a constraint under which we are working. If we want a really big area, all we have to do is make both x and y really large. What's to prevent us from having garden that is acres and acres? Simply that with 80 feet of fencing, the constraint must be satisfied, and given that neither x nor y can be negative, this means that they cannot be too large either. So, if we want to make a complete statement of the problem (which we should do in any model), it is this:

Find the maximum possible value of $A = xy$

subject to the conditions:

$2x + 2y = 80$

$x \geq 0, y \geq 0.$

This is a typical "constrained optimization" problem. We'll see later in the text a large class of such problems when we study linear programming. For now, we can use the given condition that $2x + 2y = 80$ to solve for y in terms of x. The constraint becomes

$$y = 40 - x.$$

Notice that this implies $x \leq 40$, since $y \geq 0$. Now, if we substitute this into the formula for A, we can restate the problem:

Maximize $A = x(40 - x)$

subject to:

$0 \leq x \leq 40.$

To solve this, note that $A = x(40 - x) = 40x - x^2$ is a quadratic function. Since $a < 0$, it opens downward, so the vertex must be the maximum point. Of course, the x-coordinate of this vertex must be in the interval $[0, 40]$ in order for the vertex to be the solution. To find the vertex, we complete the square:

$$\begin{aligned} A &= -x^2 + 40x \\ &= -(x^2 - 40x) \\ &= -(x^2 - 40x + 400) + 400 \\ &= -(x - 20)^2 + 400. \end{aligned}$$

Thus, the vertex is located at $(20, 400)$, and $x = 20$ is indeed in the desired interval $[0, 40]$. Thus, $y = 40 - x = 40 - 20 = 20$, and $A = 400$. The solution is the largest-possible rectangular garden enclosed by 80 feet of fencing is a 20×20 square with $A = 400$ square feet. We could plot the function $A(x) = 40x - x^2$ if desired, but that is not really necessary. We know it is a parabola opening downward from the vertex, so we know we have found the global maximum.

In solving optimization problems, one must pay attention to how the problem is stated. If the question is "where is the maximum (or minimum) value?", the solution is the x-coordinate of the global maximum (or minimum) point. If the question is "what is the maximum (or minimum) value?", the solution is the y-coordinate. In the preceding example, the question really only asked for "where" – the dimensions were to be found, but the actual maximum area of $A = 400$ was not explicitly asked for. Nevertheless, it never hurts to give the complete solution with both coordinates.

Example 1.28 Hoffs' Historical Walking Tours offers a four-day walking tour of the pyramids in Giza, Egypt. Hoffs' knows it can sell 60 tickets per month at \$3,000 per ticket. Based on a survey and market study, Hoffs' believes that each drop in price of \$100 for a ticket will generate three additional sales. What price should Hoffs' charge in order to maximize its revenue?

Since the price per ticket is the unknown we are searching for, it is tempting to let that be the variable x. But we would have to be able to express the total number of ticket sold in terms of x. This is possible, but there is a more natural approach that is easier. Since we begin with a set price of \$3,000, and we are considering dropping \$100 from this price multiple times, let x stand for the number of drops in price (of \$100). So we can express both the price per ticket and the total number of tickets sold easily in terms of x. Indeed, the price per ticket will be \$3,000 $- 100x$ by our definition of x. Also we know each drop results in three additional sales, so the total number of tickets sold is $60 + 3x$. It follows that the total revenue is

$$R(x) = (3,000 - 100x)(60 + 3x) = -300x^2 + 3,000x + 180,000$$

We note that this is a quadratic function, so it is a parabola opening downward since $a = -300 < 0$. Thus, the vertex is the maximum-possible value of R, provided the value of x that yields the vertex makes sense in terms of the original problem. Unlike in the garden problem, we do not have a definite interval on the x-axis to work with. We are told that Hoffs' believes that each drop of \$100 will generate three additional sales, but that is probably only valid over a small range. For example, if $x = 30$, the price drops to 0. It would be hard to believe that a free tour would only generate 90 additional tourists! Also, what if $x < 0$? Would adding \$100 to the price cause three fewer tickets to be sold? Nevertheless, these are questions for the interpretation stage of modeling. Let's see where the vertex is, which will complete the solving stage of modeling.

Completing the square yields

$$R(x) = -300(x^2 - 10x) + 180,000$$
$$= -300(x^2 - 10x + 25) + 180,000 + 7,500$$
$$= -300(x - 5)^2 + 187,500.$$

So the vertex is located at $(5, 187,500)$. Since $x > 0$ and is small compared to $x = 30$, we feel fairly confident that we have found the global maximum revenue. With $x = 5$, Hoffs' should drop the price to \$2,500, and the total number of sales will be $60 + 3(5) = 75$, with a maximum monthly revenue of \$187,500.

We will revisit optimization later in the text. Exercises on this will be at the end of the section.

1.6.3 Some Quadric Surfaces

Earlier in this chapter, we mentioned that the graph of first-degree equations in three variables are planes in $\mathbb{R}^3$. Such equations have the form $Ax + By + Cz = D$, and if $C \neq 0$, they can be expressed as a function of two variables $z = f(x, y) = px + qy + r$. More generally, functions $z = f(x, y)$ have graphs that are surfaces in $\mathbb{R}^3$. The case of surfaces of degree 2 are important, just as quadratic curves are important. Surfaces defined by degree 2 equations are called *quadric surfaces*. The general quadric surface has a lot of terms:

$$Ax^2 + By^2 + Cz^2 + Dxy + Exz + Fyz + Gx + Hy + Iz = J.$$

Just as there are several different types of quadratic curves, there are a number of different types of quadric surfaces, with intriguing names such as ellipsoids, paraboloids, hyperboloids, etc. The names come from the types of curves you get when you intersect these surfaces with various planes to obtain "cross sections." For example, the function $z = x^2 + y^2$ is called a *circular paraboloid* because it has horizontal cross sections that are circles and vertical ones that are parabolas. You can see both types of cross sections in Figure 1.24. The shape looks like what you would obtain by rotating a parabola around its axis of symmetry, as shown in Figure 1.24.

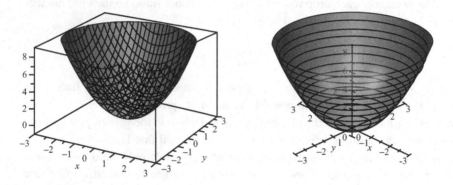

Figure 1.24 Two views of a paraboloid.

A more detailed study of quadric surfaces is usually undertaken in a multivariable calculus course. As was the case for quadratic curves, we do not intend to study these surfaces in general, but there is one special case that comes up later in the book, so we would like to be able to find a standard form in this case, just like we found a standard form for parabolas.

Consider the quadric surface $z = xy$. To get an idea of what the graph looks like, note that if $x = 0$ or if $y = 0$, then $z = 0$. This means that all points of the form $(0, y, 0)$, which is the y-axis, and all points of the form $(x, 0, 0)$, which is the x-axis, are on the graph. Furthermore, in regions where x and y have the same sign (namely, the first quadrant and the third quadrant of the xy-plane), we have $z = xy > 0$, so the surface in these quadrants is above the xy-plane (which has equation $z = 0$). In the other two quadrants, since x and y have opposite signs, we have $z = xy < 0$, so the surface drops below the xy-plane in these quadrants. So imagine a flat horizontal plane, where two opposite corners are pulled up, and the other two opposite corners are pulled down. The graph is shown in Figure 1.25.

The technical name for this surface is a *hyperbolic paraboloid*, because the horizontal cross sections have equations $xy = c$ for a constant $z = c$, and these are hyperbolas, and the vertical cross sections (which are at a 45-degree angle to the x- and y-axes) are parabolas – you can see the downward opening parabola on the bottom profile of Figure 1.25. However, don't worry about the technical name too much. It's also known as a *saddle surface*, because the picture looks like a saddle.

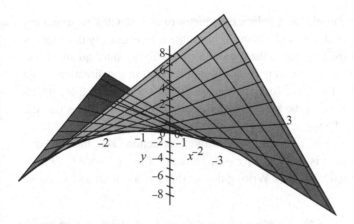

Figure 1.25 A "saddle" surface.

We remark that there are other quadric equations whose graphs are saddle surfaces. If you look it up in a calculus text, you will see that $z = x^2 - y^2$ also gives a saddle surface. The difference is that the horizontal cross sections are hyperbolas of the form $x^2 - y^2 = c$ instead of $xy = c$. In the first type, $z = xy$, the asymptotes of the hyperbolas are parallel to the coordinate axes, while in the second type, $z = x^2 - y^2$, the asymptotes of the hyperbolic cross sections are parallel to the lines $y = \pm x$. Basically, the graph of one is more or less rotated 45 degrees to obtain the graph of the other. There are other variations as well.

In this book, we are particularly interested in the ones where the two variables are multiplied together – that is, we want the "cross term" xy to be present in the equation – not the expression $x^2 - y^2$. We shall call this a *saddle surface of crossed type*. In the general equation for a quadric surface, a saddle surface of crossed type corresponds to the case when $A = B = C = E = F = 0$, and $I \neq 0$ (so that we can solve for z in terms of x and y).

In the graph of $z = xy$, the point $(0, 0, 0)$ on the graph is called a *saddle point* – that's the point where you would sit on the saddle. At that point, in two opposite directions, the saddle goes down for placement of your legs, while directly in front and behind you, the saddle goes up. That is, the saddle point has the property that it is the maximum point going in one direction, and the minimum point in a direction at right angles to that. Locating the saddle point of a saddle surface will be important. Similar to locating maxima and minima, finding saddle points on general surfaces requires calculus. However, we can locate them on quadric saddle surfaces in a way similar to what we did for locating vertices on parabolas.

Thinking in terms of transformations to the graph, if we replace x with $x - h$, then the graph of $f(x - h, y)$ is just the graph of $f(x, y)$, shifted h units in the positive x direction, similar to the case for curves. Thus, the graph of $z = (x - h)y$ is also a saddle surface, with the saddle point located at $(h, 0, 0)$ instead of $(0, 0, 0)$. Similarly, replacing $f(x, y)$ with $f(x, y - k)$ results in a horizontal shift of k units in the positive y direction, so that $z = x(y - k)$ is a saddle surface with saddle point located at $(0, k, 0)$. We can combine the shifts – $f(x - h, y - k)$ is a horizontal shift h units in the x direction and k units in the y direction; or, if you want to, you can think of it as a shift along the vector $(h, k, 0)$ in $\mathbb{R}^3$. Thus, the function $z = (x - h)(y - k)$ is a saddle surface with saddle point located at $(h, k, 0)$.

Next, consider a vertical stretch: multiply the z values by a factor of $a > 0$. Clearly, stretching vertically does not change the location of the saddle point; it just adjusts how quickly the z values increase as we move away from the saddle point in the direction of the first or third quadrant and how quickly the z values decrease as we move away from the saddle point in the direction of the second or fourth quadrant. If $a < 0$, it reverses for which quadrants z is positive and for which ones z is negative, but again, the location of the saddle point does not change. It follows that the graph of $z = a(x - h)(y - k)$ is a saddle surface with saddle point at $(h, k, 0)$ whenever $a \neq 0$.

Finally, consider a vertical translation of ℓ units in the positive z direction. This happens in a similar way to the other two variables – z is replaced by $z - \ell$. Thus, $z - \ell = a(x - h)(y - k)$ is a saddle surface with saddle point located at (h, k, ℓ). Writing this so that z is a function of x and y, we have $z = a(x - h)(y - k) + \ell$.

DEFINITION 1.29 *The* standard form *of the equation of a saddle surface of crossed type with saddle point at (h, k, ℓ) is $z = a(x - h)(y - k) + \ell$ for some $a \neq 0$.*

If you multiply out this equation, you obtain $z = a(x - h)(y - k) + \ell = axy - bx - cy + d$, where $b = ak$, $c = ah$, and $d = ahk + \ell$. Thus, any equation of the form $z = f(x, y) = axy - bx - cy + d$ represents a saddle surface of crossed type. This form of the equation we'll call the *general form*. To located the saddle point (and perhaps to plot it quickly if necessary), we must convert it from general form to standard form. This is similar to what we did for parabolas to locate the vertex. However, it is easier to do this for saddle surfaces of crossed type than for parabolas (or for saddle surfaces without the cross terms) because completing the square is not involved. It is merely a matter of using the distributive law (or you could use the formulas for b, c, and d to find h, k and ℓ).

We illustrate with an example.

Example 1.30 Put the following saddle surface into standard form and locate the coordinates of the saddle point:

$$z = \frac{1}{3}xy - 2x - 5y + 17.$$

Method 1: We know $a = \frac{1}{3}$. Grouping the first two terms, we factor out $ax = \frac{1}{3}x$:

$$z = \frac{1}{3}x(y - 6) - 5y + 17.$$

Next, we are going to want to factor out $a = \frac{1}{3}$ from the y term as well, so rewrite $-5y = \frac{-15}{3}y$. Also, since the first term involves $y - 6$, we want the second term to also have a common factor of $y - 6$. Thus, $-5y$ becomes $-5(y - 6 + 6) = -5(y - 6) - 30 = -\frac{15}{3}(y - 6) - 30$. Substituting this in, we have

$$z = \frac{1}{3}x(y - 6) - \frac{15}{3}(y - 6) - 30 + 17.$$

Factor out $\frac{1}{3}(y - 6)$ from the first two terms, and combine the last two terms:

$$z = \frac{1}{3}(x - 15)(y - 6) - 13.$$

This is standard form, so the saddle point is located at $(h, k, \ell) = (15, 6, -13)$.

Method 2: We know $a = \frac{1}{3}$, $b = 2$, $c = 5$, and $d = 17$. It follows that $c = ah$, or $5 = \frac{1}{3}h$, whence $h = 15$. Also $b = ak$, or $2 = \frac{1}{3}k$, whence $k = 6$. Finally $d = ahk + \ell$, so $17 = \frac{1}{3}(15)(6) + \ell$, or $17 = 30 + \ell$, whence $\ell = 17 - 30 = -13$. Thus, we arrive at the same standard form equation, and saddle point at $(15, 6, -13)$.

We will see how this is useful in Chapter 6. For now, we close with one last example. Find the standard form of the saddle surface with equation

$$z = -2xy + 3x - 8y + 17,$$

and locate the saddle point.

Using the first method, we prefer to group the xy term with the y term instead of with the x term, because it is easier to factor

$$
\begin{aligned}
z &= -2xy - 8y + 3x + 17 \\
&= -2y(x + 4) + 3x + 17 \\
&= -2y(x + 4) + 3(x + 4) - 12 + 17 \\
&= -2(x + 4)(y - 6) + 5.
\end{aligned}
$$

Thus, the saddle point is located at $(h, k, \ell) = (-4, 6, 5)$. We leave the second method as an exercise (see Exercise 9).

Exercises

1. Put the curve in standard form, find the coordinates of the vertex, and sketch the graph:

$$y = x^2 + 2x + 3.$$

2. Follow the same directions as in Exercise 1 for the following curve:

$$y = \frac{1}{4}x^2 - 2x.$$

3. Follow the same directions as in Exercise 1 for the following curve:

$$y = -3x^2 + 4x - 1.$$

4. Find a pair of real numbers such that the first number plus twice the second sums to 12, and the product is as large as possible.

5. Find a pair of numbers such that the first number minus twice the second is 16, and the product is as small as possible.

6. Zimmerman's Millinery is about to manufacture a run of leopard skin pillbox hats. It has been determined that if x hats are manufactured, then the cost of producing these hats is given by the formula $C = \frac{1}{10}x^2 - 80x + \$31{,}000$, and the company believes this formula to be valid for any x between 200 and 500 hats. How many hats should Zimmerman's make in order to minimize the total cost? What is the total cost? How much should the company charge for a hat to make \$40 profit per hat?

7. Yellow Brick Publishing House is selling astronaut John Nolte's new autobiography, *Rocketman across the Water*. The publisher knows it can sell $100,000$ copies at a price of $40 per book. But Yellow Brick has done a market study that suggests that for each $1 drop in price, it will generate 5,000 additional sales. What price should Yellow Brick set for the book which will maximize its total revenue? What is the total revenue it makes, assuming all the books are sold?

8. Elementary Penguin Farm sells fresh eggs. The farm has a large hen house where it keeps 300 hens. The hens are raised on organic feed and are well cared for. The average hen from this farm will produce 210 eggs in its lifetime. Recently, Walrus Animal Collective published a study suggesting that the overcrowding of hens can decrease egg production. Based on preliminary experiments that the head egg man has run, he believes that each decrease of one hen in their hen house population will result in an increase of 1.5 eggs per hen (on average) over its lifetime. What is the ideal number of hens the farm should keep in the hen house in order to maximize the total lifetime production of eggs?

9. Use method 2 to verify the location of the saddle point in Example 1.2 from Section 1.6.3.

10. Put the following saddle surface of crossed type into standard form, and locate the coordinates of the saddle point:

$$z = 3xy + 12x - 3y - 2.$$

11. Follow the same directions as Exercise 10 for the saddle surface:

$$z = -\frac{1}{5}xy - \frac{2}{5}y + 1.$$

12. Follow the same directions as Exercise 10 for the saddle surface:

$$z = 20xy + 100x + 40y + 120.$$

2 Matrix Algebra

In Chapter 1, we encountered matrices as a device to streamline the process of elimination to solve systems of linear equations. However, matrices have a life of their own beyond being a bookkeeping aid for elimination. In this chapter, the algebra of matrices and some simple applications are introduced. A large part of the rest of the book illustrates more applications of matrices. They are quite useful, and their theory is rich and deep. This book is a mere introduction, and it is not very theoretical; if the reader is interested in learning more about matrices, the next step would be to take a course in linear algebra, where the theory is developed more carefully and completely than it is here.

2.1 Matrices

A *matrix* (plural, *matrices*) is a rectangular array of numbers, called *entries* or *elements*. As we'll see later in this text, it is sometimes useful to allow the entries of a matrix to be ordered pairs instead of individual numbers. But for now, we'll stick to the given definition. The *size* or *dimensions* of a matrix are the number of (horizontal) rows and (vertical) columns (in that order), so that a 2×3 matrix is one with two rows and three columns. Capital letters will be used to stand for a matrix, and the entries inside are referred to by the corresponding lowercase letters. Different entries are distinguished by their *address*, which is a double subscript indicating in which row and column it appears. Thus, a_{ij} refers to the element of the matrix A in the ith row and jth column. For example, in the matrix

$$A = \begin{bmatrix} 2 & 1 & 3 \\ 0 & -1 & 3 \end{bmatrix},$$

we have $a_{11} = 2$, $a_{13} = 3 = a_{23}$, etc. The general pattern (for a 3×4 matrix, for example) is

$$A = A_{3 \times 4} = \begin{bmatrix} a_{11} & a_{12} & a_{13} & a_{14} \\ a_{21} & a_{22} & a_{23} & a_{24} \\ a_{31} & a_{32} & a_{33} & a_{34} \end{bmatrix}.$$

(Notice that we sometimes write the dimensions of the entire matrix A as a subscript on A.) If there are more than nine rows or columns in A, we use commas to separate the row and column subscripts. For example, a_{125} is ambiguous, but $a_{1,25}$ (the entry in the first row and 25th column) and $a_{12,5}$ (the entry in the 12th row and 5th column) are clear.

There are various special types of $m \times n$ matrices. A *square* matrix is one where $m = n$, such as a 2×2 or a 7×7 matrix. A *row matrix* is one where $m = 1$ (it consists of a single row), and a *column matrix* is one where $n = 1$ (it consists of a single column.) A 1×1 matrix is nothing more than a single number. The entries in the jj positions, running from the upper-left corner (when $j = 1$) down a 45° diagonal form the *main diagonal* of a matrix. A square matrix where all the nonzero entries appear only on the main diagonal is called a *diagonal matrix*. A diagonal matrix with every diagonal entry identical is called a *scalar matrix*, and a scalar matrix with all ones on the diagonal is called an *identity matrix*. As examples, consider the matrices:

$$B = \begin{bmatrix} 1 & 2 & 3 \\ 0 & 0 & 1 \\ 2 & 2 & 2 \end{bmatrix} \qquad C = \begin{bmatrix} -5 & 4 & 2 \end{bmatrix}$$

$$D = \begin{bmatrix} 11 \\ 23.5 \end{bmatrix} \qquad E = \begin{bmatrix} 2 & 0 & 0 \\ 0 & 3 & 0 \\ 0 & 0 & 0 \end{bmatrix}$$

$$F = \begin{bmatrix} 6 & 0 & 0 \\ 0 & 6 & 0 \\ 0 & 0 & 6 \end{bmatrix} \qquad G = \begin{bmatrix} 1 & 0 & 0 & 0 \\ 0 & 1 & 0 & 0 \\ 0 & 0 & 1 & 0 \\ 0 & 0 & 0 & 1 \end{bmatrix}.$$

Then B, E, F, and G are square matrices; C is a row matrix; D is a column matrix; E, F, and G are diagonal matrices; F and G are scalar matrices; and G is the 4×4 identity matrix. The $n \times n$ identity matrix is often denoted I_n.

Exercises

1. Give a numerical example of each of the following:

    a.   A square matrix.
    b.   A row matrix.
    c.   A diagonal matrix.

2. Is $\begin{bmatrix} 1 & 0 & 0 \\ 0 & 0 & 0 \\ 0 & 0 & 0 \end{bmatrix}$ considered to be a diagonal matrix? How about $\begin{bmatrix} 0 & 0 & 3 \\ 0 & 2 & 0 \\ 1 & 0 & 0 \end{bmatrix}$?

3. Write out the general form of a 6×4 matrix with each a_{ij} in its correct location, as we did in the text for a 3×4 matrix.

4. Write out a 5×5 matrix where $a_{ij} = i + 1$ for all i and j.

5. Write out a 5×5 matrix where $a_{ij} = i + j$ for all i and j.

6. Write out a 5×5 matrix where $a_{ij} = \max\{i, j\}$ for all i and j.

7. Write out a 5×5 matrix where $a_{ij} = |i - j|$ for all i and j.

8. Write out a 5×5 matrix where $a_{ij} = i^2 + j$ for all i and j.

9. Write out a 5×5 matrix where $a_{ij} = (-1)^{i+j}$ for all i and j.

10. Write out a 5×5 matrix where $a_{ij} = (-1)^{i+j}(i - 1)(j - 1)$ for all i and j.

11. A square matrix A is said to be *upper triangular* if $a_{ij} = 0$ whenever $i > j$. Likewise, A is said to be *lower triangular* if $a_{ij} = 0$ whenever $i < j$.

a.   Give numerical examples (3×3 or larger) of upper triangular matrices. Also give examples of lower triangular matrices.

b.   Give an example of a nonzero square matrix A (3×3 or larger) which is both upper triangular and lower triangular. What special type of matrix is A?

12.   A square matrix P with the property that each row and each column has exactly one nonzero entry, which has value 1, is called a *permutation matrix*. For example, an identity matrix I_n is a permutation matrix. Give at least two distinct examples of 3×3 permutation matrices which are not identity matrices. How many different 3×3 permutation matrices do you think there are? Can you prove your answer? How about $n \times n$ permutation matrices? Why do you think they are called permutation matrices?

2.2 Operations on Matrices I

Matrices provide a useful and efficient way to handle itemized calculations. We illustrate this assertion with examples throughout the rest of the chapter (and in later chapters) as we learn to work with matrices.

There are a number of interesting and useful operations on matrices. The treatment of matrices in this text is a mere introduction to their uses, with a goal of developing just enough to be able to use these tools in linear programming and game theory. The reader who is interested in complete, rigorous proofs of any claims we make in this chapter but do not prove is encouraged to pick up any text in Linear Algebra for further reading, such as Lay (2012), or Lang (1971).

We have already seen one way in which matrices can help us solve problems – the technique of pivoting a matrix that we learned in Chapter 1, which allows us to solve a linear system in a very efficient manner. In this chapter, we focus on other types of matrix operations besides the elementary row operations of the first chapter. The first operation is quite simple.

2.2.1 Transpose of a Matrix

Given a matrix A, it determines another matrix B called the *transpose* of A, written $B = A^T$. It is obtained from A by interchanging the rows and columns. More precisely, the elements of the transpose are given by

$$b_{ij} = a_{ji} \tag{2.1}$$

for all i, j. For example, the preceding general 3×4 matrix has transpose as follows:

$$A^T = \begin{bmatrix} a_{11} & a_{12} & a_{13} & a_{14} \\ a_{21} & a_{22} & a_{23} & a_{24} \\ a_{31} & a_{32} & a_{33} & a_{34} \end{bmatrix}^T = \begin{bmatrix} a_{11} & a_{21} & a_{31} \\ a_{12} & a_{22} & a_{32} \\ a_{13} & a_{23} & a_{33} \\ a_{14} & a_{24} & a_{34} \end{bmatrix} = (b_{ij}).$$

Visibly, the entry b_{ij} in the ij position of A^T is a_{ji} as indicated by (2.1). Notice the rows of A become the columns of A^T and the columns of A become the rows of A^T. In particular, if A has size $m \times n$, then A^T has size $n \times m$, as in the preceding example where A^T has size 4×3.

One may also think of the transpose of A as a reflection across a mirror on the main diagonal. For a numerical example, observe that

$$
\begin{bmatrix}
\underline{9} & 1 & 0 \\
1 & \underline{2} & 3 \\
-5 & 0 & \underline{8} \\
4 & 4 & 4 \\
0 & 11 & 7
\end{bmatrix}^T
=
\begin{bmatrix}
\underline{9} & 1 & -5 & 4 & 0 \\
1 & \underline{2} & 0 & 4 & 11 \\
0 & 3 & \underline{8} & 4 & 7
\end{bmatrix}.
$$

The entries on the main diagonal are underlined to call attention to the fact that these entries do not change their positions in the transpose, whereas all the other entries are reflected across the main diagonal. It is clear from the definition (or from doing a few examples) that

$$
\left(A^T \right)^T = A.
$$

Despite its simplicity, the operation of taking the transpose has some important uses. In this text, it arises in the discussion of duality in linear programming, which is covered in Section 4.5. We have one final definition to make:

DEFINITION 2.1   *A matrix A is said to be* symmetric *if $A^T = A$ and A is said to be* antisymmetric *if $A^T = -A$.*

For example, verify that A is symmetric while B is antisymmetric:

$$
A = \begin{bmatrix}
\pi & 2 & -7 \\
2 & 0 & 4 \\
-7 & 4 & 100
\end{bmatrix}
\quad
B = \begin{bmatrix}
0 & -1 & -2 \\
1 & 0 & -3 \\
2 & 3 & 0
\end{bmatrix}.
$$

2.2.2 Scalar Multiplication

Given a matrix A and a real number r, it is useful to consider the *scalar product* of r and A. It is obtained by multiplying every entry of A by a factor of r (i.e., scaling the entries up by a factor of r), resulting in a matrix, denoted rA (or Ar) that has the same dimensions as A.

For example, if $r = 3$ and $A = \begin{bmatrix} 2 & 1 & 3 \\ 0 & -1 & 3 \end{bmatrix}$, then

$$
rA = 3 \begin{bmatrix} 2 & 1 & 3 \\ 0 & -1 & 3 \end{bmatrix} = \begin{bmatrix} 6 & 3 & 9 \\ 0 & -3 & 9 \end{bmatrix}
$$

is the matrix obtained by tripling the entries of A. Scalar multiplication is a pretty trivial operation (just like scalar multiplication of vectors), and there is not much to say about it, so we move on. Scalar multiplication and transposition involve only one matrix at a time. Such operations are called *unary* operations. The next two operations are *binary* operations – they are ways to combine two matrices together to obtain a new matrix from them.

2.2.3 Matrix Addition

Given two matrices, provided they are the same size, we can add them according to Definition 2.2.

DEFINITION 2.2

$$A_{m \times n} + B_{r \times s} = \begin{cases} undefined \ if \ m \neq r \ or \ n \neq s \\ C \ if \ m = r \ and \ n = s, \end{cases} \tag{2.2}$$

where C is also of dimension $m \times n$,

and $c_{ij} = a_{ij} + b_{ij}$ for all i, j.

This definition amounts to saying that if two matrices are the same size, they are added by adding together the corresponding entries (just like vector addition.) For example,

$$\begin{bmatrix} 1 & 0 \\ 1 & 2 \\ 1 & 7 \end{bmatrix} + \begin{bmatrix} -1 & 0 \\ -1 & 1 \\ 1 & -3 \end{bmatrix} = \begin{bmatrix} 0 & 0 \\ 0 & 3 \\ 2 & 4 \end{bmatrix}.$$

On the other hand, matrices of different sizes cannot be added. For example,

$$\begin{bmatrix} 1 & 3 \\ 5 & 7 \end{bmatrix} + \begin{bmatrix} 1 & 1 & 0 \\ 1 & 1 & 0 \end{bmatrix}$$

is undefined because the dimensions do not match. Similar remarks hold regarding the subtraction of matrices – just subtract corresponding entries, provided the matrices are the same size. Thus,

$$\begin{bmatrix} 1 & 0 \\ 1 & 2 \\ 1 & 7 \end{bmatrix} - \begin{bmatrix} -1 & 0 \\ -1 & 1 \\ 1 & -3 \end{bmatrix} = \begin{bmatrix} 2 & 0 \\ 2 & 1 \\ 0 & 10 \end{bmatrix}.$$

The following theorem summarizes what happens when matrix addition is combined with transposition and/or scalar multiplication.

THEOREM 2.3   *Let A, B, and C be matrices, and let r, s be real numbers. Also, let 0 be a matrix (regardless of size) with all entries equal to 0 (called a* zero matrix*). The following rules hold:*

a. $A + B = B + A$ *(commutative law for matrix addition).*

b. $A + (B + C) = (A + B) + C$ *(associative law for matrix addition).*

c. $s(A + B) = sA + sB$ *(distributive law for scalar multiplication over matrix addition).*

d. $A + 0 = A$ *(where 0 is a zero matrix the same size as A).*

e.   *Given A, there is another matrix, denoted $-A$, with the property that $A + (-A) = 0$, where 0 is as in Part d. $-A$ is called the* additive inverse *of A.*

f. $-A = (-1)A$ *(the additive inverse of A is the same as the scalar multiple $(-1)A$).*

g. $(A + B)^T = A^T + B^T$ *(transposition distributes over matrix addition).*

h. $(sA)^T = s(A^T)$ *(transposition commutes with scalar multiplication).*

i. $r(sA) = (rs)A$ *(associative law for scalar multiplication).*

These properties will be illustrated in the exercises. We do not give formal proofs, which can be found in any linear algebra text and are quite easy in any case. Mathematically mature readers will be asked to supply some of the proofs in the exercises.

2.2.4 Matrix Multiplication

Given two matrices A and B, it may or may not be possible to multiply them together, depending on their dimensions. The definition of the matrix product is not as simple as that for the matrix sum. In order to understand how multiplication works, it is useful to consider first a special case – that of a row matrix times a column matrix. So suppose we have the following row matrix A and column matrix B:

$$A = \begin{bmatrix} a_1 & a_2 & a_3 & \cdots & a_n \end{bmatrix} \quad B = \begin{bmatrix} b_1 \\ b_2 \\ b_3 \\ \vdots \\ b_r \end{bmatrix}.$$

The *dot product* $A \cdot B$ is defined by Definition 2.3.

DEFINITION 2.4

$$A \cdot B = \begin{cases} undefined & if\ n \neq r \\ c & if\ n = r \end{cases} \tag{2.3}$$

$$where\ c = a_1 b_1 + a_2 b_2 + \cdots + a_n b_n.$$

You may regard the single number c as a 1×1 matrix or simply as a scalar. Here are examples:

$$\begin{bmatrix} 2 & 5 & 3 \end{bmatrix} \cdot \begin{bmatrix} 4 \\ -1 \\ 3 \end{bmatrix} = 2(4) + 5(-1) + 3(3) = 12$$

$$\begin{bmatrix} 2 & 5 & 3 \end{bmatrix} \cdot \begin{bmatrix} 4 \\ -1 \end{bmatrix} = undefined,\ since\ n \neq r.$$

So, in this special case, the dot product is the matrix product.

Before we go on to look at the general case, it is worth presenting a couple of examples that illustrate how such a strange way of multiplying comes up naturally. We consider two examples. In our first example, suppose the Grand Flick Railroad movie theater sells four different types of tickets: children's tickets sell for \$5 each, student tickets sell for \$7 each, adult tickets are \$10 each, and senior citizens' tickets are \$8 each. We might conveniently arrange this information in a 1×4 matrix P (for "price matrix") – we may even label the columns to help us keep track:

$$\begin{matrix} C & S & A & SC \end{matrix}$$
$$P = \begin{bmatrix} 5 & 7 & 10 & 8 \end{bmatrix}.$$

Now, on a given night, say Friday, the theater is showing the movie *Closer to Home* and will sell different amounts of each type of ticket. We might keep track of this information by listing it as a 4×1 matrix Q (for "quantity matrix"), with one row for each type of ticket:

$$Q = \begin{matrix} C \\ S \\ A \\ SC \end{matrix} \begin{bmatrix} 40 \\ 50 \\ 80 \\ 10 \end{bmatrix},$$

meaning that they sold 40 children's tickets, 50 student tickets, etc. Now, by arrangement, these matrices are the right size to be able to form their dot product $P \cdot Q$, and in this case, the product has a specific meaning:

$$P \cdot Q = \begin{bmatrix} 5 & 7 & 10 & 8 \end{bmatrix} \cdot \begin{bmatrix} 40 \\ 50 \\ 80 \\ 10 \end{bmatrix}$$

$$= 5(40) + 7(50) + 10(80) + 8(10) = 1{,}430.$$

Clearly, the dot product represents the total revenue for the theater for that night, since it is summing the total revenue made for each of the itemized types of ticket. Even the units make sense – the units of the entries of P are "dollars per ticket," and the units of the entries of Q are "tickets." When these units are multiplied together, we obtain units of "dollars" – the theater had a total revenue of \$1,430 for Friday night.

For a second example, consider an experiment with three possible outcomes. Let p_1 be the probability of the first outcome, p_2 the probability of the second outcome, and p_3 the probability of the third outcome. We can arrange this information in a "probability matrix" $P = \begin{bmatrix} p_1 & p_2 & p_3 \end{bmatrix}$. Suppose further that each outcome comes with a payoff or value, and we can arrange these values in a "value matrix":

$$V = \begin{bmatrix} v_1 \\ v_2 \\ v_3 \end{bmatrix}.$$

Then it is possible to form the dot product, and again the result has a useful interpretation:

$$P \cdot V = \begin{bmatrix} p_1 & p_2 & p_3 \end{bmatrix} \cdot \begin{bmatrix} v_1 \\ v_2 \\ v_3 \end{bmatrix}$$

$$= p_1 v_1 + p_2 v_2 + p_3 v_3.$$

Students who know some probability theory will recognize this as the *expected value* (E) of the experiment. For those who may not have encountered these ideas before, E is a weighted average of the values, where each value is weighted according to the probability that it occurs. E represents the the average payoff or average value in the long run if the experiment is repeated many times. We will revisit this idea of expected value later when we cover game theory. For now, you should at least be convinced of the utility of this rather strange definition of multiplying matrices, at least in this special case of a row matrix times a column matrix.

We remark in passing that a row matrix or a column matrix can simply be regarded as a vector (in which case there is no distinction between writing them as a row or as a column). Some readers may have encountered the dot product of vectors in a linear algebra or multivariable calculus course, where they may have learned that there is a geometric interpretation of the dot product in terms of the angle between the vectors. While this provides an important third example of the utility of the dot product concept, we will have no need of that interpretation of the dot product in this text, so we leave the interested reader to seek out this geometric interpretation on their own.

We are now ready for the general case of matrix multiplication. Let A be an $m \times n$ matrix and B an $r \times s$ matrix. The product is given by Definition 2.4:

DEFINITION 2.5

$$AB = \begin{cases} undefined \ if \ n \neq r \\ \qquad C \ if \ n = r \end{cases} \tag{2.4}$$

where C is an $m \times s$ matrix

whose entries are given by the dot products:

$$c_{ij} = (ith \ row \ of \ A) \cdot (jth \ column \ of \ B).$$

Note that the condition $n = r$ guarantees that the dot products for c_{ij} are defined.

There is no substitute for practice! Consider the example

$$\begin{bmatrix} 1 & 2 & 3 \\ 4 & 5 & 6 \end{bmatrix}_{2\times3} \begin{bmatrix} 1 & 0 \\ 2 & -1 \\ -2 & 3 \end{bmatrix}_{3\times2}$$

$$= \begin{bmatrix} (1st \ row \ of \ A) \cdot (1st \ column \ of \ B) & (1st \ row \ of \ A) \cdot (2nd \ column \ of \ B) \\ (2nd \ row \ of \ A) \cdot (1st \ column \ of \ B) & (2nd \ row \ of \ A) \cdot (2nd \ column \ of \ B) \end{bmatrix}$$

$$= \begin{bmatrix} 1(1) + 2(2) + 3(-2) & 1(0) + 2(-1) + 3(3) \\ 4(1) + 5(2) + 6(-2) & 4(0) + 5(-1) + 6(3) \end{bmatrix}$$

$$= \begin{bmatrix} -1 & 7 \\ 2 & 13 \end{bmatrix}_{2\times2}.$$

Notice that in order for the matrices to be able to multiply, the inner dimensions (n and r) must match – the threes in this case – while the size of the product C is given by the outer dimensions ($m \times s$) – the twos in this case. On the other hand, observe that

$$\begin{bmatrix} 1 & 2 & 3 \\ 4 & 5 & 6 \end{bmatrix}_{2\times3} \begin{bmatrix} 1 & 0 & 1 \\ 0 & 1 & 0 \end{bmatrix}_{2\times3} = undefined$$

because the inner dimensions do not match here.

For another example, let's return to the Grand Flick Railroad movie theater example from before. Keep the same price matrix P, but for the quantity matrix Q, let's add columns corresponding to the sales on Saturday and Sunday in addition to Friday to obtain a new 4×3 matrix:

$$Q = \begin{array}{c} \\ C \\ S \\ A \\ SC \end{array} \begin{array}{c} Fri. \quad Sat. \quad Sun. \\ \begin{bmatrix} 40 & 150 & 20 \\ 50 & 30 & 60 \\ 80 & 100 & 50 \\ 10 & 0 & 60 \end{bmatrix} \end{array}.$$

We can form the matrix product because the inner dimensions match (the four columns of P match the four rows of Q), and the answer should be a 1×3 matrix coming from the outer dimensions. The single row of PQ corresponds to the single row of P, and the three columns of PQ correspond to the three columns of Q. They can even be labeled the same:

$$PQ = \begin{bmatrix} 5 & 7 & 10 & 8 \end{bmatrix}_{1\times 4} \begin{bmatrix} 40 & 150 & 20 \\ 50 & 30 & 60 \\ 80 & 100 & 50 \\ 10 & 0 & 60 \end{bmatrix}_{4\times 3} = \$ \begin{bmatrix} Fri. & Sat. & Sun. \\ 1{,}430 & 1{,}960 & 1{,}500 \end{bmatrix}.$$

The interpretation is obvious – the entries of PQ represent the total revenue for the theater for the entire weekend in units of dollars, but itemized by day.

As you practice examples of matrix multiplication, be aware of the following difference with ordinary arithmetic: *the commutative law of multiplication can fail!* That is, for most matrices, $AB \neq BA$. This may be because one of the two products is defined and the other is not. In the movie theater example, PQ is defined and has a useful interpretation, but QP is not defined (check!).

It may also be the case that both products are defined, but result in matrices of different dimensions. In the earlier example where A is 2×3 and B is 3×2, the product AB is 2×2. However, BA is also defined, but is 3×3, so couldn't possibly be equal to AB. (The reader should check this by finding BA now.) But even if the sizes of AB and BA are the same, perhaps surprisingly, the actual matrices usually differ. For example, verify the following calculations:

$$A = \begin{bmatrix} 1 & 2 \\ 3 & 4 \end{bmatrix} \quad B = \begin{bmatrix} 1 & 0 \\ -1 & 5 \end{bmatrix}$$

$$AB = \begin{bmatrix} -1 & 10 \\ -1 & 20 \end{bmatrix}$$

$$BA = \begin{bmatrix} 1 & 2 \\ 14 & 18 \end{bmatrix}$$

So the products are quite different!

Having the commutative law fail may cause the reader wonder if other rules of arithmetic can fail for matrix multiplication. The following theorem should reassure the reader that only the commutative law is compromised.

THEOREM 2.6 *Let A, B, and C be matrices, and let r, s be real numbers. Then the following rules hold:*

a. $A(BC) = (AB)C$ *(the associative law for matrix multiplication)*
b. $A(B + C) = AB + AC$ *(left distributive law of matrix multiplication over matrix addition)*
c. $(B+C)A = BA+CA$ *(right distributive law of matrix multiplication over matrix addition)*
      To the preceding rules, we add the following rules which show how matrix multiplication combines with scalar multiplication and transposition:
d. $s(AB) = (sA)B = A(sB)$ *(scalar multiplication commutes with matrix multiplication)*
e. $(AB)^T = B^T A^T$ *(transposition distributes over matrix multiplication provided you reverse the order of the product)*

Thus, the usual basic rules of multiplication hold for matrix multiplication with the exception of the commutative law. Note that because the commutative law fails, $A(B + C)$ is different from $(B+C)A$, which is why we require two separate distributive laws. As with the theorem on matrix addition, we will not supply formal proofs of these properties, which can be found in a linear algebra text. Because of the complicated definition of matrix multiplication, the proofs of these facts may not be quite as obvious as the proofs for matrix addition. Again, these properties are illustrated in the exercises, and again, mathematically mature readers will be invited to try to prove some of them. See also Exercises 16 and 17 for examples illustrating some other familiar algebraic properties of multiplication of real numbers that do not necessarily work in the same way for matrix multiplication.

Perhaps property e is surprising at first glance. Suppose A has size 2×3 and B has size 3×5. Then AB is defined and has size 2×5, which means $(AB)^T$ has size 5×2. Now A^T has size 3×2, and B^T has size 5×3. Clearly, $A^T B^T$ is not defined. On the other hand, $B^T A^T$ is defined and has the right size 5×2. Thus, it is at least plausible that rule e holds, since they are at least the same size, whereas it would be impossible without reversing the order since the sizes differ.

Matrix operations have been advertised in this text as being particularly useful for calculations with itemized data, with the movie theater example and the expected value example illustrating this point of view. Here is another example. In Chapter 1 we studied *systems of linear equations* such as

$$2x + 3y = 7 \tag{2.5}$$
$$5x - y = 9$$

or

$$3x + y - 4z = 2 \tag{2.6}$$
$$x + 3y + 2z = 4$$
$$5x - y - z = 9.$$

This is really just another instance of itemized data – one equation for each row – so we should be able to use matrices to compactly display the same information. Given any such linear system, let A be the *coefficient matrix*: for example, in the two systems, the coefficient matrices are respectively

$$A = \begin{bmatrix} 2 & 3 \\ 5 & -1 \end{bmatrix} \text{ and } A = \begin{bmatrix} 3 & 1 & -4 \\ 1 & 3 & 2 \\ 5 & -1 & -1 \end{bmatrix}.$$

REMARK 2.7 *The coefficient matrix differs from the* augmented *coefficient matrix we encountered in Chapter 1, which also includes a column for the numbers on the right side of the equations.*

Now, arrange all the constants on the right side of the equations in a single column matrix B. Finally, make a column matrix X containing the variables of the equations. In the preceding examples,

$$X = \begin{bmatrix} x \\ y \end{bmatrix} \quad B = \begin{bmatrix} 7 \\ 9 \end{bmatrix} \text{ in the first system,}$$

$$X = \begin{bmatrix} x \\ y \\ z \end{bmatrix} \quad B = \begin{bmatrix} 2 \\ 4 \\ 9 \end{bmatrix} \text{ in the second.}$$

Notice that, in both systems, the sizes of the matrices allow us to multiply the coefficient matrix times the variable matrix, in that order. The results are as follows:

$$AX = \begin{bmatrix} 2 & 3 \\ 5 & -1 \end{bmatrix} \begin{bmatrix} x \\ y \end{bmatrix} = \begin{bmatrix} 2x + 3y \\ 5x - y \end{bmatrix}$$

$$AX = \begin{bmatrix} 3 & 1 & -4 \\ 1 & 3 & 2 \\ 5 & -1 & -1 \end{bmatrix} \begin{bmatrix} x \\ y \\ z \end{bmatrix} = \begin{bmatrix} 3x + y - 4z \\ x + 3y + 2z \\ 5x - y - z \end{bmatrix}.$$

Clearly, the entries of this product are the left-hand side of the linear systems. But if the systems are to hold, the left side must be equal to the right side. Two matrices are equal if and only if their corresponding entries are the same. Thus, in order for the systems to be true, we must have equalities of matrices:

$$\begin{bmatrix} 2x + 3y \\ 5x - y \end{bmatrix} = \begin{bmatrix} 7 \\ 9 \end{bmatrix}$$

$$\begin{bmatrix} 3x + y - 4z \\ x + 3y + 2z \\ 5x - y - z \end{bmatrix} = \begin{bmatrix} 2 \\ 4 \\ 9 \end{bmatrix}.$$

Notice that, in both cases, the single matrix equation has the form $AX = B$. It should be clear that every linear system of equations corresponds to a single matrix equation of this form and vice versa – any matrix equation of this form can be multiplied out to give a linear system. A system with m equations and n unknowns corresponds to a coefficient matrix of size $m \times n$. In short, the (single) matrix equation contains exactly the same information as the entire system. Whether or not there is any advantage to writing the system as a single matrix equation in terms of being able to solve the system is something we will address in the next section.

Note that the identity matrix I_n has the following property: for any $n \times n$ matrix A,

$$I_n A = A = A I_n. \tag{2.7}$$

For example, when $n = 3$, the reader may easily verify that

$$\begin{bmatrix} 1 & 0 & 0 \\ 0 & 1 & 0 \\ 0 & 0 & 1 \end{bmatrix} \begin{bmatrix} a & b & c \\ d & e & f \\ g & h & i \end{bmatrix} = \begin{bmatrix} a & b & c \\ d & e & f \\ g & h & i \end{bmatrix}$$

$$\begin{bmatrix} a & b & c \\ d & e & f \\ g & h & i \end{bmatrix} \begin{bmatrix} 1 & 0 & 0 \\ 0 & 1 & 0 \\ 0 & 0 & 1 \end{bmatrix} = \begin{bmatrix} a & b & c \\ d & e & f \\ g & h & i \end{bmatrix}.$$

It is this property that gives I_n its name – it's the *multiplicative identity* for $n \times n$ matrices, just as the number 1 is the multiplicative identity for real numbers. Multiplying A by I_n does not change the identity of A, just as multiplying a real number a by 1 does not change its identity.

Exercises

1. Let

$$A = \begin{bmatrix} 1 & 0 \\ 4 & -1 \end{bmatrix} \quad B = \begin{bmatrix} 2 & 3 \\ 2 & -3 \end{bmatrix} \quad C = \begin{bmatrix} 0 & 1 \\ 1 & 0 \end{bmatrix}.$$

   a. Find $2A$, $2A - 3B$, and $2A - 3B + 7C$.
   b. Find $A + B$ and $(A + B) + C$.
   c. Find $B + C$, and verify that $(A + B) + C = A + (B + C)$.
   d. Find A^T and B^T, and verify that $(A + B)^T = A^T + B^T$.

2. If T is a 4×7 matrix and ST is a 6×7 matrix, what size is S?

3. Let $A, B,$ and C be as in Exercise 1. Compute the following products, if possible:

   a. AB
   b. BA
   c. A^2
   d. C^3

4. Let A be as in Exercise 1, and let

$$D = \begin{bmatrix} 1 & 1 \\ -2 & 7 \\ 4 & 0 \end{bmatrix} \quad E = \begin{bmatrix} 1 & 2 & 3 \\ 4 & 5 & 6 \end{bmatrix}.$$

   Compute the following matrix products, if possible:

   a. AD
   b. DA
   c. DE
   d. ED

5. Let $A, B,$ and C be as in Exercise 1.

   a. Verify that $AB + AC = A(B + C)$.
   b. Verify that $BA + CA = (B + C)A$.

6. Let D and E be as in Exercise 3.

   a. Verify that $(DE)^T = E^T D^T$.
   b. Verify that $(ED)^T = D^T E^T$.

7. Let $B, C,$ and E as in Exercises 5 and 6, and let $F = \begin{bmatrix} 1 & 0 & 2 \\ 3 & 6 & 9 \\ 2 & -4 & 8 \end{bmatrix}.$

   a. Verify that $(BE)F = B(EF)$.
   b. Verify that $(CE)F = C(EF)$.

8. Let F be as in Exercise 7, and let

$$G = \begin{bmatrix} \frac{1}{2} & \frac{2}{3} & \frac{3}{4} \end{bmatrix} \quad H = \begin{bmatrix} 30 \\ 6 \\ -12 \end{bmatrix}.$$

Compute the following matrices, if possible:

a. GF
b. GH
c. HG
d. HF
e. GFH

9. a. Let $A = \begin{bmatrix} 1 & 2 & -6 \\ 0 & 5 & 1 \\ 4 & -3 & 8 \end{bmatrix}$. Show that $A + A^T$ is a symmetric matrix.

b. Prove that $A + A^T$ is always a symmetric matrix for all square matrices A.

10. a. Let $A = \begin{bmatrix} 1 & 3 & 2 & 4 \\ 3 & 0 & -1 & 5 \end{bmatrix}$. Show that AA^T and $A^T A$ are both symmetric matrices.

b. Let A be an arbitrary $m \times n$ matrix. Prove that AA^T and $A^T A$ are always symmetric matrices. What are their dimensions?

11. Consider the following system of equations:

$$3x + 3y = 21$$
$$2x - y = 12$$
$$4x + y = 33$$
$$7x + 2y = 50.$$

Write the system as a single matrix equation of the form $AX = B$, where A is the coefficient matrix of the system. What are the dimensions of the coefficient matrix?

12. Let F and H be as in Exercises 7 and 8, and let $X = \begin{bmatrix} r \\ s \\ t \end{bmatrix}$ be a variable matrix. Consider the matrix equation $FX = H$. Multiply out the left side and use it to write down the corresponding system of equations.

13. Six students are taking part in a study of the effects of vitamins on the common cold. Each student is injected with a certain number of shots of one or two serums, Serum I and/or Serum II. Matrix A indicates how many doses of each serum each student receives:

	I	II
Mark	1	0
Ariel	1	1
Tina	2	1
Robert	2	2
Ivan	1	2
Xavier	0	1

Matrix B indicates how many mg of various types of vitamins (labeled on the columns) are contained in each dose of each serum:

$$\begin{array}{c} \\ \text{Serum I} \\ \text{Serum II} \end{array} \begin{array}{cccccc} A & B_6 & B_{12} & C & D \\ \left[\begin{array}{ccccc} 300 & 100 & 0 & 1{,}200 & 200 \\ 0 & 500 & 500 & 700 & 100 \end{array} \right]. \end{array}$$

Find matrix AB, and interpret the meaning of its entries.

14. Rocky and Sadie are on a diet and on a budget. At a recent visit to the Savoy Truffle Diner for breakfast, Rocky had two eggs, three slices of bacon, two slices of buttered toast, one cup of coffee. Sadie had one egg, two slices of ham, one slice of buttered toast, two cups of coffee, and one glass of orange juice.

    a.   Put all of the preceding information into a matrix A where there is a row for Rocky, a row for Sadie, and a column for each potential breakfast item. That is, a column each for eggs, bacon slices, ham slices, buttered toast, coffee, and juice, in that order. Your matrix should be 2×6.

    b.   The following matrix B represents the breakdown of each diner item in terms of calories, milligrams cholesterol, and price (in cents):

$$\begin{array}{c} \\ \text{Egg} \\ \text{Slice of Bacon} \\ \text{Slice of Ham} \\ \text{Slice of Toast} \\ \text{Cup of Coffee} \\ \text{Glass of Juice} \end{array} \begin{array}{ccc} \text{Cal.} & \text{Chol.} & \text{Cost} \\ \left[\begin{array}{ccc} 75 & 10 & 100 \\ 150 & 6 & 75 \\ 100 & 4 & 100 \\ 125 & 1 & 50 \\ 20 & 0 & 80 \\ 80 & 0 & 150 \end{array} \right] \end{array}$$

Find the product matrix AB, and use it to answer the following questions:

       i.   What did Rocky spend on his breakfast?

     ii.   How many calories were in Sadie's breakfast?

   iii.   Who consumed more cholesterol?

    c.   Consider the 1×2 matrix $S = \left[\begin{array}{cc} 1 & 1 \end{array} \right]$. Find the matrix SAB, and interpret the meaning of its entries.

15. The Locomotive Breath Rail Company runs passenger trains between five cities. The company does not run a train between every pair of cities, but it is possible to go from one city to any of the other five if you are willing to make the journey in several steps or legs, with an overnight layover between each leg. For simplicity, we assume that all trains leave at the same time; say, 9 a.m. every day. Figure 2.1 shows how many trains go from each city to each other city. If there are no direct trains from one city to another, no arrow is shown in the diagram.

    a.   Construct a 5×5 matrix T, where the rows and columns of T are labeled with the cities A, B, C, D, and E (in that order), and where the entry t_{ij} is the number of trains running from city j to city i. Thus, the cities of departure are the columns and the cities of arrival are the rows. For example, since there are five trains from Erie to

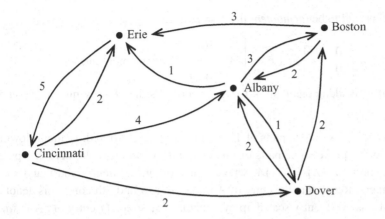

Figure 2.1 A transit graph.

Cincinnati, the five would be located in the column labeled E and the row labeled C; that is, we have $t_{35} = 5$.

b. Find $T^2 = T \cdot T$ and interpret the meaning of its entries.

c. Find $T^3 = T \cdot T \cdot T$ and interpret the meaning of its entries.

d. What is the meaning of the entries in the matrix $T + T^2 + T^3$?

16. Let A and B be as in Exercise 1.

a. Show that $(A + B)^2 \neq A^2 + 2AB + B^2$.

b. Show that $(A + B)(A - B) \neq A^2 - B^2$.

c. What conditions on square matrices M, N would be necessary and sufficient to make $(M + N)(M - N) = M^2 - N^2$?

17. Consider the matrices

$$A = \begin{bmatrix} 1 & 2 & 3 \\ 0 & 0 & -1 \\ 0 & 0 & 5 \end{bmatrix} \quad B = \begin{bmatrix} 2 & 0 & 1 \\ 1 & 1 & 1 \\ -3 & 4 & 5 \end{bmatrix}$$

$$C = \begin{bmatrix} 2 & -6 & 3 \\ 1 & 4 & 0 \\ -3 & 4 & 5 \end{bmatrix}.$$

Show that $AB = AC$, yet $B \neq C$. This demonstrates that the cancellation law does not always hold for matrices.

18. A square matrix N is said to be *nilpotent* if some power of it is the zero matrix $\mathbf{0}$; that is, if $N^k = 0$ for some positive integer k. Verify that the following two matrices are nilpotent:

$$N = \begin{bmatrix} 0 & 1 & 0 & 0 \\ 0 & 0 & 1 & 0 \\ 0 & 0 & 0 & 1 \\ 0 & 0 & 0 & 0 \end{bmatrix} \quad M = \begin{bmatrix} 0 & 0 & \sqrt{3} & \pi & 0 \\ 0 & 0 & 0 & 0 & 0 \\ 0 & 0 & 0 & 0 & 0 \\ 0 & 0 & 0 & 0 & -1 \\ 0 & 0 & 0 & 0 & 0 \end{bmatrix}.$$

19. A square matrix E is said to be *idempotent* if $E^2 = E$.

a.   Verify that $E = \begin{bmatrix} 0 & 0 \\ 0 & 1 \end{bmatrix}$ and $F = \begin{bmatrix} 1 & 0 & 4 \\ 0 & 1 & 3 \\ 0 & 0 & 0 \end{bmatrix}$ are idempotent.

b.   Prove that if E is idempotent, so is $I - E$, where I is the identity matrix the same size as E.

In some of the preceding exercises, parts of Theorem 2.3 were illustrated. In the following exercise, we ask you to prove some parts of this theorem. As an example, here is a proof of Part I of that theorem: $r(sA) = (rs)A$, where A is an arbitrary matrix and r and s are arbitrary real numbers. Recall that the entry of A in the ith row and jth column is denoted a_{ij}. The matrix sA has every entry scaled up by a factor of s, so the ij entry of sA is sa_{ij}. Similarly, the entries of $r(sA)$ are obtained from those in sA by scaling up by a factor of r, so the ij entry of $r(sA)$ is $r(sa_{ij}) = (rs)a_{ij}$. On the other hand, the entries of $(rs)A$ are obtained from those in A by scaling by a factor of rs, so the ij entry of $(rs)A$ is also $(rs)a_{ij}$. Because this is true for all i and j, it follows that as matrices, $r(sA) = (rs)A$ (because they have the exact same entries in the same positions.)

20. Using similar arguments and ideas as the preceding proof, prove the following parts of Theorem 2.3:

a.   Part a: $A + B = B + A$
b.   Part b: $(A + B) + C = A + (B + C)$
c.   Part c: $s(A + B) = sA + sB$
d.   Part g: $(A + B)^T = A^T + B^T$

21. In some of the preceding exercises, you saw parts of Theorem 2.6 illustrated. In this exercise, give proofs of the following parts of Theorem 2.6:

a.   Part b: $A(B + C) = AB + AC$
b.   Part c: $(B + C)A = BA + CA$
c.   Part d: $s(AB) = (sA)B = A(sB)$
d.   Part e: $(AB)^T = B^T A^T$.

2.3   Operations on Matrices II: Matrix Inversion

In the real number system, dividing a by b is the same as multiplying a by the reciprocal of b: $\frac{a}{b} = a \cdot \frac{1}{b}$. We sometimes write b^{-1} for $\frac{1}{b}$, and refer to it as the *multiplicative inverse* of a. In this way, we can avoid explicit mention of the process of division – if we wish to divide a by b, it is the same as the multiplication problem $a \cdot b^{-1}$. We now mimic this for matrices – we never speak of "dividing" one matrix A by another one B; instead, we seek the multiplicative inverse of B and simply multiply $A \cdot B^{-1}$. The concept of an inverse matrix of B is defined only for the case of a square matrix B.

But what exactly do we mean by "inverse matrix"? Again, we take our cue from the real number system. For a real number b, the reciprocal or inverse is characterized by the fact that the product

$bb^{-1} = 1$ – that is, the product of a number and its inverse is the multiplicative identity. This motivates the following:

DEFINITION 2.8   *Let B be a given square matrix of size n × n. Suppose there is another n × n matrix A such that*

$$AB = I_n = BA; \qquad (2.8)$$

then we say A is the inverse of B, and write $A = B^{-1}$.

REMARK 2.9   *We must specify that both products AB and BA are equal to the identity in this definition, because of the failure of the commutative law. There is no reason to expect that AB = I would be sufficient to conclude that BA = I as well.*

REMARK 2.10   *If* $A = B^{-1}$, *then the definition implies also that* $B = A^{-1}$, *so A and B are inverses of each other.*

REMARK 2.11   *Just because we made this definition, we cannot conclude that, given B, such an inverse* $A = B^{-1}$ *exists. It may or may not exist, depending on B. Some matrices have inverses (and are called* invertible *matrices), and some do not have inverses. This is exactly what happens with real numbers as well. Do all real numbers have inverses? If not, which ones don't?*

In light of our last remark, we must answer the fundamental questions:

1.   Given a square matrix B, how can we tell if it is invertible?
2.   If B is invertible, how can we find the inverse B^{-1}?

There are two approaches to the second question. One way is to set up a linear system of equations whose solution gives the desired inverse. Such a system comes from the definition of the inverse. Then solve the system by Gaussian elimination. Along the way, the first question is answered – if the system has no solution, then no inverse exists. We explore this approach in the exercises (see Exercises 10–13 at the end of this section.)

The other approach is to find a formula for the inverse, using two auxiliary concepts known as the *determinant* of a matrix and the *adjoint* of a matrix. For 2×2 matrices, both the determinant and adjoint are easy to find, so the formula will be presented momentarily. For larger square matrices, finding the determinant by hand can be a tedious chore (although its easy if you use a computer algebra system with matrix operations built in, such as Mathematica or Maple), so we defer to more advanced texts for this. We assure the reader that in this text, the 2×2 case is by far the most frequent case that appears.

Without further ado, let $A = \begin{bmatrix} a & b \\ c & d \end{bmatrix}$ be a given 2×2 matrix.

DEFINITION 2.12   *The* determinant *of A, written* det(A), *is the real number given by the formula*

$$\det(A) = ad - bc. \qquad (2.9)$$

DEFINITION 2.13   *The* adjoint matrix *of A, written* A^{adj}, *is the* 2×2 *matrix given by interchanging the two main diagonal entries and negating the other entries – that is,*

$$A^{adj} = \begin{bmatrix} d & -b \\ -c & a \end{bmatrix}. \qquad (2.10)$$

Notice that the determinant of A is a scalar while the adjoint of A is another matrix. For example, if $A = \begin{bmatrix} 3 & 1 \\ 5 & 4 \end{bmatrix}$, then $\det(A) = 3 \cdot 4 - 1 \cdot 5 = 7$; and $A^{adj} = \begin{bmatrix} 4 & -1 \\ -5 & 3 \end{bmatrix}$.

The first question is then answered by the following

THEOREM 2.14   *Let A be an 2×2 matrix. Then A is invertible (i.e., A^{-1} exists) if and only if $\det(A) \neq 0$.*

For example, the matrix $A = \begin{bmatrix} 2 & 3 \\ 5 & -1 \end{bmatrix}$ is invertible since $\det(A) = -2 - 15 - 17 \neq 0$. On the other hand, the matrix $B = \begin{bmatrix} 3 & -5 \\ -9 & 15 \end{bmatrix}$ is not invertible since $\det(B) = 45 - 45 = 0$. This theorem extends to $n \times n$ matrices, with the proper definition of determinant for larger matrices. We won't consider larger matrices, but we will explore the proof of this theorem in the 2×2 case. As for the second question, we have the following:

THEOREM 2.15   *Let $A = \begin{bmatrix} a & b \\ c & d \end{bmatrix}$ be a 2×2 matrix with $\det A \neq 0$. Then the inverse of A is given by the formula*

$$A^{-1} = \frac{1}{\det(A)} A^{adj} \tag{2.11}$$

$$= \frac{1}{ad - bc} \begin{bmatrix} d & -b \\ -c & a \end{bmatrix}.$$

That is, simply multiply the adjoint matrix by a scalar factor equal to the reciprocal of the determinant. For example, with $A = \begin{bmatrix} 2 & 3 \\ 5 & -1 \end{bmatrix}$, we have $\det(A) = -17$, so

$$A^{-1} = \frac{1}{-17} \begin{bmatrix} -1 & -3 \\ -5 & 2 \end{bmatrix} = \begin{bmatrix} \frac{1}{17} & \frac{3}{17} \\ \frac{5}{17} & -\frac{2}{17} \end{bmatrix}.$$

This checks using the definition of inverse:

$$AA^{-1} = \begin{bmatrix} 2 & 3 \\ 5 & -1 \end{bmatrix} \begin{bmatrix} \frac{1}{17} & \frac{3}{17} \\ \frac{5}{17} & -\frac{2}{17} \end{bmatrix}$$

$$= \begin{bmatrix} 1 & 0 \\ 0 & 1 \end{bmatrix} = I_2$$

as expected, and similarly for the product $A^{-1}A$, as the reader should verify. Again, with proper definition of adjoints and determinants for larger matrices, this theorem extends to $n \times n$ matrices. We eschew the larger case here, but the $n = 2$ case is easy to prove, and we do so now:

Proof   We must check that $AA^{-1} = I_2 = A^{-1}A$. But observe, assuming $\det A = ad - bc \neq 0$, we have by Theorem 2.6, Part d:

$$AA^{-1} = A\left(\frac{1}{\det A}A^{adj}\right) = \frac{1}{\det A}AA^{adj}$$

$$= \frac{1}{\det A}\begin{bmatrix} a & b \\ c & d \end{bmatrix}\begin{bmatrix} d & -b \\ -c & a \end{bmatrix}$$

$$= \frac{1}{\det A}\begin{bmatrix} ad-bc & 0 \\ 0 & ad-bc \end{bmatrix}$$

$$= \frac{1}{ad-bc}\begin{bmatrix} ad-bc & 0 \\ 0 & ad-bc \end{bmatrix} = \begin{bmatrix} 1 & 0 \\ 0 & 1 \end{bmatrix} = I_2.$$

The product in the other order is similar and left for the reader to check. This concludes the proof. ∎

As a corollary, we have a proof of part of Theorem 2.14. Indeed, if $\det A \neq 0$, then Equation (2.11) exhibits an inverse of A. The reverse implication (the "only if" part of the statement) would say that if $\det A = 0$, then no inverse exists. Clearly, Formula 2.11 on the facing page cannot work if $\det A = 0$, since we would be dividing by zero. However, that doesn't prove that some other matrix could not be the inverse in this case. In the exercises, the reader will prove the "only if" implication by using a known property of the determinant. The next theorem summarizes some properties of the inverse.

THEOREM 2.16   *Let A be an invertible $n \times n$ matrix. Then:*

a.   *The inverse of A is unique.*

b. $\left(A^{-1}\right)^{-1} = A$

c. A^T *is also invertible, and* $\left(A^T\right)^{-1} = \left(A^{-1}\right)^T$

d.   *If B is also an invertible matrix the same size as A, then AB is invertible, and* $(AB)^{-1} = B^{-1}A^{-1}$.

Proof of Part c.   Observe that $A^T\left(A^{-1}\right)^T = \left(A^{-1}A\right)^T$, by Part e of Theorem 2.6, which in turn is equal to $(I_n)^T$ since $A^{-1}A = I_n$. But I_n is a symmetric matrix so $(I_n)^T = I_n$. We have shown $A^T\left(A^{-1}\right)^T = I_n$. A similar argument shows $\left(A^{-1}\right)^T A^T = I_n$. Thus, $\left(A^{-1}\right)^T$ satisfies the definition of the inverse of A^T. Thus, A^T is invertible and $\left(A^T\right)^{-1} = \left(A^{-1}\right)^T$. The other parts of the proof are equally easy and are left for the reader in the exercises. ∎

Exercises

1.   Compute the determinant of the following 2×2 matrices:

a. $\begin{bmatrix} 1 & 2 \\ 3 & 4 \end{bmatrix}$

b. $\begin{bmatrix} 1 & -2 \\ -3 & 6 \end{bmatrix}$

c. $\begin{bmatrix} 4 & 3 \\ 5 & 5 \end{bmatrix}$

d. $\begin{bmatrix} 0 & 7 \\ 7 & 100 \end{bmatrix}$

2. For each of the matrices in Exercise 1, determine whether or not the matrix is invertible. If it is, find the inverse matrix, and verify your answer by checking that $AA^{-1} = I_2$.

3. For what value(s) of x is the following matrix not invertible?

$$\begin{bmatrix} x-1 & 3 \\ 8 & x-3 \end{bmatrix}$$

4. a. Let $A = \begin{bmatrix} 1 & 2 \\ 3 & 4 \end{bmatrix}$ and $B = \begin{bmatrix} 4 & 3 \\ 5 & 5 \end{bmatrix}$. Find AB and show that, in this example, $\det(AB) = \det A \cdot \det B$.

   b. Show that $\det(AB) = \det A \cdot \det B$ for arbitrary 2×2 matrices. (It can be shown that this is true for $n \times n$ matrices.)

5. a. It can be shown that $\det I_n = 1$ for any n. Verify it for the case $n = 2$.

   b. In one of the theorems in this section (Theorem 2.3), we showed (in the 2×2 case) that if $\det A \neq 0$, then A is invertible. This can be strengthened to say that A is invertible if and only if $\det A \neq 0$. To prove the converse, show that if $\det A = 0$, it could not possibly have an inverse. [**Hint:** Use Exercise 4b and Part a of this exercise.]

6. Suppose A is invertible, and B and C are two inverses of A. Prove that $B = C$ so that Part a of Theorem 2.16 holds. [**Hint:** One possible approach is to consider the product BAC and evaluate it in two different ways.]

7. Let $A = \begin{bmatrix} -3 & 7 \\ -2 & 1 \end{bmatrix}$. Verify Parts b and c of Theorem 2.16 for A.

8. a. Prove Part c of Theorem 2.16 for 2×2 matrices using Equation (2.11).

   b. Prove Part d of Theorem 2.16.

9. Show by example that $A + B$ need not be invertible if both A and B are invertible.

   In the text, it was mentioned that in lieu of Equation (2.11), an alternate approach exists for finding the inverse of a matrix based on solving systems of equations. The next few exercises explore that approach.

10. Let $A = \begin{bmatrix} 3 & 6 \\ 2 & 5 \end{bmatrix}$. Since $\det A \neq 0$, A is invertible. Let $B = \begin{bmatrix} w & x \\ y & z \end{bmatrix}$ be the unknown inverse, so that $AB = I$.

   a. Multiply out AB and set it equal to I.

   b. Notice that in your answer to Part a, the first column of AB only involves the variables w and y, while the second column of AB only involves the variables x and z. Thus, by setting the first column of AB to the first column of I, you should obtain a system of two linear equations involving w and y. Write out this system of equations. Similarly, write out the system involving x and z that is obtained from setting the second column of AB equal to the second column of I.

   c. Solve the first system by pivoting the augmented matrix of the system $\begin{bmatrix} 3 & 6 & 1 \\ 2 & 5 & 0 \end{bmatrix} \longrightarrow$

   $\begin{bmatrix} 1 & 0 & w \\ 0 & 1 & y \end{bmatrix}$ as we learned in Chapter 1.

d.   Notice that the coefficient matrix of both systems is the same matrix – namely A. This means that solving the second system by pivoting, you would use exactly the same elementary row operations in the same order that you used in Part c to solve the first system. Therefore, in yet another instance of using matrices to efficiently handle itemized calculations, you can solve both systems at once by annexing a fourth column $\begin{bmatrix} 0 \\ 1 \end{bmatrix}$ to the augmented matrix you used in Part c. That is, pivot

$$\begin{bmatrix} 3 & 6 & 1 & 0 \\ 2 & 5 & 0 & 1 \end{bmatrix} \longrightarrow \begin{bmatrix} 1 & 0 & w & x \\ 0 & 1 & y & z \end{bmatrix}$$ using the same steps you used in Part c.

Perform the pivoting and thereby find all four variables w, x, y, and z. Thus, you have found B, the inverse of A. Check your answer using Formula 2.11.

11. In Problem 10, you showed that in order to find the inverse of $A = \begin{bmatrix} 3 & 6 \\ 2 & 5 \end{bmatrix}$, you annex an identity matrix I_2 to form an augmented matrix and pivot until the two left columns become I_2, at which point A^{-1} automatically appears in the right two columns.

a.   In this exercise, show that you can do the same thing for any invertible 2×2 matrix $A = \begin{bmatrix} a & b \\ c & d \end{bmatrix}$. That is, to find A^{-1}, row reduce (pivot) $\begin{bmatrix} A & I_2 \end{bmatrix} \longrightarrow \begin{bmatrix} I_2 & A^{-1} \end{bmatrix}$. [**Hint:** treat a, b, c, and d as known and follow the steps you did in Exercise 10. You will need to know that $\det A = ad - bc \neq 0$.]

b.   Use this method to find the inverse of $\begin{bmatrix} 3 & 5 \\ 4 & 7 \end{bmatrix}$ and check your answer either by using Equation (2.11) or by multiplying AA^{-1} to obtain the identity.

c.   What happens if you try to apply this method to a matrix that is not invertible, such as $\begin{bmatrix} 2 & -4 \\ 9 & -18 \end{bmatrix}$?

12. The method of Exercise 11 actually works for any $n \times n$ matrix, and the reason is essentially the same as the one given in Exercises 10 and 11 for the case $n = 2$. Use this method to find the inverses of the following 3×3 matrices, and check your answers by multiplying AA^{-1} to obtain I_3.

a. $\begin{bmatrix} 3 & 1 & -4 \\ 1 & 3 & 2 \\ 5 & -1 & -1 \end{bmatrix}$

b. $\begin{bmatrix} 4 & 6 & 2 \\ 2 & 3 & 2 \\ 1 & 2 & \frac{1}{2} \end{bmatrix}$

13. Find, if possible, the inverse of the following 4×4 matrices, using the method of Exercises 11 and 12.

a. $\begin{bmatrix} 1 & -1 & 0 & 0 \\ 0 & 1 & -1 & 0 \\ 0 & 0 & 1 & -1 \\ 0 & 0 & 0 & 1 \end{bmatrix}$

$$\begin{bmatrix} 1 & 2 & 3 & 4 \\ 0 & 1 & 0 & 1 \\ 2 & 4 & 6 & 8 \\ 0 & 0 & 0 & 3 \end{bmatrix}$$

b.

14. Recall from Exercise 15 of the last section that a square matrix N is *nilpotent* if $N^k = \mathbf{0}$ for some $k > 0$.

   a. Suppose N is nilpotent with $k = 4$. Show that $I - N$ is invertible by direct computation as follows:

   $$(I - N)\left(I + N + N^2 + N^3\right) = I,$$

   and so $(I - N)^{-1} = I + N + N^2 + N^3$.

   b. What matrix N should be used to show that Part a gives an alternate way to find the inverse of the matrix in Part a of Exercise 13?

   c. Generalize Part a to an arbitrary $k > 0$. What is the inverse of $I - N$ if $N^k = \mathbf{0}$?

15. Recall from Exercise 16 of the last section that a square matrix E is called *idempotent* if $E^2 = E$. Explain why the only invertible idempotent $n \times n$ matrix is I_n.

2.4 Solving Linear Systems by Matrix Inversion

At this point, we haven't addressed the fundamental question of whether there is any advantage to writing a linear system as a matrix equation, in terms of finding the solution to the system. Recall from Chapter 1 that every linear system had one of three possibilities:

1. There might be no solution (called an *inconsistent* system).
2. There might be infinitely many solutions (called a *redundant* system).
3. There might be a unique solution (called a *non-degenerate* system).

One usually hopes to be in the third case, that of a unique solution, which means in particular that the number of equations is the same as the number of unknowns – thus, the coefficient matrix is square. But even in the case of a square matrix, the system might degenerate to one of the other two cases. In Chapter 1, we explored the solution of systems by elimination, which we streamlined and made efficient by the use of elementary row operations on the (augmented) coefficient matrix. This method (Gaussian elimination) worked in any of the three cases just mentioned.

In this section, we explore a second method of using matrices to solve systems that is based on writing the system as a single matrix equation $AX = B$. So there will turn out to be some advantage to writing a system in this form. However, it only works in the case of a non-degenerate system (case 3) because it requires finding the inverse of the coefficient matrix A.

To motivate the solution, consider what we do to solve a single linear equation in just one unknown of the form $ax = b$. For example, how do we solve $3x = 7$? Most people would instantly give the solution as $x = \frac{7}{3}$, obtained by dividing both sides of the equation by 3 in order to isolate the variable x. In reality, this is actually a several-step process, with most of the steps suppressed because of their familiarity:

$$3x = 7$$

$$\frac{1}{3}(3x) = \frac{7}{3}\left(\text{multiply both sides by } \frac{1}{3}\right)$$

$$\left(\frac{1}{3} \cdot 3\right)x = \frac{7}{3} \text{ (associative law of multiplication)}$$

$$1 \cdot x = \frac{7}{3}\left(3 \text{ and } \frac{1}{3} \text{ are multiplicative inverses}\right)$$

$$x = \frac{7}{3} \text{ (1 is the multiplicative identity).}$$

Because of the theorems in the last section, all the justifications given in the string of steps are valid also for invertible matrices, so the matrix equation $AX = B$ should be solvable the same way:

$$AX = B$$

$$A^{-1}(AX) = A^{-1}B \text{ (multiply both sides by } A^{-1} \text{ on the left)}$$

$$\left(A^{-1}A\right)X = A^{-1}B \text{ (associative law of matrix multiplication)}$$

$$IX = A^{-1}B \text{ (since } A^{-1} \text{ and } A \text{ are multiplicative inverse)}$$

$$X = A^{-1}B \text{ (since } I \text{ is the multiplicative identity).}$$

So, provided A^{-1} exists, the solution to the entire system can be found by a simple matrix multiplication $X = A^{-1}B$.

For example, consider the system (2.5) from Section 2.3:

$$2x + 3y = 7$$
$$5x - y = 9,$$

which we have already written in matrix form:

$$\begin{bmatrix} 2 & 3 \\ 5 & -1 \end{bmatrix}\begin{bmatrix} x \\ y \end{bmatrix} = \begin{bmatrix} 7 \\ 9 \end{bmatrix}.$$

Since this has the required form $AX = B$ and A is invertible, we know the (unique) solution is $X = A^{-1}B$. Since we have already calculated A^{-1} in Section 2.4 (Exercise 2), we have

$$X = \begin{bmatrix} x \\ y \end{bmatrix} = A^{-1}\begin{bmatrix} 7 \\ 9 \end{bmatrix} = \begin{bmatrix} \frac{1}{17} & \frac{3}{17} \\ \frac{5}{17} & -\frac{2}{17} \end{bmatrix}\begin{bmatrix} 7 \\ 9 \end{bmatrix} = \begin{bmatrix} 2 \\ 1 \end{bmatrix}.$$

Thus, $x = 2$ and $y = 1$ is the solution, as the reader can easily verify.

Similarly, the system (2.6) can be written in $AX = B$ form as

$$\begin{bmatrix} 3 & 1 & -4 \\ 1 & 3 & 2 \\ 5 & -1 & -1 \end{bmatrix}\begin{bmatrix} x \\ y \\ z \end{bmatrix} = \begin{bmatrix} 2 \\ 4 \\ 9 \end{bmatrix},$$

so the solution is given by

$$X = \begin{bmatrix} x \\ y \\ z \end{bmatrix} = A^{-1}B = \begin{bmatrix} 3 & 1 & -4 \\ 1 & 3 & 2 \\ 5 & -1 & -1 \end{bmatrix}^{-1} \begin{bmatrix} 2 \\ 4 \\ 9 \end{bmatrix}.$$

Since A is 3×3, we have no nice formula for the inverse and can only use the technique indicated in Exercises 10–11 of Section 2.4. It can be shown (see Exercise 12a) that

$$A^{-1} = \begin{bmatrix} -\frac{1}{72} & \frac{5}{72} & \frac{14}{72} \\ \frac{11}{72} & \frac{17}{72} & -\frac{10}{72} \\ -\frac{16}{72} & \frac{8}{72} & \frac{8}{72} \end{bmatrix}$$

so that

$$\begin{bmatrix} x \\ y \\ z \end{bmatrix} = \begin{bmatrix} -\frac{1}{72} & \frac{5}{72} & \frac{14}{72} \\ \frac{11}{72} & \frac{17}{72} & -\frac{10}{72} \\ -\frac{16}{72} & \frac{8}{72} & \frac{8}{72} \end{bmatrix} \begin{bmatrix} 2 \\ 4 \\ 9 \end{bmatrix} = \begin{bmatrix} 2 \\ 0 \\ 1 \end{bmatrix}.$$

That is, $x = 2$, $y = 0$, and $z = 1$. The reader should verify that this is a solution (the unique one, which follows from the fact that A is invertible).

Next, consider the example:

$$3x - 8y = 22$$
$$-9x + 24y = 55.$$

In this case, we have $\det A = 72 - 72 = 0$, so A is not invertible. Therefore, the system cannot be solved by the method of matrix inversion. We know there must be either no solution or infinitely many, but matrix inversion cannot distinguish between these two cases when A^{-1} does not exist. In general, such systems must be solved by elimination to tell which degenerate case we are in. However, in this example, it is easy to see that the lines represented by the two equations are parallel, and so there is no solution.

Our final examples illustrate one of the advantages of solving a system by the method of matrix inversion – namely that changes in the right side of the equations are easily and efficiently dealt with via this technique. Figure 2.2 is a picture of a set of bongo drums.

As can be seen, bongo drums consist of a set of two wooden drums, connected by a wooden block or bridge. They originated in Cuba, brought there from Africa during the slave trade. Traditional bongos had animal skins for drumheads, attached to the drum using heat. Although some bongo players still prefer animal skin heads because of the warm sound, most modern bongos have a synthetic material for the drumhead, which has the advantage over animal skins that it is impervious to changes in pitch due to variations in temperature and humidity. Modern bongos have metal tuning hardware attached to the drums to make tuning easier and also to brace the drums and make them stronger. The larger of the two drums is called the *hembra* (female) and the smaller is called the *macho* (male). The wood that is used is typically oak or mahogany, although modern bongos can be made of fiberglass in lieu of wood.

Consider a hypothetical drum manufacturing company called Bongo Fury, Inc. They manufacture three types of drums – a set of bongo drums similar to those shown above (dubbed the Yellow Shark model), a set of conga drums that are similar to bongos but larger (dubbed the Jawaka

Figure 2.2 Typical bongo drums.

model), and a set of timbales (dubbed the Chunga model.) Timbales are smaller and shallower than bongos and were also first developed in Cuba. The three models use the same raw materials: mahogany for the drums and connecting bridge, the synthetic material (referred to as *tar shot* in our example) for the drum heads, and steel for the lugs and tuning hardware. Each set of Yellow Shark bongos consists of 4 pounds mahogany, 2 square feet of tar shot, and 1 pound of steel. Each set of Jawaka congas consists of 6 pounds of mahogany, 3 square feet of tar shot, and 2 pounds of steel. Each set of Chunga timbales consists of 2 pounds of mahogany, 2 square feet of tar shot, and $\frac{1}{2}$ pound of steel.

The question is how many sets of each model should Bongo Fury produce this week if the company has available 1,400 lbs mahogany, 800 square feet of tar shot, and 400 pounds of steel hardware? We assume that all the raw materials are to be used up.

For the solution, let x denote the number of sets of Yellow Shark bongos, y denote the number of sets of Jawaka congas, and z denote the number of sets of Chunga timbales. Each resource (raw material) leads to an equation:

$$4x + 6y + 2z = 1{,}400 \text{ (pounds mahogany)}$$
$$2x + 3y + 2z = 800 \text{ (square feet of tar shot)}$$
$$x + 2y + \frac{1}{2}z = 400 \text{ (pounds steel).}$$

To solve this, we rewrite it in matrix form and obtain

$$AX = B$$

$$\begin{bmatrix} 4 & 6 & 2 \\ 2 & 3 & 2 \\ 1 & 2 & \frac{1}{2} \end{bmatrix} \begin{bmatrix} x \\ y \\ z \end{bmatrix} = \begin{bmatrix} 1{,}400 \\ 800 \\ 400 \end{bmatrix}.$$

It can be shown (Exercise 12b of the last section) that

$$A^{-1} = \begin{bmatrix} \frac{5}{4} & -\frac{1}{2} & -3 \\ -\frac{1}{2} & 0 & 2 \\ -\frac{1}{2} & 1 & 0 \end{bmatrix}.$$

We can now solve the system via matrix inversion:

$$
\begin{bmatrix} x \\ y \\ z \end{bmatrix} = X = A^{-1}B = \begin{bmatrix} \frac{5}{4} & -\frac{1}{2} & -3 \\ -\frac{1}{2} & 0 & 2 \\ -\frac{1}{2} & 1 & 0 \end{bmatrix} \begin{bmatrix} 1,400 \\ 800 \\ 400 \end{bmatrix} = \begin{bmatrix} 150 \\ 100 \\ 100 \end{bmatrix}.
$$

Thus, Bongo Fury should produce $x = 150$ sets of Yellow Shark bongos, $y = 100$ sets of Jawaka congas, and $z = 100$ sets of Chunga timbales.

So far, it looks as if it would take the same amount of work to solve the problem this way as it would to solve it by elimination, since we needed elimination to find A^{-1}. However, suppose that the following week, the supply of mahogany increased to 1,500 pounds, while the other resources remained at the current levels. How should Bongo Fury adjust its production plans to reflect the change in supplies? Once A^{-1} is in hand, no further work is necessary to adapt to these market changes. Changes in the resources only affect the matrix B. So we can find the modified solution with a simple matrix multiplication with the new B:

$$
X = A^{-1}B = \begin{bmatrix} \frac{5}{4} & -\frac{1}{2} & -3 \\ -\frac{1}{2} & 0 & 2 \\ -\frac{1}{2} & 1 & 0 \end{bmatrix} \begin{bmatrix} 1,500 \\ 800 \\ 400 \end{bmatrix} = \begin{bmatrix} 275 \\ 50 \\ 50 \end{bmatrix}.
$$

So for the following week, Bongo Fury should produce $x = 275$ sets of Yellow Shark bongos, $y = 50$ sets of Jawaka Congas, and $z = 50$ sets of Chunga timbales. For the week after that, suppose there is a strike in the shipping industry, so the supply of steel goes down to 200 pounds and the supply of mahogany goes down to 800 pounds. However, the tar shot is produced locally so the supply is unaffected by the shipping strike. Now how should they adjust their production? Again, we merely compute $A^{-1}B$ with the new B. The reader should check that this time,

$$
X = A^{-1} \begin{bmatrix} 800 \\ 800 \\ 200 \end{bmatrix} = \begin{bmatrix} 0 \\ 0 \\ 400 \end{bmatrix},
$$ so during the strike, Bongo Fury should produce 400 sets of

Chunga timbales and none of the other two types of drum. Compare this to the amount of work we would have to do if we solved the system using elimination. Every time a change in the last column was made, we would have to start all over again and perform all of the pivoting operations from the beginning. Surely, the reader can appreciate how much more efficient the method of matrix inversion is in this situation.

Exercises

1. Solve the following system using matrix inversion:

$$
5x + y = 8
$$
$$
2x - y = 13.
$$

2. Solve the following system using matrix inversion:

$$
x + 6y = 4
$$
$$
5x - 2y = 4.
$$

3. Solve the following system using matrix inversion:

$$-x + y = 3$$
$$9x + 3y = 13.$$

4. Solve the following system using matrix inversion:

$$x + 2y = 1$$
$$5x + 9y = 16.$$

5. a. Given $A = \begin{bmatrix} 10 & -1 & 2 \\ 5 & 0 & 3 \\ 7 & -2 & -2 \end{bmatrix}$, verify that

$$A^{-1} = \begin{bmatrix} \frac{6}{9} & -\frac{6}{9} & -\frac{3}{9} \\ \frac{31}{9} & -\frac{34}{9} & -\frac{20}{9} \\ -\frac{10}{9} & \frac{13}{9} & \frac{5}{9} \end{bmatrix}$$

by checking that $AA^{-1} = I_3$.

b. Solve the following system using matrix inversion:

$$10x - y + 2z = 34$$
$$5x + 3z = 31$$
$$7x - 2y - 2z = 0.$$

6. In the Bongo Fury Drum company example from the text, how many sets of each type of drum should the company produce if the supply of tar shot increases to 1,000 square feet, while the mahogany and steel supplies remain fixed at 1,400 pounds and 400 pounds, respectively?

7. The Green Onions Natural Food grocery store orders all their dairy products from Maggie's Farm and from Dixie's Chicken Farm. On any given day, Maggie's Farm can deliver 50 dozen eggs and 100 gallons of milk. On any given day, Dixie's Chicken Farm can deliver 40 dozen eggs and 70 gallons of milk.

    a. Each week, the store can sell 350 dozen eggs and 650 gallons of milk. How many days per week should each farm make a delivery to the store?
    b. Suppose the demand increases to 400 dozen eggs and 750 gallons of milk. Now how many days per week should each farm make a delivery?
    c. How many days per week should each farm make a delivery if the demand changed to 380 dozen eggs and 690 gallons of milk per week.

8. The Aoxomoxoa Chemical Company mixes nitric acid for industrial use. In the warehouse, the company keeps a large supply of nitric acid in two concentrations – one vat has a 50% concentration and one vat has a 20% concentration.

    a. They receive an order for 240 gallons of nitric acid at 30% concentration. How many gallons from each vat does Aoxomoxoa Chemical Company need to mix to fill the order?

b.   The company also receives a second order for 240 gallons at 40% concentration. How many gallons from each vat are needed to mix to fill this order?

c.   The company also receives an order for 180 gallons at 25% concentration. How many gallons from each vat are needed need to mix to fill this order?

d.   Finally, the company receives a fourth order for 400 gallons at 40% strength. Set up a single matrix multiplication that will solve for all four orders (from Parts a–d) at once.

9.   The Low Spark Body Shop has a bay that specializes in repairs to sedans and to pickup truck bodies due to rust. Each sedan repair takes 4 hours on the lift and 7 hours in the paint room. Each truck repair requires 5 hours on the lift and 6 hours in the paint room.

a.   Suppose the shop has 33 hours available on the lift and 55 hours available in the paint room. How many repairs of each type should it schedule?

b.   Suppose the shop has 35 hours available on the lift and 53 hours available in the paint room. How many repairs of each type should it schedule?

c.   Suppose the shop has 44 hours available on the lift and 66 hours available in the paint room. How many repairs of each type should it schedule?

d.   One week, the body shop repaired six sedans and three pickup trucks. How many hours did it use on the lift in total and how many hours in the paint room? (Use the matrix multiplication!)

10.   Solve the following system by matrix inversion:

$$w - x = 10$$
$$x - y = 20$$
$$y - z = 40$$
$$z = 100.$$

[**Hint**: See Exercise 13a or 14 from the last section to invert the 4×4 coefficient matrix.]

2.5   Some Applications to Cryptography and Economics

2.5.1   A Matrix Cipher

There has been an interest in transmitting coded or secret messages at least since the days of Caesar, and probably earlier. A scheme for encoding a message is called a *cipher*. The simplest cipher for encoding a message (actually called the *Caesar cipher*) is a substitution cipher where each letter in the message is replaced by a substitute letter. Such messages are not too difficult to decode, and form the basis for cryptograms as popular word puzzles today. In 1929, Lester Hill developed a way to encode messages using invertible matrices. Messages encoded by a *Hill cipher* are much more difficult to decode than those encoded by a simple substitution cipher, so the secrecy of the message is less likely to be compromised.

Naturally, in the digital age of the Internet, encryption of information has become extremely important and many more sophisticated ciphers have been developed which are even more difficult to decode than the Hill cipher. In this section, we focus on a relative of the Hill cipher because it provides a nice application of invertible matrices.

The idea is to assign a number $(1-26)$ to each letter and also assign a 0 to a blank space between words. Then, break the message up into blocks of size n, and arrange them into column matrices (or n-tuples). Group the columns together to form a message matrix M that has n rows. Next, multiply (on the left) the entire message by some invertible $n \times n$ matrix with integer entries A. The person who receives the coded message can only decode it if he or she knows what A is. He or she can recover the message by simply multiplying the coded message matrix by A^{-1}. We'll do an example. Begin with the following substitution table:

Table 2.1 Converting between letters and numbers

blank space	A	B	C	D	E	F	G	H
0	1	2	3	4	5	6	7	8
I	J	K	L	M	N	O	P	Q
9	10	11	12	13	14	15	16	17
R	S	T	U	V	W	X	Y	Z
18	19	20	21	22	23	24	25	26

Next, suppose $A = \begin{bmatrix} 1 & -1 & 2 \\ -2 & 3 & -1 \\ 0 & 1 & 2 \end{bmatrix}$, which has inverse $A^{-1} = \begin{bmatrix} -7 & -4 & 5 \\ -4 & -2 & 3 \\ 2 & 1 & -1 \end{bmatrix}$. The reader should verify that these are indeed inverse matrices.

REMARK 2.17 *Notice that the entries of A^{-1} are also integers. An invertible matrix with integer entries does not necessarily have an inverse with integer entries. However, if the determinant of A is ± 1, then the entries of the inverse are guaranteed to be integers as well, because of the formula $A^{-1} = \frac{1}{\det A} A^{adj}$. For this cipher, it is not necessary that the inverse A^{-1} have integer entries for it to work, but it is a lot more convenient.*

Now select a message to encode. We will encode the following: EQUAL RIGHTS FOR ALL. First, break up the message into blocks of size 3, including the spaces between words (because A is 3×3) : EQU AL_ RIG HTS _FO R_A LL_. Now convert each symbol to a number using the substitution table (Table 2.1), and write the blocks of three as column matrices in order. For example, EQU becomes $\begin{bmatrix} 5 \\ 17 \\ 21 \end{bmatrix}$, etc. The entire message converts to a matrix M (the "message" matrix) by putting the columns together:

$$M = \begin{bmatrix} 5 & 1 & 18 & 8 & 0 & 18 & 12 \\ 17 & 12 & 9 & 20 & 6 & 0 & 12 \\ 21 & 0 & 7 & 19 & 15 & 1 & 0 \end{bmatrix}.$$

Notice we fill out the last column with enough zeros to make sure we have a rectangular array for M. To scramble the message, multiply on the left by A to obtain

$$AM = \begin{bmatrix} 1 & -1 & 2 \\ -2 & 3 & -1 \\ 0 & 1 & 2 \end{bmatrix} \begin{bmatrix} 5 & 1 & 18 & 8 & 0 & 18 & 12 \\ 17 & 12 & 9 & 20 & 6 & 0 & 12 \\ 21 & 0 & 7 & 19 & 15 & 1 & 0 \end{bmatrix}$$

$$= \begin{bmatrix} 30 & -11 & 23 & 26 & 24 & 20 & 0 \\ 20 & 34 & -16 & 25 & 3 & -37 & 12 \\ 59 & 12 & 23 & 58 & 36 & 2 & 12 \end{bmatrix}.$$

This last matrix is what you send your friend/lover/employer/fellow spy. The receiver of the message can only decode it if he or she knows what A^{-1} is. Simply multiply by A^{-1} to recover M, since $A^{-1}(AM) = M$:

$$\begin{bmatrix} -7 & -4 & 5 \\ -4 & -2 & 3 \\ 2 & 1 & -1 \end{bmatrix} \begin{bmatrix} 30 & -11 & 23 & 26 & 24 & 20 & 0 \\ 20 & 34 & -16 & 25 & 3 & -37 & 12 \\ 59 & 12 & 23 & 58 & 36 & 2 & 12 \end{bmatrix}$$

$$= \begin{bmatrix} 5 & 1 & 18 & 8 & 0 & 18 & 12 \\ 17 & 12 & 9 & 20 & 6 & 0 & 12 \\ 21 & 0 & 7 & 19 & 15 & 1 & 0 \end{bmatrix}.$$

The receiver can now use the substitution table to replace the numbers with letters and read the message:

$$\begin{bmatrix} E & A & R & H & & R & L \\ Q & L & I & T & F & & L \\ U & & G & S & O & A & \end{bmatrix}.$$

The real Hill cipher differs from what we did in only one way. The numbers of the coded matrix AM are replaced by letters, so the receiver receives a string of letters instead of numbers. In order for that to work, the numbers in AM must be in the range $0-26$ so the substitution table (Table 2.1) can be used. To make that happen, each number is replaced by its remainder when divided by 27 (this is called *modular arithmetic*), and the matrix A must be invertible over the integers modulo 27. To avoid dealing with modular arithmetic, in our simplified version of the Hill cipher, we will be content to use the matrix AM with numerical entries as we did before rather than try to convert to letters.

Let's finish with one more example. Captain Daltrey is a scout for the Royal Army and is stationed on the north side of Entwistle Ridge overlooking Moon Valley, where the army is expecting a rebel attack. Daltrey receives an inquiry from Field Marshall Townshend about the extent of Daltrey's ability to see the entire valley from his perch on the ridge so he can warn the troops before the attack. Daltrey sends the following coded reply:

$$AM = \begin{bmatrix} 15 & -13 & 24 & 24 & 44 & 7 & 21 & 10 & 28 & -14 \\ -21 & 40 & -28 & 3 & -49 & 13 & -39 & -16 & -11 & 47 \\ 6 & 14 & 15 & 36 & 26 & 22 & 2 & 4 & 33 & 19 \end{bmatrix}.$$

What did Daltrey say? Knowing the same matrix A as above was used to code the message, Townshend can decode it by multiplying by A^{-1}. The reader should verify that $A^{-1}(AM) = M$ is given by the following matrix:

$$M = \begin{bmatrix} 9 & 1 & 19 & 0 & 18 & 9 & 19 & 14 & 13 & 5 \\ 0 & 14 & 5 & 6 & 0 & 12 & 0 & 4 & 9 & 19 \\ 3 & 0 & 5 & 15 & 13 & 5 & 1 & 0 & 12 & 0 \end{bmatrix}.$$

Now, using the substitution table, we replace each number with its letter:

$$\begin{bmatrix} I & A & S & & R & I & S & N & M & E \\ & N & E & F & & L & & D & I & S \\ C & & E & O & M & E & A & & L & \end{bmatrix}.$$

Thus, Captain Daltrey's reply is "I can see for miles and miles." Note that when we are encoding or decoding a message, we ignore all apostrophes, commas, periods, and other punctuation marks. If desired, you could certainly build these symbols into the substitution table by going higher than the number 26, and you would thereby have the ability to encode grammatically correct messages.

2.5.2 Leontief Input–Output Models

In 1973, Harvard economist Wassily Leontief earned a Nobel Memorial in Economic Sciences for his work showing how different sectors of the economy are interrelated and how some of the outputs of one sector are used as inputs for other sectors. Leontief's ideas are today known as *input–output models* of the economy, and they provide a nice application of matrices and matrix inversion in particular. Furthermore, in the 1940s when Leontief first used his model to analyze the economy of the United States, he divided the economy into 500 sectors, so the size of the matrices made hand calculations impractical. Leontief used the primitive Mark II computer at Harvard in 1946 to complete the analysis. Not only was it a beautiful application of matrices, but it also was one of the first-ever significant uses of a computer for mathematical modeling.

In this section, we illustrate Leontief's model, but in order to keep the size of the problem manageable, our hypothetical economy will be divided into just four sectors instead of 500. Our fictional country is called Sunhillow, and we represent the four sectors of its economy (chemicals, electricity, services, and steel) as vertices in the directed graph in Figure 2.3.

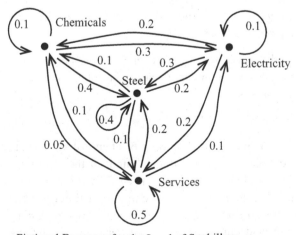

Fictional Economy for the Land of Sunhillow

Figure 2.3 Fictional economy for the land of Sunhillow.

The arrows in this graph indicate flow of production between the sectors. The fractions that label the arrows are interpreted as telling what fraction of that sector's output (measured in dollars) is used as input for the various other sectors over one time period (which could be a month or a year, for example) for the other sectors to produce one dollar of output. For example, the 0.05 on the arrow pointing from chemicals to services indicates that for each time period (say, a month), each dollar of output from the services sector will require 0.05 of a dollar input from the chemicals sector. There are arrows from each sector to themselves (such arrows are called *loops* in graph theory) because some of the output from each sector is used as input into the same sector. For example, the .1 on the loop connected the electricity sector to itself indicates that 10% of the output from that sector is funneled back into the same sector; that is, each dollar of output produced by the electricity sector uses $0.10 of the sector's output to produce it.

Consider the sum of all the percentages or fractions of outputs from a fixed sector. That sum represents the total output from that sector which is required for all the industries in all the sectors for a month of the economy. For example, leading away from the chemicals sector, we find the sum: $0.1 + 0.3 + 0.05 + 0.4 = 0.85$. This tells us that 85% of the output from the chemicals sector is required by the industries of the entire economy (including the chemicals sector itself) any given month. The remaining 15% is available for sale to consumers. A model such as this example, where the sum of all the outgoing percentages is less than 100% (thereby leaving a surplus of output, which can be used to satisfy consumer demand) is called an *open Leontief model*.

By contrast, if we assume there is no consumer demand and, therefore, that all of the output from each sector of the economy is reabsorbed by the various sectors of the economy, the sum of all the output percentages for each sector will be exactly 100%. This is called a *closed Leontief model*. Both types of models are useful, but in this section, we will only consider open models.

In an open model, let d_i denote the consumer demand (in dollars) for the ith industry sector for one time period. and let x_i be the total production of the ith sector (in dollars) for one time period. By assumption, each sector produces enough output to provide for the needs of all the other sectors and still have some remaining output to be use to satisfy consumer demand. It seems intuitively clear, then, that by producing enough (in other words, by making the x_i large enough), we should be able to satisfy any demand. This is a true fact, whose proof we omit.

PROPOSITION 2.18   *In an open Leontief model, (where the sum of the output percentages for each sector is strictly smaller than 100 %), any consumer demand can be met.*

How do we go about finding the correct production levels? In our example, consider the production x_1 from the chemicals sector of the Sunhillow economy. Clearly, the total demand for chemicals output for the four sectors of the economy is

$$0.1x_1 + 0.3x_2 + 0.05x_3 + 0.4x_4.$$

If we add the consumer demand d_1 to this, we obtain the total amount of chemical sector output needed, which is the unknown x_1. For example, if x_2 is the dollar output of the electricity sector, $0.3x_2$ represents the total dollar output of the chemicals sector needed by the electricity sector, since each dollar output from the electricity sector requires $0.30 output from the chemicals sector, and similarly for the other sectors. Each sum such as the preceding one – before adding the consumer demand looks like a dot product, and our experience with itemized calculations suggests that we should be able to do all the relevant calculations at once with a suitable matrix multiplication. Indeed, let $A = (a_{ij})$ be the matrix where a_{ij} represents the percentage of dollar output from the ith sector which is required by the jth sector (the numbers that label the arrows in the diagram).

This is called the *input–output matrix*. With the sectors listed in the order chemicals, electricity, services, steel, the input–output matrix for the Sunhillow economy is

$$A = \begin{array}{c} \\ Chemicals \\ Electricity \\ Services \\ Steel \end{array} \begin{array}{cccc} C & E & S & St \\ \left[\begin{array}{cccc} 0.1 & 0.3 & 0.05 & 0.4 \\ 0.2 & 0.1 & 0.2 & 0.3 \\ 0.1 & 0.1 & 0.5 & 0.2 \\ 0.1 & 0.2 & 0.1 & 0.4 \end{array}\right] \end{array}.$$

We define the *production matrix* to be $X = \begin{bmatrix} x_1 \\ x_2 \\ x_3 \\ x_4 \end{bmatrix}$ and the *demand matrix* to be: $D = \begin{bmatrix} d_1 \\ d_2 \\ d_3 \\ d_4 \end{bmatrix}$. Observe that the matrix AX represents the outputs of the four sectors that are needed by the industries in the four sectors. If we add to this the demand, we obtain the total production needed. That is, in an open Leontief model, the following equation governs the production:

$$AX + D = X. \tag{2.12}$$

In the Sunhillow example, this becomes

$$\begin{bmatrix} 0.1 & 0.3 & 0.05 & 0.4 \\ 0.2 & 0.1 & 0.2 & 0.3 \\ 0.1 & 0.1 & 0.5 & 0.2 \\ 0.1 & 0.2 & 0.1 & 0.4 \end{bmatrix} \begin{bmatrix} x_1 \\ x_2 \\ x_3 \\ x_4 \end{bmatrix} + \begin{bmatrix} d_1 \\ d_2 \\ d_3 \\ d_4 \end{bmatrix} = \begin{bmatrix} x_1 \\ x_2 \\ x_3 \\ x_4 \end{bmatrix}.$$

To solve this system, observe that

$$AX + D = X$$
$$X - AX = D$$
$$(I - A)X = D.$$

Notice that $I - A$ is a square matrix. Assuming it has an inverse, we can then solve by multiplying through by this inverse to obtain

$$X = (I - A)^{-1}D. \tag{2.13}$$

Our assumption that we are working with an open model will imply that not only does $I - A$ have an inverse but that this inverse has all nonnegative entries (proof omitted), and therefore, we can find X, and X will also have nonnegative entries, no matter what demand D we have. (This is how the preceding proposition is proved.) Thus, the mathematical solution of the model does indeed make sense when interpreted in terms of the original real world problem; obviously, we seek nonnegative solutions, as negative production levels are meaningless.

We conclude with an illustration with the Sunhillow economy. Suppose that the monthly consumer demand, in millions of dollars, for the four sectors is

$$D = \begin{bmatrix} 25 \\ 100 \\ 80 \\ 40 \end{bmatrix}.$$

Find the correct monthly production levels to meet this demand. For the solution, we have

$$I - A = \begin{bmatrix} 1 & 0 & 0 & 0 \\ 0 & 1 & 0 & 0 \\ 0 & 0 & 1 & 0 \\ 0 & 0 & 0 & 1 \end{bmatrix} - \begin{bmatrix} 0.1 & 0.3 & 0.05 & 0.4 \\ 0.2 & 0.1 & 0.2 & 0.3 \\ 0.1 & 0.1 & 0.5 & 0.2 \\ 0.1 & 0.2 & 0.1 & 0.4 \end{bmatrix}$$

$$= \begin{bmatrix} 0.9 & -0.3 & -0.05 & -0.4 \\ -0.2 & 0.9 & -0.2 & -0.3 \\ -0.1 & -0.1 & 0.5 & -0.2 \\ -0.1 & -0.2 & -0.1 & 0.6 \end{bmatrix}.$$

Finding the inverse of this matrix by hand is possible but tedious. Using Mathematica, the inverse can be found quickly. To five significant figures, it is approximately

$$(I - A)^{-1} = \begin{bmatrix} 1.7089 & 1.1421 & 1.0391 & 2.0567 \\ 0.77286 & 1.9236 & 1.2237 & 1.8849 \\ 0.76428 & 1.0133 & 2.8768 & 1.9751 \\ 0.66982 & 1.0004 & 1.0605 & 2.9669 \end{bmatrix}.$$

Therefore, by (2.13), we have

$$X = (I - A)^{-1}D$$

$$= \begin{bmatrix} 1.7089 & 1.1421 & 1.0391 & 2.0567 \\ 0.77286 & 1.9236 & 1.2237 & 1.8849 \\ 0.76428 & 1.0133 & 2.8768 & 1.9751 \\ 0.66982 & 1.0004 & 1.0605 & 2.9669 \end{bmatrix} \begin{bmatrix} 25 \\ 100 \\ 80 \\ 40 \end{bmatrix}$$

$$= \begin{matrix} \text{Chemicals} \\ \text{Electricity} \\ \text{Services} \\ \text{Steel} \end{matrix} \begin{bmatrix} 322.33 \\ 384.97 \\ 429.59 \\ 320.3 \end{bmatrix}.$$

Thus, the monthly productions necessary to meet the consumer demand are approximately given by the entries of X, in units of millions of dollars.

Exercises

In Exercises 1–3, messages (which are all famous quotes) were encoded with the given matrices A. In each case, A^{-1} is given. Verify that $AA^{-1} = I$, and then use A^{-1} to decode the message.

1. Attributed to Jack Kerouac:

$$A = \begin{bmatrix} 1 & -1 & 2 \\ -2 & 3 & 1 \\ 0 & 1 & 2 \end{bmatrix}.$$

A^{-1} is given in the text in this section.

$$AM = \begin{bmatrix} 22 & 13 & -13 & 44 & 12 & -10 & 30 & 6 \\ -21 & 31 & 51 & -7 & -16 & 39 & 18 & -9 \\ 18 & 43 & 25 & 60 & 8 & 19 & 59 & 3 \end{bmatrix}.$$

2. Attributed to Oscar Wilde:

$$A = \begin{bmatrix} 5 & -5 & 3 & 0 & 2 & 0 \\ 12 & -13 & 7 & 3 & 6 & 0 \\ -10 & 11 & -6 & -2 & -5 & 0 \\ -2 & 2 & -1 & 0 & -1 & 0 \\ 21 & -22 & 13 & 2 & 10 & 0 \\ 0 & 0 & 0 & 0 & 0 & 1 \end{bmatrix}$$

$$A^{-1} = \begin{bmatrix} 1 & 2 & 3 & -1 & 0 & 0 \\ -1 & 2 & 4 & 0 & 1 & 0 \\ -1 & 0 & 1 & 3 & 1 & 0 \\ 0 & 1 & 1 & 1 & 0 & 0 \\ -3 & 0 & 1 & -2 & 1 & 0 \\ 0 & 0 & 0 & 0 & 0 & 1 \end{bmatrix}$$

and $AM =$

$$\begin{bmatrix} 59 & -43 & 1 & 16 & 167 & 75 & 136 & 11 & 147 & 75 & 75 \\ 139 & -113 & 27 & 98 & 428 & 246 & 393 & 86 & 380 & 246 & 180 \\ -113 & 97 & -18 & -67 & -357 & -192 & -320 & -57 & -317 & -192 & -150 \\ -21 & 21 & 0 & -8 & -66 & -29 & -53 & -6 & -58 & -29 & -30 \\ 248 & -184 & 35 & 107 & 744 & 359 & 630 & 86 & 660 & 359 & 315 \\ 3 & 8 & 5 & 5 & 0 & 7 & 5 & 5 & 0 & 7 & 0 \end{bmatrix}.$$

3. Attributed to Bertrand Russell:

$$A = \begin{bmatrix} 1 & -1 & 2 & 0 & 0 & 0 \\ -2 & 3 & -1 & 0 & 0 & 0 \\ 0 & 1 & 2 & 0 & 0 & 0 \\ 0 & 0 & 0 & 1 & -1 & 2 \\ 0 & 0 & 0 & -2 & 3 & -1 \\ 0 & 0 & 0 & 0 & 1 & 2 \end{bmatrix}$$

$$A^{-1} = \begin{bmatrix} -7 & -4 & 5 & 0 & 0 & 0 \\ -4 & -2 & 3 & 0 & 0 & 0 \\ 2 & 1 & -1 & 0 & 0 & 0 \\ 0 & 0 & 0 & -7 & -4 & 5 \\ 0 & 0 & 0 & -4 & -2 & 3 \\ 0 & 0 & 0 & 2 & 1 & -1 \end{bmatrix}$$

$$AM = \begin{bmatrix} 58 & -14 & 6 & 32 & -7 & 23 & 25 & -7 & 19 \\ -61 & 47 & 7 & -13 & 29 & -16 & 0 & 29 & -15 \\ 37 & 19 & 14 & 37 & 15 & 23 & 38 & 15 & 17 \\ 26 & 39 & 51 & 51 & -10 & -12 & 71 & -10 & 20 \\ -3 & -3 & -43 & -33 & 39 & 44 & -73 & 39 & -40 \\ 34 & 55 & 41 & 46 & 19 & 20 & 46 & 19 & 0 \end{bmatrix}.$$

4. Let $B = \begin{bmatrix} 2 & 1 & 0 & 0 & 3 \\ 1 & 1 & 1 & 0 & 0 \\ 0 & -1 & 1 & -1 & 0 \\ 0 & 0 & 1 & 1 & 2 \\ 0 & 0 & 0 & -1 & -2 \end{bmatrix}$.

Verify that $B^{-1} = \begin{bmatrix} -2 & 5 & 3 & -8 & -11 \\ 2 & -4 & -3 & 7 & 10 \\ 0 & 0 & 0 & 1 & 1 \\ -2 & 4 & 2 & -6 & -9 \\ 1 & -2 & -1 & 3 & 4 \end{bmatrix}$ by checking that $BB^{-1} = I_5$.

The following messages in Exercises 5–11 were all coded using the matrix B from Exercise 4. Use B^{-1} to decode them. (They are all quotes from rock song lyrics.)

5.

$$BM = \begin{bmatrix} 89 & 66 & 71 & 27 \\ 25 & 36 & 9 & 27 \\ 0 & -35 & -13 & -19 \\ 62 & 24 & 47 & 47 \\ -50 & -24 & -47 & -32 \end{bmatrix}$$

6.

$$BM = \begin{bmatrix} 73 & 78 & 29 & 59 & 13 & 111 & 44 & 53 \\ 19 & 13 & 23 & 21 & 24 & 50 & 35 & 28 \\ -10 & -14 & -6 & -27 & -1 & 4 & -5 & 16 \\ 42 & 47 & 64 & 36 & 41 & 74 & 31 & 55 \\ -38 & -47 & -43 & -36 & -22 & -51 & -16 & -33 \end{bmatrix}$$

7.

$$BM = \begin{bmatrix} 87 & 60 & 42 & 60 & 77 & 67 & 27 & 55 \\ 33 & 10 & 36 & 33 & 30 & 18 & 26 & 28 \\ 0 & -30 & 1 & -3 & -7 & 1 & 12 & -13 \\ 50 & 55 & 29 & 13 & 49 & 35 & 38 & 58 \\ -38 & -55 & -14 & -8 & -38 & -30 & -18 & -43 \end{bmatrix}$$

8.

$$BM = \begin{bmatrix} 51 & 101 & 15 & 81 & 14 & 117 & 32 \\ 33 & 56 & 28 & 50 & 16 & 33 & 27 \\ -3 & -1 & 10 & -1 & -5 & -3 & -10 \\ 7 & 64 & 23 & 60 & 23 & 51 & 4 \\ -2 & -41 & -4 & -37 & -14 & -46 & 0 \end{bmatrix}$$

9.

$$BM = \begin{bmatrix} 53 & 49 & 10 & 89 & 75 & 36 & 22 & 51 & 17 & 38 & 33 & 100 & 14 \\ 27 & 25 & 28 & 42 & 38 & 43 & 34 & 46 & 21 & 14 & 36 & 57 & 7 \\ -45 & 11 & 18 & 5 & 8 & -14 & 18 & 8 & 5 & 4 & -8 & -17 & 0 \\ 36 & 55 & 28 & 60 & 68 & 41 & 28 & 23 & 18 & 46 & 29 & 61 & 0 \\ -36 & -35 & -5 & -40 & -45 & -21 & -5 & -5 & -6 & -33 & -14 & -43 & 0 \end{bmatrix}$$

10.

$$BM = \begin{bmatrix} 108 & 41 & 32 & 79 & 14 & 33 & 61 & 95 & 21 & 120 & 41 \\ 33 & 28 & 24 & 42 & 20 & 20 & 36 & 35 & 29 & 56 & 33 \\ -3 & -19 & -13 & 13 & 8 & -22 & -17 & -40 & 1 & -1 & -25 \\ 45 & 24 & 13 & 60 & 18 & 19 & 19 & 50 & 18 & 66 & 0 \\ -40 & -19 & -8 & -37 & -5 & -18 & -14 & -50 & -4 & -46 & 0 \end{bmatrix}$$

11.

$$BM = \begin{bmatrix} 59 & 94 & 105 & 17 & 1 & 63 & 98 & 25 & 60 & 4 \\ 29 & 25 & 28 & 14 & 15 & 33 & 33 & 31 & 12 & 19 \\ -25 & 4 & -17 & -11 & 9 & -21 & -1 & -2 & 2 & 11 \\ 25 & 42 & 47 & 43 & 18 & 33 & 69 & 21 & 32 & 15 \\ -24 & -37 & -47 & -31 & -4 & -28 & -55 & -7 & -29 & 0 \end{bmatrix}$$

12. The following message (also a lyric from a rock song) was coded using the matrix A given. Check the given inverse and decode the message.

$$A = \begin{bmatrix} 3 & 4 & 0 & 0 & 0 & 0 & 0 \\ 4 & 5 & 0 & 0 & 0 & 0 & 0 \\ 0 & 0 & 1 & -1 & 2 & 0 & 0 \\ 0 & 0 & -2 & 3 & -1 & 0 & 0 \\ 0 & 0 & 0 & 1 & 2 & 0 & 0 \\ 0 & 0 & 0 & 0 & 0 & 1 & 2 \\ 0 & 0 & 0 & 0 & 0 & 3 & 7 \end{bmatrix}$$

$$A^{-1} = \begin{bmatrix} -5 & 4 & 0 & 0 & 0 & 0 & 0 \\ 4 & -3 & 0 & 0 & 0 & 0 & 0 \\ 0 & 0 & -7 & -4 & 5 & 0 & 0 \\ 0 & 0 & -4 & -2 & 3 & 0 & 0 \\ 0 & 0 & 2 & 1 & -1 & 0 & 0 \\ 0 & 0 & 0 & 0 & 0 & 7 & -2 \\ 0 & 0 & 0 & 0 & 0 & -3 & 1 \end{bmatrix}$$

$$AM = \begin{bmatrix} 102 & 32 & 60 & 51 & 31 & 122 & 137 & 135 & 131 & 59 \\ 131 & 40 & 80 & 64 & 40 & 156 & 176 & 174 & 166 & 74 \\ 30 & -14 & 3 & -13 & 59 & 37 & 7 & 63 & 14 & 0 \\ -25 & 47 & -1 & 51 & -58 & -56 & 4 & -69 & -7 & 0 \\ 26 & 19 & 5 & 25 & 40 & 12 & 13 & 38 & 17 & 0 \\ 28 & 37 & 28 & 24 & 33 & 45 & 54 & 19 & 26 & 0 \\ 91 & 126 & 89 & 77 & 104 & 153 & 182 & 65 & 91 & 0 \end{bmatrix}$$

13. In the economy for Sunhillow in the example in the text, suppose the demand matrix is as follows:

$$D = \begin{bmatrix} 5 \\ 30 \\ 15 \\ 10 \end{bmatrix}.$$

Find the production levels for each sector that will meet the demand.

14. The fictional country of Mekano hired the Mirror and Door Consulting Firm to study its use of energy. For the purposes of the study, the firm only considers two sectors of the economy – fossil fuels, denoted F (which includes energy produced from gas, oil, and coal) and green energy, denoted G (which includes energy produced by solar, wind, and geothermal sources.) All other sectors of the economy are considered to be part of the consumer demand. The result is the following 2×2 input–output matrix for the annual dollar output of the economy:

$$A = \begin{matrix} & \begin{matrix} F & \ G \end{matrix} \\ \begin{matrix} F \\ G \end{matrix} & \begin{bmatrix} 0.5 & 0.05 \\ 0.2 & 0.1 \end{bmatrix} \end{matrix}$$

a. Draw the directed graph showing the flow of economic output between these two sectors of the economy. Explain why this is an open Leontief model.

b. Find the matrix $(I - A)^{-1}$.

c. Suppose the annual demand (in millions of dollars) from consumers and all other economic sectors is given by

$$D = \begin{bmatrix} 500 \\ 200 \end{bmatrix}.$$

Find the production levels that will meet this demand.

d. In the near future, Mekano estimates that its total annual demand for energy will increase by 200 units (that is, by $200,000,000). Determine the production levels that will meet this demand under two hypothetical scenarios – first assuming that the additional demand is placed entirely on the fossil fuel sector F, so that the demand becomes $D = \begin{bmatrix} 700 \\ 200 \end{bmatrix}$. Secondly, assuming the additional demand is placed entirely with the green energy sector G, so that the demand becomes $D = \begin{bmatrix} 500 \\ 400 \end{bmatrix}$. If Mekano could control where the additional demand could be placed, which scenario would be better for Mekano and why?

15. Suppose you are given the following input–output matrix A and the demand matrix D. Assuming an open Leontief model, determine the production levels that will meet the demand.

$$A = \begin{bmatrix} 0.1 & 0.25 \\ 0.5 & 0.3 \end{bmatrix} \text{ and } D = \begin{bmatrix} 400{,}000 \\ 500{,}000 \end{bmatrix}$$

16. Suppose you are given the following input–output matrix A and the demand matrix D. Assuming an open Leontief model, determine the production levels that will meet the demand.

$$A = \begin{bmatrix} 0.1 & 0.5 & 0.2 \\ 0.3 & 0.4 & 0.2 \\ 0 & 0.2 & 0.3 \end{bmatrix} \text{ and } D = \begin{bmatrix} 500 \\ 200 \\ 400 \end{bmatrix}$$

(You will need to find the inverse of a 3×3 matrix. You may use the technique of Exercise 11 from Section 2.4, or you may use a computer algebra system such as Mathematica.)

17. a. Any mathematical model has assumptions built into it. Consider Leontief's original analysis of the US economy in the 1940s broken down into 500 sectors. Would you expect the predictions of Leontief's model to remain valid for more than a year or two into the future? Why or why not? What assumptions are made in the model?

b. Suppose that during the 1940s, a similar analysis was made of the economy of the USSR (Soviet Union). Assume similar levels of accuracy and the same number of sectors of the economy – 500. Speculate on whether the predictions of Leontief's model would be more accurate for farther into the future than the predictions for the US economy or whether they would be less accurate. Give reasons for your answers.

3 Graphical Linear Programming

3.1 Introduction and Graphical Solutions

Linear programming problems are special cases of *optimization problems*. There is an *objective function* whose values change according to the values of input variables called *decision variables*. The goal is to choose the values of the decision variables that will produce the optimal value of the objective function. "Optimal" means either the largest possible value (this is called a *maximization* problem) or the smallest possible value (this is called a *minimization* problem.)

One is usually not free to specify just any values of the decision variables; typically there are *constraints* that limit the choices. What distinguishes a linear programming problem from a general constrained optimization problem is that both the constraints and the objective function are determined by linear equations; that is, the decision variables appear only to the first power in these equations. In practice, the constraints often arise from a limited supply of resources, so linear programming can be viewed as a tool to determine the *best way to allocate scarce resources*.

However, we are getting ahead of ourselves. In this section, we present a very simplified scenario that shows how a linear programming problem might arise naturally. We begin with a simple algebra problem of the type we have seen in Chapter 1.

Example 3.1 *Angie and Kiki's Lemonade Stand.* Angie and her sister Kiki want to set up a roadside lemonade stand in order to save money for college. They sell two types of lemonade: a glass of sweet lemonade uses three lemons and 2 tablespoons of sugar, and it sells for $1.25. A glass of tart lemonade uses four lemons and 1 tablespoon of sugar, and it sells for $1.50. How many glasses of each type should be mixed if their mother has given them 60 lemons and 30 tablespoons of sugar to use?

In solving a problem like this, we make some simplifying assumptions, such as assuming the children can sell all of the lemonade they make. We also assume that the other ingredients (water, ice, cups, etc.) are in unlimited supply, so the only constraints come from the limited amounts of lemons and sugar. We can model this as a system of linear equations. Let x stand for the number of glasses of sweet lemonade and y stand for the number of glasses of tart lemonade.

We can now obtain an equation (i.e., a constraint) from each limited resource, just as we did with the bongo drum example in Chapter 2:

$$3x + 4y = 60 \text{ (lemons)}$$
$$x + 2y = 30 \text{ (tablespoons sugar)}$$

We recognize this as a system of two linear equations in two unknowns, and we have several methods at our disposal to solve it. (Before solving it, think for a moment and see if you can spot another, unstated assumption that we made in setting up this system. We'll come back to this point later.) For now, let's solve the system using matrix inversion, as a review of of what was covered in Chapter 2. The matrix form of the system is

$$\begin{bmatrix} 3 & 4 \\ 2 & 1 \end{bmatrix} \begin{bmatrix} x \\ y \end{bmatrix} = \begin{bmatrix} 60 \\ 30 \end{bmatrix}.$$

The inverse of the coefficient matrix is

$$\frac{1}{-5} \begin{bmatrix} 1 & -4 \\ -2 & 3 \end{bmatrix}.$$

Therefore the solution is

$$\begin{bmatrix} x \\ y \end{bmatrix} = -\frac{1}{5} \begin{bmatrix} 1 & -4 \\ -2 & 3 \end{bmatrix} \begin{bmatrix} 60 \\ 30 \end{bmatrix} = \begin{bmatrix} 12 \\ 6 \end{bmatrix}.$$

That is, the solution is $(x, y) = (12, 6)$, so our heroines should mix 12 glasses of sweet lemonade and six glasses of tart lemonade. The reader may have noticed that we did not use the prices at all in finding the solution to the system, but if you want to know the total revenue R earned from selling the lemonade, it is given by

$$R = 1.25x + 1.5y = 1.25(12) + 1.5(6) = \$24,$$

a modest addition to the girls' college funds.

Of course, one may also find the solution by substitution, elimination, or even by graphing it, as shown in Figure 3.1.

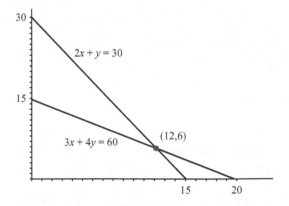

Figure 3.1 Graphical solution to the lemonade stand problem.

The solution point $(12, 6)$ is the point of intersection of the two lines – the only point to satisfy both constraints simultaneously.

As noted before, this is nothing more than a simple algebra problem, not (yet) a linear programming problem. But now consider the following twist on the problem. Kiki decides they should modify their recipes by including a third ingredient: limes. The reasons for doing so aren't important – perhaps she wants to make their lemonade a bit more exotic so they can effectively compete with a second lemonade stand that opened down the street, or perhaps she just suddenly developed a taste for lime juice. Regardless of the reasons, the modified recipes are as follows: a glass of sweet lemonade uses three lemons, one lime, and 2 tablespoons sugar. A glass of tart lemonade uses four lemons, two limes, and 1 tablespoon sugar. Assume the price they charge for each glass remains the same as above. If their mother provides them with 28 limes in addition to the other ingredients, solve the modified problem.

A moment's reflection should convince the reader that the only change to the system of equations is the addition of a new equation coming from the limited supply of limes. The modified system (with resources listed in aphabetical order) is the following:

$$3x + 4y = 60 \text{ (lemons)}$$
$$x + 2y = 28 \text{ (limes)}$$
$$2x + y = 30 \text{ (tablespoons sugar)}$$

Of course, this is a 3 × 2 system – the coefficient matrix is not square. Thus, it has no inverse, which means the system likely is not one with a unique solution. Recall from Chapter 1 that this means either there is no solution or there are infinitely many. To see which case we are actually in, we could apply the method of elimination as covered in Chapter 1, but it is easier just to observe the graph of the three lines in Figure 3.2.

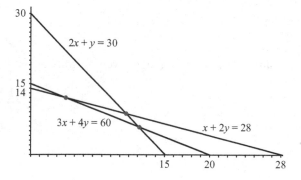

Figure 3.2 The three constraint lines corresponding to the resources.

Now, a solution to the system must be a common solution to all three equations – that is, a point simultaneously on the graph of all three lines. Visually, it is clear that there is no such point (it is a rare system of three lines that all meet concurrently at a single point!). Thus, the system has no solution at all.

Don't forget, when modeling a real life problem, after obtaining a mathematical solution, we must interpret the solution (or lack thereof) in terms of the original problem. In other words, what do the children do about this lack of a solution? Do they close the lemonade stand and go home? Do they reject Kiki's idea to add limes and revert to the boring old lemonade recipes? Surely, there must be a way for them to persevere with Kiki's innovation.

When a model is not helping us solve a problem, perhaps it's time to take a look at the model itself. What are the underlying assumptions? Is the model too simplistic? Can we modify it to be a better reflection of reality? In this example, we can indeed modify the model to make it more realistic, and the resulting modification will prove to be fruitful (pun intended!). The needed change has something to do with the unstated assumption mentioned at the beginning of the example.

If you haven't spotted it yet, here's how we can coax it out. Look again at the equation for lemons. Like any equation, it has three parts – a left-hand side, a right-hand side, and an equal sign that separates them. Let's consider carefully what each part means. The left side, $3x + 4y$, represents the total amount of lemons used to mix the lemonade if we make x glasses of sweet and y glasses of tart. The right side, 60, represents the total amount of lemons available for use. There is no problem with either side. The problem lies in the equals sign! By placing this sign between $3x + 4y$ and 60, we are insisting that these two expressions are exactly the same. In other words, we are insisting that we use every lemon we have. This is the unstated assumption that we built into our model by writing the equations – that we use up all of our resources. Clearly, the amount of lemons used cannot be greater than the number we have, but there is no reason why it cannot be less, which would entail having some leftover lemons at the end of the day. The same holds for all three resources. In each case, the problem only requires that the left side is less than or equal to the right side – therefore, each equation should be replaced by an inequality. This is the modified model, which is more realistic:

$$3x + 4y \leq 60 \text{ (lemons)}$$
$$x + 2y \leq 28 \text{ (limes)}$$
$$2x + y \leq 30 \text{ (tablespoons sugar)}.$$

Each inequality says that the amount of resource used is no more than the amount available. We are encouraged by having discovered a more realistic model for our problem, but also concerned about maybe having missed other unstated assumptions. Perhaps this has already occurred to the reader, but negative values of x and y do not make sense in terms of the original problem. As long as we are listing inequalities that must be satisfied, then we should also list these:

$$x \geq 0$$
$$y \geq 0.$$

What effect does this relaxation of our constraints have on our problem? Whereas before our solution set was empty, we shall find that now it has become quite large – in fact, infinite (assuming fractional values for x and y are allowed, which we do assume for the moment). For example, it is easy to verify that the ordered pair $(4, 3)$ satisfies all the constraints simultaneously, since it entails using 24 lemons, 10 limes, and 11 tablespoons sugar, and it consists of positive values for both x and y. Other ordered pairs that satisfy all five inequalities include $(1, 6)$, $(0, 10)$, $(5, 5)$, and many others as well. In fact, we can depict the entire solution set graphically using the following ideas.

Every line of the form $ax + by = c$ divides the Cartesian plane into three regions. On one side of the line, we have $ax + by < c$; on the other, we have $ax + by > c$; and, of course, on the line itself, we have $ax + by = c$. So whenever we want to graph the solution set of of a linear inequality, we may do so by shading in the proper side of the line and including the points on the line if it is an inequality that is not strict (that is, if it involves $\leq$ or $\geq$ instead of $<$ or $>$).

The easiest way to determine which side to shade in is to plug in a *test point* not on the line itself. If the test point satisfies the inequality, so does every point on the same side of the line as the test point, and these can be shaded in as solutions. If the test point does not satisfy the inequality, then shade in the points on the side opposite of the line from the test point.

A rigorous justification of this procedure is based on a theorem from calculus called the *intermediate value theorem*. Readers who know the intermediate value theorem might want to prove for themselves that this test point procedure works. For readers who do not know calculus, the intermediate value theorem is a way of making the following statement precise: "since the expression $ax + by$ varies continuously, it cannot go from a value less than c to a value greater than c without passing through c itself; that is, without crossing the graph of the line itself." For obvious reasons, the solution set of a linear inequality in the plane is called a *half-plane*, and in higher dimensions, it is called a *half-space*.

We can use any of the previously named points (or many others) as a test point. In practice, the point $(0,0)$ is an easy point to use for the first three inequalities since it is not on any of those lines. Based on this method, we see the solution set for each of the first three inequalities is the points below the lines, while the solution set for the nonnegativity inequalities is clearly the points in the first quadrant. Since we seek common solutions to all the inequalities at once, we are interested in the intersection of all five half-planes (the points in the first quadrant that are below all of the other three lines), which is the shaded region shown in Figure 3.3.

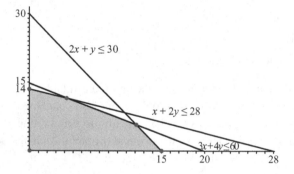

Figure 3.3 The possible solutions to the modified model.

So, indeed, there are now infinitely many solution points. Our modification transformed the problem from one with no solution to one with many solutions. The collection of all these points that are common solutions to all the constraints are called the *feasible points* or *feasible region*. We are now almost finished converting this into a true linear programming problem.

All that remains to do is note that while we may have many feasible solutions, some solutions are better than others. To see what we mean by "better," consider the feasible point $(x, y) = (0,0)$. What is wrong with this solution? Well, since $x = 0 = y$, the children are not making any lemonade at all and, therefore, no revenue at all. Making money to save for college was the main goal of the children in the first place. After all, higher education is not inexpensive! The feasible point $(x, y) = (1,6)$ corresponds to a revenue of $1.25(1) + 1.5(6) = \$10.25$. On the other hand, the feasible point $(5,5)$ corresponds to a revenue of $1.25(5) + 1.5(5) = \$13.75$. Since this is more revenue for Angie and Kiki, $(5,5)$ is clearly a superior solution to $(1,6)$.

The point is, when you have many solutions available, it is possible to compare them and see how they measure up according to some goal or *objective* that you have. Some solutions will better serve your objective than others. At last, we begin to see clearly how to state the problem. In this example, the objective is to maximize the revenue, but we must stick to feasible solutions.

The transformation of our problem is complete - we can now restate it as a bona fide linear programming problem:

Let x be the number of glasses of sweet lemonade.

Let y be the number of glasses of tart lemonade.

We must

$$\text{Maximize: } R = 1.25x + 1.5y$$

subject to the constraints:

$$3x + 4y \leq 60 \text{ (lemons)}$$
$$x + 2y \leq 28 \text{ (limes)}$$
$$2x + y \leq 30 \text{ (sugar)}$$

and the nonnegativity constraints:

$$x \geq 0$$
$$y \geq 0.$$

In any linear programming problem, you will see the same four components, and it's best to write them in the order shown:

- Label the meaning of the *decision variables* (the x and y in our example).
- Express the *objective function* in terms of the decision variables (what we want to optimize – the revenue R in our example).
- Express the constraints in terms of the decision variables as a system of linear inequalities (we call these the *structural constraints* – in our example, these inequalities come from the limited resources and the recipes for the lemonade).
- Record the *nonnegativity constraints*.

What makes this a linear model is that the objective function as well as the left-hand side of all the constraints are given by linear functions in the decision variables (x and y appear only to the first power, and no products such as xy appear). This might be a good time to go back and reread the first paragraph of this section and see if it makes more sense now. So this explains the word "linear" in "linear programming." What about the word "programming"?

To see where that comes from, it is helpful to assess exactly where we are in the process. We have now clarified the statement of the problem but have not addressed how to find the (optimal) solution. The alert reader will recall that there are now infinitely many feasible points, and may wonder if that is an obstruction to finding the optimal one. After all, we cannot check every feasible point to find the one with maximum revenue. And perhaps you may wonder about another potential problem – how do we know in advance that there even is a point with maximum revenue?

Fortunately, in the following sections, we give a prescription (several, actually) for some short-cuts. It turns out that it is not necessary to check every single feasible point. The steps that you follow in these prescriptions or *algorithms* are what is meant by a *program*. The word is simply a synonym for "algorithm" or "technique" – a series of steps to follow, like a recipe, in order to solve the problem. The word has nothing at all to do with, for example, the notion of a computer program, even though linear programming problems can be solved by using a computer. It's just a matter of who or what is following the steps in the recipe – us or our computer. In a later section of the text, we will illustrate how to solve linear programming problems with a computer.

The lemonade stand problem is an example of what is known as a *standard form maximization problem*. It is important to recognize problems in this form when we learn the *simplex algorithm* in a later chapter, since problems of this form are the easiest type of problem to solve. To be standard form, all the structural constraints must have the form:

(linear combination of decision variables) $\leq$ (nonnegative constant).

In a maximization problem, if all the constraints have this form, except that the constant on the right side of one or more of the constraints is negative, we shall say that the problem is in *semi-standard form*.

Sometimes, your objective is to minimize something, such as cost, instead of maximizing something. In a minimization problem, to be in standard form, all the structural constraints must have the form:

(linear combination of decision variables) $\geq$ (nonnegative constant).

Similarly, in a minimization problem, if all the constraints have the this form, except that the constant on the right-hand side of one or more of the constraints is negative, we say the problem is in *semi-standard form*.

Our next example is a standard form minimization problem.

Example 3.2 *O'Casek's Used Cars*. Mr. O'Casek is a used car dealer who just opened a second lot in a nearby town. He needs to ship some of his existing inventory from his first lot to his new one, a drive of 30 miles. He decides that he should transport at least 24 SUVs, at least 30 sedans, and at least 40 compact cars. He calls *Moving in Stereo, Inc.*, who has two types of vehicles to ship the cars – automobile transport vans and flatbed trucks. Each automobile transport van can hold one SUV, two sedans, and four compact cars (simultaneously), while each flatbed truck can hold 2 SUVs, one sedan, and one compact car (simultaneously). Between fuel and driver expenses, the cost for a transport van amounts to $6 per mile, while the cost for flatbed truck is $4 per mile. How many of each vehicle should O'Casek hire in order to minimize the total cost of the move?

The setup: Let x stand for the number of transport vans and y stand for the number of flatbed trucks. Minimizing the total cost is the same as minimizing the cost per mile (since the difference between the cost per mile and the total cost is merely a factor of 30). So let C be the cost per mile. Then we must

$$\text{minimize: } C = 6x + 4y$$

subject to:

$$x + 2y \geq 24 \text{ (SUVs)}$$
$$2x + y \geq 30 \text{ (sedans)}$$
$$4x + y \geq 40 \text{ (compacts)}$$

and

$$x \geq 0, y \geq 0.$$

Observe that it is indeed a standard form minimization problem, since each inequality is of the greater-than type. For practice, the reader should graph the feasible region for this problem. Notice that the feasible region does not contain the origin and that it is *unbounded* – very large values of x and y do not violate the constraints, unlike the lemonade stand problem. One final observation to make is that while we might make the case for fractional values of x and y in the lemonade stand problem, the same cannot be said for this problem. It may be possible to sell $\frac{2}{3}$ of a glass of lemonade, but you certainly cannot hire $\frac{1}{2}$ of a flatbed truck! Thus, this problem has an additional unstated constraint – that the solution x and y be integers. Such problems are called *integer programming* problems and are discussed in Chapter 9. For the moment, we will ignore this requirement, which would cause unnecessary complications at this point, and just hope that the answers come out to be whole numbers. *Therefore, until the section on integer programming, we assume all variables in this book are allowed to take on fractional values.*

Our final example is already set up for us:

$$\text{maximize: } z = 10x + 3y + 6z$$

subject to:

$$2x + 3y + 5z \leq 150$$

$$x + 4y + 2z \geq 25$$

$$x \geq 0, \ y \geq 0, \ z \geq 0.$$

This problem differs from the previous problems in two ways. First, notice it has more than two decision variables. One of the approaches we take to solving a linear programming problem is to draw a graph of the feasible region, as we did earlier. Since this problem has three decision variables, the graph would be three-dimensional, which makes it difficult to visualize and draw. Problems with more than three variables would require even more dimensions, making it impossible to visualize the feasible set. Therefore, graphical methods of solution will be of limited use in the real world, where problems often have many variables. Graphical methods are very useful, however, to help us understand the nature of linear programming problems, so we will focus on these methods first.

Second, notice that the problem is not in standard form, because the second constraint has the inequality sign pointing the wrong way for a maximization problem. A problem like this is said to have *mixed constraints*. Notice that this problem can be put into semi-standard form if we multiply the offending inequalities through by a factor of -1, which will reverse the direction of the inequality sign at the expense of making the constant on the right negative. It's also worth pointing out that it is possible to have a linear programming problem with a constraint that is an exact equality instead of an inequality. Such problems are also not considered to be in standard form.

We close this introduction by noting that each linear programming problem has a size associated with it – if it has m constraints (not counting the nonnegativity constraints) and n decision variables, we say that it is an $m \times n$ problem. Thus, the lemonade stand problem and the used car problem are both 3×2 problems, while the third example is a 2×3 problem.

Exercises

The first six exercises deal with the following system of equations and accompanying graph (Figure 3.4).

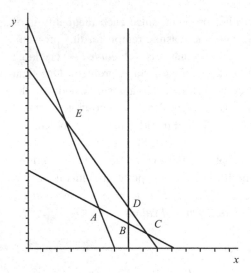

Figure 3.4 A system of four equations.

$$5x + 2y = 60 \tag{3.1}$$
$$3x + 6y = 60 \tag{3.2}$$
$$5x + 4y = 90 \tag{3.3}$$
$$x = 14 \tag{3.4}$$

1. Find the coordinates of the seven x and y intercepts of the four lines in the system, and thereby identify which equation goes with each line in Figure 3.4.
2. Find the coordinates of the five labeled points of intersection A, B, C, D, and E in Figure 3.4.
3. Determine which region of the graph should be shaded in to indicate the solution set of the system of inequalities:

$$5x + 27 \geq 60$$
$$3x + 6y \geq 60$$
$$5x + 4y \leq 90$$
$$x \leq 14$$
$$x \geq 0, \ y \geq 0$$

4. Determine which region of the graph should be shaded in to indicate the solution set to the system of inequalities:

$$5x + 2y \leq 60$$
$$3x + 6y \geq 60$$
$$5x + 4y \leq 90$$
$$x \leq 14$$
$$x \geq 0, \ y \geq 0$$

5. Determine which region of the graph should be shaded in to indicate the solution set to the system of inequalities:

$$5x + 2y \leq 60$$
$$3x + 6y \geq 60$$
$$5x + 4y \leq 90$$
$$x \geq 14$$
$$x \geq 0, \ y \geq 0$$

6. Determine which region of the graph should be shaded in to indicate the solution set to the system of inequalities:

$$5x + 2y \geq 60$$
$$3x + 6y \leq 60$$
$$5x + 4y \leq 90$$
$$x \leq 14$$
$$x \geq 0, \ y \geq 0$$

7. Graph the system of inequalities, shade in the solution set, and find the coordinates of the points of intersection:

$$10x + 16y \leq 240$$
$$15x + 8y \leq 240$$
$$x \geq 0, \ y \geq 0$$

8. Graph the system of inequalities, shade in the solution set, and find the coordinates of the points of intersection:

$$5x + 27 \geq 30$$
$$5x + 7y \geq 70$$
$$x \geq 0, \ y \geq 0$$

9. Graph the system of inequalities, shade in the solution set, and find the coordinates of the points of intersection:

$$4x + 2y \geq 36$$
$$4x + 3y \leq 48$$
$$x \geq 0, \ y \geq 0$$

10. Graph the system of inequalities, shade in the solution set, and find the coordinates of the points of intersection:

$$4x + 7y \leq 120$$
$$20x + 25y \leq 500$$
$$x + y \geq 12$$
$$y \geq 6$$
$$x \geq 0, \ y \geq 0$$

Set up the following problems as linear programming problems. Be sure to label the decision variables, clearly state the objective function, and record all the constraints. (Do not graph the feasible region or try to solve the problems.)

11. Bruce Jax is responsible for constructing end tables and kitchen tables for the White Room Woodshop. Each end table uses 2 square yards of $\frac{3}{4}$-inch oak boards and takes 2 hours to complete. Each kitchen table uses 4 square yards of oak board and takes 3 hours to complete. This week he has available 36 square yards of oak boards and 32 hours of time. Other resources are unlimited. How many of each item should he make if he is paid $70 for each end table and $100 for each kitchen table?

12. Joni is putting her designing skills to work at the Court and Sparkle Jewelry Emporium. She makes two signature design bracelet models. Each Dawntreader bracelet uses 1 ruby, 6 pearls, and 10 opals. Each Hejira bracelet uses 3 rubies, 3 pearls, and 15 opals. She has 54 rubies, 120 pearls, and 300 opals to work with. If either model results in a profit of $1,800 for Joni, how many of each type should she make?

13. Toys in the Attic, Inc. operates two workshops to build toys for needy children. Mr. Tyler's shop can produce 36 Angel dolls, 16 Kings and Queens board games, and 16 Back in the Saddle rocking horses each day it operates. Mr. Perry's shop can produce 10 Angel dolls, 10 Kings and Queens board games, and 20 Back in the Saddle rocking horses each day it operates. It costs $144 to operate Mr. Tyler's shop for one day and $166 to operate Mr. Perry's shop for one day. Suppose the company receives an order from the Kids Dream On charity foundation for at least 720 Angel dolls, at least 520 Kings and Queens board games, and at least 640 Back in the Saddle rocking horses. How many days should they operate each shop in order to fill the order at the least possible total cost?

14. The ocean liner Gigantic is being refitted with new lifeboats and safety equipment. There are three types of lifeboats available. Each wooden lifeboat holds 40 people, weighs 1,000 pounds, and comes equipped with six first aid kits and 40 life jackets. The cost for a wooden boat is $6,000. Each large rubber raft holds 24 people, weighs 400 pounds, and comes equipped with four first aid kits and 30 life jackets. Each small rubber raft holds 10 people, weighs 250 pounds, and comes equipped with one first aid kit and eight life jackets. The cost for a large rubber raft is $4,000, and the cost for a small rubber raft is $1,500. The Gigantic has 40 storage berths for these lifeboats. Each berth can hold one wooden lifeboat, or two large rubber rafts, or four small rubber rafts. According to commercial marine regulations, all of the following conditions must be met:

- The total weight of the lifeboats must not exceed 38,000 lbs.
- The 40 storage berths must be filled completely.
- They must have the ability to float at least 1,600 people in the lifeboats.
- They must carry a total of at least 1,500 life jackets.
- They must carry a total of at least 200 first aid kits.

In addition, due to contract agreements with CSK Wooden Ships, the manufacturer, the ocean liner Gigantic obligated to purchase

- At least 12 wooden lifeboats.
- At least a total of 80 rubber rafts (regardless of size).

How many of each type of lifeboat should be purchased in order to minimize the total cost?

15. Arnold and Penny Layne have up to $60,000 to invest for a one-year period. There are three investment options they are considering: certificates of deposit, which have an expected annual return of 3%; municipal bonds (the proceeds of which go to the cleaning and upkeep of local buses, police cars, and fire engines), which have an expected annual return of 5%; and stocks on women's apparel, with an expected annual return of 11%. They would like to maximize their expected return for the year but are adhering to the following guidelines suggested by their financial adviser:

   • The amount invested in municipal bonds should be at least $10,000.
   • Because of stock volatility, the amount invested in stocks should be at most $10,000 more than the amount invested in bonds.
   • Because of the reliability of certificates of deposit, at least $\frac{1}{3}$ of the total investment should be in certificates of deposit.

   Be sure you have all of the constraints.

   The next three exercises all relate to the following scenario. The buffet bar at the Acquiring the Taste Bistro offers five different items, which the bistro sells individually by weight. Customers may add whatever they want to their plate, but they pay for each item separately. Because the bistro customers are health conscious, the nutritional information in Table 3.1 is posted at the buffet bar.

Table 3.1 Acquiring the Taste nutritional information.

Per ounce	Garden salad w/dressing	Grilled vegetables	Pasta w/tomato sauce	Meatballs	Curried chicken salad
Fat (g)	1	1	3	6	4
Protein (g)	0	1	2	10	12
Carbohydrates (g)	3	4	12	6	3
Sodium (mg)	10	12	15	25	20
Sugar (g)	12	5	6	4	1
Calories	15	20	40	60	50
Cost (cents)	25	40	30	50	60

16. Derek just moved to town and has a new job, so he is most concerned with saving money. When he eats at the bistro, his objective is to minimize cost. However, he does have some dietary requirements, namely the following:

   • He would like to limit his fat intake to 40 grams or less.
   • He wants a minimum of 80 grams protein.
   • He wants a minimum of 60 grams carbohydrates.
   • Because he has high blood pressure, Derek must limit his sodium intake to at most 200 milligrams.
   • He would like to limit his calorie intake to at most 700 calories.

17. Ray is a body builder, so his objective is to maximize protein intake. Ray's other requirements are the following:

- He would like to limit his fat intake to at most 60 grams.
- He wants a minimum of 120 grams of carbohydrates.
- He would like to have between 90 and 200 grams of sugar.
- He is on a budget and cannot spend more than $7.50 for lunch.

18.   Phil is on a weight-loss diet, and his objective is to minimize calorie intake. Phil's other requirements are the following:

- He would like to limit his fat intake to at most 20 grams.
- He would like to limit his carbohydrate intake to at most 50 grams.
- He would like to limit his sugar intake to at most 88 grams.
- He would like to limit his cost to at most $6.00.
- He needs to eat at least 12 ounces of food altogether, in order to avoid being hungry before his next meal.

19.   Green's Heavy Metal Foundry mixes three different alloys composed of copper, zinc, and iron. Each 100-pound unit of Alloy I consists of 50 pounds of copper, 50 pounds of zinc, and no iron. Each 100-pound unit of Alloy II consists of 30 pounds of copper, 30 pounds of zinc, and 40 pounds of iron. Each 100-pound unit of Alloy III consists of 50 pounds of copper, 20 pounds of zinc, and 30 pounds of iron. Each unit of Alloy I generates $100 profit, each unit of Alloy II generates $80 profit, and each unit of Alloy III generates $40 profit. There are 12,000 pounds of copper, 10,000 pounds of zinc, and 12,000 pounds of iron available. The foundry has hired Gary Giante, an outside consultant, to help them maximize its profit. What problem does Mr. Giante need to solve in order to advise the foundry?

20.   The Creative Thought Matters Biological Research Company has developed a promising new antibiotic that has been shown to be effective in fighting strains of hepatitis infection that are resistant to other antibiotic agents. They have stockpiled a supply of this new drug at two research hospitals. The Huey News Hospital in Atlanta, GA, has 330 kilograms of the agent available to ship (beyond what they will keep there), and the Lewis Memorial Hospital in Boston, MA, has 450 kilograms of it available to ship (beyond what they will keep there). In order to fight a growing national problem of resistant infection, the company wants to ship the available units of antibiotic to three other hospital clinics across the country – one in Chicago, IL; one in Denver, CO; and one in San Francisco, CA. Chicago has requested at least 300 kilograms, Denver has requested at least 120 kilgrams, and San Francisco has requested at least 360 kilograms. The drug must be kept cold during shipment, so shipping is expensive. Table 3.2 indicates the cost (in dollars) of shipping one kilogram of the drug from each supply hospital to each of the three clinics at the recipient hospitals.

Table 3.2 Shipping costs and capacities.

	Chicago	Denver	San Francisco	Shipping Capacities
Huey News Hospital (Atlanta)	$30	$45	$80	330 kg
Lewis Memorial Hospital (Boston)	$40	$40	$70	450 kg
Clinic Requests	300 kg	120 kg	360 kg	

If the Creative Thought Matters Company wants to ship the drugs at minimum cost, how many kilograms should it ship from each research hospital to each clinic? Set this up as a linear programming problem.

REMARK 3.3 *Problems like this are called* transportation problems. *The problems that the US military (in particular, the US Air Force) faced at the end of World War II in finding the cheapest way to move massive amounts of personnel and equipment from supply points to desired locations in battle were among the very first linear programming problems that were formulated and solved. This led directly to the development of linear programming as an important area of study.*

21. Set up the following brainteaser as a linear programming problem. Find the largest three-digit number N such that the sum of the digits of N is at most 12, the hundreds digit minus the units digit is no larger than the tens digit plus four, and the number N minus the number M obtained from N by reversing the digits is no larger than 693. Be sure to include all of the constraints.

3.2 The Decision Space: $m \times 2$ Problems

We now discuss techniques for solving linear programming problems. In general, methods of solution involving graphs and pictures are limited to small problems. The most well-known method works for problems with only two decision variables, but there is no limit on the number of constraints; in other words, it works for problems of size $m \times 2$. In most texts, this method is referred to as *the graphical method*, but a more precise name would be *graphing in the decision space*, because the axes of the coordinate system correspond to the two decision variables.

There is another graphical method, which we call *graphing in the constraint space*. This method works for problems of size $2 \times n$, because the axes of the coordinate system correspond to the two constraints, while there is no limit on the number of decision variables. In this section, we cover graphing in the decision space, and later in the chapter, we cover graphing in the constraint space.

One advantage to the method of graphing in the decision space, besides the fact that many realistic problems are $m \times 2$, is that it is easy to derive from this technique the *fundamental theorem of linear programming*, also known as the *corner point theorem*.

To illustrate this method, we continue with the example of Angie and Kiki's lemonade stand. For convenience, we restate the problem here and reproduce the graph of the feasible region, along with some special points labeled and new lines in Figure 3.5.

Let x denote the number of glasses of sweet lemonade.

Let y denote the number of glasses of tart lemonade.

We must

$$\text{maximize: } R = 1.25x + 1.5y$$
$$\text{subject to:}$$
$$3x + 4y \le 60 \text{ (lemons)}$$
$$x + 2y \le 28 \text{ (limes)}$$

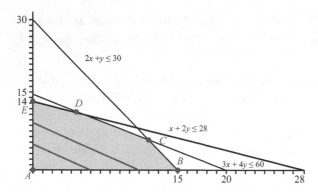

Figure 3.5 The feasible set for the lemonade stand problem.

$$2x + y \leq 30 \text{ (tablespoons sugar)}$$

$$x \geq 0, \ y \geq 0$$

Consider the feasible point $(6,0)$, where six glasses of sweet lemonade is made, but no tart lemonade is made. The revenue for that point is $R = 1.25(6) + 1.5(0) = \$7.50$. There are other feasible points that also produce a revenue of \$7.50; for example, $(0,5)$ works. In fact, we claim that every feasible point on the line segment connecting $(6,0)$ to $(0,5)$ will produce the same revenue. Why is that? If we take our revenue function $R = 1.25x + 1.5y$ and set it equal to \$7.50 to find all the points that generate the revenue of \$7.50, we obtain

$$1.25x + 1.5y = 7.50,$$

which we recognize as the equation of a straight line, and since it contains the above two points, it is therefore the unique line passing through those two points. The portion of this line that lies in the shaded region (it is the short line segment in the lower-left corner of the graph) will be all the feasible points generating that revenue. Because every point on this line corresponds to the same value of the objective function, we shall refer to it as a *level objective line*, or perhaps by the fancier name, *iso-objective line*, or, at least in this example, *iso-revenue line* (the prefix "iso" means "same").

Let's find another level objective line – say, the one with revenue \$15. It is not hard to see that both $(12,0)$ and $(0,10)$ are on this line. The portion of this line that lies inside the feasible region is the second, longer line segment cutting across the feasible region in Figure 3.5. It appears in the graph that the two level objective lines might be parallel. Indeed, they are, because the equation of this second level objective line is

$$1.25x + 1.5y = 15.00.$$

In any line of the form $Ax + By = C$, the slope m is determined by the coefficients A and B; indeed, we have $m = -\frac{B}{A}$. Clearly, the various level objective lines differ only in the constant C – they all have the same values of A and B (namely, 1.25 and 1.5, respectively). Thus, we see that all level objective lines have the same slope, so they are all parallel to each other.

Furthermore, because the A and B in this problem are both positive (as they are in many problems in practice), larger and larger values of x and y lead to larger and larger values of R.

In other words, the iso-objective lines that correspond to the larger values of R are those to the right and up in the first quadrant, in a direction away from the origin. So, to find the feasible points with the largest possible values of R, we should steadily move the level objective line farther out from the origin, keeping it parallel to itself as it moves. Eventually, because the feasible region is bounded, these level objective lines will lie entirely outside of the feasible region. Thus, the last feasible point these lines pass through just before they leave the feasible region must be the optimal point.

That implies that the optimal point cannot be an interior point of the feasible region – it must be on the boundary. In fact, in this example, it looks as if the optimal point is actually a corner point of the region. (Can you tell which corner by studying Figure 3.5?) We'll see in the next section that this fact – that the optimal point must be a corner point – is always true, assuming the feasible region is nonempty and bounded.

Let's assume this *corner point theorem* is true for the moment. In light of that, our seemingly impossible problem now seems tractable. Instead of checking the revenue at an infinite number of feasible points, we need only check the corner points, which are always finite in number (in this problem, there are five corners.) The first step is to find the coordinates of all the corner points. The point A is the origin with coordinates $(0,0)$. We already know this point is not the optimal point, but we include all the corner points in our discussion for completeness. The points B and E are intercepts of the sugar line and the lime line, so their coordinates are easy to find. They are $B = (15,0)$ and $E = (0,14)$. The point C we already know from the problem (Example 3.1) we solved in the last section – it is the intersection of the sugar line and the lemon line, so has coordinates $C = (12,6)$. That leaves only point D, which is the intersection of the lemon line and the lime line. We find the coordinates for D by solving that particular system as we did before to find the coordinates for C. The system is

$$3x + 4y = 60$$
$$x + 2y = 28.$$

In matrix form, we can solve by inverting the coefficient matrix as usual:

$$\begin{bmatrix} 3 & 4 \\ 1 & 2 \end{bmatrix} \begin{bmatrix} x \\ y \end{bmatrix} = \begin{bmatrix} 60 \\ 28 \end{bmatrix}$$

$$\begin{bmatrix} x \\ y \end{bmatrix} = \begin{bmatrix} 3 & 4 \\ 1 & 2 \end{bmatrix}^{-1} \begin{bmatrix} 60 \\ 28 \end{bmatrix}$$

$$= \frac{1}{2} \begin{bmatrix} 2 & -4 \\ -1 & 3 \end{bmatrix} \begin{bmatrix} 60 \\ 28 \end{bmatrix}$$

$$= \begin{bmatrix} 4 \\ 12 \end{bmatrix}.$$

Thus, the coordinates of D are $(4, 12)$.

Now, in the Table 3.3, we compute the revenue for each corner point and simply choose the largest one:

Table 3.3 Revenue at each corner point.

Corner	(x, y)	$R = 1.25x + 1.5y$
A	$(0, 0)$	$0
B	$(15, 0)$	$18.75
C	$(12, 6)$	$24.00
D	$(4, 12)$	$23.00
E	$(0, 14)$	$21.00

From the table it is clear that C is the optimal point, so the solution to the problem is

$$x = 12 \text{ glasses sweet lemonade}$$
$$y = 6 \text{ glasses tart lemonade}$$
$$R = \$24.00 \text{ (maximized)}.$$

Coincidentally, the solution $(12, 6)$ is the same point it was before Kiki introduced limes into the mix, but there is no way we could have known that before we started.

Thus, at the end of the day, the children have $24.00 that they can put away for college, assuming they can agree on how to split it. Should they split it 50/50 – that sounds fair, right? What if Kiki claims it was her idea to include limes in the recipe that made the lemonade stand so successful and so she is entitled to 60%? What if Angie claims it was her charm and promotional skills that actually led to the sales, so she is entitled to 60%? This is another type of problem that has nothing to do with linear programming. Problems like this are called *fair division problems*, and they also have been studied from a mathematical standpoint. But that's for another text, not here, so we leave the girls to sort it out for themselves.

Before we leave this problem, we might ask ourselves if it is important for the girls to know anything else besides x, y, and R. We will see in a later section that there are quite a few interesting and useful questions the girls could ask at this point after an optimal solution has been found. For now, let us mention just one simple question. Since our modified model with inequalities allowed for leftover resources, wouldn't they like to know how much of each resource will be leftover? We introduce three "hidden" variables to this problem, one for each resource:

Let S_1 denote the number of leftover lemons.
Let S_2 denote the number of leftover limes.
Let S_3 denote the number of leftover tablespoons sugar.

Because information is valuable, we take the position that the final solution to any linear programming problem should include not only the values of the decision variables (x and y in our problem) and the value of the objective function (R in our problem), but also the values of these S_i (the leftover resources in our problem).

In the lemonade stand problem, it is clear that since the optimal point C is the intersection of the lemon line and the sugar line, there will be no leftover lemons or sugar. Indeed, to find the coordinates of C, we solved the system of equations – not inequalities – which is tantamount to insisting that these resources are used up. On the other hand, the point C does not lie on the the lime line but below it. Therefore, there will be leftover limes. It is not hard to find out how many.

Recall that $x + 2y$ is the expression that represents the number of limes used, as it is the left-hand side of the lime constraint. At the point $C = (12, 6)$, we see this becomes $x + 2y = 12 + 2(6) = 24$, so we have used 24 limes. Because we began with 28 limes, there will be four leftover limes at point C. Thus, a more complete solution to the lemonade stand problem is written as follows:

$$x = 12 \text{ glasses sweet lemonade}$$
$$y = 6 \text{ glasses tart lemonade}$$
$$R = \$24.00 \text{ (maximized)}$$
$$S_1 = 0 \text{ leftover lemons}$$
$$S_2 = 4 \text{ leftover limes}$$
$$S_3 = 0 \text{ leftover tablespoons sugar.}$$

We'll see later on that there is actually more information we can add to the final solution, but for the purposes of this section (and its exercises), we'll regard this as the complete solution.

We now have our first method for solving linear programming problems, which works for all $m \times 2$ problems. In principle, it works for all $m \times n$ problems, but when $n > 2$, it is too difficult to carry out because the feasible region is n-dimensional. Note that in this method, it does not matter if the problem is in standard form or not, unless the feasible set is unbounded (see step 5 of Algorithm 3.1). Here is a summary of the program when $n = 2$:

ALGORITHM 3.4 *Decision Space Graphical Solution to $m \times 2$ linear programming problems.*

1. *Carefully plot the graph of the feasible region in the decision space (that is, in the xy-plane, where x and y are the decision variables.)*
2. *If the feasible set is empty, there are no feasible solutions and, therefore, no optimal solution.*
3. *If the feasible set is nonempty, determine the coordinates of all the corner points.*
4. *If the feasible set is bounded, test every corner in the objective function. The corner with the largest value of the objective function is the optimal solution if it is a maximization problem, and the corner with the smallest value of the objective function is the optimal solution if it is a minimization problem.*
5. *If the feasible set is an unbounded region in the first quadrant and the problem is a standard form maximization problem with all coefficients of the objective function being positive, then there is no largest value and, hence, no optimal point. If it is a standard form minimization problem with all coefficients of the objective function being positive, then there is an optimal point at one of the corners. If any of the coefficients of the objective function are negative, or if it is a problem with mixed constraints (so not in standard form), then there may or may not be an optimal solution. The best way to find out is to plot some of the level objective lines and see in what direction the values of the objective function are increasing and decreasing. We can tell if there is an optimal solution by moving these level objective lines in the desired direction to see if they eventually leave the feasible region (if they do, the last feasible point they touch will be a corner point, and this is the optimal solution.)*

To illustrate the case of a minimization problem, let's use this algorithm to solve O'Casek's Used Car problem (Example 3.2). Recall the setup: Let x stand for the number of transport vans and y stand for the number of flatbed trucks. if we let C stand for the cost per mile, then we must

$$\text{minimize } C = 6x + 4y$$
subject to:
$$x + 2y \geq 24 \text{ (SUVs)}$$
$$2x + y \geq 30 \text{ (sedans)}$$
$$4x + y \geq 40 \text{ (compact cars)}$$
$$x \geq 0, \ y \geq 0.$$

Earlier, we asked the reader to graph the feasible region for this problem. Your graph should look like Figure 3.6.

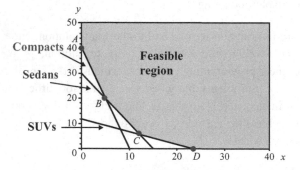

Figure 3.6 The feasible set for Example 3.2.

In this problem, the feasible region is unbounded, but since it is a standard form minimization problem, there is an optimal solution by point 5 of the algorithm, and it must occur at one of the corners labeled A, B, C, or D. (If a few of the level objective lines are plotted this will become obvious.) We leave it as an exercise to find the coordinates of the four corners. The results are displayed in Table 3.4, along with the values of the objective function.

Table 3.4 Corner point data for Example 3.2.

Corner	(x, y)	$C = 6x + 4y$
A	$(0, 40)$	$160
B	$(5, 20)$	$110
C	$(12, 6)$	$96
D	$(24, 0)$	$144

Clearly, the minimum cost is at $C = (12, 6)$, so $x = 12$ transport vans and $y = 6$ flatbed trucks is just what O'Casek needed to ship his inventory at least cost. The total cost is $96 per mile times 30 miles, or $2,880, which is minimized. To find the complete solution, we need to compute the value of all S_i, too. Recall there is one S_i for each constraint. In the lemonade stand, they were interpreted as leftover resources, but in this problem, they are interpreted as surplus vehicles shipped beyond the minimum required by the constraint. Thus:

S_1 is the number of surplus SUVs shipped.

S_2 is the number of surplus sedans shipped.

S_3 is the number of surplus compact cars shipped.

Since point $C = (12,6)$ is the intersection of the SUV line and the sedan line, those two constraints are met exactly, and there is no surplus of either type. But the point C is above the compact car line, so there will be surplus compact cars shipped (assuming the transport vans and flatbed trucks are all filled to capacity). To determine how many, we evaluate the left side of the compact car constraint to obtain the number shipped: $4x + y = 4(12) + 6 = 54$. Since the constraint only required 40, we see there is a surplus of 14 compact cars beyond the requirement. So the complete solution is as follows:

$$x = 12 \text{ (transport vans)}$$

$$y = 6 \text{ (flatbed trucks)}$$

$$C = \$96 \text{ per mile} \times 30 \text{ miles } = \$2,880 \text{ (minimized)}$$

$$S_1 = 0 \text{ surplus SUVs}$$

$$S_2 = 0 \text{ surplus sedans}$$

$$S_3 = 14 \text{ surplus compact cars.}$$

Exercises for this section and the next are deferred until the end of Section 3.4. The next two sections are devoted to tying up a few loose ends first.

3.3 Convex Sets and the Corner Point Theorem

The purpose of this section is to elaborate on the *corner point theorem*, which might also be called the *fundamental theorem of linear programming*. None of the information in this section is needed to solve any of the exercises that appear at the end of Section 3.4, if the reader is willing to accept the truth of this theorem and the algorithm of the last section. Recall from Chapter 1 that the space of all n-tuples of real numbers is denoted $\mathbb{R}^n$.

DEFINITION 3.5   *A subset X of $\mathbb{R}^n$ is said to be* convex *if it has the property that whenever two points $P, Q \in X$, then the entire line segment joining P to Q is contained in X.*

Figure 3.7 illustrates the idea, at least for plane figures (when $n = 2$).

We'll see that a feasible region for a linear programming problem is always a convex set and cannot have curved boundaries (like the upper-right figure pictured). Something with a hole, like the lower-left figure pictured, cannot be convex. Figures without holes are called *simple*. In the Euclidean plane, that makes the feasible region for any linear programming problem a simple, convex polygon, like the upper-left figure pictured. In higher dimensions, the generalization of a polygon is called a *polytope*, and a simple, convex polytope is called a *simplex*. The point of all these definitions is that the feasible region for a linear programming problem is always a simplex.

Further examples of convex sets, aside from simplices (the plural of "simplex") include points, lines, line segments, planes, half-planes, hyperplanes, and half-spaces. Recall that half-planes and half-spaces in higher dimensions describe the solutions set to a linear inequality. The proof of the next result is quite easy, so we include it:

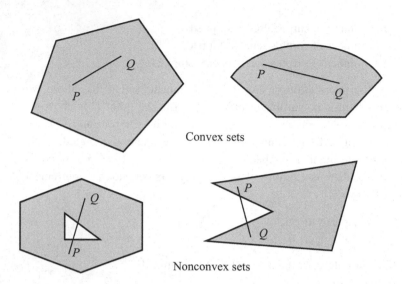

Convex sets

Nonconvex sets

Figure 3.7 Planar convex and nonconvex sets.

PROPOSITION 3.6   *Let* $X_1, X_2, \ldots, X_m$ *be convex sets in* $\mathbb{R}^n$. *Then the intersection* $X = X_1 \cap X_2 \cap \cdots \cap X_m$ *is also a convex set.*

Proof Let P and Q be arbitrary points in X. To prove the proposition, we need to verify that the entire line segment PQ lies in X. Since $P \in X$, we know that $P \in X_1, P \in X_2$, etc.; that is, $P \in X_i$ for each $i, 1 \leq i \leq m$, by definition of intersection. Similarly, $Q \in X_i$ for each i, $1 \leq i \leq m$. Choose an arbitrary i. Since $P \in X_i$ and $Q \in X_i$, and since X_i is convex, the entire segment PQ lies in X_i. Since i is arbitrary, this shows the segment PQ lies in each X_i. Then, by definition of intersection, this means PQ lies in X. But that is exactly what we wanted to show. Thus, X is convex.   ■

REMARK 3.7   *The preceding proof does not depend on m being finite. The reader who is comfortable working with indexed collections of sets can easily extend the proof to arbitrary intersections. However, we only need the finite case in this text.*

We are now ready to state the fundamental theorem of linear programming:

THEOREM 3.8   Corner Point Theorem.   *Consider an* $m \times n$ *linear programming problem with a nonempty feasible region* X. *Then,*

1.   *X must be a convex set.*
2.   *If X is bounded, then X is a simplex, and the optimal solution always exists and can be located at one of the corner points of X.*
3.   *If X is unbounded, then there may or may not be an optimal solution, but if there is, it can be located at one of the corners of X.*

We can give a proof of Part 1 easily. Regardless of whether the ith constraint is a linear equation or a linear inequality, its solution set X_i is a convex set in $\mathbb{R}^n$, since it is a hyperplane or a half-space, respectively. The feasible set X is by definition the common set of solution to the m constraints. Thus, $X = X_1 \cap X_2 \cap \cdots \cap X_m$, and by Proposition 3.14, X is a convex set.

For Parts 2 and 3, we indicate a heuristic outline of the proof here. We already outlined the steps when we solved the lemonade stand problem. For the moment, suppose $n = 2$ and the feasible region X is bounded, so it is a simple convex polygon. If P is an optimal point, we argue first that P cannot be an interior point as follows. If P were an interior point, surround P with a small open disk D, centered at P, lying entirely in X (in fact, this property can be used as a definition of what it means to be an interior point of X). Now consider the level objective line ℓ passing through P, and suppose the equation of ℓ is $ax + by = c$. Since ℓ is a straight line passing through the center P of the disk D, it actually must be a diameter of D. Therefore, D contains points on both sides of ℓ. We have already observed in Section 3.2 that one side of ℓ contains points where $ax + by < c$, and the other contains points where $ax + by > c$. Since there are points inside of D on both sides of ℓ, there are points in particular where $ax + by > c$. But all the points of D are feasible since D is contained in X. So there are feasible points where $ax + by > c$, and since this is a higher value of the objective function than the value at P, we see P cannot be a maximum point. Similarly, since D contains points where $ax + by < c$, there are feasible points where the value of the objective function is smaller than the value at P, so P cannot be a minimum point either. Thus, an interior point cannot be optimal.

Now, suppose that P is not at a corner, but somewhere along an edge k of the polygon X. Again, let ℓ be the level objective line through P. Thus ℓ and the line m containing the segment k are two straight lines containing P. There are two possibilities – either ℓ and m intersect only at P, or else they are concurrent, in which case the edge k is entirely contained in ℓ. In the first case, when they intersect only at P, we proceed in a way similar to the case of an interior point. That is, we note that since k crosses ℓ, then k contains points on both sides of ℓ. That is, k contains points where $ax + by < c$ and points were $ax + by > c$. But every point of k is feasible, so P can neither be a minimum nor a maximum point. In this case, the only possibility left for P is that P is a corner of X.

In the second case, where k is entirely contained in ℓ, every point of k satisfies $ax + by = c$, since ℓ is a level objective curve. Let Q and Q' be the corners of X that are the endpoints of the edge k. Then, in particular, $ax + by = c$ is satisfied at both Q and Q' as well as at P. Therefore, if P is optimal, so are Q and Q', so again we see there must be an optimal point at a corner point of X. This line of reasoning covers Part 2 of the proof, at least when $n = 2$. Note that it shows that an optimal point always occurs at a corner, but it does not say that every optimal point is at a corner of X. As we saw in the second case, it can happen that if the level objective curve is parallel to one of the edges k of X, then every point on that edge k is optimal. In particular, a linear programming problem can have more than one solution (in fact, infinitely many!).

What about higher dimensions, when $n > 2$? Well, the same ideas apply, but there are more steps. For example if there are $n = 3$ decision variables, x, y, z, and the feasible set is a simplex X in $\mathbb{R}^3$ (a simple, convex polyhedron in space), the objective function has the form $w = ax + by + cz$. When this is set to a constant $ax + by + cz = d$, this is the equation of a plane instead of a line. Thus, we have level objective planes instead of level objective lines. Now follow the same steps as before. First assume the optimal point P is an interior point of X. Then surround it by a small ball B centered at P, lying entirely in X. Then the level objective plane through P cuts this ball in half; thus, there are feasible points on both sides of the level objective plane. Therefore, P can neither be a maximum point nor a minimum point.

Thus, P must lie on the boundary of X, which consists of simple polygon faces. If P lies in the interior of one of these polygonal faces, again consider the level objective plane through P. If this plane is not coincident with the plane of the polygonal face, then the polygonal face (which consists of feasible points) will contain points on both sides of the level objective plane, leading to the same contradiction as before. If the level objective plane is coincident with the plane of the polygonal face, then every point in this face has the same value of the objective function, so they are all optimal points, including the corners, so an optimal point can be found at a corner of X. Thus, P cannot lie on the interior of any polygonal face of X unless every point on that face is optimal. Next, assume P lies on an edge of X. But again, either this edge will contain feasible points on both sides of the level objective plane through P or else the edge lies entirely within this level objective plane. In the first case, we have a contradiction, and in the second, the entire edge (including the endpoints of the edge, which are corners of X) are optimal points. The conclusion is that in all cases, either P is a corner point or else it shares the same value of the objective function as some corner point Q of X, whence there is always an optimal point at a corner, and there may be more than one optimal point. Similar ideas apply in higher dimensions as well.

These arguments should give you an idea why Part 2 of the corner point theorem holds, but this should not be construed as a complete, rigorous proof. Once you are comfortable with using the corner point theorem and have had a bit more mathematical training, then you can read a complete proof and check all the details.

The same ideas hold for Part 3 of the theorem. Notice that we assumed that P exists, and proceeded to show an optimal corner point also exists (which may or may not be P itself.) In Part 3 of the theorem, we can't assume that an optimal point exists in the first place. In Part 2, we can, because the domain of the objective function is a closed, bounded domain, and this implies that the objective function assumes both its maximum and minimum values. Readers who have taken calculus will recognize that this statement is merely a version of the *extreme value theorem*. There are ways to show this without using the extreme value theorem as well. For example, the proof in Calvert and Voxman (1989) shows this by using something called *convex combinations* of corner points. This concludes our discussion of the corner point theorem.

3.4 Problems with No Solution or Infinitely Many Solutions

The discussion in Section 3.3 made it clear that it is possible for a linear programming problem to have infinitely many solutions. But it is also possible that there is no solution if the feasible region is empty or unbounded. Suppose that a linear programming problem has two distinct solution points P and Q. Then the level objective line (or plane or hyperplane) through P contains Q since they both have the same value of the objective function. Consider the line segment PQ joining P to Q. This segment lies entirely within that level objective line (or plane, etc.), so every point on that segment has the same value of the objective function as P and Q. But since the feasible set X is convex by Theorem 3.8, Part 1, every one of the points on this segment is feasible. The conclusion is that if P and Q are optimal, so is every point on PQ, so there are actually infinitely many solutions.

We have just shown that, like the systems of linear equations in Chapter 1, there are exactly three possibilities for a linear programming problem: there is no solution, a unique solution, or infinitely many solutions. We've seen examples of the case of a unique solution (the lemonade stand problem and the used car problem both had unique solutions). In this section, we give examples of the other two cases.

Example 3.9 A problem with an empty feasible set.

$$\text{Maximize: } w = 10x + 3y + 6z$$

subject to:

$$2x + 3y + 5z \leq 150$$
$$2x + 3y + 5z \geq 225$$
$$x \geq 0, \; y \geq 0, \; z \geq 0.$$

Clearly, no point $P = (x, y, z)$ can have the property that $2x+3y+5z$ is less than or equal to 150 and simultaneously greater than or equal to 225. It is not even necessary to see the graph, which is three-dimensional. It is not hard to visualize the graph, in any case. The equations $2x+3y+5z = c$, for different values of c, are parallel planes in space. The solution set to the first inequality is the half-space containing the origin, of the plane closer to the origin. The solution set to the second inequality is the half-space not containing the origin, of the plane farther from the origin. There is no intersection of these half-spaces because their boundary planes are parallel. Thus, there are no feasible points.

Example 3.10 A problem with unbounded feasible set and no solution.

$$\text{Maximize: } P = 2x + y$$

subject to:

$$y \leq 10$$
$$x + y \geq 5$$
$$x \geq 0, \; y \geq 0$$

Figure 3.8 shows a graph of the feasible region in the decision space.

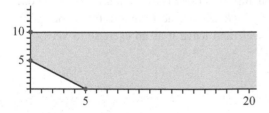

Figure 3.8 An unbounded feasible set.

This region is unbounded in the positive x direction. In particular, points of the form $(x, 0)$ are all feasible if $x \geq 5$. The value of the objective function at $(x, 0)$ is $2x$. Since x can be chosen arbitrarily large, we see the objective function becomes arbitrarily large also. P grows without bound as you move to the right. Thus, there is no maximum point, and the problem has no solution.

On the other hand, if the problem had been to minimize P, a solution would exist, which can be easily seen by plotting a few level objective lines. According to the algorithm in Section 3.2 (or determined by viewing the level objective lines), the solution would be one of the three corners

$(0, 10)$, $(0, 5)$, or $(5, 0)$. By checking the objective function, we see $P = 10$ at $(0, 10)$ or $(5, 0)$, but $P = 5$ at $(0, 5)$, so $(0, 5)$ would be the optimal minimum point.

Example 3.11 A problem with infinitely many solutions.

$$\text{Maximize: } P = 35x + 10y$$

subject to:

$$2x + y \le 18$$
$$7x + 2y \le 42$$
$$x + 2y \ge 12$$
$$x \ge 0, \ y \ge 0$$

The reader might practice graphing the feasible set, as demonstrated in Figure 3.9.

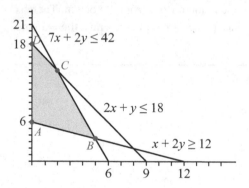

Figure 3.9 Feasible set for this Example 3.5.

Note that the problem is not in standard form – it has mixed constraints, so $(0, 0)$ is not a feasible point. However, the feasible set is bounded and nonempty, so there is a solution, and at least one optimal point must be at one of the four corners $A, B, C,$ and $D,$ according to the corner point theorem. In Table 3.5, complete the problem as usual (and leave it as an exercise to verify the coordinates of B and C).

Table 3.5

Corner	(x, y)	
A	$(0, 6)$	60
B	$(5, 3.5)$	210
C	$(2, 14)$	210
D	$(0, 18)$	180

We see that the maximum value of P occurs at two distinct corners B and C. This is because the level objective line is parallel to the edge $BC,$ as the reader should verify. We can write down the complete solution at both corners, including the values of all S_i. At corner $B,$ we have

$$x = 5$$
$$y = 3.5$$
$$P = 210 \text{ (maximized)}$$
$$S_1 = 4.5$$
$$S_2 = 0$$
$$S_3 = 0,$$

and at corner C we have

$$x = 2$$
$$y = 14$$
$$P = 210 \text{ (maximized)}$$
$$S_1 = 0$$
$$S_2 = 0$$
$$S_3 = 18.$$

As we observed at the beginning of this section, every point along the segment BC is actually an optimal point, so this problem has infinitely many solutions. Observe that every point along the segment BC has $S_2 = 0$, because constraint 2 is exactly satisfied along this line. However, if you are at any point along the edge other than the endpoints B or C, then no other S_i will have a value of 0, since none of these points lie on the lines defining any of the other constraints.

When solving a linear programming problem like this, where there are infinitely many optimal points, one might narrow it down to a unique solution if there are secondary objectives, where you would prefer one of the optimal corners over the other optimal points. For example, if it is more important to satisfy the first constraint exactly to the third; that is, if you prefer S_1 to be zero over S_3, then you would choose C over B, even though they have the same value of the primary objective function. This might happen, for example, if S_1 represents a perishable resource, so one would prefer not to have any of it leftover to store. This concludes our discussion of graphing in the decision space.

Exercises

1. Consider Angie and Kiki's lemonade stand problem (3.1). Suppose the girls decide to increase the prices of their lemonade to $1.50 for a glass of sweet lemonade and $2.50 for a glass of tart lemonade. With no other changes to the problem, determine the optimal lemonade production plan now. Be sure to indicate the leftover resources in your complete solution.

2. Solve the following linear programming problem. Be sure to give the complete solution.

a.

$$\text{maximize: } R = 10x + 3y$$

subject to:

$$5x + 2y \geq 60$$
$$3x + 6y \geq 60$$
$$5x + 4y \leq 90$$
$$x \leq 14$$
$$x \geq 0, \ y \geq 0$$

(See Exercise 3 from Section 3.1.)

b.   Keep the same objective function and constraints as in Part a, but now minimize R.

3.   Solve the maximization problem (see Exercise 7 from Section 3.1):

$$\text{maximize: } P = x + 3y$$

subject to:

$$10x + 16y \leq 240$$
$$15x + 8y \leq 240$$
$$x \geq 0, \ y \geq 0$$

4.   Solve the minimization problem (see Exercise 8 from Section 3.1):

$$\text{minimize: } C = 20x + 12y$$

subject to:

$$5x + 2y \geq 30$$
$$5x + 7y \geq 70$$
$$x \geq 0, \ y \geq 0$$

5.   a.   Solve the maximization problem (see Exercise 10 from Section 3.1):

$$\text{maximize: } z = 500x + 440y$$

subject to:

$$4x + 7y \leq 120$$
$$20x + 25y \leq 500$$
$$x + y \geq 12$$
$$y \geq 6$$
$$x \geq 0, \ y \geq 0.$$

b.   Minimize z, subject to the same constraints.

6.   Solve the problem that was set up in Exercise 11 from Section 3.1, which we reproduce here for convenience: Bruce Jax is responsible for constructing end tables and kitchen tables for the White Room Woodshop. Each end table uses 2 square yards of $\frac{3}{4}$-inch oak boards and takes 2 hours to complete. Each kitchen table uses 4 square yards of oak board and takes 3

hours to complete. This week he has available 36 square yards of oak boards and 32 hours of time. Other resources are unlimited. How many of each item should he make if he is paid $70 for each end table and $100 for each kitchen table? Be sure to include the complete data set for the solution.

7. Solve the following problem (See Exercise 12 of Section 3.1). Joni is putting her design skills to work at the Court and Sparkle Jewelry Emporium. She makes two bracelet designs. Each Dawntreader design uses 1 ruby, 6 pearls, and 10 opals. Each Hejira design uses 3 rubies, 3 pearls, and 15 opals. She has 54 rubies, 120 pearls, and 300 opals to work with. If either model yields a profit of $1,800, how many of each type of bracelet should she make in order to maximize her profit? Be sure to include the complete data set for the solution.

8. Solve the following problem (see Exercise 13 of Section 3.1). Toys in the Attic, Inc., operates two workshops to build toys for needy children. Mr. Tyler's shop can produce 36 Angel dolls, 16 Kings and Queens board games, and 16 Back in the Saddle rocking horses each day it operates. Mr. Perry's shop can produce 10 Angel dolls, 10 Kings and Queens board games, and 20 Back in the Saddle rocking horses each day it operates. It costs $144 per day to operate Mr. Tyler's shop and $166 per day to operate Mr. Perry's shop. Suppose the company receives an order from the Kids Dream On charity foundation for at least 720 Angel dolls, at least 520 Kings and Queens board games, and at least 640 Back in the Saddle rocking horses. How many days should they operate each shop in order to fill the order at least possible total cost?

9. Solve the following diet mix problem. Kerry's Kennel is mixing two commercial brands of dog food for its canine guests. A bag of Dog's Life Canine Cuisine contains 3 pounds of fat, 2 pounds of carbohydrates, 5 pounds of protein, and 3 ounces of vitamin C. A giant-size bag of Way of Life Healthy Mix contains 1 pound of fat, 5 pounds of carbohydrates, 10 pounds of protein, and 7 ounces of vitamin C. The requirements for a week's supply of food for the kennel are that there should be at most 21 pounds of fat, at most 40 pounds of carbohydrates, and at least 21 ounces of vitamin C. How many bags of each type of food should be mixed in order to design a diet that maximizes protein?

10. Solve the following scheduling problem. The Poseidon's Wake Petroleum Company operates two refineries. The Cadence Refinery can produce 40 units of low-grade oil, 10 units of medium-grade oil, and 10 units of high-grade oil in a single day. (Each unit is 1,000 barrels.) The Cascade Refinery can produce 10 units of low-grade oil, 10 units of medium-grade oil, and 30 units of high-grade oil in a single day. Poseidon's Wake receives an order from the Mars Triangle Oil retailers for at least 80 units of low-grade oil, at least 50 units of medium-grade oil, and at least 90 units of high-grade oil. If it costs Poseidon's Wake $1,800 to operate the Cadence Refinery for a day and $2,000 to operate the Cascade Refinery for a day, how many days should Poseidon's Wake operate each refinery to fill the order at lowest cost?

11. Consider the following linear programming problem:

$$\text{minimize: } R = 10x - 3y$$
$$\text{subjectto:}$$
$$x + 4y \geq 30$$
$$27x - 3y \geq 33$$
$$x \geq 0, \ y \geq 0.$$

Show that the feasible region is unbounded and that R has neither a maximum nor a minimum over this region.

12. a.   Graph the feasible region X given by the constraints:

$$2x + 3y \leq 48$$
$$x + y \leq 18$$
$$3x + y \geq 12$$
$$x \geq 0, \; y \geq 0.$$

  b.   Find the minimum value of $C = 14x + 21y$ over the region X. List the complete solution for the optimal point.

  c.   Find the maximum value of C over the region. List the complete optimal data.

  d.   Suppose you wanted to maximize C over the region X, and you also had a secondary objective (less important than maximizing C) of maximizing $z = x + y$. What point would you consider to be optimal now?

13.   Solve the following linear programming problem. The Jefferson Plastic Fantastic Assembly Corporation manufactures gadgets and widgets for airplanes and starships. Each case of gadgets uses 2 kilograms of steel and 5 kilograms of plastic. Each case of widgets uses 2 kilograms of steel and 3 kilograms of plastic. The profit for a case of gadgets is $360, and the profit for a case of widgets is $200. Suppose Jefferson Plastic Fantastic Assembly Corporation has 80 kilograms of steel and 150 kilograms of of plastic available on a daily basis, and can sell everything it manufactures. How many cases of each should it manufacture if it is obligated to produce at least 10 cases of widgets per day? The objective is to maximize daily profit.

14.   Mr. Cooder, a farmer in the Purple Valley, has at most 400 acres to devote to two crops: rye and barley. Each acre of rye yields $100 profit per week, while each acre of barley yields $80 profit per week. Due to local demand, Mr. Cooder must plant at least 100 acres of barley. The federal government provides a subsidy to grow these crops in the form of tax credits. They credit Mr. Cooder four units for each acre of rye and two units for each acre of barley. The exact value of a "unit" of tax credit is not important. Mr. Cooder has decided that he needs at least 600 units of tax credits in order to be able to afford to his loan payments on a new harvester. How many acres of each crop should he plant in order to maximize his profit?

15.   Joanne's Antique Restoration Emporium refurbishes Victorian furniture. Each dining room set requires 16 hours stripping and sanding time, 4 hours refinishing time, and 4 hours in the upholstery shop. Each bedroom set requires 36 hours stripping and sanding time, 4 hours refinishing time, and 2 hours in the upholstery shop. Each month, they have available 720 person-hours in the stripping and sanding shop, 100 person-hours in the refinishing shop, and 80 person-hours in the upholstery shop. Each dining room set generates $600 profit, and each bedroom set generates $900 profit. How many of each type of furniture set should Joanne accept to work on each month in order to maximize her profit?

16.   Solve the following linear programming problem:

$$\text{minimize: } z = 15x + 51y$$
$$\text{subject to:}$$
$$x + y \leq 100$$

$$3x + 10y \geq 300$$
$$18x + 10y \geq 900$$
$$-6x + 50y \geq 240$$
$$x \geq 0, \ y \geq 0.$$

17. As a result of a federal discrimination lawsuit, the town of Yankee, NY, is required to build a low-income housing project. The outcome of the lawsuit specifies that Yankee should build enough units to be able to house at least 44 adults and at least 72 children. Yankee also must meet a separate requirement to be able to house at least 120 people altogether. Yankee has available up to 54,000 square feet on which to build. Each townhouse requires 1,800 square feet and can house six people (two adults and four children). Each apartment requires 1,500 square feet and can house four people (two adults and two children). Each townhouse costs $100,000 to build, and each apartment costs $80,000 to build. How many of each type of housing unit should Yankee build in order to minimize the total cost?

18. Solve the linear programming problem:

$$\text{maximize: } z = 3x + 4y$$
$$\text{subject to:}$$
$$x + y \leq 12$$
$$5x + 2y \geq 36$$
$$7x + 4y \geq 14$$
$$x \geq 0, \ y \geq 0.$$

Do you notice something unusual about one of the constraints? What is it?

19. Solve the following linear programming problem:

$$\text{maximize: } z = 5x + 2y$$
$$\text{subject to:}$$
$$x + y \geq 11$$
$$2x + 3y \geq 24$$
$$x + 3y \leq 18$$
$$x \geq 0, \ y \geq 0.$$

20. The Topographic Starship Tour Company has been offering extended tours of the dark side of the moon's surface ever since the Roundabout Interplanetary Mining Company set up a lunar base in the year 2025. They have two type of vehicles – a land rover (called a Relayer) and a flying jet shuttle (called a Khatru). Each Relayer can hold six first-class passengers, 10 economy-class passengers, and 600 kilograms of luggage and supplies. Each Khatru can hold 10 first-class passengers, four economy-class passengers, and 400 kilograms of luggage and supplies. Each Relayer costs $8,000 to run a two-week tour, and each Khatru costs $11,500 to run a two-week tour. A group of research geologists and potential mining investors want to take a tour.

    a.   There are 58 first-class tourists and 40 economy-class tourists in the group, and they have a total of 3,400 kilograms of luggage and supplies to bring along. How many of each vehicle should they charter in order to minimize the total cost?

    b.   Suppose there were 46 economy-class tourists instead of 40, with no other changes from Part a. Find the optimal solution now. What is different about the feasible set in Part b? [**Remark**: When this happens, the problem is said to be *degenerate*. A more formal definition is given later in the chapter.]

3.5　The Constraint Space: 2 × *n* Problems

Consider the following standard form maximization problem:

Example 3.12 A farmer has 150 acres available to split between two crops: asparagus and beans. Each bag of asparagus seed is sufficient to plant 1 acre and requires 8 pounds of fertilizer. Each bag of bean seeds is sufficient to seed 2 acres and requires 10 pounds of fertilizer. She has a total of 1,000 pounds of fertilizer to use on her two crops. The profit from a bag of asparagus seeds is $1,500, and the profit from a bag of bean seeds is $2,000. How many bags of each should she plant in order to maximize profit?

We will introduce a second graphical method in this section, called *graphing in the constraint space*, and use it to solve this problem. For comparison purposes, we solve it first by the usual method of graphing in the decision space.

    Solution: Let x stand for the number of bags of asparagus seed and y for the number of bags of bean seed. We must do the following:

$$\text{maximize: } P = 1,500x + 2,000y$$
$$\text{subject to:}$$
$$x + 2y \leq 150 \text{ (acres)}$$
$$8x + 10y \leq 1,000 \text{ (lbs. fertilizer)}$$
$$x \geq 0, \ y \geq 0$$

Figure 3.10 shows the feasible region in the decision space.

    Since the feasible set is bounded, the corner point theorem guarantees that there is a maximum point at one of the four corners, as shown in Table 3.6. [**Exercise:** verify the coordinates of the corner points]:

Table 3.6 Corner point data for Example 3.6.

Corner	(x, y)	$P = 1,500x + 2,000y$
A	$(0, 0)$	0
B	$(125, 0)$	$187,500
C	$\left(\frac{250}{3}, \frac{100}{3}\right)$	$191,666.67
D	$(0, 75)$	$150,000

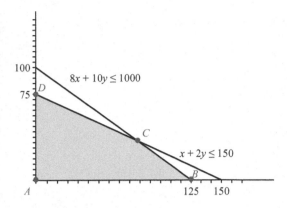

8x + 10y ≤ 1000

x + 2y ≤ 150

Figure 3.10 Feasible set for Example 3.6.

So the complete solution is the following:

$$x = \frac{250}{3} = 83\frac{1}{3} \text{ bags asparagus seed}$$

$$y = \frac{100}{3} = 33\frac{1}{3} \text{ bags bean seed}$$

$$P = \$191,666.67 \text{ profit (maximized)}$$

$$S_1 = 0 \text{ leftover acres land}$$

$$S_2 = 0 \text{ lbs. leftover fertilizer.}$$

We now illustrate the second method, graphing in the constraint space.

REMARK 3.13 *In the following discussion, we will be making several parenthetical statements referring to topics in linear algebra. The purpose of these remarks is merely to make connections for those students who have already studied linear algebra. If you have not taken a linear algebra course, you can safely ignore these parenthetical remarks without affecting your ability to apply this graphical technique.*

In this method, the axes correspond to the constraints, not the decision variables. So in our example, the horizontal axis will be land (in acres), and the vertical axis will be fertilizer (in pounds). Because we are limited to at most 150 acres and 1,000 pounds of fertilizer, the only part of this constraint space that is relevant to our problem is the rectangle in the first quadrant with one corner at the origin in Figure 3.11.

We construct a vector for each of the decision variables, with components given by how much of each resource they use up. Thus, since a bag of asparagus seed implies using 1 acre and 8 pounds of fertilizer, we represent it by the following column vector:

$$A = \begin{bmatrix} 1 \\ 8 \end{bmatrix}.$$

Similarly, a bag of bean seed is represented in the constraint space by a column vector indicating how much of the constraints (resources) it uses up:

$$B = \begin{bmatrix} 2 \\ 10 \end{bmatrix}.$$

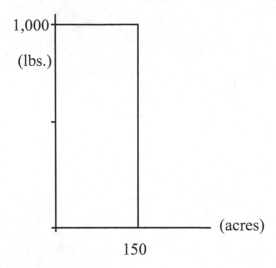

Figure 3.11 The resource space.

If we plot A in our rectangle in Figure 3.11, we obtain a visual representation of how a bag of asparagus seed uses up our resources. Ignoring B for the moment, what happens if we plot the scalar multiple $10A$? Well,

$$10A = 10 \begin{bmatrix} 1 \\ 8 \end{bmatrix} = \begin{bmatrix} 10 \\ 80 \end{bmatrix}$$

clearly represents how much of our resources 10 bags of asparagus seeds will use. In the graph in Figure 3.11, $10A$ lies along the ray from the origin determined by A. In general, if only asparagus is planted, the resulting point will be of the form xA, where x is the number of bags of seed planted, and this vector lies along the ray determined by A. We can let x increase until the ray hits one of the edges of the rectangle, which means one of the resources is used up. (Which one?)

There is a similar ray determined by B, which contains vectors of the form yB, where y is the number of bags of bean seed planted, assuming no asparagus is planted. If some of each crop is planted – say, x bags of asparagus seed and y bags of bean seed – then the resources used up is obviously the sum of the resources used for each crop. In other words, the total amount of resources used by both crops is given by the linear combination $xA + yB$. Indeed,

$$xA + yB = x \begin{bmatrix} 1 \\ 8 \end{bmatrix} + y \begin{bmatrix} 2 \\ 10 \end{bmatrix} = \begin{bmatrix} x + 2y \\ 8x + 10y \end{bmatrix}$$

is a vector whose components are given by the left-hand side of the constraint inequalities, which we already know represents the amount of resources used. In order for the structural constraints to be satisfied, the plot of this vector $xA + yB$ must lie within the rectangle plotted above. But we must also consider the nonnegativity constraints. Linear combinations of the vectors $\{A, B\}$ can be located anywhere in the plane. However, if both x and y are nonnegative, it is clear from the way vectors add (parallelogram law, as we saw in Chapter 1) that the linear combination $xA + yB$ will lie in the first quadrant and between the rays determined by A and B, and conversely. Thus, if we want to consider all linear combinations that are feasible, they would lie in the shaded region within the rectangle between the rays determined by A and B, as shown in Figure 3.12 (the vectors A and B are not drawn to scale in the picture).

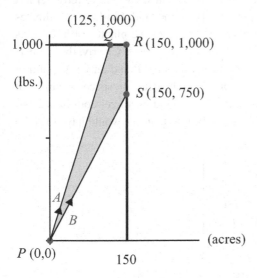

Figure 3.12 Feasible linear combinations of vectors A and B.

Now, just as in the previous graphical method, we can plot level objective lines. For example, along the ray for asparagus, there lies the point $20A$, with a corresponding profit of $P = 1,500x + 2,000y = 1,500(20) + 2,000(0) = \$30,000$. To find the point along the ray for beans corresponding to the same profit, we must solve $30,000 = 1,500(0) + 2,000y$, whence $y = 15$. Thus, the vector $15B$ also corresponds to a profit of \$30,000. The line segment connecting these points ($20A$ and $15B$) would lie on the level objective line for $P = \$30,000$. Moving the line out from the origin, keeping it parallel to itself as it moves, we see that the last feasible point that it touches, just before leaving the feasible region entirely, is again one of the corners, then that must be the optimal point. So once again, to find the optimal point, all we need to do is to test the objective function at each corner and choose the largest one, just as we did when we graphed in the decision space.

There is one wrinkle, however, which is that the coordinates listed in Figure 3.12 for P, Q, R, and S are land and fertilizer coordinates, not x and y. We need to find the (x, y) coordinates for P, Q, R, and S first, and then we would see a direct correspondence with the points A, B, C, and D in the feasible region in the decision space.

For example, if we move along the ray for asparagus, we would eventually hit the point Q when we run out of fertilizer. Each bag of asparagus seed requires an acre of land, and since we have 150 acres available, we would not run out of land until we planted 150 bags of seed. However, before we ran out of land, we'd run out of fertilizer. Indeed, each bag uses 8 pounds, and since we have 1,000 pounds, available, we'd run out of fertilizer as soon as we planted $x = 125$ bags. Since we must stop planting asparagus as soon as the first resource runs out, we must stop when we have planted 125 bags, whence $x = 125$. At this point, all the fertilizer is used up, so $y = 0$. Thus, the land, fertilizer coordinates for point Q are $(125, 1,000)$ as shown, but the (x, y) coordinates are $(125, 0)$, which is point B in the decision space.

Similarly, if we planted only beans and moved up along the ray for beans until we hit point S, it is clear we run out of land first. Since each bag is enough for 2 acres, and we have 150 acres, we would run out of land after planting 75 bags of bean seed. Note that we still have fertilizer left over at this point since we have used $75 \times 10 = 750$ pounds. Thus, the land, fertilizer coordinates of the point S are $(150, 750)$; but the (x, y) coordinates are $(0, 75)$ since we planted no asparagus and 75 bags of bean seed. This corresponds to point D in the decision space. Clearly, the point P corresponds to the point A in the decision space, since x and y are both 0 if and only if no land or fertilizer is used. Finally, the point R corresponds to the point C in the decision space because that is the one feasible point where both resources are used up. So the land, fertilizer coordinates are $(150, 1,000)$ as shown, but what about the (x, y) coordinates? To find that, we would have to solve this linear combination equation for x and y:

$$xA + yB = \begin{bmatrix} 150 \\ 1,000 \end{bmatrix}$$

$$\begin{bmatrix} x + 2y \\ 8x + 10y \end{bmatrix} = \begin{bmatrix} 150 \\ 1,000 \end{bmatrix}.$$

This corresponds to the linear system:

$$x + 2y = 150$$
$$8x + 10y = 1,000.$$

But solving this system is exactly what we had to do to find the (x, y) coordinates of point C in the decision space. So we know the answer will be the same. Thus, the (x, y) coordinates of R are $\left(\frac{250}{3}, \frac{100}{3}\right)$. So, R indeed corresponds to C. The reader should find it very easy to believe that the two feasible regions (one in the decision space and one in the constraint space) are just two views of the same thing. (Readers who have taken a course in linear algebra might guess that there is a linear transformation that goes from the decision space to the constraint space, which maps the first feasible region onto the other in a bijective fashion. Those readers might find it instructive to find the matrix of that linear transformation. Not only will it look quite familiar, but it is also a good example of a linear transformation arising quite naturally in the context of an applied problem.)

To complete the solution by graphing in the constraint space, in Table 3.7, we create a table similar to Table 3.6, which we gave for the decision space.

Table 3.7 Corner point data from the resource space.

Corner	(Land, fertilizer)	(x, y)	$P = 1,500x + 2,000y$
P	$(0, 0)$	$(0, 0)$	0
Q	$(125, 1,000)$	$(125, 0)$	\$187,500
R	$(150, 1,000)$	$\left(\frac{250}{3}, \frac{100}{3}\right)$	\$191,666.67
S	$(150, 750)$	$(0, 75)$	\$150,000

We arrive at the correct solution – point R. One slight advantage of this technique is that we have for each point the land and fertilizer coordinates in addition to the (x, y) coordinates, which makes it very easy to determine the values of the S_i. For example, at point Q, since we have used

125 acres and all 1,000 pounds of the fertilizer, it's obvious that $S_1 = 25$ acres and $S_2 = 0$ pounds of fertilizer. Of course, we really only care about the optimal point R, and the complete solution here is

$$x = \frac{250}{3} \text{ bags asparagus seed}$$

$$y = \frac{100}{3} \text{ bags bean seed}$$

$$P = \$191,666.67 \text{ (maximized)}$$

$$S_1 = 0 \text{ leftover acres land}$$

$$S_2 = 0 \text{ leftover lbs. fertilizer.}$$

Of course, this is the same optimal data we obtained by graphing in the decision space.

Now, this particular problem was 2×2, so the feasible region could be graphed using either method. The claim we are making, however, is that graphing in the constraint space can be used for $2 \times n$ problems for any n. While this is true, having more than two decision variables adds a layer of complexity to the technique. We now introduce some new terminology that will help us deal with the added complexity coming from this plurality of decision variables, and moreover, this new terminology will also be useful when we introduce the *simplex algorithm* in Chapter 5.

Our starting point in the linear combination of A and B that corresponds to the corner point R is the following:

$$xA + yB = \begin{bmatrix} 150 \\ 1,000 \end{bmatrix}.$$

We contend that there is some missing information here – not only are the vectors A and B important, but so are the the two unit vectors on the two axes:

$$U_1 = \begin{bmatrix} 1 \\ 0 \end{bmatrix}$$

$$U_2 = \begin{bmatrix} 0 \\ 1 \end{bmatrix}.$$

Since the first coordinate is land and the second is fertilizer, we can associate U_i with the variable S_i. The reason that the vectors U_i appear to be missing from the equation is that at the point R, both of the S_i have value 0. The complete linear combination should involve all four of the relevant vectors:

$$xA + yB + S_1U_1 + S_2U_2 = \begin{bmatrix} 150 \\ 1,000 \end{bmatrix}. \tag{3.5}$$

Now, there are infinitely many solutions to this vector equation. One way to see that is to write it out as a system of equations, as was done in Chapter 1, and notice that there are only two equations but four unknowns. (Readers who have studied linear algebra know another reason why there are non-unique solutions – because the set of vectors $\{A, B, U_1, U_2\}$ is a *linearly dependent* set.)

Furthermore, there are infinitely many solutions even if we restrict attention to those solutions where all four variables are nonnegative. Indeed, every feasible point corresponds to such a solution. The reason is that to be feasible, we already know $x \geq 0$ and $y \geq 0$. But if the point is feasible, the amount of resources must be less than or equal to what we have available, which means the values of S_i are also nonnegative, since they measure the leftover resources. To see a specific example, 30 bags of asparagus seed and 20 bags of bean seed correspond to the point $(x, y) = (30, 20)$. But this is a feasible point since it uses 70 acres and 440 pounds of fertilizer. So it corresponds to the following nonnegative solution to (Eq. (3.5)):

$$30A + 20B + 80U_1 + 560U_2 = \begin{bmatrix} 150 \\ 1{,}000 \end{bmatrix}$$

$$30 \begin{bmatrix} 1 \\ 8 \end{bmatrix} + 20 \begin{bmatrix} 2 \\ 10 \end{bmatrix} + 80 \begin{bmatrix} 1 \\ 0 \end{bmatrix} + 560 \begin{bmatrix} 0 \\ 1 \end{bmatrix} = \begin{bmatrix} 150 \\ 1{,}000 \end{bmatrix}.$$

So every feasible point is such a nonnegative solution. Of course, the point $(30, 20)$ is feasible but not optimal – in fact, it is not even a corner point. Is there a way we can recognize which nonnegative solutions represent a corner point in the decision space? Indeed, we can. We already noted that the corner point R (which happens to be optimal) corresponds to the solution when $S_1 = S_2 = 0$ (and x and y are nonnegative). But this is true of all the corners – each corner corresponds to setting two of the four variables in Eq. (3.5) equal to 0. The reason is that a corner, after all, is determined by the fact that two of the lines that define the feasible set (including the axes $x = 0$ and $y = 0$) intersect, and that, in turn, means two of the variables are 0. The rest of the variables should have nonzero (so strictly positive) values, unless another line from the list of constraints passes through the corner in question.

We want to refer to these variables easily, so we give them names. The two variables that determine where two constraint lines cross (and so define a "corner" point), and whose values are 0, are called *nonbasic variables*. The rest of the variables (typically with positive values) will be called *basic variables*. For example, at the optimal point R, x and y are the basic variables (with positive values $\frac{250}{3}$ and $\frac{100}{3}$, respectively), while S_1 and S_2 are the nonbasic variables. The reader can verify that the notion of a *basic solution*, which was introduced in Chapter 1, is exactly what we need here. A solution to Eq. (3.5) with two variables set equal to 0 is a basic solution precisely because the vectors corresponding to the basic variables are a basis of the constraint space.

In general, in an $m \times n$ problem, there are $m + n$ variables. A solution to the vector equation analogous to Eq. (3.5) in that case is called a basic solution if there are m basic variables and n nonbasic variables. Incidentally, if three or more of the constraint lines all pass through the same corner, this will lead to some of the basic variables also having a value of 0 since "by coincidence" three or more of the constraints are all satisfied exactly at the same corner. In this case, the problem is said to be *degenerate*. For an example, see Exercise 20b from Section 3.4.

In this problem, we have four variables and we must select two of them to be nonbasic – there are six different such selections, so six different basic solutions (see Table 3.8). But our feasible region only has four corners. This should not cause any confusion since, as Table 3.8 shows, the other two basic solutions are actually "corner points" outside the feasible region, so we don't care about them. All six basic solutions to Eq. (3.5) are shown in Table 3.8.

Notice the two last rows correspond to infeasible points since one of the basic variables is negative. To see how we found the values of the variables in the second column, for each row, we set the nonbasic variables equal to 0 in Eq. (3.5). This reduces it to a system of equations with a unique solution for the basic variables, which we solve as usual.

So, to summarize, feasible points correspond to nonnegative solutions of Eq. (3.5), while basic solutions of Eq. (3.5) are those where two of the lines cross, whence two of the variables in

Table 3.8 Basic solutions.

Basic variables	Basic solution to Eq. (3.5)	Corner
x, y	$\frac{250}{3}A + \frac{100}{3}B + 0U_1 + 0U_2 = \begin{bmatrix} 150 \\ 1{,}000 \end{bmatrix}$	R
x, S_1	$125A + 0B + 25U_1 + 0U_2 = \begin{bmatrix} 150 \\ 1{,}000 \end{bmatrix}$	Q
y, S_2	$0A + 75B + 0U_1 + 250U_2 = \begin{bmatrix} 150 \\ 1{,}000 \end{bmatrix}$	S
S_1, S_2	$0A + 0B + 150U_1 + 1{,}000U_2 = \begin{bmatrix} 150 \\ 1{,}000 \end{bmatrix}$	P
x, S_2	$150A + 0B + 0U_1 + (-200)U_2 = \begin{bmatrix} 150 \\ 1{,}000 \end{bmatrix}$	Not feasible
y, S_1	$0A + 100B + (-50)U_1 + 0U_2 = \begin{bmatrix} 150 \\ 1{,}000 \end{bmatrix}$	Not feasible

Eq. (3.5) must be 0. The corner points of the feasible region are those solutions that are both feasible and basic. In Table 3.8, the last two rows are not feasible, so we simply discard them. The four remaining basic solutions in the first four rows of the table (regardless if you are thinking of the decision space or of the constraint space) are the corner points of the feasible region. It may be instructive to the reader to look more carefully at both the decision space graph and the constraint space graph and see exactly where these "infeasible corners" are located (see Exercise 2 at the end of this section).

Before we consider a larger example, is there any way we could have predicted in advance which choices of basic variables would lead to feasible basic solutions (i.e., corners) and which would not? Yes, indeed, there is a way. Recall that each of the four variables corresponds to one of the four vectors in $\{A, B, U_1, U_2\}$. If one deletes from this set the two vectors corresponding to the nonbasic variables, the two vectors that are left form a *basis* of the constraint space, as noted earlier. Thus, for example, the point Q corresponds to the basis $\{A, U_1\}$ because the basic variables are x and S_1 at this corner, while the point R corresponds to the basis $\{A, B\}$ because the basic variables are x and y at this corner.

Now, every choice of two basis vectors from the set $\{A, B, U_1, U_2\}$ leads to one of the six rows in Table 3.8. The key to determining which basic solutions are feasible is to remember we want nonnegative solutions. So, for example, if we select $\{A, U_2\}$ as our basis, we must solve this equation:

$$xA + S_2U_2 = \begin{bmatrix} 150 \\ 1{,}000 \end{bmatrix}.$$

But it is clear from simply looking at the graph of the feasible region in the constraint space that there are no nonnegative solutions to this. Recall from Chapter 1 that nonnegative linear combinations of A and U_2 would lie between the ray determined by A and the ray determined by U_2 because vectors add by the parallelogram law. But the ray for U_2 is the positive vertical axis (it corresponds to a unit of fertilizer), and points between this axis and the ray determined by A are not located in the shaded region! This explains why the basic solution in the fifth row of the table is not feasible.

Similarly, the ray for U_1 is the positive horizontal axis (it corresponds to a unit of land), and no vectors lying between this axis and the ray for B are in the shaded region. This explains why the basic solution in the sixth row is not feasible. The other four choices of a basis all must lead to a feasible point because the vector $R = \begin{bmatrix} 150 \\ 1{,}000 \end{bmatrix}$ does lie in the shaded region (it is the diagonal of the shaded rectangle), and *it also lies between the two rays corresponding to the two basis vectors.* Since R is the diagonal of the rectangle, it follows that in order to find a basis of two vectors that has the property that R lies between the two rays, we see one of the rays must meet the top edge of the rectangle (where the number of units of fertilizer is 1,000), and the other must meet the right edge of the rectangle (where the number of units of land is 150). Clearly, these four choices result in the first four rows of Table 3.8 so are in fact our corner points.

Armed with this insight about predicting feasible basic solutions, we are now prepared to look at a larger example. Let's make the preceding problem a 2×3 problem by adding a third possible crop – corn. Suppose a bag of corn seed is enough to plant 1 acre of land, uses 20 pounds of fertilizer, and generates \$3,700 profit. The complete statement of the modified problem is as follows:

Let x be the number of bags of asparagus seed.

Let y be the number of bags of bean seed.

Let z be the number of bags of corn seed.

Maximize: $P = 1{,}500x + 2{,}000y + 3{,}700z$

subject to:

$x + 2y + z \le 150$ (acres land)

$8x + 10y + 20z \le 1{,}000$ (lbs. fertilizer)

$x \ge 0,\ y \ge 0,\ z \ge 0.$

Since there are three decision variables, if we tried to solve this problem by the method of graphing in the decision space, we would require a three-dimensional graph. This is not impossible (see Exercise 8 at the end of this section), but it is substantially more difficult than a two-dimensional graph, so instead, we solve it by graphing in the constraint space. Since there are only two resources, land and fertilizer, this graph is still two-dimensional.

In fact, since we didn't change the amounts of resources available, we still work with the same rectangle as above in the constraint space. The vectors A, B, U_1, U_2 are still present (as are the rays associated with them); they still have the same coordinates as before, and the same interpretations. What's new in the modified problem is the presence of a fifth vector for corn and its corresponding ray. Let C be the corn vector, whose coordinates represent the amount of resources used up by one bag of corn seed:

$$ C = \begin{bmatrix} 1 \\ 20 \end{bmatrix} $$

Then, in addition to the rays for asparagus and beans (and the positive axes, which are the rays for land and fertilizer), the vectors of the form zC determine a third product ray that we add to the previous diagram and we shade in everything between the (outermost) product rays, just as we did in the 2×2 case. The resulting diagram is shown in Figure 3.13.

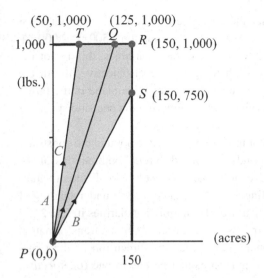

Figure 3.13 Feasible linear combinations of vectors A, B, and C.

You should observe the following changes. First, there is a new point T labeled where the corn ray meets the top edge of the rectangle. We find that point by taking $x = 50$ bags of corn seed, which will use up all 1,000 pounds of fertilizer (since corn requires 20 pounds per bag), and 50 acres of land (since corn uses 1 acre per bag). Thus, the land, fertilizer coordinates of T are $(50, 1,000)$. Note that the (x, y, z) coordinates of T must be $(0, 0, 50)$, since after 50 bags of corn seed are planted, there is no fertilizer left to be able to plant either of the other two crops. Second, the shaded region consists of (the intersection of the rectangle with) all nonnegative linear combinations of $\{A, B, C\}$, by the way vectors add using the parallelogram law. However, unlike the 2×2 problem, the points in the shaded region do *not* have unique expressions as linear combinations of the product vectors $\{A, B, C\}$. (Readers familiar with linear algebra will recognize that the reason for this is that the set $\{A, B, C\}$ is linearly dependent.) This means we have to take care before identifying a point in the shaded region as feasible or not.

To illustrate this complication, consider the point T. When we think of writing $T = 0A + 0B + 50C$, it is definitely a feasible point. However, there are other linear combinations of $\{A, B, C\}$ that sum to $T = \begin{bmatrix} 50 \\ 1,000 \end{bmatrix}$. For instance, we could also write $T = xA + yB + 0C = xA + yB$ for some x, y, since we know any point in the plane can be expressed in terms of the basis $\{A, B\}$. However, since T is not between the rays for A and B, such a linear combination could not possibly be a nonnegative one – either x or y would have to be negative, and therefore, it would not be a feasible point!

The lesson to be gleaned from this example is that when $n > 2$ in a $2 \times n$ problem, there are many different ways to express any vector as a linear combination of the *decision vectors* (corresponding to the decision variables). Consequently, just because the coordinates in the constraint space put a point in the shaded region, we cannot conclude it represents a feasible point until we specify exactly which linear combination of the decision vectors we use to represent the point.

There is a further complication when $n > 2$. Set the objective function to a constant – for illustrative purposes, let's use \$100,000 in our example. We obtain the equation $1,500x + 2,000y + 3,700z = 100,000$. Do you see the problem? Because there are three variables, this is not the equation of a line! Thus, when $n > 2$, there is no such thing as a level objective line in the constraint space. (Of course, this equation is the equation of a level objective plane in the three-dimensional decision space, but this does not help us since we are ignoring the decision space for the moment.)

This difficulty is really connected to the previous complication – the same point in the constraint space (whether in the shaded region or not) corresponds to many different combinations of the product vectors (that is, with different points in the decision space), possibly with different values of the objective function. (Again, a familiarity with linear algebra can help one's understanding of the reasons for this. This time, it's the concept of a linear transformation that clarifies it. In a 2×3 problem, the coefficient matrix of the structural constraints is the matrix of a linear transformation from a three-dimensional space to a two-dimensional space. Since the dimension of the target is less than the dimension of the domain, the transformation cannot be one to one (or *injective*, as it is often called). Thus, different points in the domain correspond to the same point in the target. In other words, in the 2×2 example, we made the remark that the shaded region in the decision space and the shaded region in the constraint space were really two views of the same thing. However, in the 2×3 example, the shaded region in the constraint space is a two-dimensional "projection" or "shadow" of the feasible region in the three-dimensional decision space. When you make such a projection, you lose information and distinct points have the same projection in the target.)

That appears to be an obstacle in solving $2 \times n$ problems when $n > 2$ using the technique of graphing in the constraint space. If we cannot associate each point in the shaded region of our rectangle with either a unique set (x, y, z) of coordinates in terms of the decision variables or with a unique value of the objective function, how do we determine which points are optimal? Will they still be corner points of the shaded region? We contend that the optimal point need not be an actual geometric corner of the shaded region in the constraint space. In the example at hand, for instance, it is possible that a feasible linear combination of A, B, C that identifies with the point Q might be optimal, yet Q is clearly not a corner of the shaded region in the constraint space. Of course, there is something special about Q that separates it from other shaded points. It's the new language of basic solutions and basic variables that is the way out of the dilemma.

PROPOSITION 3.14   *The optimal point must be at a basic feasible solution.*

Proof   We know from the corner point theorem that the optimal point must be at a corner of the feasible set in the decision space. But those corners of the simplex are defined by the fact that (in a $2 \times n$ problem) n of the $n + 2$ total variables must be value 0, so the point in question exactly satisfies n of the equations. Therefore, there are n nonbasic variables at the optimal point, and two basic variables. That means one can express the vector R (which is the vector whose components are the right side of the structural constraints) as a linear combination of exactly two vectors (and the other n vectors have coefficients of 0). But this is exactly the definition of a basic solution of a vector equation and concludes the proof.   ∎

So let's mimic what we did in the 2×2 case, using the proposition. In this case, we have $m + n = 2 + 3 = 5$ variables (corresponding to three decision vectors A, B, C and two unit vectors U_1 and U_2 (corresponding to the two constraints) in the constraint space. According to the proposition, we seek basic feasible solutions to the vector equation:

$$xA + yB + zC + S_1U_1 + S_2U_2 = \begin{bmatrix} 150 \\ 1{,}000 \end{bmatrix} = R.$$

Each basic solution corresponds to a choice of two out of the five vectors $\{A, B, C, U_1, U_2\}$ to be a basis, and the corresponding two variables will be basic variables. The other three variables will be nonbasic and have a value of 0. The number of ways of selecting two elements from a set of five is 10. (This is a *binomial coefficient* or a *combination* $C(5,2)$. See Appendix.) However, not all 10 choices will yield a feasible solution. To be feasible, the vector $R = \begin{bmatrix} 150 \\ 1{,}000 \end{bmatrix}$ must be between the rays generated by the basic vectors. So one of the basis vectors must lie above R – which is the diagonal of the rectangle, and so its ray meets the top edge – and the other must be below the diagonal, so its ray meets thee right edge of the rectangle. A glance at the graph of the rectangle in Figure 3.13 shows that $\{U_2, A, C\}$ are the three above and $\{B, U_1\}$ are below R. Thus, only $3 \times 2 = 6$ of the 10 basic solutions will actually be feasible. In Table 3.9, we have omitted the four infeasible cases and have listed only the six feasible points, one of which must be optimal by the proposition. We also list the corresponding basic and nonbasic variables and values of the objective function (rounded to the nearest dollar) in Table 3.9.

Table 3.9 Basic feasible solutions.

Basis	Basic variables	Nonbasic variables	Basic solution (suppressing nonbasic vectors)	Profit
$\{A, B\}$	x, y	z, S_1, S_2	$\frac{250}{3}A + \frac{100}{3}B = R$	\$191,667
$\{A, U_1\}$	x, S_1	y, z, S_2	$125A + 25U_1 = R$	\$187,500
$\{B, C\}$	y, z	x, S_1, S_2	$\frac{200}{3}B + \frac{50}{3}C = R$	\$195,000
$\{C, U_1\}$	z, S_1	x, y, S_2	$50C + 100U_1 = R$	\$185,000
$\{B, U_2\}$	y, S_2	x, z, S_1	$75B + 250U_2 = R$	\$150,000
$\{U_1, U_2\}$	S_1, S_2	x, y, z	$150U_1 + 1{,}000U_2 = R$	\$0

The third row has the highest profit, so it is the optimal point. The complete solution is as follows:

$$x = 0 \text{ bag asparagus seed}$$
$$y = \frac{200}{3} \text{ bags bean seed}$$
$$z = \frac{50}{3} \text{ bags corn seed}$$
$$P = \$195{,}000 \text{ (maximized)}$$
$$S_1 = 0 \text{ leftover acres}$$
$$S_2 = 0 \text{ lbs. leftover fertilizer.}$$

Just to reiterate what was stated earlier, the choice of $\{A, C\}$, for example, as a basis would yield a basic solution; however, the vector $R = \begin{bmatrix} 150 \\ 1{,}000 \end{bmatrix}$ would not be between the asparagus ray and corn ray. So this basic solution would not be feasible, and we do not include it in Table 3.9. Thus, we have successfully solved a 2×3 problem using graphing in the constraint space. You may have noticed that without the availability of level objective lines, the shading in the picture wasn't so relevant. All that really matters is the relative positions of the product rays compared to the vector R, the diagonal of the rectangle.

Extending these ideas into general $2 \times n$ problems, we can now state the complete algorithm for this method.

ALGORITHM 3.15 *Constraint space graphical solution of $2 \times n$ linear programming problems (for standard form maximization problems).*

1. *In the plane representing the constraint space, draw the rectangle corresponding to the two constraints.*
2. *Plot a "decision" vector (also known as a "product" vector) A_i for each of the n decision variables x_i. The components of A_i are the elements of the ith column of coefficients of the system of inequalities that describe the structural constraints.*
3. *Draw the ray through each A_i and extend it until it hits either the top edge or the right edge of the rectangle. (If desired, you may shade in the region between the outermost rays, but this isn't really necessary.)*
4. *Let R be the vector whose components are the right-hand side of the structural constraints, the total available resources, so that R is the diagonal of the rectangle, and let U_1 and U_2 be the unit vectors on the horizontal and vertical axes of the constraint space. Consider the vector equation obtained from setting R equal to a linear combination of the $n + 2$ vectors $\{A_1, A_2, \ldots, A_n, U_1, U_2\}$. It has the form:*

$$x_1 A_1 + x_2 A_2 + \cdots + x_n A_n + S_1 U_1 + S_2 U_2 = R. \tag{3.6}$$

5. *Determine which pairs of vectors correspond to feasible basic solutions. In each pair, the vector R must be between the rays determined by the chosen pair of basis vectors.*
6. *The optimal point is the feasible basic solution found in step 5 with the largest value of the objective function. (It need not be unique.)*

We leave it to the reader to consider how this algorithm may be modified to handle minimization problems or problems that are not standard form. At this point, we can solve any $m \times 2$ problem (graphing in the decision space) and any $2 \times n$ problem (graphing in the constraint space). However, if we wish to be able to efficiently solve problems that are 3×3 or larger, neither graphical method will be very helpful. We need to develop a more algebraic technique that will not rely on graphs and diagrams so much. The best and most widely used algebraic approach is the simplex algorithm, which we present in Chapter 5.

It is interesting to note that in a later chapter, when we discuss zero-sum games, a similar phenomenon occurs. We will discover that there are some fairly simple methods that will solve $m \times 2$ or $2 \times n$ games, but a more powerful method is needed for games of size 3×3 or larger. Surprisingly, it turns out that this more powerful method again involves the simplex algorithm!

Exercises

1. Consider the following vector equation:

$$x\begin{bmatrix} 2 \\ 7 \end{bmatrix} + y\begin{bmatrix} 4 \\ 1 \end{bmatrix} + S_1\begin{bmatrix} 1 \\ 0 \end{bmatrix} + S_2\begin{bmatrix} 0 \\ 1 \end{bmatrix} = \begin{bmatrix} 44 \\ 50 \end{bmatrix}.$$

    a.   Find the basic solution corresponding to the basic variables x, y.
    b.   Find the basic solution corresponding to the basic variables x, S_1.
    c.   Find the basic solution corresponding to the basic variables y, S_1.
    d.   Find the basic solution corresponding to the basic variables S_1, S_2.
    e.   If this vector equation described the solution to a standard form maximization linear programming problem, one of the preceding basic solutions would not be feasible. Which one and why not?

2. In the table of basic solutions to the 2 × 2 farming problem discussed in this section, two of the basic solutions were not feasible. In the graph of this problem in the constraint space, determine exactly where these infeasible points are located. Do the same thing for the graph of this problem in the decision space.

3. Solve the following 2 × 2 problem (see Exercise 3 from Section 3.4) using the method of graphing in the constraint space:

$$\text{maximize: } P = x + 3y$$
$$\text{subject to:}$$
$$10x + 16y \le 240$$
$$15x + 8y \le 240$$
$$x \ge 0, \ y \ge 0.$$

4. Solve the following 2 × 2 problem (see Exercise 4 from Section 3.4) using the method of graphing in the constraint space. Notice that it is a minimization problem. What adjustments must be made in Algorithm 3.15 to handle minimization problems?

$$\text{Minimize: } C = 20x + 12y$$
$$\text{subject to:}$$
$$5x + 2y \ge 30$$
$$5x + 7y \ge 70$$
$$x \ge 0, \ y \ge 0$$

5. Solve the following 2 × 2 problem (see Exercise 6 from Section 3.4) using the method of graphing in the constraint space. Bruce Jax is responsible for constructing end tables and kitchen tables for the White Room Woodshop. Each end table uses 2 square yards of $\frac{3}{4}$-inch oak boards and takes 2 hours to complete. Each kitchen table uses 4 square yards of oak board and takes 3 hours to complete. This week he has available 36 square yards of oak boards and 32 hours of time. Other resources are unlimited. How many of each item should he make if he is paid $70 for each end table and $100 for each kitchen table?

6. The Glass House is a shop that produces three types of specialty drinking glasses. The Runaway is a 12-ounce water tumbler, the Reunion is a 16-ounce beer glass, and the Experience is an elegant stemmed champagne flute glass. A case of Runaway glasses takes 1 hour on the molding machine and 0.1 hour to pack and generates a profit of $40. A case of Reunions glasses takes 1.5 hours on the molding machine and 0.1 hours to pack and generates a profit of $60. A case of Experience glasses takes 1.5 hours on the molding machine and 0.2 hours to pack and generates a profit of $70. Each week, there are 720 hours of time available on the molding machine and 60 hours available to pack. How many cases of each type should they manufacture in order to maximize profit?

7. The Spooky Boogie Costume Salon makes and sells four different Halloween costumes: the witch, the ghost, the goblin, and the werewolf. Each witch costume uses 3 yards of material and takes 3 hours to sew. Each ghost costume uses 2 yards of material and takes 1 hour to sew. Each goblin costume uses 2 yards of material and takes 3 hours to sew. Each werewolf costume uses 2 yards of material and takes 4 hours to sew. The profits for each costume are as follows: $10 for the witch, $8 for the ghost, $12 for the goblin, and $16 for the werewolf. If they have 600 yards of material and 510 sewing hours available before the holiday, how many of each costume should they make in order to maximize profit, assuming they sell everything they make?

8. Consider the three crop farming problem discussed in this section of the text. Solve the problem by graphing in the decision space. [**Remark**: This problem is for readers who have basic experience making plots in space. Since it is a 2×3 problem, the decision space is three dimensional, and each inequality determines a half-space with a plane as a boundary.]

9. Consider the three-crop farming problem discussed in this section of the text. In addition to asparagus, beans, and corn, now the farmer is considering a possible fourth crop – eggplant. Each bag of eggplant seed is enough to seed 3 acres and requires 10 pounds of fertilizer. Each bag of eggplant seed will generate $2,800 profit. There are no other changes to the problem – the other three crops still have the same resource requirements and generate the same profit. With the addition of this new crop, solve the problem again.

10. Consider the following variation on Angie and Kiki's lemonade stand. Suppose Kiki's idea was to include a third type of lemonade instead of a third ingredient? Thus the resources are just lemons and sugar (no limes). They have 60 lemons and 30 tablespoons of sugar available. In addition to the sweet lemonade (three lemons and 2 tablespoons of sugar per glass) and the tart lemonade (four lemons and 1 tablespoon of sugar per glass), they also make a lite lemonade (two lemons and $\frac{4}{3}$ tablespoons of sugar per glass). The prices are $1.25 for a glass of sweet, $1.50 for a glass of tart, and $1.00 for a glass of lite. How many glasses of each should they make in order to maximize their revenue. Use the method of graphing in the constraint space.

11. Touch of Grey Art Supplies, Ltd., produces and sells artists' supplies, including acrylic paint. In any acrylic paint, there are two basic ingredients: pigment (which is what gives the paint its color) and a synthetic resin binder (which forms a base to hold the particles of pigment). Touch of Grey has five different recipes for red paint of various hues, values, and intensities. The amounts of pigment and binder in one tube of each of the five colors, as well as the profit generated, is indicated in Table 3.10. There are other ingredients in the paint, such as fillers and dyes, but these are in unlimited supply and do not affect the ability for the company to make a profit.

Table 3.10 Ingredients and profits for various red paints.

	Cadmium red	Napthol crimson	King crimson	Venetian red	Providence red
Pigment (g)	50	40	40	40	50
Binder (g)	80	100	60	50	60
Profit ($)	4	3	5	4	2

Suppose the company has available 10 kilograms of pigment and 15 kilograms of binder. The problem is to determine how many tubes of each color should the company mix if it wants to maximize profit, assuming it sells everything it makes.

a.   Why can we conclude, before we even solve the problem, that there will be at most two different colors of paint mixed at the optimal point?

b.   Solve the problem by graphing in the constraint space.

c.   In real life, the company would probably mix some of each color instead of just two. Speculate on how our model might be modified to make it more realistic.

4   Sensitivity Analysis and Duality

In Chapter 3, we learned how to solve some simple linear programming problems. The solution of a problem is not the end of the story in real life. In this chapter, we introduce two important ideas that go beyond the solution of a problem. As we suggested in Chapter 1, once we learn how to use a model, we can seek to improve it. One type of improvement for a model is something to make it a better reflection of the physical problem we want to solve – that is, make the model more realistic. An example of this that was mentioned then was the study of trajectories of falling bodies. Once we understand how it works in a vacuum, we might then modify the model to include air resistance.

The first main idea of this chapter is *sensitivity analysis*. The idea is to make a small change in the original problem and see how this change affects the solution. Why one would want to do this and how it is done is explored in Sections 4.1–4.4, as well as in later chapters. Sensitivity analysis is not about making a model more realistic, but it is a different sort of improvement to a model that makes the model more flexible and broadly applicable.

In Section 4.5, we introduce the second main idea of the chapter – *duality*. It is an interesting fact that linear programming problems come in pairs called dual problems. For each maximization problem, there is a corresponding minimization problem that is related to it in a certain way, and vice versa. It turns out to be quite useful to know how these pairs of problems are related. In subsequent chapters, we will exploit this relationship in a way that will ultimately save us a lot of work.

4.1   Introduction to Sensitivity Analysis

Consider the lemonade stand problem from Chapter 3. We know the optimal solution to the problem is given by the following:

$$x = 12 \text{ glasses sweet lemonade}$$
$$y = 6 \text{ glasses tart lemonade}$$
$$R = \$24.00 \text{ (maximized)}$$
$$S_1 = 0 \text{ leftover lemons}$$
$$S_2 = 4 \text{ leftover limes}$$
$$S_3 = 0 \text{ leftover tablespoons sugar.}$$

Sensitivity analysis is the study of how sensitive this solution is to minor changes in the original problem. For example, suppose we change the price of a glass of tart lemonade from $1.50 to

$1.55, with no other changes to the problem. What will the new solution be? Will $(12, 6)$ still represent the optimal solution, or will the optimal values of the decision variables change to a different corner of the feasible set? How will the values of the slack variables change, and what will the optimal revenue change to?

We can ask similar questions with regard to any small change to the problem. Instead of changing the price of a glass of lemonade, we can ask what would happen to the solution if our supply of lemons changed from 60 to 65, or if our supply of sugar decreased from 30 tablespoons to 20. What if it was possible to increase the supply of just one of the resources – lemons, limes, or sugar? How can we measure which resource is the most valuable?

We could even address small changes in the recipes of the lemonade itself if we so desired. Suppose we used 3.5 lemons per glass of tart lemonade instead of four? How would this affect the optimal solution? Certainly, it would be useful for Angie and Kiki to know all of this information, in case they are considering changes to their lemonade stand operation.

Now, to answer any of these questions, one way to proceed would be to simply solve the problem again with the changed information. But that can be rather tedious, and the point of sensitivity analysis is to extract useful information without having to solve the entire problem over again. The ease of extracting this type of useful information depends on the technique used to solve the problem. Doing sensitivity analysis with algebraic methods such as the simplex algorithm (which we cover in the next chapter) is generally more effective and easier than with graphical methods. So we will revisit these ideas later after the simplex algorithm is introduced.

Let us divide the type of questions we could ask into several distinct categories:

- Changes in the objective coefficients. Here we consider what happens to the optimal solution if the only change we make is to one coefficient of the objective function. In the lemonade stand problem, this means changing the price of a glass of sweet lemonade or the price of a glass of tart lemonade.
- Changes in the capacity. This refers to what happens when we change one of the numbers on the right-hand side of one of the inequalities that define the feasible set. In the lemonade stand problem, this means changing the amount of one of the resources – lemons, limes, or sugar. The capacity to make lemonade obviously depends on how much of these resources we have available to allocate, explaining why we call this "changes in the capacity."
- The marginal value associated with a constraint. The exact definition of marginal value appears in Section 4.3. In the lemonade stand problem, the constraints arise from the resources – lemons, limes, and sugar. So each resource will have a marginal value that helps answer the question we raised earlier of which resource is the most valuable at a given corner point.
- Structural changes. This refers to what happens when we change one coefficient on the left-hand side of one of the constraints. In the lemonade stand problem, this means we are changing the amount of one of the resources in the recipe of one of the types of lemonade. Another possible structural change would be the addition of a new decision variable. In the lemonade stand problem, this refers to situation where the girls start selling a third type of lemonade – for example, a lite lemonade with low sugar content.

The first three of these types of questions are commonly studied in most textbooks. Structural changes, the fourth type, are less frequently addressed, but we mention these changes in the context of sensitivity analysis for the sake of completeness.

We close this introduction with some general remarks. Some models, and the physical phenomena they represent, have the property that tiny changes in the problem can lead to drastic changes in the answer. Sometimes called the "butterfly effect," this is one of the hallmarks of chaos theory, an interesting branch of mathematics that we do not explore in this text. Linear programming is not chaotic. A "small" change in one of our coefficients leads to a predictable "small" change in the solution.

However, a "larger" change in one of the coefficients can lead to a more drastic change in the solution. Therefore, with each of the types of changes mentioned before, there is a question that needs to be answered – how small is a "small" change? How far can we go in changing a coefficient before we induce a "drastic" change in the solution? As we shall see, there will be a range of values for each coefficient we change which qualify as a "small" change. We call this the *stable range* of that coefficient. If you go outside of the stable range, the solution changes in a more drastic way than if you stay inside the range. What we mean by a "more drastic" change depends on the coefficient we are changing and will be clarified in the sections that follow. Part of sensitivity analysis is to determine those stable ranges.

4.2   Changes in the Objective Coefficients

In this section, we consider what happens to the solution when we change one of the coefficients of the objective function. A key observation is that the feasible set is completely determined by the constraints of the problem, and the objective function has nothing to do with this set. It follows that no matter how we change the objective function, the feasible set remains unchanged. What can change, of course, is which corner point is the optimal one. It turns out that a small change in the objective coefficient will not change which point is optimal, but a larger change in this coefficient can induce a change to a different corner, the more drastic change referred to at the end of the last section. Therefore, the *stable range* of the coefficient is, in this case, the set of all values of the coefficient that retain the original optimal corner.

We illustrate this with our favorite example, the lemonade stand problem. Recall the setup of the problem:

Let x denote the number of glasses of sweet lemonade.
Let y denote the number of glasses of tart lemonade.

We must

maximize: $R = 1.25x + 1.5y$

subject to:

$3x + 4y \leq 60$ (lemons)

$x + 2y \leq 28$ (limes)

$2x + y \leq 30$ (tbsp. sugar)

$x \geq 0, \ y \geq 0.$

We begin with a simple question: Suppose we change the price of a glass of sweet lemonade from \$1.25 to \$1.30. What is the revised optimal data? We ask the same question if we change it to \$1.00 per glass.

Since the feasible set is unchanged, the corners have the same coordinates as in the original problem. So there is no need to draw the graph of the feasible set. All we need do is evaluate the revised objective function (revenue) at each corner, creating a table similar to the one in Section 3.2 in Table 4.1.

Table 4.1 Corner point data for modified objective functions.

Corner	(x, y)	$R = 1.3x + 1.5y$	$R = x + 1.5y$
A	$(0,0)$	\$0	0
B	$(15,0)$	\$19.50	\$15.00
C	$(12,6)$	\$24.60	\$21.00
D	$(4,12)$	\$23.20	\$22.00
E	$(0,14)$	\$21.00	\$21.00

Here we have added a column for each of the changes. In the case when the price \$1.30 per glass, observe that point C is still the optimal point. Thus, we know that all the optimal data is the same as it was before the change, except for the additional revenue. Thus, $(x, y) = (12, 6)$, and we have four leftover limes, with a maximum revenue of \$24.60.

In the case when the price is \$1.00 per glass, notice that the optimal point has changed – point D is now optimal. Thus, $(x, y) = (4, 12)$ and $R = \$22.00$ is the maximum revenue. Since the optimal point has changed, we need to compute the leftover resources. Since point D is the intersection of the lemon and lime lines (see Figure 3.5 in Section 3.2), we have used up all the lemons and limes. But we do have leftover sugar: $2x + y = 2(4) + 12 = 20$, and since we started with 30 tablespoons, we have $S_3 = 10$ tablespoons of sugar left over.

Finding the revised optimal data is then quite easy. What about the stable range for the price of a glass of sweet lemonade? Notice that with the first change to \$1.30 per glass, the optimal point remained at point C, whereas the second change to \$1.00 per glass triggered a switch to a different corner (D in this case.) This means that 1.3 is within the stable range, but 1 is not.

The key to finding the stable range is to observe that when the objective coefficient is changed, the slope of the level objective lines (the iso-revenue lines) is changed. Recall that the slope of the level objective line is exactly what determines which corner is optimal. For example, in the original problem, when $R = 1.25x + 1.5y$, the slope is $-\frac{1.25}{1.5} = -\frac{5}{6}$. As we move to larger and larger values of R, the level objective lines move away from the origin, and the last corner touched depends on the slope. See Figure 3.5 in Section 3.2.

Clearly, The point C is optimal because the slope $-\frac{5}{6}$ of the level objective lines is between the slope of segment BC, which is -2, and the slope of segment CD, which is $-\frac{3}{4}$. Also, the graph makes it clear that for a small change in the slope of the level objective line, the slope will still be between these values so that C will remain the optimal corner. For example, when we change the price to \$1.30, the slope becomes $-\frac{1.3}{1.5} = -0.86667$, which still lies between -2 and $-\frac{3}{4} = -0.75$, explaining why C is still the optimal corner. However, when we change the price to \$1, the slope becomes $-\frac{1}{1.5} = -0.66667$, which is not in the interval $[-2, -0.75]$, explaining why the optimal corner moves to a different corner.

It should now be clear how to determine the stable range for the price of a glass of sweet lemonade. If we replace the coefficient 1.25 with a variable a, the objective function becomes

$$R = ax + 1.5y.$$

The slope of this line is given by $m = -\frac{a}{1.5}$. If we want point C to remain optimal, this slope must lie in the interval $[-2, -0.75]$. Thus, we set up inequalities

$$-2 \le -\frac{a}{1.5} \le -0.75,$$

which we solve for a:

$$-3 \le -a \le -\frac{9}{8} = -1.125$$

$$1.125 \le a \le 3.$$

Thus, we have found the stable range. In order for the point C to be optimal (assuming the price of a glass of tart lemonade remains fixed at \$1.50), the price for a glass of sweet lemonade must be between \$1.125 and \$3.00. Having this knowledge is useful for the girls selling the lemonade. If they want to increase the price of sweet lemonade, or have a sale and decrease the price, then as long as the new price is in the stable range, they do not have to change their optimal program of 12 glasses of sweet lemonade and six glasses of tart lemonade.

In a similar way, we can find the stable range for the price of a glass of tart lemonade (assuming the price of sweet remains fixed at \$1.25). We replace the coefficient \$1.50 with a variable b to obtain the objective function:

$$R = 1.25x + by,$$

which has slope $-\frac{1.25}{b}$, and set this slope to be in the critical interval $[-2, -0.75]$:

$$-2 \le -\frac{1.25}{b} \le -0.75.$$

Solving for b yields

$$0.75 \le \frac{1.25}{b} \le 2$$

$$\frac{4}{3} \ge \frac{b}{1.25} \ge \frac{1}{2}$$

$$\frac{5}{3} \ge b \ge \frac{5}{8}.$$

Therefore, the stable range for the price of a glass of tart lemonade is

$$\$0.625 \le b \le \$1.6666\ldots$$

So the girls know that raising the price to \$1.67 or more, or lowering it to \$0.62 or lower, will require switching to a new optimal corner. Otherwise, the point $C = (12, 6)$ remains the optimal point.

In general, if the problem is in standard form, to find the stable range without referring to the graph of the feasible set, all one has to do is list the slopes of each constraint equation in increasing order and find which two values on the list the slope of the objective function lie between. That will give the critical interval that plays the role of $[-2, -0.75]$ in this example. We remark that the same procedure works for a minimization problem in standard form. If the problem has mixed constraints so it is not in standard form, it is best to refer the graph of the feasible set, although one could derive an algorithm that works in all cases.

For example, consider the following problem with mixed constraints:

$$\text{maximize: } P = 30x + 10y$$

subject to:

$$2x + y \le 18$$

$$7x + 2y \le 42$$

$$x + 2y \ge 12$$

$$x \ge 0, \ y \ge 0.$$

The slope of the level objective lines is $m = -3$. The slopes of the constraints are, in order, $-2, -\frac{7}{2}$, and $-\frac{1}{2}$. However, the third constraint, with slope $-\frac{1}{2}$, is irrelevant if we assume that the objective coefficients are nonnegative. To see why, consider the graph in Figure 3.9.

Because the coefficients of the objective function are nonnegative, it is clear from that the point A could never be the optimal point. Only B, C, or D could be the last point touched by the level objective lines as P increases. Thus, we do not need to consider the slope of the segment AB, which is the edge of the feasible set lying on the greater-than-type constraint. Only the slopes of BC and CD are relevant to determine the optimal point.

Notice that the slope $m = -3$ lies between these two slopes:

$$-\frac{7}{2} \le m \le -2,$$

which suggests that the point C is the optimal point. We verify this in the usual way (compare with Example 3.11) in Table 4.2.

Table 4.2 Values of the objective function at each corner point.

Corner	(x, y)	$P = 30x + 10y$
A	$(0, 6)$	60
B	$(5, 3.5)$	185
C	$(2, 14)$	200
D	$(0, 18)$	180

So the point C is indeed optimal. Should m change to a value that is less than $-\frac{7}{2}$, then the level objective curves would be quite steep and the optimal point would switch to corner B. Should m change to a value greater than -2, then the optimal point would switch to corner D. These are the only possibilities, and no information is obtained by comparing m to $-\frac{1}{2}$, the slope of the greater-than-type constraint. Had this been a minimization problem (with positive objective coefficients), then, likewise, we could ignore all the greater-than-type constraints, as the only possible optimal minimum points are A, B, or D. There would be no need to compare m to the slopes of segments BC and CD.

We can now find the stable ranges of the objective coefficients. For the coefficient of x, replace 30 with a variable a, and solve the inequalities

$$-\frac{7}{2} \le -\frac{a}{10} \le -2$$

for a. We obtain $-35 \le -a \le -20$, or

$$20 \le a \le 35.$$

It is worth observing that in Example 3.11 from the last chapter, the value of a was an endpoint 35 of this stable range. This explains why in that example there were two optimal corners B and C. When a is at the endpoint of a stable range, it is always the case that the level objective curve is parallel to one of the constraint lines.

For the stable range of the y coefficient, replace 10 with a variable b an solve the inequalities

$$-\frac{7}{2} \le -\frac{30}{b} \le -2$$

for b. We obtain $2 \le \frac{30}{b} \le \frac{7}{2}$, and (assuming $b > 0$) taking reciprocals, we obtain

$$\frac{2}{7} \le \frac{b}{30} \le \frac{1}{2},$$

or

$$\frac{60}{7} \le b \le 15,$$

the stable range for b that we seek. Following similar procedure, we can find the stable range of the objective coefficients in any $m \times 2$ problem. For problems with more than two decision variables, one would have to rely on an algebraic approach, so we defer discussion of this until after we cover the simplex algorithm.

Exercises

The following exercises appeared previously in the exercises of Section 3.4, where they were to be solved by graphing in the decision space.

1. Joni is putting her designing skills to work at the Court and Sparkle Jewelry Emporium. She makes two signature design bracelet models. Each Dawntreader bracelet uses 1 ruby, 6 pearls, and 10 opals. Each Hejira bracelet uses 3 rubies, 3 pearls, and 15 opals. She has 54 rubies, 120 pearls, and 300 opals to work with. If either model results in a profit of $1,800 for Joni, how many of each type should she make? .

    a. Suppose the profit for a Dawntreader bracelet increases to $2,400. Find the new optimal data.
    b. Find the stable range for the profit of a Dawntreader bracelet.
    c. Find the stable range for the profit of a Hejira bracelet.

2. Toys in the Attic, Inc., operates two workshops to build toys for needy children. Mr. Tyler's shop can produce 36 Angel dolls, 16 Kings and Queens board games, and 16 Back in the Saddle rocking horses each day it operates. Mr. Perry's shop can produce 10 Angel dolls, 10 Kings and Queens board games, and 20 Back in the Saddle rocking horses each day it operates. It costs $144 per day to operate Mr. Tyler's shop and $166 per day to operate Mr. Perry's shop. Suppose the company receives an order from the Kids Dream On charity foundation for at least 720 Angel dolls, at least 520 Kings and Queens board games, and at

least 640 *Back in the Saddle* rocking horses. How many days should they operate each shop in order to fill the order at least possible total cost?

a.   Suppose Mr. Perry can streamline operations in his shop so that the daily cost comes down to $120 per day. Find the new optimal data.

b.   Find the stable range for the costs of daily operations for both of the shops.

3.   Kerry's Kennel is mixing two commercial brands of dog food for its canine guests. A bag of Dog's Life Canine Cuisine contains 3 pounds of fat, 2 pounds of carbohydrates, 5 pounds of protein, and 3 ounces of vitamin C. A giant-size bag of Way of Life Healthy Mix contains 1 pound of fat, 5 pounds of carbohydrates, 10 pounds of protein, and 7 ounces of vitamin C. The requirements for a week's supply of food for the kennel are that there should be at most 21 pounds of fat, at most 40 pounds of carbohydrates, and at least 21 ounces of vitamin C. How many bags of each type of food should be mixed in order to design a diet that maximizes protein?

a.   Suppose the recipe for Dog's Life Canine Cuisine changes so that the protein content of each bag increases to 12 pounds Find the new optimal data.

b.   Find the stable range for the protein in a bag of Dog's life Canine Cuisine.

c.   Find the stable range for the protein in a bag of Way of Life Healthy Mix.

4.   The Poseidon's Wake Petroleum Company operates two refineries. The Cadence Refinery can produce 40 units of low-grade oil, 10 units of medium-grade oil, and 10 units of high-grade oil in a single day. (Each unit is 1,000 barrels.) The Cascade Refinery can produce 10 units of low-grade oil, 10 units of medium-grade oil, and 30 units of high-grade oil in a single day. Poseidon's Wake receives an order from the Mars Triangle Oil retailers for at least 80 units of low-grade oil, at least 50 units of medium-grade oil, and at least 90 units of high-grade oil. If it costs Poseidon's Wake $1,800 to operate the Cadence Refinery for a day and $2,000 to operate the Cascade Refinery for a day, how many days should Poseidon's Wake operate each refinery to fill the order at least cost?

a.   Suppose that the Cadence refinery daily cost can be reduced to $1,200. Find the new optimal data.

b.   Find the stable range for the daily cost of operating each refinery.

c.   Suppose that *both* operating costs are cut in half. Explain in a short sentence or two why this has no effect on which corner is optimal.

5.   The Jefferson Plastic Fantastic Assembly Corporation manufactures gadgets and widgets for airplanes and starships. Each case of gadgets uses 2 kilograms of steel and 5 kilograms of plastic. Each case of widgets uses 2 kilograms of steel and 3 kilograms of plastic. The profit for a case of gadgets is $360, and the profit for a case of widgets is $200. Suppose Jefferson Plastic Fantastic Assembly Corporation has 80 kilograms of steel available and 150 kilograms of plastic available on a daily basis, and can sell everything it manufactures. How many cases of each should it manufacture if it is obligated to produce at least 10 cases of widgets per day? The objective is to maximize daily profit.

Find the stable range for the profit of a case of gadgets and for the profit of a case of widgets.

6.   As a result of a federal discrimination lawsuit, the town of Yankee, NY, is required to build a low-income housing project. The outcome of the lawsuit specifies that Yankee should build enough units to be able to house at least 44 adults and at least 72 children. Yankee also must meet a separate requirement to be able to house at least 120 people altogether. Yankee has available up to 54,000 square feet on which to build. Each townhouse requires 1,800 square feet and can house six people (two adults and four children). Each apartment requires 1,500 square feet and can house four people (two adults and two children). Each townhouse costs $100,000 to build, and each apartment costs $80,000 to build. How many of each type of housing unit should Yankee build in order to minimize the total cost?

   Find the stable range of the cost of building a townhouse and for the cost of building an apartment.

4.3   Marginal Values Associated with Constraints

Consider again Angie and Kiki's lemonade stand problem. Imagine that the girls are excited about the $24 they made from their optimal solution of $x = 12$ glasses of sweet lemonade and $y = 6$ glasses of tart lemonade, and they feel they can sell even more lemonade the next day. So they ask their mother to increase the amount of lemons, limes, and sugar for the next day. Mom replies that she cannot afford to increase all the ingredients, but she makes the following offer as a compromise: "I'll either increase your lemon supply by five lemons or your lime supply by five limes or your sugar supply by 5 tablespoons. But you have to decide which ingredient you want to increase. Which one do you want?"

   How do the girls decide which ingredient to increase? They suspect that an increase in any of them will result in more revenue, but in order to choose, they should try to measure exactly how much revenue they would obtain from each additional lemon and similarly for the other ingredients. This leads us directly to Definition 4.1.

DEFINITION 4.1   *The* marginal value *of a resource is the additional revenue obtained by increasing that resource by one unit while making no other changes to the problem.*

   Let us introduce a new variable M_i to represent the marginal value of the ith resource. So M_1 is the marginal value of lemons, M_2 is the marginal value of limes, and M_3 is the marginal value of sugar. If the girls knew these marginal values, they could make an informed decision and choose the resource with the highest marginal value as the resource to increase.

   Now, even with no previous mathematical training, the girls immediately reply to their mom: "Well, we certainly do not want to increase our supply of limes!" Do you see why not? At the current optimal solution, recall that there are $S_2 = 4$ leftover limes. If they started with even more than 28 limes, it simply means there would be even more limes left over at the end. The optimal solution would still be at the point $C = (12, 6)$. They could not increase their revenue at all because they have run out of the other ingredients, and additional limes would not help them at this point. The girls intuitively understand that the marginal value of any resource that is leftover is 0. That is, if $S_i > 0$, then $M_i = 0$. This phenomena is called *complementary slackness* and is discussed further in Section 4.5. A constraint for which there is no slack ($S_i = 0$) is said to be a *binding* constraint, while one where there is slack ($S_i > 0$) is said to be *non-binding*.

So, having ruled out additional limes, the girls still must choose between additional lemons or additional sugar. So how do we compute the marginal value of these resources? According to the definition, we need to find the additional revenue generated if that resource is increased by one unit. Consider the case of lemons first, where the supply is increased from 60 to 61 lemons. This means that the the constraint for lemons is replaced by the modified constraint:

$$3x + 4y \leq 61.$$

Unlike the changes in the objective coefficients that we studied in the last section, changes in the right-hand-side values of the constraints do change the feasible set. Therefore, the coordinates of some of the corners will change. Let's take a look at the feasible set in Figure 4.1.

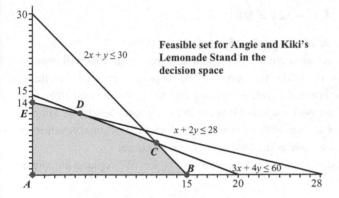

Figure 4.1 Feasible set before we modify the constraint for lemons.

The effect of changing 60 to 61 is that the the line representing the lemon constraint is moved slightly farther away from the origin. In the graph in Figure 4.1, it is clear that this will move slightly the locations of points C and D, so they will have new coordinates. Points A, B, and E are unaffected by this change, so their coordinates remain unchanged. So, in theory, what one should do is calculate the new coordinates of the points C and D and compute the revenue at all five corners in order to find the new optimal point.

However, we can cut down on the work involved, since it should be clear by considering the level objective lines that the new optimal point will still be the (modified) point C. Indeed, none of the slopes of the constraints have changed, so the slope of the level objective lines still lies between the slope of the segment BC and the (modified) segment CD, whence the modified point C is still optimal. Thus, all we really need to do it find the modified point C and compute the revenue for that one corner.

We do so now. The important fact about corner C is that it is the intersection of the lemon constraint line and the sugar constraint line. So we can find the coordinates by solving the system:

$$3x + 4y = 61$$
$$2x + y = 30.$$

Writing this as a matrix equation $AX = B$,

$$\begin{bmatrix} 3 & 4 \\ 2 & 1 \end{bmatrix} \begin{bmatrix} x \\ y \end{bmatrix} = \begin{bmatrix} 61 \\ 30 \end{bmatrix},$$

we can solve it quickly by matrix inversion, obtaining

$$\begin{bmatrix} x \\ y \end{bmatrix} = A^{-1}B = \frac{1}{-5}\begin{bmatrix} 1 & -4 \\ -2 & 3 \end{bmatrix}\begin{bmatrix} 61 \\ 30 \end{bmatrix}$$

$$= \frac{1}{-5}\begin{bmatrix} -59 \\ -32 \end{bmatrix} = \begin{bmatrix} 11.8 \\ 6.4 \end{bmatrix}.$$

Thus, the revenue at the modified optimal point C is

$$1.25(11.8) + 1.50(6.4) = \$24.35.$$

Therefore, the increase in revenue generated by the additional lemon is

$$\$24.35 - \$24 = \$0.35.$$

We have shown that the marginal value of a lemon is 35 cents; $M_1 = 0.35$.

In summary, to find the marginal value, we increase that resource by one unit. This will change the coordinates of some of the corners, including the current optimal point, provided that there are no leftover units of resource at the optimal point. Assuming that the optimal point does not switch to a different corner (which is the case for a small change in the resource level), we simply compute the revenue for the revised coordinates of that corner and subtract the original revenue to obtain the increase in revenue generated by one additional unit of that resource.

Let's illustrate this to find the marginal value of a tablespoon of sugar. We replace the sugar constraint with the modified constraint:

$$2x + y \leq 31.$$

This will move the sugar constraint line slightly in the graph, resulting in new coordinates for points B and C. To find the new coordinates of C, we solve the system:

$$\begin{bmatrix} 3 & 4 \\ 2 & 1 \end{bmatrix}\begin{bmatrix} x \\ y \end{bmatrix} = \begin{bmatrix} 60 \\ 31 \end{bmatrix},$$

resulting in

$$\begin{bmatrix} x \\ y \end{bmatrix} = \frac{1}{-5}\begin{bmatrix} 1 & -4 \\ -2 & 3 \end{bmatrix}\begin{bmatrix} 60 \\ 31 \end{bmatrix}$$

$$= \frac{1}{-5}\begin{bmatrix} -64 \\ -27 \end{bmatrix} = \begin{bmatrix} 12.8 \\ 5.4 \end{bmatrix}.$$

We then compute the associated revenue by substituting this point into the objective function:

$$R = 1.25(12.8) + 1.50(5.4) = \$24.10.$$

This is an increase in 10 cents, so $M_3 = 0.10$, the marginal value of a tablespoon of sugar. To answer the question originally posed by the girls, sugar increases the revenue by $0.10 per tablespoon, so presumably an increase of 5 tablespoons will result in an increase of $0.50 in their revenue. Lemons will increase their revenue by $0.35 per lemon, so presumably an increase of five lemons will generate an additional $1.75 revenue. If they can increase their supply of only one resource, they should choose lemons, as they are the most valuable resource at the current corner point C.

Notice that we can verify that complementary slackness holds in this way. If we increased the limes (or any resource that has leftovers at corner C), the line for the lime constraint moves slightly, changing the coordinate for points D and E. However, the coordinates for the optimal point C do not change at all, because the line for limes is not part of the system of equations we solved to determine the point C. Thus, the optimal point remains unchanged at values $(12, 6)$, and the revenue remains unchanged at $24.00, resulting in a marginal value of 0 for limes. The same holds for any constraint that is non-binding at the optimal point.

We remark that we have made a few unstated assumptions in the preceding calculations. First of all, while we correctly calculated that the additional revenue for one additional lemon is $0.35, notice that we assumed that it will be 35 cents for *each* additional lemon, not just the first one. The justification for this relies on two facts. First of all, the revenue function is linear. If the revenue were given by a nonlinear function, one would need calculus to compute these marginal values (for those who know calculus, the marginal values are approximated by derivatives – actually, by partial derivatives, to be precise – and for nonlinear functions, these derivatives are not constant – they vary from point to point). In a linear programming problem, the objective function is always linear, so the marginal values are indeed constant.

Secondly, marginal values are examples of what we call "local" properties. They are valid for small changes but not necessarily for large changes. To see this, just imagine increasing the supply of lemons to a very large amount. At some point, we will have too many lemons, and there will some leftover lemons at the optimal point. That is, at some point, the lemon constraint becomes non-binding. Once this happens, the marginal value of a lemon suddenly drops to 0 by complementary slackness. There is a stable range involved for the changes in the resources, which is discussed in the next section. If we increase the number of lemons from 60 to 65, then as long as 65 is in the stable range, the conclusion that we gain $5 \times 0.35 = \$1.75$ in revenue is correct. Each additional lemon generates $0.35 additional revenue as long as the total number of lemons stays within this stable range. The details of how to find this stable range are presented in Section 4.4. In this section, we merely wanted to define marginal values and show how to find them.

The lemonade stand problem is a standard form maximization problem, but we now want to generalize these ideas to any linear programming problem. The first thing to realize is that if we increase the value of the number on the right-hand side of a constraint by one unit, the exact same ideas lead to a change in the value of the objective function regardless of whether or not the constraint was created from a "resource." Thus, the concept of a marginal value applies to each constraint. Also, the reason our units for marginal values were dollars in this example because the objective function was revenue measured in dollars. But marginal values are not always a monetary concept. The units are whatever the units of the objective function are. For example, if we are trying to design an optimal diet that maximizes protein, then the marginal value of a constraint is the additional protein obtained when we increase the right-hand side of that constraint by one unit.

Thus, for a maximization problem with objective function P, if $P_{optimal}$ represents the optimal value and P_{new} represents the new optimal value obtained after raising the constraint by one unit, then the marginal value is the change in P: $\Delta P = P_{new} - P_{optimal}$.

What happens in a maximization problem when we increase the right-hand side of a greater-than-type constraint? In the lemonade stand example when we increased the value of the lemon constraint (or any of the less-than-type constraints), we move the line slightly farther away from the origin, resulting in some new feasible points. In other words, the feasible set of with the modified constraint contains the original feasible set as subset. It stands to reason that with a larger feasible set, it is possible that the objective function might increase, making ΔP positive. However, with a greater-than-type constraint, moving the line farther from the origin will typically reduce the size of the feasible set – that is, the revised feasible set X' is a subset of the original feasible set X. It follows that the optimal value of the objective function on X' is less than or equal to the maximum value on X. This shows that for a greater-than-type constraint in a maximization problem, $P_{new} \leq P_{optimal}$, so that the marginal value ΔP is negative.

Finally, for a minimization problem, the argument works in reverse. For a standard form minimization, all the constraints are of the greater-than type, and the feasible set X is the unbounded region above the lines in the first quadrant (think of an example such as O'Casek's used car problem). Increasing the value of the right-hand side of a constraint again results in a new feasible set X', which is a subset of X. It follows that the minimum value of the objective C function on X' is greater than or equal to the minimum value on the larger set X, so $C_{new} \geq C_{optimal}$; therefore, the marginal value $\Delta C = C_{new} - C_{optimal} \geq 0$. But if we consider a minimization problem with mixed constraints and increase the value of the right side of a less-than-type constraint, the revised feasible set X' contains X as a subset, and consequently, the minimum value of C on X' is less than or equal to the minimum value on X. It follows that the marginal value $\Delta C \leq 0$.

Thus, the general definition is as stated in Definition 4.2.

DEFINITION 4.2   *In any linear programming problem, each constraint has a marginal value associated with it, defined as follows. If the objective function is z, then the marginal value of a constraint is defined as $\Delta z = z_{new} - z_{optimal}$, where z_{new} is the optimal value of z on the revised feasible set obtained when we increase the value of the constant on the right-hand side of that constraint by one unit.*

Also, we have shown the following:

COROLLARY 4.3   *For any constraint of a standard form problem (maximization or minimization), the marginal value is nonnegative. If the problem has mixed constraints, the marginal value is nonpositive for any constraint that points in the "wrong" direction. That is, the marginal value is nonpositive for a greater-than-type constraint in a maximization problem or a less-than-type constraint in a minimization problem.*

We remark that it can be shown that if a constraint is an exact equality, then the marginal value is unrestricted (it could be positive, negative, or zero.)

We illustrate Corollary 4.3 with the following problem with mixed constraints:

$$\text{maximize: } P = 10x + 3y$$

subject to:

$$2x + 3y \geq 45$$
$$2x + y \leq 35$$
$$x \geq 0, y \geq 0.$$

The feasible set in the decision space has three corner points, as shown in Figure 4.2.

The reader should verify that the corner B is the optimal point with maximum value of $P = 10(15) + 3(5) = 165$.

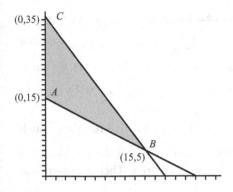

Figure 4.2 Feasible set.

To compute the marginal value M_1, we increase the right side by 1. To find the coordinates of the modified optimal point B, we must solve

$$\begin{bmatrix} 2 & 3 \\ 2 & 1 \end{bmatrix}\begin{bmatrix} x \\ y \end{bmatrix} = \begin{bmatrix} 46 \\ 35 \end{bmatrix}$$

$$\begin{bmatrix} x \\ y \end{bmatrix} = \frac{1}{-4}\begin{bmatrix} 1 & -3 \\ -2 & 2 \end{bmatrix}\begin{bmatrix} 46 \\ 35 \end{bmatrix} = \begin{bmatrix} \frac{59}{4} \\ \frac{11}{2} \end{bmatrix}.$$

The corresponding value of the objective function is $P_{new} = 10\left(\frac{59}{4}\right) + 3\left(\frac{11}{2}\right) = 164$. Thus, the marginal value is $M_1 = P_{new} - P_{old} = 164 - 165 = -1$.

To compute the marginal value M_2, we raise the right side by 1. To find the coordinates of the modified optimal point B, we must solve

$$\begin{bmatrix} 2 & 3 \\ 2 & 1 \end{bmatrix}\begin{bmatrix} x \\ y \end{bmatrix} = \begin{bmatrix} 45 \\ 36 \end{bmatrix}$$

$$\begin{bmatrix} x \\ y \end{bmatrix} = \frac{1}{-4}\begin{bmatrix} 1 & -3 \\ -2 & 2 \end{bmatrix}\begin{bmatrix} 45 \\ 36 \end{bmatrix} = \begin{bmatrix} \frac{63}{4} \\ \frac{18}{4} \end{bmatrix}.$$

The corresponding value of the objective function is $P_{new} = 10\left(\frac{63}{4}\right) + 3\left(\frac{18}{4}\right) = 171$. Thus, the marginal value is $M_2 = 171 - 165 = 6$.

Suppose we change the problem to a minimization problem: minimize $C = 5x + 12y$, with the exact same constraints as before. So the feasible set is identical to the pictured set in Figure 4.2. The reader should verify that the optimal point is again point $B = (15, 5)$, with minimum value of $C = 5\,(15) + 12\,(5) = 135$. To compute the marginal values, note that again we raise the first constraint to 46 to find M_1, and we raise the second constraint to 36 to find M_2, just as in the maximization problem. Thus, the coordinates of the modified optimal points are the same as before.

It follows that this time, for the first constraint, we have $C_{new} = 5\left(\frac{59}{4}\right) + 12\left(\frac{11}{2}\right) = \frac{559}{4}$, whence $M_1 = \frac{559}{4} - 135 = \frac{19}{4} = 4.75$. For the second constraint, we have $C_{new} = 5\left(\frac{63}{4}\right) + 12\left(\frac{18}{4}\right) = \frac{531}{4}$, whence $M_2 = \frac{531}{4} - 135 = -\frac{9}{4} = -2.25$.

We conclude this section by developing a more streamlined method to compute marginal values. Recall the computations in the lemonade stand problem to compute the marginal value of lemons.

The optimal point C was the intersection of the lemon and sugar line constraints. So the modified optimal point is obtained by solving $AX = B$, where $B = \begin{bmatrix} 61 \\ 30 \end{bmatrix}$ and A is the coefficient matrix of the system. Of course, the original coordinates of point C are obtained in the same way, using $B = \begin{bmatrix} 60 \\ 30 \end{bmatrix}$ instead. Thus, the formal solution is given by $X = A^{-1}B$ in each case. Once $X = \begin{bmatrix} x \\ y \end{bmatrix}$ is obtained, we substitute the values into the objective function, which is $R = 1.25x + 1.5y$. But this calculation is also a matrix multiplication (or dot product). Let $F = [1.25, 1.5]$ be the 1×2 matrix with the objective coefficients as entries. Then the revenue at the point $X = \begin{bmatrix} x \\ y \end{bmatrix}$ is simply given by $R = FX$. So the marginal value is given by computing the two revenues and subtracting: $M_1 = FX_{new} - FX_{old}$. Now, by using the properties of matrix multiplication from Chapter 2, we have

$$
\begin{aligned}
M_1 &= FX_{new} - FX_{old} \\
&= F(X_{new} - X_{old}) \\
&= F(A^{-1}B_{new} - A^{-1}B_{old}) \\
&= FA^{-1}(B_{new} - B_{old}) \\
&= FA^{-1}\left(\begin{bmatrix} 61 \\ 30 \end{bmatrix} - \begin{bmatrix} 60 \\ 30 \end{bmatrix} \right) \\
&= FA^{-1} \begin{bmatrix} 1 \\ 0 \end{bmatrix}.
\end{aligned}
$$

In other words, we have $M_1 = FA^{-1}U_1$, where U_1 is the unit vector with a 1 in the first coordinate and 0s in the other coordinates. This makes sense because M_1 involves changing the number of lemons available by 1 and not changing anything else. Of course, $M_2 = 0$ by complementary slackness. And similarly, $M_3 = FA^{-1}U_2$, where $U_2 = \begin{bmatrix} 0 \\ 1 \end{bmatrix}$ because we change the sugar supply by 1 and make no other changes. (Recall that the matrix A is 2×2, and the rows correspond to the first and third constraints – lemons and sugar.)

These formulas allow one to compute the marginal values as a simple matrix multiplication, without having to think about the details. But we can simplify even further. Exploiting the use of matrix algebra as a way of doing itemized calculations, let's compute both M_1 and M_3 at once, by multiplying FA^{-1} by a 2×2 matrix with U_1 in the first column and U_2 in the second column. In this way, M_1 will appear in the first column of our answer and M_3 will appear simultaneously in the second column. But this matrix has the form $[U_1, U_2] = \begin{bmatrix} 1 & 0 \\ 0 & 1 \end{bmatrix}$, which is the identity matrix! It follows that when we multiply FA^{-1} by this matrix, we just obtain FA^{-1} as the result.

The conclusion is that FA^{-1}, a 1×2 matrix, has M_1 in its first coordinate and M_3 in its second; that is, $FA^{-1} = [M_1, M_3]$. We illustrate this for the lemonade stand problem. We have

$$
A = \begin{bmatrix} 3 & 4 \\ 2 & 1 \end{bmatrix},
$$

and so

$$A^{-1} = \frac{1}{-5} \begin{bmatrix} 1 & -4 \\ -2 & 3 \end{bmatrix} = \begin{bmatrix} -\frac{1}{5} & \frac{4}{5} \\ \frac{2}{5} & -\frac{3}{5} \end{bmatrix}.$$

Thus,

$$FA^{-1} = [1.25, 1.5] \begin{bmatrix} -\frac{1}{5} & \frac{4}{5} \\ \frac{2}{5} & -\frac{3}{5} \end{bmatrix} = [\ 0.35 \quad 0.10\ ],$$

and we recognize the entries as the marginal values we computed earlier: $0.35 per lemon and $0.10 per tablespoon sugar.

Let's try this for a maximization problem with mixed constraints. For example, in the maximization problem with mixed constraints, we have

$$A = \begin{bmatrix} 2 & 3 \\ 2 & 1 \end{bmatrix}$$

$$A^{-1} = \frac{1}{-4} \begin{bmatrix} 1 & -3 \\ -2 & 2 \end{bmatrix} = \begin{bmatrix} -\frac{1}{4} & \frac{3}{4} \\ \frac{1}{2} & -\frac{1}{2} \end{bmatrix}.$$

Thus:

$$FA^{-1} = [\ 10 \quad 3\ ] \begin{bmatrix} -\frac{1}{4} & \frac{3}{4} \\ \frac{2}{4} & -\frac{2}{4} \end{bmatrix} = [\ -1 \quad 6\ ],$$

which agrees with the marginal values we computed earlier.

For the minimization problem above with mixed constraints, we obtain

$$FA^{-1} = [\ 5 \quad 12\ ] \begin{bmatrix} -\frac{1}{4} & \frac{3}{4} \\ \frac{2}{4} & -\frac{2}{4} \end{bmatrix} = [\ \frac{19}{4} \quad -\frac{9}{4}\ ],$$

which again agrees with the values we obtained earlier $[4.75, -2.25]$.

We have shown the following:

PROPOSITION 4.4 *Consider an $m \times 2$ linear programming problem, where the optimal point is the intersection of two lines given by the 2×2 system $AX = B$. Let F be the row matrix with entries given by the two objective coefficients. Then for the two constraints that are binding, the marginal values are given by the two entries in FA^{-1}, and all the other marginal values (for the non-binding constraints) have value 0 (by complementary slackness).*

We remark that this proposition can be generalized to all $m \times n$ problems. However, if there are more than two decision variables, one would not bother using the graphical approach to solve the problem, but would use the simplex algorithm instead. Then the proposition is not needed since the simplex algorithm automatically computes the marginal values, as we shall see.

Exercises

The following exercises appeared previously in the exercises to Section 3.4, where they were to be solved by graphing in the decision space.

1. Joni is putting her designing skills to work at the Court and Sparkle Jewelry Emporium. She makes two signature design bracelet models. Each Dawntreader bracelet uses 1 ruby, 6 pearls, and 10 opals. Each *Hejira* bracelet uses 3 rubies, 3 pearls, and 15 opals. She has 54 rubies, 120 pearls, and 300 opals to work with. If either model results in a profit of $1,800 for Joni, how many of each type should she make? .

    a. Explain why the marginal value of a ruby is 0.
    b. Find the marginal value of a pearl by increasing the pearl supply to 121, finding the coordinates of the modified optimal point, and computing the net increase in profit.
    c. Find the marginal value of an opal using the same approach as in Part c. Increase the opal supply to 301, find the coordinates of the modified optinal point, and compute the net increase in profit.
    d. Instead, find the marginal values of both pearls and opals simultaneously using Proposition 4.4.

2. Kerry's Kennel is mixing two commercial brands of dog food for its canine guests. A bag of Dog's Life Canine Cuisine contains 3 pounds of fat, 2 pounds of carbohydrates, 5 pounds of protein and 3 ounces of vitamin C. A giant-size bag of Way of Life Healthy Mix contains 1 pound of fat, 5 pounds of carbohydrates, 10 pounds of protein, and 7 ounces of vitamin C. The requirements for a week's supply of food for the kennel are that there should be at most 21 pounds of fat, at most 40 pounds of carbohydrates, and at least 21 ounces of vitamin C. How many bags of each type of food should be mixed in order to design a diet that maximizes protein?

    Compute the marginal value associated with each constraint. For each one, interpret exactly what it means using the correct units.

3. The Jefferson Plastic Fantastic Assembly Corporation manufactures gadgets and widgets for airplanes and starships. Each case of gadgets uses 2 kilograms of steel and 5 kilograms of plastic. Each case of widgets uses 2 kilograms of steel and 3 kilograms of plastic. The profit for a case of gadgets is $360, and the profit for a case of widgets is $200. Suppose Jefferson Plastic Fantastic Assembly Corporation has 80 kilograms of steel available and 150 kilograms of of plastic available on a daily basis and can sell everything it manufactures. How many cases of each should the company manufacture if it is obligated to produce at least 10 cases of widgets per day? The objective is to maximize daily profit.

    Compute the marginal value associated with each constraint.

4. Toys in the Attic, Inc., operates two workshops to build toys for needy children. Mr. Tyler's shop can produce 36 Angel dolls, 16 Kings and Queens board games, and 16 Back in the Saddle rocking horses each day it operates. Mr. Perry's shop can produce 10 Angel dolls, 10 Kings and Queens board games, and 20 Back in the Saddle rocking horses each day it operates. It costs $144 per day to operate Mr. Tyler's shop and $166 per day to operate Mr. Perry's shop. Suppose the company receives an order from the Kids Dream On charity foundation for at least 720 Angel dolls, at least 520 Kings and Queens board games, and at least 640 Back in the Saddle rocking horses. How many days should the company operate each shop in order to fill the order at least possible total cost?

    Compute the marginal value associated with each constraint. For each one, interpret exactly what it means using the correct units.

5. The Poseidon's Wake Petroleum Company operates two refineries. The Cadence Refinery can produce 40 units of low-grade oil, 10 units of medium-grade oil, and 10 units of

high-grade oil in a single day. (Each unit is 1,000 barrels). The Cascade Refinery can produce 10 units of low-grade oil, 10 units of medium-grade oil, and 30 units of high-grade oil in a single day. Poseidon's Wake receives an order from the Mars Triangle Oil retailers for at least 80 units of low-grade oil, at least 50 units of medium-grade oil, and at least 90 units of high-grade oil. If it costs Poseidon's Wake $1,800 to operate the Cadence Refinery for a day and $2,000 to operate the Cascade Refinery for a day, how many days should the company operate each refinery to fill the order at least cost?

Compute the marginal value associated with each constraint.

6. Mr. Cooder, a farmer in the Purple Valley, has at most 400 acres to devote to two crops: rye and barley. Each acre of rye yields $100 profit per week, while each acre of barley yields $80 profit per week. Due to local demand, Mr. Cooder must plant at least 100 acres of barley. The federal government provides a subsidy to grow these crops in the form of tax credits. The government credits Mr. Cooder four units for each acre of rye and two units for each acre of barley. The exact value of a "unit" of tax credit is not important. Mr. Cooder has decided that he needs at least 600 units of tax credits in order to be able to afford to his loan payments on a new harvester. How many acres of each crop should he plant in order to maximize his profit?

Compute the marginal value associated with each constraint.

7. Joanne's Antique Restoration Emporium refurbishes Victorian furniture. Each dining room set requires 16 hours stripping and sanding time, 4 hours refinishing time, and 4 hours in the upholstery shop. Each bedroom set requires 36 hours stripping and sanding time, 4 hours refinishing time, and 2 hours in the upholstery shop. Each month, Joanne's Antique Restoration Emporium has available 720 person-hours in the stripping and sanding shop, 100 person-hours in the refinishing shop, and 80 person-hours in the upholstery shop. Each dining room set generates $600 profit and each bedroom set generates $900 profit. How many of each type of furniture set should Joanne accept to work on each month in order to maximize her profit?

Compute the marginal value associated with each constraint.

8. As we saw in Section 4.2, when an objective coefficient is changed to a value within its stable range, the coordinates of the optimal point do not change because the feasible set is unchanged. However, because of the modified objective coefficient, the value of the objective function changes at the optimal point. Also, because of the modified objective coefficient, the marginal values also can change. Consider Angie and Kiki's lemonade stand problem. We know that the stable range for the cost of a glass of sweet lemonade is the interval $[1.125, 3]$.

    a. Suppose the price for a glass of sweet lemonade is raised from $1.25 to $1.40. Compute the new optimal revenue. Also, compute the new marginal values for lemons and sugar. Use the matrix multiplication FA^{-1} of the proposition at the end of the section.

    b. Answer the same question as Part a if the price for a glass of sweet lemonade is lowered from $1.25 to $1.15.

9. Consider the Court and Sparkle Jewelry Emporium problem (Problem 1). Suppose the profit for a Dawntreader bracelet increases to $2,400. Compute the new marginal values for the three resources.

4.4   Other Changes: Drawbacks of the Graphical Methods

We have observed that marginal values are a local phenomenon, meaning that they remain valid for a range of values as we change the number in the right-hand side of the relevant constraint. A large change in this number, going outside this range, will result in a more drastic change. To see what kind of change, consider again the lemonade stand problem with its optimal solution at point C in the graph of the feasible set in Figure 4.1.

In computing the marginal value of a lemon, we increased the lemons supply from 60 to 61, which had the effect on the graph of moving the line for the lemon constraint slightly farther away from the origin. The new position of this line is parallel to the one in Figure 4.1, because the slope does not change when we increase the lemon supply to 61. If the reader visualizes what happens to this line as we increase lemons even further, we have already observed that the coordinates of the optimal point C change slightly (as do the coordinates of the non-optimal point D). Let's call the (infeasible) point of intersection of the lime line and the sugar line F. The modified point for C moves slightly up the sugar constraint line toward F as we increase the supply of lemons, and the modified point D also moves to the right along the lime line toward F.

At some point, if we increase the supply of lemons enough, the modified lemon constraint line will pass through F (and points C and D converge to and become coincident with F.) At this level of lemon supply, since all three lines pass though the same point F, the problem is degenerate. But more importantly, think of what happens if we increase the lemon supply further. If we keep moving the lemon line past this point, it completely misses the feasible set – that is, it become superfluous. This is because we have run out of the other two resources, so additional lemons no longer help. Thus, the optimal point would remain at point F, and the optimal solution will have leftover lemons – the lemon constraint becomes non-binding. We already know by complementary slackness that at this point, the marginal value of a lemon suddenly becomes 0. Another way to say it is that the optimal point F has a different set of basic variables than does the point C because $S_1 > 0$ and now $S_2 = 0$. (Basic variables were defined in Section 3.5.)

Similar reasoning applies if we lower the supply of lemons. The lemon constraint line now moves slowly toward the origin, and the points C and D move away from F instead of toward it. In particular, the point C moves down the segment BF toward B on the sugar line. As soon as this line hits point B, the problem degenerates, and any further lowering of the supply of lemons would cause a switch of the basic variables in the optimal point and, at the same time, a change in the marginal value of a lemon. This leads us to Definition 4.5.

DEFINITION 4.5   *For each constraint, the* stable range *of the capacity number (the value of the right-hand side of the constraint) is the range of values for which the marginal value associated with that constraint is unchanged. Equivalently, it is the range of values for which the basic variables in the optimal solution are unchanged.*

Clearly, we can find the range as follows. We know the equation of any line parallel to the lemon constraint line has equation

$$3x + 4y = d.$$

As long as d is chosen so that the point C lies between B and F on the sugar line, then the basic variables of the optimal solution are unchanged. For this line to pass through B, we substitute in the coordinates of B into this line:

$$3(15) + 4(0) = d$$
$$45 = d.$$

Similarly, for this line to pass through F, we would substitute the coordinates of F into this line. Although we have not previously computed the coordinates of F, the reader should verify that the coordinates are $\left(\frac{32}{3}, \frac{26}{3}\right)$, obtained by solving the system of equations consisting of the lime line and the sugar line. Thus,

$$3\left(\frac{32}{3}\right) + 4\left(\frac{26}{3}\right) = d$$
$$\frac{200}{3} = d.$$

Thus, we have found the stable range for d:

$$45 \le d \le \frac{200}{3} = 66\frac{2}{3}.$$

The conclusion is that as long as the supply of lemons is between 45 and $\frac{200}{3}$, then the optimal point has the same basic variables as point C (namely, x, y, and S_2), and the marginal value of a lemon calculated in the last section, \$0.35, is valid throughout this range. If the supply of lemons exceeds $\frac{200}{3}$, then S_1 becomes a basic variable (and M_1 changes to 0). If the supply of lemons drops below 45, then again the basic variables change. In this case, y becomes nonbasic, and both S_2 and S_3 become basic because both limes and sugar are left over at the modified point B. Again, the marginal value of a lemon will change, but it won't become 0 in this case since there are no leftover lemons. We leave it to the reader to find what it changes to if $d < 45$.

Similar reasoning can be applied to find the stable range for sugar. This time it is the sugar line that moves, and point C moves along the segment of the lemon line connecting point D to the (unlabeled) point $(20, 0)$ on the x-axis. Any line parallel to the sugar line has equation

$$2x + y = d.$$

To pass through D, we must have

$$2(4) + 12 = 20 = d.$$

To pass through $(20, 0)$, we have

$$2(20) + 0 = 40 = d.$$

Thus, the stable range for sugar is the interval $[20, 40]$. If the supply of sugar is kept between 20 and 40 tablespoons, then the modified optimal point C will have the same basic variables x, y, and S_2, and the marginal value of a tablespoon of sugar will be \$0.10, as we computed in Section 3.5. Going outside this range will change both the basic variables of the optimal point and the marginal value of sugar.

Finally, what about the stable range for limes? Well, since S_2 is a basic variable, there are leftover limes at point C and $M_2 = 0$. As the girls already complained to their mother, increasing the limes won't help at all; we will simply have even more leftover limes. So there is no upper limit to the stable range for limes! What about a lower limit? Well, if we lower the supply of limes, there will be a point when the lime line passes through C, the problem degenerates, and any further lowering causes a change in basic variables. But we already know exactly how much to lower the supply of limes in order for that to happen. Since $S_2 = 4$ (there are four limes left over), we simply need lower the supply by four limes in order for the lime line to pass through the optimal point C! Thus, the stable range for the supply of limes is the interval $[24, \infty]$.

In summary, for an $m \times 2$ problem, we can, with some effort, find the stable range for the capacity of each constraint. But it may involve finding coordinates of infeasible basic solutions (such as the points F and $(20, 0)$ in the lemonade stand problem). Furthermore, if one actually does change the supply of any of the resources, even if it is in the stable range, one still has to compute the values of the revised optimal point. Both of these tasks require some effort and extra algebraic work. We'll present an example or two shortly. For problems with more than two decision variables, it seems daunting to try to extend what we have done here to higher dimensions. We'll see in a later chapter that the simplex algorithm is more efficient at making both of these types of calculations, so we will not attempt to apply what we did in this section to larger problems.

Before doing our example, it is worth comparing the notions of a stable range for the objective coefficients to that of a capacity coefficient (right-hand side of a constraint). In the former case, throughout the stable range the coordinates of the optimal point remain unchanged. In the latter case, the coordinates of the optimal point change, but the set of basic variables remains unchanged (and, as we have observed, the marginal values also remain unchanged in this range).

We introduced sensitivity analysis as an attempt to find modified optimal solutions in a problem when we make a small change in the original problem without having to solve the entire problem over. Let's see how this applies in the lemonade stand problem.

Example 4.6 Suppose the girls' mother is short on lemons and one day only gives them 48 lemons to work with instead of 60. Find the revised optimal data.

To solve this, note that 48 is within the stable range $[45, 66\frac{2}{3}]$. Therefore, we know that at the revised optimal point, we have the same basic variables x, y, S_2 and the same nonbasic variables S_1 and S_3. So we expect to have leftover limes, but no leftover lemons or sugar. But to find the revised values of the basic variables, we need to solve the revised system:

$$3x + 4y = 48$$
$$2x + y = 30$$

While it is not that difficult to do this by other methods, if you recall the discussion in Section 2.5 regarding the bongo drum example, changes in the right side are easiest to deal with if you use matrix inversion. Presumably, one would find the inverse of this coefficient matrix when we solve the original problem, in order to find the coordinates $(12, 6)$ of point C. So, either by going back to look up the inverse or recomputing it, we solve this system to obtain

$$\begin{bmatrix} x \\ y \end{bmatrix} = \frac{1}{-5} \begin{bmatrix} 1 & -4 \\ -2 & 3 \end{bmatrix} \begin{bmatrix} 48 \\ 30 \end{bmatrix} = \begin{bmatrix} \frac{72}{5} \\ \frac{6}{5} \end{bmatrix}.$$

What is the revised revenue R? We can evaluate the objective function to obtain

$$R = 1.25\left(\frac{72}{5}\right) + 1.5\left(\frac{6}{5}\right) = 19.8.$$

There is another way to find the revenue. Recall that the original optimal revenue was $24.00. The marginal value of a lemon is $0.35, which means it's the additional revenue if we had one more lemon. But that also means it's a loss of $0.35 if we had one fewer lemon. Since having 48 lemons means we have 12 fewer lemons, the revised optimal revenue must be

$$24 - 12(0.35) = 19.8,$$

a valid approach since 48 is in the stable range where the marginal value of 0.35 per lemon applies.

What about the value of S_2, the other basic variable? Just as in the original problem, we compute this by finding out how many limes we actually used (which is $x + 2y$) and subtract this form the available 28:

$$S_2 = 28 - \left(\frac{72}{5} + 2\left(\frac{6}{5}\right)\right) = \frac{56}{5} = 11.2.$$

In summary, here is the revised optimal data (remember, we allow fractional answers for the moment):

$$x = \frac{72}{5} = 14.4 \text{ glasses sweet lemonade}$$

$$y = \frac{6}{5} = 1.2 \text{ glasses tart lemonade}$$

$$R = \$19.80 \text{ (maximized)}$$

$$S_1 = 0 \text{ leftover lemons}$$

$$S_2 = 11.2 \text{ leftover limes}$$

$$S_3 = 0 \text{ leftover tablespoons sugar.}$$

Since the marginal values are unchanged in the stable range, we also know $M_1 = .35$, $M_2 = 0$, and $M_3 = 0.1$.

As a second example, suppose we keep the lemons supply fixed at 60 but increase the sugar supply from 30 to 35 tablespoons. Now what is the optimal data? Since 35 is in the stable range of $[20, 40]$ for sugar, and the marginal value is 0.1 per tablespoon, we expect $\$24.50$ at the revised optimal point.

Indeed, we have

$$\begin{bmatrix} x \\ y \end{bmatrix} = \frac{1}{-5} \begin{bmatrix} 1 & -4 \\ -2 & 3 \end{bmatrix} \begin{bmatrix} 60 \\ 35 \end{bmatrix} = \begin{bmatrix} x \\ y \end{bmatrix} = \begin{bmatrix} 16 \\ 3 \end{bmatrix},$$

and the number of limes used is $x + 2y = 16 + 2(3) = 22$, so the there are $28 - 22 = 6$ leftover limes. Thus,

$$x = 16 \text{ glasses sweet lemonade}$$

$$y = 3 \text{ glasses tart lemonade}$$

$$R = 1.25(16) + 1.5(3) = \$24.50 \text{ (maximized)}$$

$$S_1 = 0 \text{ leftover lemons}$$

$$S_2 = 6 \text{ leftover limes}$$

$$S_3 = 0 \text{ leftover tablespoons sugar.}$$

This seems to be somewhat less work than solving the entire problem over, but not by much. Furthermore, what if we changed the capacity to a value outside of the stable range? It may be possible to continue our analysis of the moving lines in the feasible set to figure out what the basic variables are in the revised solution, but that would be even more tedious than what we have done so far. It seems unlikely to be able to save much work and one may as well just solve the whole problem over. But again, when we learn the simplex algorithm, we shall see that even in the case we stray outside of the stable range, we can deal with the revised problem without having to start over and solve the whole problem again.

This concludes our discussion of marginal values and changes to the capacity coefficients (right side of the constraints). What about the structural changes we mentioned in Section 4.1? That is, what happens if we change one coefficient on the left side of one of the constraints. This means that the slope of the constraint line has changed, so it becomes more difficult to do a graphical analysis than when we changed the right-hand side. We'll do one example to illustrate the kinds of issues that come up, but after that, we will abandon any attempt at a systematic solution to this question using graphical methods.

For example, in the lemonade stand problem, suppose the girls are considering changing the recipe of sweet lemonade from using three lemons per glass to using 2.5 lemons per glass. Thus, the equation of the lemon line becomes

$$2.5x + 4y = 60.$$

What changes in the graph of the feasible set? The x intercept of the lemon line becomes $\frac{60}{2.5} = 24$ instead of 20. Since the y intercept remains unchanged, it is easy to plot the modified line. Clearly, this means that the coordinates of points D and C have changed. Notice also that the slope of this line increased from $-\frac{3}{4}$ to $-\frac{5}{8}$. In Section 4.2, we observed that the relationship between the slopes of the constraint lines and the slope of the level objective lines determines which corner is optimal. Recall that for this problem, we have $m_{lemons} = -\frac{3}{4}$, $m_{limes} = -\frac{1}{2}$, $m_{sugar} = -2$, and $m_{revenue} = -\frac{5}{6}$. In order, we have

$$m_{sugar} < m_{revenue} < m_{lemons} < m_{limes}.$$

The fact that the slope of the level objective lines $m_{revenue}$ is between the slope of the sugar constraint line and the slope of the lemon constraint line is why the optimal point is the point C, which is the intersection of those two constraint lines. In Section 4.2, we saw what happens when we vary $m_{revenue}$. However, here, it is m_{lemons} that has changed. Since the slope $m_{revenue}$ is still between the slopes of the sugar and revised lemon constraint lines, we expect that the revised point C is still optimal.

So we must compute the revised coordinates for point C, using the revised lemon equation in the system:

$$2.5x + 4y = 60$$
$$2x + y = 30.$$

The matrix formulation of the system is

$$\begin{bmatrix} 2.5 & 4 \\ 2 & 1 \end{bmatrix} \begin{bmatrix} x \\ y \end{bmatrix} = \begin{bmatrix} 60 \\ 30 \end{bmatrix}$$

and has solution

$$\begin{bmatrix} x \\ y \end{bmatrix} = \frac{2}{-11} \begin{bmatrix} 1 & -4 \\ -2 & \frac{5}{2} \end{bmatrix} \begin{bmatrix} 60 \\ 30 \end{bmatrix} = \begin{bmatrix} \frac{120}{11} \\ \frac{90}{11} \end{bmatrix}.$$

This solution uses $\frac{300}{11} = 27\frac{3}{11}$ limes, so there is $\frac{8}{11}$ of a lime left over. The revised revenue is $R = 1.25\left(\frac{120}{11}\right) + 1.5\left(\frac{90}{11}\right) = 25.909 \approx \$25.91$. So the revised optimal data is

$$x = \frac{120}{11} \approx 10.91 \text{ glasses sweet lemonade}$$

$$y = \frac{90}{11} \approx 8.18 \text{ glasses tart lemonade}$$

$$R \approx \$25.91 \text{ (maximized)}$$

$$S_1 = 0 \text{ leftover lemons}$$

$$S_2 = \frac{8}{11} \text{ leftover lime}$$

$$S_3 = 0 \text{ leftover tablespoons sugar.}$$

Assessing what we have done, in this example, the point C was still optimal because the revised coordinates allowed for $S_2 \geq 0$. A larger change in this coefficient might move the point C up past the point F (see Section 4.2) along the sugar line, requiring a change in basic variables to the optimal point. It is not hard to see to what the coefficient must change. The slope of the revised sugar line $ax + 4y = 60$ is $m = -\frac{a}{4}$, and for the problem to degenerate, we require this line pass through F. Since we already know this line also passes through $(0, 15)$, we can find the slope of the segment connecting $(0, 15)$ to F, set it equal to $-\frac{a}{4}$ and solve for a. In this way, we can find a stable range for this coefficient. We leave the details to the interested reader.

On the down side, observe that to find the revised coordinates of C, we had to solve a system with a different coefficient matrix than the original because we changed a structural coefficient. In this problem, that was not hard to do because it is a 2×2 system. However, in a problem with more variables, this is a formidable task. It is almost as much work as solving the entire problem again.

Finally, what about the other type of structural change mentioned in Section 4.1 – adding a new decision variable? Well, that changes an $m \times 2$ problem into an $m \times 3$ problem, and with the added dimension, we would not solve the modified problem graphically anyway. So we will not attempt to study these types of changes here. Overall, it seems that structural changes are the most work-intensive changes to deal with in a linear programming problem. We suspect that is why these types of changes are not generally considered in many texts.

In summary, we have considered all the types of changes outlined in Section 4.1 for $m \times 2$ problems, plus we have defined the important notion of marginal value associated with a constraint. Of these types of changes, only the changes in the objective coefficients left the feasible set unchanged. This means it was easy to study these changes using a graphical analysis. All of the other types of changes, including the computation of the marginal values, required a change of coordinates in the optimal point. For the marginal values computations, as well as changes in the capacity, this was not too hard because there were only two variables. However, it would be difficult to mimic these techniques for problems with more variables. Even for changes in the structural constraints, the $m \times 2$ problems could be handled, but for larger problems, dealing with such changes would be almost as much work as just solving the problem over, as noted before.

In summary, there are a number of drawbacks to using graphical methods. They include the following:

• Extra work is required to compute the values of the slack variables S_i when graphing in the decision space, and extra work is required to compute the values of the decision variables when graphing in the constraint space.

• Extra work is required to compute the marginal values M_i (much more work than for the slack variables!).

• Finding stable ranges for capacity changes is a bit tedious (but not as tedious as for structural changes).

• If there are more than two decision variables and more than two constraints (i.e., if the problem is 3×3 or larger), the graphical methods do not apply.

Are there any advantages to the graphical methods, other than being helpful to define important concepts? Well, there are a couple. Finding stable ranges in the objective coefficients was easy for $m \times 2$ problems. Also, graphical methods do not depend on whether the problem is maximization or minimization, or whether the problem is standard form or has mixed constraints. But that's about it. For almost anything else, we'll see that the simplex algorithm, which we begin to study in the next chapter, is superior to the graphical methods. We'll see that when the simplex algorithm is applied, it automatically computes the values of all the variables, including the slack variables and the marginal values, with no extra work. Plus it handles $m \times n$ problems for all m, n. Finally, we'll see in Chapter 9 that it is fairly easy to find stable ranges for capacity changes, as well as modified optimal data for changes in the capacity, with very little additional work.

Exercises

The following exercises appeared previously in the exercises of Section 3.4, where they were to be solved by graphing in the decision space.

1. Joni is putting her designing skills to work at the Court and Sparkle Jewelry Emporium. She makes two signature design bracelet models. Each Dawntreader bracelet uses 1 ruby, 6 pearls, and 10 opals. Each Hejira bracelet uses 3 rubies, 3 pearls, and 15 opals. She has 54 rubies, 120 pearls, and 300 opals to work with. If either model results in a profit of $1,800 for Joni, how many of each type should she make? .

   a. Compute the stable range of each resource.
   b. Find the revised optimal data if the number of pearls is increased to 150.
   c. Find the revised optimal data if the number of opals in increased to 330.

2. Toys in the Attic, Inc., operates two workshops to build toys for needy children. Mr. Tyler's shop can produce 36 Angel dolls, 16 Kings and Queens board games, and 16 Back in the Saddle rocking horses each day it operates. Mr. Perry's shop can produce 10 Angel dolls, 10 Kings and Queens board games, and 20 Back in the Saddle rocking horses each day it operates. It costs $144 per day to operate Mr. Tyler's shop and $166 per day to operate Mr. Perry's shop. Suppose the company receives an order from the Kids Dream On charity foundation for at least 720 Angel dolls, at least 520 Kings and Queens board games, and at least 640 Back in the Saddle rocking horses. How many days should the company operate each shop in order to fill the order at least possible total cost?

a. Find the stable range for each of the constraints.

b. Suppose the request for at least 640 rocking horses was lowered to at least 608 rocking horses. Find the revised optimal data.

3. Mr. Cooder, a farmer in the Purple Valley, has at most 400 acres to devote to two crops: rye and barley. Each acre of rye yields $100 profit per week, while each acre of barley yields $80 profit per week. Due to local demand, Mr. Cooder must plant at least 100 acres of barley. The federal government provides a subsidy to grow these crops in the form of tax credits. The government credits Mr. Cooder two units for each acre of rye and two units for each acre of barley. The exact value of a 'unit' of tax credit is not important. Mr. Cooder has decided that he needs at least 600 units of tax credits in order to be able to afford to his loan payments on a new harvester. How many acres of each crop should he plant in order to maximize his profit?

a. Find the stable range for land in acres.

b. Suppose that Mr. Cooder can buy 50 more acres. Find the revised optimal data.

c. Suppose that one year he decides to let half of his farm to lie fallow (that is, he plants no crops in half of his land) so that the soil can regain nutrients. What is the revised optimal data in this case?

4. In the discussion of the lemonade stand problem in this section, we suggested that a stable range can be found for the coefficient, which is the number of lemons used in a glass of sweet lemonade. Find this stable range.

5. In the lemonade stand problem, suppose the girls decrease the amount of sugar used in a glass of sweet lemonade from 2 tablespoons to $\frac{5}{3}$ tablespoons. Find the revised optimal data.

4.5 Duality

In Chapter 1, we considered the following optimization problem: With 80 feet of fencing forming the perimeter, what dimensions $x \times y$ form the rectangular garden of maximum area? We saw that the solution is when $x = y = 20$ feet, so the solution is a square. This is not a linear programming problem, of course, since the area is given by the nonlinear function $A = xy$.

Some reflection might convince the reader that this problem can be phrased as a minimization problem instead: if a rectangular garden has 400 square feet area, what dimensions yield the least amount of fencing (i.e., smallest perimeter)? The solution is again $x = y = 20$ feet, so the solution is a square. Even though the two problems are expressed differently, they both express the idea that a square is in some sense the most efficient rectangle. Among all rectangles of a fixed perimeter, the square has the maximum area; or, equivalently, among all rectangles of a fixed area, the square has the minimum perimeter.

This example suggests that there may be two ways to express a single optimization problem, as a maximization or as a minimization, depending on what part of the problem you view as the constraint(s) and what part you view as the objective function. When the optimization problem is a linear programming problem, we can make this idea very precise. Given a linear programming problem (which we refer to as the *primal* problem), there is another linear programming problem that is closely related to the primal problem, called the *dual* problem. When the primal is a maximization problem, the dual is a minimization, and when the primal is a minimization problem, the dual is a maximization. The problems are so closely related that, in a sense, they can be regarded as two ways to view the same problem, and have the "same" solution, just as in the previous example the solution to both problems about rectangles is a square.

The concept of a dual problem has important theoretical uses, and it also provides a tool to save work when doing certain computations. We are not concerned in this book with the theoretical aspects, leaving that for more advanced texts in programming and optimization. We do intend to exploit the computational aspects of duality. In Section 4.5.1, we present a technique for finding the dual problem, without worrying about what the decision variables of the dual problem mean. This is all one really needs to know about the dual in order to exploit the computational shortcuts mentioned before. The technique for finding the dual is mechanical and very easy, but it gives no hint as to why the two problems are connected conceptually.

The more interesting aspect of the theory is explaining how the two problems relate to one another. This is done in Section 4.5.2, where we show how to give an interpretation of the decision variables in the dual problem. To make an analogy, in Section 4.5.1, you learn how to drive a car, which you can easily do without knowing anything about combustion engines, while in Section 4.5.2, you learn how the car works. In subsequent sections of this book, only knowing how to drive is required.

4.5.1   The Construction of the Dual

We illustrate the technique of finding the dual problem using a maximization problem in standard form as the primal problem. Then the dual is always a minimization problem also in standard form. In Section 3.5, we discussed the following problem (a modified version of Example 3.12):

A farmer has 150 acres available to split between three crops: asparagus, beans, and corn. Each bag of asparagus seed is sufficient to plant 1 acre and requires 8 pounds of fertilizer. Each bag of bean seeds is sufficient to seed 2 acres and requires 10 pounds of fertilizer. Each bag of corn seed will plant 1 acre and requires 20 pounds of fertilizer. She has a total of 1,000 lbs. fertilizer to use on her two crops. The profit from a bag of asparagus seeds is $1,500, and the profit from a bag of bean seeds is $2,000. How many bags of each should she plant in order to maximize profit?

> Let x be the number of bags of asparagus seed.
>
> Let y be the number of bags of bean seed.
>
> Let z be the number of bags of corn seed.
>
> Maximize: $P = 1,500x + 2,000y + 3,700z$
>
> subject to:
>
> $x + 2y + z \leq 150$ (acres land)
>
> $8x + 10y + 20z \leq 1,000$ (lbs. fertilizer)
>
> $x \geq 0,\ y \geq 0,\ z \geq 0.$

Observe that this is a 2×3 problem in standard form. We begin by arranging all the information in this problem in a table or matrix. It is a modified version of the augmented coefficient matrix of the system of equations (if we replaced the inequalities by equations.) See the following table.

	x	y	z	Capacity
S_1	1	2	1	150
S_2	8	10	20	1,000
P	1,500	2,000	3,700	0

The middle two rows are the augmented coefficient matrix of the system. We have added a row on the bottom representing the objective function, and filled in the last entry of this row with a 0 (the significance of the 0 will be explained in the next chapter). We also have added a row on top with the label of the appropriate variable for each column. Finally, we have added a column on the left with labels for the rows. We have labeled the row with the slack variable corresponding to that constraint. (S_1 is leftover acres of land, and S_2 is leftover pounds of fertilizer.) The reason for labeling the rows with these variables is explained in the next chapter. Finally, the last row is labeled with profit P, the objective function.

This modified augmented matrix contains all the important information of the problem and has a specific name – it is called the *condensed tableau* for the problem. The word "tableau" is French for matrix or table, and what it is "condensed" from is explained in the next chapter. Step 1 in finding the dual problem of a given primal problem is writing down the condensed tableau. Step 2 is simple: take the transpose of the condensed tableau – except we must change the names of all the variables in the labels. The idea is the transpose should be the condensed tableau of the dual problem in standard form. Since x, y, z, S_1, S_2, and P already have a meaning, we should use different letters for the variables of the dual problem. In our example, the transpose is as shown in the following table.

	w_1	w_2	
T_1	1	8	1,500
T_2	2	10	2,000
T_3	1	20	3,700
C	150	1,000	0

What letters you use for the new variables are immaterial since in this section we do not even bother trying to interpret the variables' meaning. All you need to know is that w_1 and w_2 are the decision variables for the dual problem, and T_1, T_2, and T_3 are the slack variables for the dual, or, rather, the surplus variables, since the dual is a minimization problem in standard form. Finally, C is the objective function (which is to be minimized). Notice that the dual problem is a 3×2 problem because taking transpose reverses the number of rows and columns.

Step 3 is to simply read off the problem from the dual condensed tableau. So the dual problem is as follows:

$$\text{minimize } C = 150w_1 + 1,000w_2$$

subject to:

$$w_1 + 8w_2 \geq 1,500$$
$$2w_1 + 10w_2 \geq 2,000$$
$$w_1 + 20w_2 \geq 3,700$$
$$w_1 \geq 0, \ w_2 \geq 0.$$

Notice that the constraints are all greater-than-type constraints because the dual problem is a standard form minimization. Again, in this section, we do not care about the meaning of the w_i or the meaning of C. We address that in the next section.

Notice also that if this minimization problem had been what we started with as the primal problem, then the dual of this problem is the original maximization, because for any matrix we know that $(A^T)^T = A$. Thus, the two problems are considered to be duals to each other. So the process of finding the dual problem for any standard form problem is mechanical and easy:

Step 1. Construct the condensed tableau of the primal problem.

Step 2. Take the transpose of the tableau (being careful to change the names of the variables) to obtain the condensed tableau of the dual problem.

Step 3. Read off the problem from its condensed tableau found in step 2.

Let's illustrate the process starting with a minimization problem. Consider the following problem from the preceding exercise sets:

Toys in the Attic, Inc., operates two workshops to build toys for needy children. Mr. Tyler's shop can produce 36 Angel dolls, 16 Kings and Queens board games, and 16 Back in the Saddle rocking horses each day it operates. Mr. Perry's shop can produce 10 Angel dolls, 10 Kings and Queens board games, and 20 Back in the Saddle rocking horses each day it operates. It costs \$144 per day to operate Mr. Tyler's shop and \$166 per day to operate Mr. Perry's shop. Suppose the company receives an order from the Kids Dream On charity foundation for at least 720 Angel dolls, at least 520 Kings and Queens board games, and at least 640 Back in the Saddle rocking horses. How many days should the company operate each shop in order to fill the order at least possible total cost?

By now, the reader has surely set up this problem, and it is indeed a standard form minimization. If x is the number of days to operate Mr. Tyler's shop, and y is the number of days to operate Mr. Perry's shop, then we must

$$\text{minimize } C = 144x + 166y$$
$$\text{subject to:}$$
$$36x + 10y \geq 720 \text{ (Angel dolls)}$$
$$16x + 10y \geq 520 \text{ (Kings and Queens board games)}$$
$$16x + 20y \geq 640 \text{ (Back in the Saddle rocking horses)}$$
$$x \geq 0, \ y \geq 0.$$

Step 1 is to write down the condensed tableau.

	x	y	Requirements
S_1	36	10	720
S_2	16	10	520
S_3	16	20	640
C	144	166	0

Step 2 is to take the transpose and relabel variables.

	u	v	w	Capacity
T_1	36	16	16	144
T_2	10	10	20	166
P	720	520	640	0

Step 3 is to read off the dual problem as a standard form maximization:

$$\text{maximize: } P = 720u + 520v + 640$$

$$\text{subject to:}$$

$$36u + 16v + 16w \leq 144$$

$$10u + 10v + 20w \leq 166$$

$$u \geq 0, \ v \geq 0, \ w \geq 0.$$

This is the dual problem we seek.

Earlier we hinted that the solution to both problems was "the same." Since the variables are different, what does this actually mean? In dual problems, the following property holds, which is called the strong duality principle.

Strong Duality Principle. In a pair of dual linear programming problems, the maximum value of P equals the minimum value of C; that is, the optimal values of the two objective functions agree.

We remark that there is also something known as *weak duality* principle, which deals with comparing the values of the objective functions of the primal and the dual problem when we plug in a feasible solution (not necessarily optimal) to each. It says in this case that the value of P is less than or equal to the value of C. We will not need this more general principle in this text, but the interested reader can learn more about it in chapter 6 of Calvert and Voxman (1989).

Finally, we remark that in case the problem has mixed constraints or a constraint that is an equation instead of an inequality, there are additional modifications in the formulation of the dual problem. We may explore some of these in the exercises and in later chapters. The complete general case is outlined in chapter 6 of Calvert and Voxman (1989).

4.5.2   Economic Interpretation of the Dual

We now turn to the issue of how to interpret what the variables in the dual problem mean. We do so, as usual, by working through examples and then generalizing. We begin with Angie and Kiki's lemonade stand problem. Recall the setup:

Let x denote the number of glasses of sweet lemonade.

Let y denote the number of glasses of tart lemonade.

We must

$$\text{maximize: } R = 1.25x + 1.5y$$

$$\text{subject to:}$$

$$3x + 4y \leq 60 \ \text{(lemons)}$$

$$x + 2y \leq 28 \text{ (limes)}$$
$$2x + y \leq 30 \text{ (tb.sp. sugar)}$$
$$x \geq 0, \ y \geq 0$$

Now consider the following hypothetical scenario: On a particular day, Angie and Kiki cannot be present at the lemonade stand because they are playing in a soccer tournament. But they do not want to miss out on the $24 they would make for the day, and they already have all the ingredients and supplies for a day at the stand. Fortunately, their cousins Lisa and Nicole are visiting. Lisa and Nicole offer to buy the resources from Angie and Kiki and offer to run the stand for that day. The question is, what prices should they offer for the resources that would be fair prices? And which would entice Angie and Kiki to sell them?

Let w_1 be the price they offer to buy a lemon, let w_2 be the price they offer to buy a lime, and let w_3 be the price they offer to buy a tablespoon of sugar. Now Lisa and Nicole would be paying a total of

$$C = 60w_1 + 28w_2 + 30w_3$$

to buy all the resources, and they would prefer that this cost C is as small as possible. So Lisa and Nicole's goal is to minimize C. But under what constraints? Well, we stipulated that Angie and Kiki should be inclined to agree to the offer. Let's consider a glass of sweet lemonade. The resources used in making that glass are three lemons, one lime, and 2 tablespoons sugar. That means, at the given prices, they would be paying Angie and Kiki a total of

$$3w_1 + w_2 + 2w_3.$$

On the other hand, if Angie and Kiki stayed and sold the lemonade themselves, they would be getting $1.25 for that glass of lemonade; that is, for that mix of resources. So if the total amount Lisa and Nicole paid was less than $1.25, the offer wouldn't be fair in the eyes of Angie and Kiki, so they would not agree. Thus, to be fair, the prices must be chosen so that

$$3w_1 + w_2 + 2w_3 \geq 1.25.$$

A similar analysis with the tart lemonade says the girls would feel cheated unless

$$4w_1 + 2w_2 + w_3 \geq 1.50.$$

As long as both of these requirements are met, then Angie and Kiki would feel like it was a fair offer and would be inclined to agree. Finally, since the w_i are prices, it is clear that $w_i \geq 0$ for all i. In short, Lisa and Nicole must solve the following problem:

minimize $C = 60w_1 + 28w_2 + 30w_3$

subject to:

$3w_1 + w_2 + 2w_3 \geq 1.25$

$4w_1 + 2w_2 + w_3 \geq 1.50$

$w_1 \geq 0, \ w_2 \geq 0, \ w_3 \geq 0.$

The reader should recognize at once that Lisa and Nicole's problem is exactly the dual of Angie and Kiki's problem! There are a few observations to make. First of all, and most importantly, look carefully at the definition of w_1. It is the amount of money (revenue) that Angie and Kiki will make per lemon. Similarly, w_2 is the revenue from a lime, and w_3 is the revenue for a tablespoon of sugar. But we have seen these quantities before – they are precisely the marginal values of the resources! This tells us how to interpret the variables of the dual problem:

CONCLUSION 4.7   *The decision variables in the dual problem are the same thing as the marginal values of the constraints in the primal problem. Similarly, the decision variables in the primal are the same as the marginal values in the dual problem.*

The next observation to make is that strong duality indeed holds. We already know the marginal values in the lemonade stand problem. They are $M_1 = w_1 = \$0.35$ per lemon, $M_2 = w_2 = \$0$ per lime, and $M_3 = w_3 = \$0.10$ per tablespoon sugar. If the girls agree to this offer at these prices, then at the optimal solution $(x, y) = (12, 6)$, we know the maximum possible revenue for Angie and Kiki is $R = \$24.00$. But this is also the minimum value of the total cost to Lisa and Nicole:

$$C = 60w_1 + 28w_2 + 30w_3$$
$$= 60(0.35) + 28(0) + 30(0.1) = \$21.00 + \$0 + \$3.00 = \$24.00.$$

Thus, Lisa and Nicole end up paying them $24, the amount they would have made if they sold the lemonade on their own anyway. You might wonder why they don't just offer to give Angie and Kiki "$24.00 for the whole package" rather than specifying prices for each resource. In one sense, that seems like what they are doing, and if the offer was anything less than $24, the girls wouldn't accept it, thus suggesting why strong duality should hold. But it is a bit more subtle than that. What Lisa and Nicole are buying are the resources – not the finished product (lemonade), so this is what the prices should reflect.

Furthermore, what if, hypothetically, the prices w_i were chosen so that one of the constraints was satisfied – say, the first one – but second was not, *even though the total came out to be $24*. This is not a fair offer. The girls might agree to let them make the sweet lemonade since they are getting $1.25 (at least) per glass, but they would not allow them to make the tart lemonade since they could make more revenue per glass by just making it and selling it themselves. That is, if Angie and Kiki were smart, they would agree to outsource the manufacture of sweet lemonade to Lisa and Nicole, but not to outsource to the manufacture of tart lemonade (at least not at the offered prices). In this way, they stand to make more than the $24 they would make on their own. For example, if the prices were chosen so that Lisa and Nicole would pay them $1.40 for the ingredients needed for each glass of sweet lemonade, that means they make $12(1.40) + 6(1.50) = \$25.80$ by outsourcing just one of the products they make, an increase of $1.80 over what they would earn making everything themselves.

Thus, we see that for the prices to fair and to entice the girls to sell, each constraint must be met individually. Incidentally, these fair prices for the resources have another name in some texts – they are called *shadow prices* as well as marginal values.

The next observation to make is that it is fair for Nicole and Lisa to ask for all the limes for free! Indeed, we have $w_2 = M_2 = 0$, since at the optimal point, we have leftover limes. If that seems strange, recall that lemonade also involves other ingredients - cups, ice, straws – which we assume are unlimited. Clearly, Lisa and Nicole will need all those ingredients as well, but they do not offer to pay for them as they are not constraining the production of lemonade, being in unlimited supply. Given that, with a surplus, limes are also not constraining the production of lemonade, it makes sense not to offer a positive price for them either. In essence, Nicole and Lisa offer to pay a positive price only for the ingredients – lemons and sugar – which are limiting the production of lemonade, as they are binding constraints.

Finally, we observe that the amount Lisa and Nicole pay Angie and Kiki – $24.00 – is exactly what they make at the end of the day once they sell all the lemonade. So they make all their money back, according to the strong duality principle. However, Lisa and Nicole don't make a profit. So why would they even make such an offer to Angie and Kiki? Well, they could be just doing this as a favor to help their cousins and support their budding careers as soccer players. However, another reason to make such an offer is to suppose they think that they can raise the prices of the lemonade and still sell all the lemonade they produce. For example, if they raise the cost of a glass of sweet to $1.50 and the cost of a glass of tart to $1.80, then at the end of the day, their revenue is

$$12\,(1.5) + 6\,(1.8) = \$28.80.$$

The offer to Angie and Kiki was to pay them $24.00 for the resources. If Lisa and Nicole can actually sell all the lemonade at the higher prices, then they can pocket the $4.80 difference as their own profit.

We now outline an example with a minimization problem to make sure our interpretations are still valid. Consider the Toys in the Attic, Inc., problem from the last subsection. The scenario is similar to the previous example – Mr. Tyler and Mr. Perry have to be away for several weeks (perhaps they are on tour with a rock and roll band), so they cannot run their shops. Zander's Tricky Outsourcing Company offers to run the shops to fill the order received (they believe they can run the shops more cheaply and efficiently than Toys in the Attic can). They offer to sell the constructed finished products back to Toys in the Attic. The question is, what are fair prices? Let w_1 be the price they set for an Angel doll, let w_2 be the price they set for a Kings and Queens board game, and let w_3 be the price they set for a Back in the Saddle rocking horse. Then the total amount that Zander makes by selling the products to Toys in the Attic is

$$R = 720w_1 + 520w_2 + 640w_3.$$

Obviously, Zander would like this to be maximized. But the constraints are determined by how much it costs Toys in the Attic to run each shop for a day. For example, consider Mr. Tyler's shop. In a day of operation, they produce 36 dolls, 16 board games, and 16 rocking horses. Thus, at the prices set by Zander, Toys in the Attic will end up paying Zander

$$36w_1 + 16w_2 + 16w_3.$$

Unless this is at most $144, Mr. Tyler would prefer to just run the shop himself or hire someone else. He would not accept the offer. Similarly, Mr. Perry would only be likely to agree if the total cost for running his shop for a day is at most $166. (In fact, in this example, by having different people in charge of the two shops, it illustrates in a very compelling way why Zander must satisfy each constraint individually.) In short, Zander must solve this problem in order to set his prices:

maximize $R = 720w_1 + 520w_2 + 640w_3$

subject to:

$36w_1 + 16w_2 + 16w_3 \le 144$

$10w_1 + 10w_2 + 20w_3 \le 166$

$w_1 \ge 0,\ w_2 \ge 0,\ w_3 \ge 0.$

We recognize this as the dual problem of the Toys in the Attic problem. Again, the variables w_i give the value of the objective function which are generated by one unit of the ith constraint of the primal problem, so again $w_i = M_i$, the ith marginal value in the primal problem. In this case, the ith constraint corresponds to a finished product, not a resource, but that is because we started with the primal problem being the minimization. Recall that the way we interpret M_i in this case is how much Toys in the Attic would save if the ith constraint were lowered by one unit. For example, if the order was for only 719 Angel dolls instead of 720, how much savings they would realize? Whatever that value is, it should be the price w_1 that Zander sets for an Angel doll, and similarly for the other prices. This verifies Conclusion 4.7: in all cases, the decision variables of the dual problem are the marginal values associated with the constraints in the primal problem.

4.5.3 Complementary Slackness

We already have observed that if $S_i > 0$, then $M_i = 0$ (leftover resources have no marginal value), a principle to which we referred as *complementary slackness*. Also, the reverse holds: if $M_i > 0$, then $S_i = 0$ (if a resource has a positive marginal value, then that constraint is binding – all of the resource is used up).

In particular, in either of these two cases, we have $S_i \cdot M_i = 0$, since one or the other of the variables is 0. Are these two cases exhaustive? According to our conventions in this text, none of the slack variables are negative at the optimal point. So we need not worry about the case $S_i < 0$. Of course, it is possible that $M_i < 0$. But again, since $M_i \neq 0$, that constraint must be binding so that $S_i = 0$. But there is one other case – can both $S_i = 0 = M_i$ simultaneously?

If $S_i = 0$, the constraint is binding, but if $M_i = 0$, having additional units would not increase the value of the objective function. This can happen, although it is rare. For example, in the lemonade stand problem, suppose that the supply of limes is 24 instead of 28. Then the reader can verify that all three constraint lines pass through the point $C = (12, 6)$, which is the optimal point. Recall that when this happens, we say the problem is degenerate. At this point, all the limes are used up, so $S_2 = 0$. However, as it is easily verified, if we increase the amount of limes, then any additional limes become left over, and the optimal point does not change from point C. Therefore, no new additional revenue is obtained and the marginal value $M_2 = 0$. We remark that even though the problem is degenerate, neither M_1 nor M_3 has value 0, so while it is possible that both the slack and the marginal value are simultaneously 0 in a degenerate problem, it is not necessarily true for all the constraints.

In summary, for non-degenerate problems, either $S_i > 0$ and $M_i = 0$, or $S_i = 0$ and $M_i \neq 0$. For degenerate problems, it is possible that $M_i = 0 = S_i$ simultaneously. These cases exhaust the possibilities. But notice that, in all the cases, $S_i \cdot M_i = 0$, since at least one of the variables must be 0. This shows the following:

PROPOSITION 4.8   Complementary Slackness. *In any linear programming problem, $S_i \cdot M_i = 0$ for all i.*

Because M_i is the ith decision variable in the dual problem, this says if the i'th constraint in the primal has slack, then the ith decision variable in the dual has optimal value 0. By applying Proposition 4.8 to the dual, it also says if the ith constraint of the dual has slack, then the ith decision variable in the primal has optimal value 0. It's called "complementary" because one is comparing slackness in the primal with something that happens in the dual problem. In other words, slackness in the primal can be detected by looking at the dual (and conversely).

Exercises

For the following Problems 1–6, find the dual problem following the method used in Section 4.5.1.

1. Joni is putting her designing skills to work at the Court and Sparkle Jewelry Emporium. She makes two signature design bracelet models. Each Dawntreader bracelet uses 1 ruby, 6 pearls, and 10 opals. Each Hejira bracelet uses 3 rubies, 3 pearls, and 15 opals. She has 54 rubies, 120 pearls, and 300 opals to work with. If either model results in a profit of $1,800 for Joni, how many of each type should she make? .

2. The Poseidon's Wake Petroleum Company operates two refineries. The Cadence Refinery can produce 40 units of low-grade oil, 10 units of medium-grade oil, and 10 units of high-grade oil in a single day. (Each unit is 1,000 barrels.) The Cascade Refinery can produce 10 units of low-grade oil, 10 units of medium-grade oil, and 30 units of high-grade oil in a single day. They receive an order from the Mars Triangle Oil retailers for at least 80 units of low-grade oil, at least 50 units of medium-grade oil, and at least 90 units of high-grade oil. If it costs Poseidon's Wake $1,800 to operate the Cadence Refinery for a day and $2,000 to operate the Cascade Refinery for a day, how many days should Poseidon's Wake operate each refinery to fill the order at least cost?

3. Joanne's Antique Restoration Emporium refurbishes Victorian furniture. Each dining room set requires 16 hours stripping and sanding time, 4 hours refinishing time, and 4 hours in the upholstery shop. Each bedroom set requires 36 hours stripping and sanding time, 4 hours refinishing time, and 2 hours in the upholstery shop. Each month, Joanne's Antique Restoration Emporium has available 720 person-hours in the stripping and sanding shop, 100 person-hours in the refinishing shop, and 80 person-hours in the upholstery shop. Each dining room set generates $600 profit, and each bedroom set generates $900 profit. How many of each type of furniture set should Joanne accept to work on each month in order to maximize her profit?

4. Bruce Jax is responsible for constructing end tables and kitchen tables for the White Room Woodshop. Each end table uses 2 square yards of $\frac{3}{4}$-inch oak boards and takes 2 hours to complete. Each kitchen table uses 4 square yards of oak board and takes 3 hours to complete. This week, he has available 36 square yards of oak boards and 32 hours of time. Other resources are unlimited. How many of each item should he make if he is paid $70 for each end table and $100 for kitchen table?

5. The Glass House is a shop that produces three types of specialty drinking glasses. The Runaway is a 12-ounce water tumbler, the Reunion is a 16-ounce beer glass, and the Experience is an elegant stemmed champagne flute glass. A case of Runaway glasses takes 1 hour on the molding machine and 0.1 hour to pack, and it generates a profit of $40. A case of Reunions glasses takes 1.5 hours on the molding machine and 0.1 hours to pack, and it generates a profit of $60. A case of Experience glasses takes 1.5 hours on the molding machine and 0.2 hours to pack, and it generates a profit of $70. Each week, there are 720 hours of time available on the molding machine and 60 hours available to pack. How many cases of each type should The Glass House manufacture in order to maximize profit?

6. The Spooky Boogie Costume Salon makes and sells four different Halloween costumes: the witch, the ghost, the goblin, and the werewolf. Each witch costume uses 3 yards of material and takes 3 hours to sew. Each ghost costume uses 2 yards of material and takes 1 hour to sew. Each goblin costume uses 2 yards of material and takes 3 hours to sew. Each werewolf costume uses 2 yards of material and takes 4 hours to sew. The profits for each costume are as follows: $10 for the witch, $8 for the ghost, $12 for the goblin, and $16 for the werewolf. If the Spooky Boogie Costume Shop has 600 yards of material and 510 sewing hours available before the holiday, how many of each costume should it make in order to maximize profit, assuming it sells everything they make?

7. Observe that in Exercise 4, the dual problem has just two decision variables. Solve the dual problem using the method of graphing in the decision space. Compare your answer to the solution of the primal problem (which was assigned in Section 3.4). Verify that the principle of strong duality holds as well as complementary slackness.

8. Observe that in Exercise 5, the dual problem has just two decision variables. Solve the dual problem using the method of graphing in the decision space. In this problem, the primal has too many variables to solve by using the decision space. However, in Section 3.5, this problem was assigned to be solved using the method of graphing in the constraint space. Compare the solutions of the primal and the dual. Verify strong duality and complementary slackness.

9. Observe that in Exercise 6, the dual problem has just two decision variables. Solve the dual problem using the method of graphing in the decision space. In this problem, the primal has too many variables to solve by using the decision space. However, in Section 3.5, this problem was assigned to be solved using the method of graphing in the constraint space. Compare the solutions of the primal and the dual. Verify strong duality and complementary slackness.

10. In Section 3.4, the following problem was assigned in the exercises:

$$\text{maximize: } P = x + 3y$$

subject to:

$$10x + 16y \le 240$$
$$15x + 8y \le 240$$
$$x \ge 0, \ y \ge 0.$$

Find the dual problem and solve it. Compare the solutions of the primal and the dual. Verify strong duality and complementary slackness.

11. Follow the same directions as in Exercise 10 for this problem:

$$\text{minimize: } C = 20x + 12y$$

subject to:

$$5x + 2y \ge 30$$
$$5x + 7y \ge 70$$
$$x \ge 0, \ y \ge 0.$$

12. In this exercise, we explore the dual problem in case there are mixed constraints. Consider the following problem:

$$\text{maximize } R = 120x + 200y$$
$$\text{subject to:}$$
$$10x + 6y \leq 120$$
$$10x + 4y \geq 100$$
$$x \geq 0, \; y \geq 0.$$

a.   Solve the problem by graphing in the decision space. Your answer should have the values of x, y, R, and the slack variables.

b.   Use matrix multiplication FA^{-1} to find the marginal values for the two constraints.

c.   Our construction of the dual is based on having a problem in standard form. For a maximization problem, this means all constraints should be of the less-than type. Convert the second constraint to the less-than type by multiplying through by -1. Then construct the condensed tableau as usual, and write down the dual problem.

d.   Solve the dual problem you obtained in Part c by graphing in the decision space. Verify that strong duality holds. Do the optimal values of the dual agree with the marginal values you computed in Part b?

13.   Following the steps in Part c of Exercise 12, find the dual problem to the following problem: Mr. Cooder, a farmer in the Purple Valley, has at most 400 acres to devote to two crops: rye and barley. Each acre of rye yields \$100 profit per week, while each acre of barley yields \$80 profit per week. Due to local demand, Mr. Cooder must plant at least 100 acres of barley. The federal government provides a subsidy to grow these crops in the form of tax credits. The government credits Mr. Cooder four units for each acre of rye and two units for each acre of barley. The exact value of a "unit" of tax credit is not important. Mr. Cooder has decided that he needs at least 600 units of tax credits in order to be able to afford to his loan payments on a new harvester. How many acres of each crop should he plant in order to maximize his profit?

14.   Consider the following variation of the O'Casek used car problem from Section 3.2 in the text:

$$\text{minimize } C = 6x + 4y$$
$$\text{subject to:}$$
$$x + 2y \geq 24 \text{ (SUVs)}$$
$$2x + y \geq 30 \text{ (sedans)}$$
$$4x + y \geq 54 \text{ (compact cars)}$$
$$x \geq 0, \; y \geq 0.$$

In this variation, the only difference from the original problem is that now we are requiring at least 54 compact cars to be shipped, instead of at least 40.

a.   Find the optimal solution and verify that the problem is degenerate.

b.   Compute the marginal values associated with each constraint. Verify that complementary slackness holds and that for one constraint, both S_i and M_i are simultaneously 0.

5   The Simplex Algorithm

In Chapter 3, we considered two distinct geometric approaches to solving linear programming problems. Since geometry is easiest to visualize in two dimensions, these approaches were limited in scope to $m \times 2$ or $2 \times n$ problems. However, while these techniques do not apply directly to larger problems, during our discussion of the corner point theorem, we did glean enough from these techniques to say that the following points hold true for a general linear programming problem:

- The set of feasible points is always a simple, convex polytope in the decision space, also known as a *simplex X*.
- An optimal point, if one exists, can always be found at one of the corners of the simplex X.
- The corners of X correspond to the basic, feasible solutions to Eq. (3.6), or, rather, to a version of Eq. (3.6) that has m unit vectors U_j instead of two, for the general $m \times n$ problem:

$$x_1 A_1 + x_2 A_2 + \cdots + x_n A_n + S_1 U_1 + S_2 U_2 + \cdots + S_m U_m = R. \tag{5.1}$$

In order to be able to solve larger linear programming problems, we need a technique that is more algebraic than the techniques of Chapter 3. We might imagine the problem as a two-step process. First, we somehow find a feasible point P (that is any point in X), and from there, we seek to improve our choice until it becomes optimal. If our initial choice of a feasible point P happens to be a basic solution to Eq. (5.1), there is a beautiful technique developed by George Dantzig in 1948, which today is known as the *simplex algorithm*. It is the most widely applied method for solving linear programming problems, and it is very efficient.

If the initial choice P is feasible but not a basic solution to Eq. (5.1), there are several techniques for moving from this choice toward the optimal point. Such techniques are termed *interior point methods*. Most interior point methods use some sophisticated mathematics, such as multivariable calculus or projective geometry. By contrast, the simplex algorithm is entirely algebraic. Since in this textbook, we do not assume the reader knows calculus, we will limit our discussion only to the simplex algorithm.

Incidentally, why should the simplex algorithm be entirely algebraic? The reason is the third bullet point – and, as we saw in Chapter 1, we can travel from one basic solution of a vector equation to another by pivoting the augmented coefficient matrix. Thus, you might expect that the simplex algorithm involves pivoting, and it does. We must make two modifications to the pivoting we learned in Chapter 1. First, we modify the method for deciding where to place the pivot positions. Second, we impose rules to ensure that, once we are inside the feasible set X, we do not accidentally stray outside of X.

So, where do we start? Well, we need a basic solution of Eq. (5.1) to be our initial point P. Consider the origin $P = (0, 0, \ldots, 0)$. We have assumed that every decision variable x_i must be nonnegative. That is, we always assume our linear programming problem has the nonnegativity constraints $x_i \geq 0$. So, if we select the nonbasic variables to be the x_i, that corresponds to the basic solution of equation Eq. (5.1), where $x_i = 0$ for all i (which is the origin), and all the slack variables S_i are basic variables. So, in the simplex algorithm, we always start at the origin.

Of course, the astute reader will object that the origin, while it may always be a basic solution of Eq. (5.1), will not always be a feasible solution! That is correct; the origin is not always a feasible point – see Example 3.11 from Section 3.4 for an example. To deal with this, the simplex algorithm has two distinct phases. Phase I of the simplex algorithm is a set of rules for pivoting until we obtain a feasible basic solution (that is, until we "land on" a corner of X). At that point, Phase II takes over, which is a different set of rules for pivoting that applies only when we start at a basic feasible point and continue until the optimal point is reached. Phase II has rules built in to ensure that we do not leave the feasible region X.

Of course, if the origin happens to be feasible, we don't need Phase I at all, and we can start with Phase II. And there is a large class of problems for which the origin is feasible – the standard form maximization problems. So in this chapter, we will begin with those problems, and we will not need to discuss Phase I until Chapter 9.

5.1   Standard Form Maximization Problems

We could, at this point, just present the rules for Phase II pivoting, and the reader could learn them by rote. Since you are already familiar with pivoting from Chapter 1, it would not be difficult to learn the simplex algorithm. However, we would like to offer some motivation for these rules and explain some of the reasoning behind the algorithm. So, instead, we will take a longer route to our destination, where we are able to appreciate the sights along the way. And what better way to motivate the algorithm than by starting with a familiar standard form maximization problem?

So let us review the lemonade stand problem from Chapter 3. Recall the statement of the problem:

> Let x be the number of glasses of sweet lemonade.
>
> Let y be the number of glasses of tart lemonade.
>
> Maximize: $R = 1.25x + 1.5y$
>
> subject to:
>
> $3x + 4y \leq 60$ (lemons)
>
> $x + 2y \leq 28$ (limes)
>
> $2x + y \leq 30$ (tbsp. sugar)
>
> $x \geq 0,\ y \geq 0.$

The first thing we want to do, in preparation for an algebraic technique, is to replace the system of inequalities that describe the feasible set into a system of equations. But we already know how to do that – we must include the *slack variables* S_1, S_2, and S_3. Recall from Chapter 3 that these represent the leftover resources of lemons, limes, and tablespoons of sugar, respectively. For each resource, if you add the leftovers to what you've used up the result is the total amount of that resource available. Thus, for example, the amount of lemons used is $3x + 4y$, and S_1 is the number of leftover lemons; therefore, the first inequality becomes

$$3x + 4y + S_1 = 60.$$

Repeating the process for each structural constraint leads us to the following system of equations:

$$3x + 4y + S_1 = 60$$
$$x + 2y + S_2 = 28$$
$$2x + y + S_3 = 30.$$

Of course, we must have $x \geq 0, y \geq 0$, and $S_i \geq 0$ for $i = 1, 2, 3$ as well in order to obtain a feasible solution. We know from Chapter 1 that we can view a system of equations in three ways. Besides the system itself, there is augmented coefficient matrix, a version of which is incorporated into the simplex algorithm, but there is also the vector point of view. If we write the preceding system of equations as a vector equation, the result is

$$x \begin{bmatrix} 3 \\ 1 \\ 2 \end{bmatrix} + y \begin{bmatrix} 4 \\ 2 \\ 1 \end{bmatrix} + S_1 \begin{bmatrix} 1 \\ 0 \\ 0 \end{bmatrix} + S_2 \begin{bmatrix} 0 \\ 1 \\ 0 \end{bmatrix} + S_3 \begin{bmatrix} 0 \\ 0 \\ 1 \end{bmatrix} = \begin{bmatrix} 60 \\ 28 \\ 30 \end{bmatrix}.$$

We hope that the reader recognizes this equation – it is precisely Eq. (5.1) for this problem:

$$xA_1 + yA_2 + S_1U_1 + S_2U_2 + S_3U_3 = R.$$

In this way, since the simplex algorithm involves pivoting the system, which is equivalent to this vector equation, we may view the simplex algorithm as a natural algebraic extension of the method of graphing in the constraint space. Indeed, by Proposition 3.2, we know the optimal point is a basic, feasible solution. The method of graphing in the constraint space was a brute force approach, where *every* basic, feasible solution was checked (in the case when $m = 2$). The simplex algorithm is a way to generalize this to the case $m > 2$, together with the improvement that we need not check every basic, feasible solution of Eq. (5.1) – we can usually arrive at the optimal solution after checking just a few of them.

Now that we know the connection with the vector equation and the method of graphing in the constraint space, we will return to the point of view of writing the system of equations by using the augmented coefficient matrix, and base the simplex algorithm on pivoting this matrix. In order to motivate the steps of the pivoting process, it will be instructive to refer to the graph of the feasible set (in the decision space, of course, since this is two-dimensional) in Figure 4.1.

For comparison purposes, we also modify Table 3.3 in Table 5.1, where the coordinates of each corner were listed, along with the value of the objective function there, and we have added two columns.

The reader who is familiar with the section on graphing in the constraint space already knows that each corner is associated with a choice of so-called *basic variables* (and a corresponding choice of basis of the constraint space). But even if you haven't read that section, the table indicates the correspondence between the corners of X and a selection of basic variables. In Figure 4.1, each corner is specified by the intersection of two lines. For example, the corner point A is the intersection of the two axes which have equations $x = 0$ and $y = 0$. The corner point B is the

Table 5.1 Corner point data for the lemonade stand problem, including basic variables at each corner.

Corner	(x, y)	$R = 1.25x + 1.5y$	Basic variables	Nonbasic variables
A	$(0, 0)$	\$0	S_1, S_2, S_3	x, y
B	$(15, 0)$	\$18.75	x, S_1, S_2	y, S_3
C	$(12, 6)$	\$24.00	x, y, S_2	S_1, S_3
D	$(4, 12)$	\$23.00	x, y, S_3	S_1, S_2
E	$(0, 14)$	\$21.00	y, S_1, S_3	x, S_2

intersection of the x-axis and the sugar line. Note that the equations of these lines are $y = 0$ and $S_3 = 0$, the latter one because to be on the sugar line indicates there is no leftover sugar. And corner point C is the intersection of the lemon and sugar lines, with equations $S_1 = 0$ and $S_3 = 0$, etc. In each case, two of the five variables are set equal to 0 (these are the nonbasic variables), which gives an intersection point of two lines, and the remaining three variables are (strictly) positive, and these are the basic variables.

Again, to tie this in with what we did in the section on graphing in the constraint space, consider Eq. (5.1) when the nonbasic variables are set to 0. For example, at point C, we set $S_1 = S_3 = 0$ (because all the lemons and sugar is used up) to obtain

$$xA + yB + S_2U_2 = R.$$

The remaining variables x, y, S_2 are called basic variables because they correspond to the set of vectors $\{A, B, U_2\}$, which is a basis of the constraint space. Students who have studied linear algebra know that when the vectors on the left side of the equation form a basis, there is a unique solution to this equation for any given R (that is essentially the definition of what a basis is). In this case, because we already solved the lemonade stand problem, we know what that solution is for the corner C: $(x, y) = (12, 6)$, with $S_2 = 4$ leftover limes, so the equation becomes

$$12A + 6B + 4U_2 = R.$$

At a different corner, the equation would be different, with a different basis. For example, at corner D, which is not optimal, the solution is $(x, y) = (4, 12)$, with 10 leftover tablespoons of sugar, so the equation becomes

$$4A + 12B + 10U_3 = R,$$

which is the unique solution using the basis $\{A, B, U_3\}$. In general, of course, it is precisely these solutions that we will be looking to find via the simplex algorithm.

What we did in this example would hold in any standard form maximization problem. If the problem is size $m \times n$, then there are n decision variables $x_1, x_2, \ldots, x_n$ (with corresponding decision vectors as we called them in Chapter 3, $A_1, A_2, \ldots, A_n$). There are also m slack variables $S_1, S_2, \ldots, S_m$ coming from the m structural constraints (with corresponding unit vectors $U_1, U_2, \ldots, U_m$) for a total of $m + n$ variables altogether in Eq. (5.1):

$$x_1A_1 + x_2A_2 + \cdots + x_nA_n + S_1U_1 + \cdots + S_mU_m = R.$$

A feasible solution requires all $m + n$ variables to be nonnegative, while a basic solution is obtained by setting n of the variables to 0. The m vectors that remain are always a basis of the constraint space. The optimal point must always be at a basic, feasible solution (a corner point of the simplex X).

Now, let's switch over to the point of view of the augmented coefficient matrix. Returning to the lemonade stand problem, the augmented coefficient matrix of the system is

$$\begin{bmatrix} 3 & 4 & 1 & 0 & 0 & 60 \\ 1 & 2 & 0 & 1 & 0 & 28 \\ 2 & 1 & 0 & 0 & 1 & 30 \end{bmatrix}.$$

Before we start any pivoting, we need to add some information to this matrix. It is common to use the word *tableau* to refer to these augmented matrices with the additional information. The additional information takes the following form. We need labels for each column; we use the name of the variable for the label, except the final column, which is the right side of the equation. We label this column with the letter C (for "capacity" since the levels of the resources determine in what capacity we can manufacture lemonade). We also need labels for each row. For the initial tableau, we will label the rows with the slack variables, but these labels will change as we pivot, and it will soon become clear what they represent. Finally, we need to build into this tableau the information contained in the objective function $R = 1.25x + 1.5y$. We do this in the following way. Note that we obtain no revenue from leftover resources, so we can rewrite the objective function as

$$R = 1.25x + 1.5y + 0S_1 + 0S_2 + 0S_3.$$

We can now add a new row to the bottom of the matrix representing the objective function (and label the row with R for revenue), except in the bottom row, we negate each coordinate first (reasons for this will be explained later). Also, note that the bottom row has nothing in the C column. For the initial tableau, we place a 0 there, but this will change as we pivot (and what it represents will be explained later). The result is called the *initial tableau*.

	x	y	S_1	S_2	S_3	Capacity
S_1	3	4	1	0	0	60
S_2	1	2	0	1	0	28
S_3	2	1	0	0	1	30
R	-1.25	-1.5	0	0	0	0

Compare this to the *condensed tableau* we encountered in Chapter 4, and carefully note the differences. In the condensed tableau, the numbers in the last row are not negated, and there are no columns for the slack variables. That more or less explains why it is called a condensed tableau – it is condensed from the initial tableau by removing the slack variable columns (and ignoring the negatives in the last row).

Now, each tableau in (Phase II of) the simplex algorithm corresponds to being at one of the corners of the simplex X. We have already stated that we begin at the origin, so this initial tableau must somehow correspond to the origin. Here's how. The labels for the rows (except for the bottom row) are the *basic variables* for the corner in question. The values of these basic variables appear in the right-hand column (the capacity column), and the variables that are not listed as row labels must have value 0 since they are the nonbasic variables. Finally, in the lower-right-hand corner (in the objective row R and the capacity column C) is the value of the objective function. So to read off the information from the tableau, we have $x = y = 0$ since they are nonbasic variables (so this is indeed the origin or corner point A). We also have $S_1 = 60$, $S_2 = 28$, and $S_3 = 30$ from reading off the last column (which makes sense since we have not made any lemonade, so all of the resources are leftover), and, finally, $R = 0$ from the lower-right corner (since no lemonade entails no revenue).

So comparing this to the preceding table, we see the initial tableau always corresponds to the origin. The negative numbers in the last row are what tell us that we are not yet at an optimal point. Each pivot that we perform will move us from one corner to an adjacent corner on the simplex X. Each such pivot will also increase the value of the objective function R. We stop when there are no more negative numbers in the bottom row. That corner will be the optimal point.

So, to see what to do next, imagine being at point A (the origin) and moving to an adjacent corner along an edge of the feasible set X. That means either moving along the x-axis (to head to point B) or moving along the y-axis (to head for point E). Moving along the x-axis means x is increasing; that is, it means we are now making some sweet lemonade. Similarly, moving along the y-axis means y is increasing, corresponding to producing tart lemonade. Which should we do? Well, we already have solved this problem graphically, and we know the optimal point is point C. If we go along the x-axis, we arrive at point C in two steps – that is, two pivots. If we go along the y-axis, we need to pass through points E and D before arriving at point C, so it means three pivots.

Since we would like to be efficient, you might think the best choice is to go along the x-axis. The problem is when we use the simplex algorithm, we usually do not have a graph to refer to, and we don't know which corner is optimal ahead of time. So we need to make the decision whether to introduce x (sweet lemonade) or y (tart lemonade) based entirely on the initial tableau with no outside help. So, although it seems counterintuitive for this problem, we actually should choose y, because in the absence of better information, we'd rather make \$1.50 than \$1.25. In other words, the first step is to choose a column in the initial tableau that corresponds to the most negative entry in the bottom row, because each unit of the product with the larger negative coefficient stands to increase the objective function the most. (This is an example of what is known as a *greedy* algorithm – one makes the choice that stands to give you the most.) You should somehow indicate that column in your work – perhaps by circling it or drawing an arrow pointing to it (we'll use shading in the textbook) – it's called the *incoming column* or *pivot column*, and the variable it falls under is called the *incoming variable*. So our tableau now looks like the following table, and y is the incoming variable.

	x	y	S_1	S_2	S_3	Capacity
S_1	3	4	1	0	0	60
S_2	1	2	0	1	0	28
S_3	2	1	0	0	1	30
R	-1.25	-1.5	0	0	0	0

The next step is to choose a row (which is never the last row), called the *outgoing row* or *pivot row*, with corresponding *outgoing variable*. To see how that works, go back to the graph and visualize traveling up the y-axis, which means mixing more and more tart lemonade. Can we do this forever? No, we must stop when we run out of resources. Since each glass of tart lemonade uses 1 tablespoon of sugar, we have enough sugar to make 30 glasses of tart lemonade. But, of course, we cannot actually make 30 glasses because we run out of both of the other ingredients first. We have enough lemons to make 15 glasses of tart lemonade, since each glass uses four lemons, and we have 60 lemons. But we only have enough limes for 14 glasses of tart lemonade, since each glass uses two limes, and we started with 28 limes. On the graph, this is quite evident. The numbers 14, 15 and 30 are obvious; they are the y-intercepts of the lemon, lime, and sugar lines, respectively. But to stay within the feasible region, we must stop making tart lemonade as soon as the first ingredient – limes – runs out, after we make 14 glasses of tart lemonade. But again, we must somehow see this from within the tableau itself, since we will not have the graph around to refer to in most linear programming problems. So do the numbers 14, 15, and 30 appear in the tableau?

Well, not directly, but if you take each number in the capacity column and divide it by the corresponding number in the pivot column, you obtain these numbers as ratios: $\frac{60}{4} = 15$, $\frac{28}{2} = 14$, and $\frac{30}{1} = 30$. (Performing these divisions is actually how we found the y-intercepts in the first place when we graphed these lines.) Since we must stop when we run out of the first ingredient, we choose the smallest ratio; this determines the pivot row. If any ratio is undefined due to a 0 in the shaded column, just ignore that row – it means that the constraint line is actually parallel to the y-axis and has no intercept. You also should ignore any rows where the ratio comes out negative. This means the y-intercept is negative, so the corresponding corner is not in the first quadrant and, hence, is not a feasible corner.

Thus, the pivot row is the second row, and the outgoing variable is S_2. Again, in your work, you should indicate this row by circling it; we will again use shading. Here is the initial tableau with the pivot row and pivot column indicated. The position of the matrix where they cross is called the *pivot position*, and the entry there is called the *pivot number*.

	x	y	S_1	S_2	S_3	R	Ratio
S_1	3	4	1	0	0	60	$\frac{60}{4} = 15$
S_2	1	[2]	0	1	0	28	$\frac{28}{2} = 14$
S_3	2	1	0	0	1	30	$\frac{30}{1} = 30$
R	-1.25	-1.5	0	0	0	0	

Here, we have boxed off the pivot number, which is 2 in the a_{22} position. We are now ready to do the first pivot. The idea is to replace the pivot column with a unit column, so we want a 1 where the pivot position is, and we want all other entries in that column cleared, just as we did in Chapter 1. At the same time, the outgoing variable S_2 is replaced by the incoming variable y, so the new set of basic variables is S_1, y, S_3. If you consult the preceding table, that choice of basic variables does correspond to corner point E, and this is what we expected by looking at the graph – march up the y-axis until we run out of limes at point E.

Now, there are several choices for elementary row operations that will place a 1 in the pivot position. For example, one could divide the second row by 2. Or we might replace the second row by itself minus the third row. In Chapter 1, we didn't care much which row operations we used, as long as the 1 resulted in the a_{22} position. However, in the simplex algorithm, we will prescribe exactly which row operations to use, because we want to be able to attach a meaning or interpretation to every entry in the tableau. We will discuss the meanings of the entries in the paragraphs that follow. To pivot, the first thing we do is replace the pivot row with itself divided by the pivot number 2 (which will put a 1 in the pivot position). At the same time, the label of the row is replaced by then incoming variable y. We will be using this row in subsequent operations, so it is convenient to give it a name to be able to refer to it. Let's call it the *normalized pivot row* ("normalized" because we have placed the 1 in the pivot position), to which we refer most often by its abbreviation NPR.

Now, no other labels of rows or columns are changed (in particular, the labels on the columns never change), but each non-pivot row r_j (including the last row) is replaced by the following combination of rows:

$$(\text{row } j) \text{ is replaced by } (\text{itself} - c_j \cdot \text{NPR}),$$

where c_j is the number in the jth row of the pivot column. The effect of this operation is to place a 0 in the jth row in the pivot column. The reader is strongly advised to get out some scrap paper and practice this to make sure the details of the technique are understood!

We will go through this example in detail. Dividing the pivot row by 2 (the pivot number) yields the NPR.

| $\frac{1}{2}$ | 1 | 0 | $\frac{1}{2}$ | 0 | 14 |

Now for row r_1, the number c_1 in the pivot column is 4, so the rule above tells us to replace r_1 by $r_1 - 4 \cdot \text{NPR}$. For notational convenience, we write this as $r_1 \to r_1 - 4 \cdot \text{NPR}$. The details are as follows:

$$
\begin{array}{ccccccc}
 & 3 & 4 & 1 & 0 & 0 & 60 \\
-4 & (\frac{1}{2} & 1 & 0 & \frac{1}{2} & 0 & 14), \\
\hline
 & 1 & 0 & 1 & -2 & 0 & 4 \\
\end{array}
$$

and the result replaces r_1. Similarly, we replace the other non-pivot rows r_3 and r_4 and obtain the next tableau.

	x	y	s_1	s_2	s_3	Capacity	Pivot operation
s_1	3	4	1	0	0	60	$r_1 \to r_1 - 4 \cdot \text{NPR}$
s_2	1	2	0	1	0	28	
s_3	2	1	0	0	1	30	$r_3 \to r_3 - 1 \cdot \text{NPR}$
R	-1.25	-1.5	0	0	0	0	$r_4 \to r_4 - (-1.5)\,\text{NPR}$

	x	y	S_1	S_2	S_3	Capacity
S_1	1	0	1	-2	0	4
y	$\frac{1}{2}$	1	0	$\frac{1}{2}$	0	14
S_3	$\frac{3}{2}$	0	0	$-\frac{1}{2}$	1	16
R	-0.5	0	0	0.75	0	21

The reader should stop and verify all the arithmetic in the pivot operations. This completes our first pivot, and the new tableau should indicate that we are at corner point E. Does it? Well, the basic variables are listed as the row labels: S_1, y, and S_3. From the preceding table, those are the correct basic variables for point E. What are the values of all the variables? Well, x and S_2 are nonbasic (since they do not appear as row labels), so they have values of 0. The values of the basic variables, as noted before, are read off from the last column. In particular, $y = 14$, so we are at the point with coordinates $(x, y) = (0, 14)$, which is indeed point E. The revenue there is \$21, as can be read off for the lower-right corner. And we already see an advantage of the simplex algorithm over the graphical methods because the slack variables are automatically computed, without any extra work, as we see $S_1 = 4$ (leftover lemons), $S_2 = 0$ (leftover limes), because S_2 is nonbasic, and $S_3 = 16$ leftover tablespoons of sugar.

How do we know we are still at a feasible point? Well, to be a feasible solution, every variable must be nonnegative. The nonbasic variables are all 0, so they are nonnegative. The basic variables have values which you read off from the last column. In this case, they are all positive, so point E is indeed feasible. If you follow the rules of Phase II pivoting correctly, you should never obtain a negative entry in the last column – that would indicate an infeasible solution. This is why you ignore negative ratios when choosing the pivot row: if you choose a row with a negative ratio, the next tableau would have a negative number in the right-hand column, which means leaving the feasible region, and it would then require Phase I pivoting to get back into the feasible region.

To be more precise, it is the number in the pivot column that should not be negative. Since all the numbers in the capacity column are positive, the ratios have the same signs as the numbers in the pivot column. If there is a 0 in the capacity column (this is discussed further later), the ratio will be the signless number 0, but if the number in the pivot column is negative, that row should still not be chosen as the pivot row.

Finally, because there are still negative entries in the bottom row, we stand to improve our value of the objective function by another pivot, so this corner E is not the optimal one.

Now we simply iterate this procedure until all the negative entries in the bottom row are gone, at which point we will be at the optimal point. We finish this problem and do other examples in the next section, where we summarize the simplex algorithm. Exercises will follow at the end of the next section.

REMARK 5.1 *Some textbooks place positive entries in the bottom row of the initial tableau, instead of negative ones, as we have done. If you construct your initial tableau with positive entries in the bottom row, you must pivot until all the positive entries are gone from the last row – that is, until only negative entries and zeros appear in the last row. It's just a convention, and it doesn't really matter which convention you follow as long as you are consistent. In this text, we only use the convention illustrated in the preceding text, where we start with negative entries in the bottom row and pivot until there are no more negative entries there. We prefer this convention because there is a certain symmetry between the right-hand column and the bottom row: all nonnegative entries in the right-hand column indicate a feasible point, and all nonnegative entries in the last row indicate an optimal point.*

5.2   Phase II Pivoting

Summarizing where we are in the process, we have observed that whenever we have a standard form maximization problem, the origin is a feasible point. We set up the initial tableau, as indicated in the last section, to reflect that we are starting at the origin. The rules for setting up the *initial tableau* are as follows:

1.   Start with the augmented coefficient matrix of the system of equations obtained by adding the slack variables to the inequalities the describe the feasible simplex. Annex a row at the bottom of this matrix with the coefficients, which are the negatives of the coefficients in the objective function. Fill out the rest of this row with zeros.

2.   Label each column with its variable (decision variables first, followed by slack variables), and label the last column with C or with the word "capacity." These labels never change throughout the pivoting process.

3.   Label each row of the tableau with the name of a slack variable, except the bottom row, which is labeled with a variable representing the objective function.

Now, whenever a tableau represents a feasible point (regardless of whether it is the initial tableau or not), we use Phase II pivoting rules to pivot the tableau repeatedly until the optimal point is reached. Here are the rules we need, based on the example we started earlier:

ALGORITHM 5.2   Phase II pivoting algorithm for feasible tableaux in a maximization problem.

1.   *Determine the* incoming *variable and* pivot column *first. If there are no negative entries in the bottom row, then stop – you are at an optimal point. Otherwise, choose the column with the most negative entry in the bottom row. If there is a tie, pick either column; it doesn't matter which one you choose in most cases (we'll discuss the rare exceptions later). Circle the pivot column.*

2.   *Determine the* outgoing *variable and* pivot row *next. For each row (except the last row – never pivot on the objective function row!) with a (strictly) positive entry in the pivot column, compute the ratio of the number in the capacity column divided by the corresponding number in the pivot column. If there are no positive entries in the pivot column, then stop – this problem has no solution. (That is, it means the feasible set is unbounded, and since it is a standard form maximization problem, there is no maximum point.) Choose the row with the smallest ratio. If there is a tie, pick either row; it doesn't matter which one you choose in most cases (we'll discuss the rare exceptions later.) Circle the pivot row.*

3.   *The number at the intersection of the pivot row and pivot column is called the* pivot number. *Now perform the pivot, using the following rules:*

   a.   *In the new tableau, put the same labels on each column as before. Also put the same labels on each row, except for the pivot row. Replace the outgoing variable with the new incoming variable as the label for the pivot row.*

b.   *Replace the entries in the pivot row with the entries in the* normalized *pivot row (NPR), which is obtained by dividing each entry in the pivot row by the pivot number. (The effect is to place a 1 in the pivot position.)*

c.   *Replace each non-pivot row r_i according to the rule*

$$r_i \rightarrow r_i - c_i \cdot \text{NPR};\qquad(5.2)$$

that is, each non-pivot row (including the bottom row) is replaced by itself minus a multiple of the NPR, where the multiple c_i is the number in that row r_i and in the pivot column. (The effect is to replace the c_i with a in the pivot column.)

4.   *Go to step 1 and repeat the process with the new tableau you obtain.*

We illustrate this algorithm by finishing the lemonade stand problem. So far, in the previous section, we set up the initial tableau corresponding to the origin (corner A in the graph) and pivoted once, to arrive at the second tableau, corresponding to the corner E on the graph.

	x	y	S_1	S_2	S_3	Capacity	Ratios	
S_1	1	0	1	-2	0	4	4	
y	$\frac{1}{2}$	1	0	$\frac{1}{2}$	0	14	28	$r_2 \rightarrow r_2 - \frac{1}{2}\cdot\text{NPR}$
S_3	$\frac{3}{2}$	0	0	$-\frac{1}{2}$	1	16	$\frac{32}{3}$	$r_3 \rightarrow r_3 - \frac{3}{2}\cdot\text{NPR}$
R	-0.5	0	0	0.75	0	21		$r_4 \rightarrow r_4 - (-0.5)\text{NPR}$

The only negative entry in the bottom row is in the first column, so x is the incoming variable (step 1). The ratios in step 2 are $\frac{4}{1} = 4$ for the first row, $\frac{14}{\frac{1}{2}} = 28$ for the second row, and $\frac{16}{\frac{3}{2}} = \frac{32}{3}$ for the third row. The smallest ratio is 4, so the first row is the pivot row, and the outgoing variable is S_1. Notice that the pivot number is 1, so the pivot row is already normalized; that is, the NPR is the same row as the pivot row (except for the label x for the row). This explains the shaded row and column in the tableau. To complete step 3, we make the pivot operations listed to the right of the tableau. This will bring us to the third tableau, which we will present a few paragraphs later, and which we expect from the graph to correspond to corner point D.

Before we actually complete the pivoting to obtain the third tableau, it is worthwhile to look at the some of the entries to see if we can interpret their meaning. For example, the -0.5 in the bottom row indicates a further pivot is necessary because it is negative, but what does the (absolute) value of 0.5 mean? At point E, the value of x is 0, so we are not making any sweet lemonade. By pivoting on the first column, x is the incoming variable (meaning after the pivot, the value of x will be positive). Since the value of the entry in the bottom row is related to the objective function, is it possible that the -0.5 is telling us something about how the revenue R will increase if we bring x into solution? After all, in the initial tableau, that is true: the -1.25 in that entry indicates that, starting at the origin A, if we create (and sell) one glass of sweet lemonade, the revenue will increase by \$1.25. If our mathematical feelers are working correctly, perhaps the -0.5 in this tableau indicates that, starting at the point E, if we create (and sell) one glass of sweet lemonade, our revenue will increase by \$0.50?

At first, that may sound incorrect. If we create a glass of sweet lemonade and sell it, why wouldn't the revenue go up by \$1.25, not \$0.50? Hopefully, it is clear because at the point E, we have run out of one of the resources – namely limes. The value of y is 14, so we are mixing

14 glasses of tart lemonade to obtain the $21 revenue. We cannot hold y at the value 14 as we increase x, as that would correspond to moving horizontally away from point E on the graph, which clearly takes us out of the feasible set (because it requires the use of limes we do not have). In order to stay in the feasible set as we increase x and move away from E, we must simultaneously decrease y so that we stay along the edge ED of the feasible simplex X. Otherwise put, in order to obtain the limes we need to produce some sweet lemonade, we would have to give up some of the tart lemonade we had planned on making. According to the recipes for lemonade, each glass of sweet lemonade requires one lime, and each glass of tart requires two, so in order to create one glass of sweet, we'd have to give up $\frac{1}{2}$ glass of tart lemonade. (This just says that the slope of the line segment ED, which is on the lime line, is $-\frac{1}{2}$). Now what does that do to our revenue? Well, we gain $1.25 from selling the glass of sweet lemonade, but at the same time, we lose the revenue for $\frac{1}{2}$ glass of tart – that is, we lose $0.75 – so our net gain in revenue is exactly $1.25 - 0.75 = 0.50$.

Thus, our interpretation is correct: the -0.50 in the bottom signifies that each glass of newly created sweet lemonade (from the point E) generates $0.50 additional revenue. The same interpretation holds for all the numbers in the bottom row. Look at the 0.75 under S_2. The correct interpretation is this – at point E, we have no leftover limes, but if we want to bring S_2 into solution – that is, if we want to move to a feasible point with one unit of leftover limes (so, with $S_2 = 1$) – it will cost us $0.75 to do so (cost us rather than increase the revenue, because it is positive rather than negative). That is because, as we have just seen, the only way to gain a lime at point E is to give up half a glass of tart lemonade, which explains the $0.75 loss in revenue.

The variables S_1, y, and S_3 are already in solution since they are basic variables, so these generate no additional profit, explaining the 0 in the bottom row under these variables. The best way to think of it is that in order to create one more unit of y or S_1 or S_3 in solution, we'd have to give up that exact amount from what is already in the solution; that is, we'd have to decrease the value of y by the exact amount to cancel out in terms of revenue what we'd gain by creating the extra unit, so it is an even exchange. Indeed, suppose we wish to create an additional unit of leftover sugar. We already have 16 units, so we want to make it 17. In order to do that, we'd have to give up one glass of tart lemonade – or $1.50 in revenue. But what does that extra tablespoon of sugar give us? The ability to create exactly one more glass of tart lemonade, resulting in an increase of $1.50 in revenue to exactly balance with what we lost by recovering the sugar! The upshot is the following.

CONCLUSION 5.3   *The numbers in the last row of a tableau have the following interpretation: they are the negatives of the increase in revenue generated when one more unit of the variables at the head of their column are brought into solution from the corner point represented by the tableau.*

In fact, we can make a similar interpretation for every entry of the tableau (except the right-hand column). Consider the entire first column under the variable x.

	x
S_1	1
y	$\frac{1}{2}$
S_3	$\frac{3}{2}$
R	-0.5

In order to create one unit of x in solution (that is, in order to mix one glass of sweet lemonade), we already observed that we would need to sacrifice $\frac{1}{2}$ glass of tart lemonade. Consider the effect of that on each of the basic variables. The half glass of tart lemonade allows us to recover two lemons and $\frac{1}{2}$ tablespoon of sugar in addition to the one lime it gives us. But to make a glass of sweet lemonade, we need three lemons, so in addition to the two that we recover, we need to trade in one more additional lemon from the four leftover lemons we had at point E. This explains the 1 in the S_1 row. We already know the reason for the $\frac{1}{2}$ in the y row: in order to gain one glass of sweet lemonade, we sacrifice $\frac{1}{2}$ glass of tart lemonade. Finally, each glass of sweet uses 2 tablespoons of sugar, so in addition to the $\frac{1}{2}$ tablespoon we recovered from giving up the tart lemonade, we need an additional $\frac{3}{2}$ tablespoons from the 16 leftover tablespoons we already had leftover at point E. So Conclusion 5.3 generalizes to the following:

CONCLUSION 5.4   *In any feasible tableau, in order to create one additional unit of any of the variables x_j in the solution, we would have to exchange it for c_{ij} units of the ith row variable for each i, (including the bottom row, which gives the effect on the objective function of this additional unit of x_j). This holds whether x_j is a decision variable or a slack variable.*

We remark that while Conclusions 5.3 and 5.4 are valid for any variable, it really is only useful for nonbasic variables, since basic variables are already in solution. (Note that the column of any basic variable is a unit vector, so the c_{ij} are all 0 in this case except for the 1 that occurs in the row labeled with that basic variable. So, like the argument for revenue, bringing a basic variable into solution is an even exchange and rather pointless.)

With that interpretation of the entries of a tableau being understood, let's return to pivoting the lemonade stand tableau for corner point E and see if we can make some predictions based on our interpretation of the entries. We know that each additional glass of sweet lemonade we make will generate and additional \$0.50 revenue, but we cannot keep adding sweet lemonade forever to our solution. Eventually, we will run out of one of the resources. The 1 under the x column and in the S_1 row tells us, by Conclusion 5.4, that each additional x we create will cost us one lemon. Since we only have four leftover lemons at point E, we can only create four glasses of sweet lemonade before running out of lemons. (The reader may want to check that lemons are indeed the first resource that will run out as we travel along the segment ED.) Since each glass of sweet lemonade comes at the expense of $\frac{1}{2}$ glass of tart, in order to create four glasses sweet lemonade, we lose two glasses of tart from the 14 we had at point E. Thus, the coordinates of D should be $(x, y) = (4, 12)$. We already know that is correct from when we solved this problem graphically, so this confirms our understanding of how to interpret the entries in the tableau. Moreover, the -0.5 in the bottom row tells us that each additional glass of sweet lemonade generates \$0.50 revenue, and since we are making four glasses, that represents a total increase of \$2.00 – from \$21.00 at point E to a predicted \$23.00 at point D. That we also know is correct. What about leftover sugar? Each additional glass of sweet lemonade costs us $\frac{3}{2}$ tablespoons of sugar by Conclusion 5.3, so that is a total of 6 tablespoons for four glasses. Since we started with 16 leftover tablespoons at point E, and we use 6, we should be left with 10 tablespoons of sugar left over at point D. This is also correct, and all of this will be evident when we perform the pivot to obtain the tableau for point D.

Of course, one does not need to worry about how to interpret the entries of the tableaux while pivoting – all one must do is perform the algorithm correctly, which can be done in a mechanical way (in fact, it easily be programmed on a computer). As we shall see, the interpretation of the entries in the right-hand column and in the bottom row are far more important than the rest of the numbers in the tableau. The result of the pivot of the tableau for point E is the following tableau for point D. Since point D is also not optimal, we compute the ratios and shade in the pivot row and column for the subsequent (and final) pivot which will take us to point C.

	x	y	S_1	S_2	S_3	C	Ratios	Pivot operations
x	1	0	1	-2	0	4	ignore	$r_1 \to r_1 - (-2)$ NPR
y	0	1	$-\frac{1}{2}$	$\frac{3}{2}$	0	12	8	$r_2 \to r_2 - \frac{3}{2}$ NPR
S_3	0	0	$-\frac{3}{2}$	$\frac{5}{2}$	1	10	4	
R	0	0	0.5	-0.25	0	23		$r_4 \to r_4 - (-\frac{1}{4})$ NPR

Looking at the tableau, we read off the following information: $(x, y) = (4, 12)$ as expected; there are 10 tablespoons of leftover sugar, as expected; and the revenue is \$23, as expected. The variables S_1 and S_2 are both nonbasic, so all the lemons and limes are used up. This is clearly the tableau for the point D. Notice also how the interpretation of the entries changed. Every number in the first row is exactly the same as in the last tableau, but the row label has changed. Thus, for example, the 4 in the last column means four glasses sweet lemonade, whereas in the previous tableau, it meant four leftover lemons. The tableau is not optimal since there is a negative in the last row, so the S_2 column becomes the pivot column next. Ignoring the first row because of the negative entry in the pivot column, the smallest positive ratio is the 4 in the third row, so S_3 becomes the outgoing variable, to be replaced by the incoming variable S_2 in the next tableau. The NPR is obtained by dividing the entries in the pivot row by $\frac{5}{2}$, leading to the following.

NPR $= 0$	0	$-\frac{3}{5}$	1	$\frac{2}{5}$	4

Performing the next pivot leads to this tableau.

	x	y	S_1	S_2	S_3	Capacity
x	1	0	$-\frac{1}{5}$	0	$\frac{4}{5}$	12
y	0	1	$\frac{2}{5}$	0	$-\frac{3}{5}$	6
S_2	0	0	$-\frac{3}{5}$	1	$\frac{2}{5}$	4
R	0	0	0.35	0	0.10	24

There are no more negative entries in the bottom row, so this is the *final tableau*, representing an optimal point. We read off the data from the final tableau:

$$x = 12 \text{ glasses sweet lemonade}$$
$$y = 6 \text{ glasses tart lemonade}$$
$$R = \$24 \text{ (maximized)}$$
$$S_1 = 0 \text{ leftover lemons}$$
$$S_2 = 4 \text{ leftover limes}$$
$$S_3 = 0 \text{ leftover tbsp. sugar.}$$

Observe that this agrees with our previous solution. However, more can be gleaned from the final tableau. Look at the numbers in the bottom row. What does the 0.35 mean in the S_1 column?

Well, from Conclusion 5.3 or 5.4, we know it means that if we wanted to allow leftover lemons into solution, each one would cost us $.35 in revenue. But that is because we'd need to reserve them from the original 60 lemons we were allotted. Isn't that the same thing as saying if we were alloeted an *additional* lemon right from the beginning, it would generate an *additional* $0.35 in revenue? In other words, the 0.35 means that each unit of additional resource (lemons) would generate $0.35 in revenue. But we have seen this definition (and quantity) before! In Section 4.3, this was precisely the definition of the *marginal value* of the resource lemons. Similarly, each additional tablespoon of sugar would generate an additional $0.10 revenue, while each additional lime generates no additional revenue (since we already have leftover limes at the point C). Thus, we have another way of interpreting Conclusion 5.3:

CONCLUSION 5.5   *In the final tableau, the marginal values of each constraint appear in the bottom row in the column corresponding to the slack variable for that constraint.*

This holds for any tableau, but we single out the final tableau, as that is the important case. It takes a little practice (and somewhat more patience) to master the simplex algorithm, so we will do more examples in this section as well as future sections. Before looking at more examples, however, we can already list some of the advantages of the simplex algorithm over the other methods we have seen:

- At each stage in the process, and in particular at the final tableau, automatically and with no extra work, the method gives us the values of all the slack variables as well as the decision variables (remember we had to compute these separately when we solved a problem graphically).
- In the final tableau, not only do the values of the decision and slack variables appear (in the right-hand column), but also, automatically and with no extra work, the values of the marginal values of the constraints appear (in the bottom row).
- Not every feasible corner need be checked (for example, with the simplex algorithm, we know that C is the optimal corner for the lemonade stand problem, even though we never visited the corner point B while we pivoted. (This is one contributing factor to why the simplex algorithm is so efficient.)
- It works for any size problem, $m \times n$, without changes to the algorithm (unlike either graphical approach, for example).

And this is just the beginning. In Chapter 9, we'll see that the simplex algorithm makes certain aspects of sensitivity analysis quite easy – much easier than than the graphical methods.

Are there any disadvantages of the simplex algorithm? Well, one has to be very careful doing arithmetic – a mistake in the first pivot could lead to much wasted effort before it is discovered. Of course, since the algorithm can be programmed on a computer, this isn't really a disadvantage so much as it is an inconvenience. Another issue is what we learned so far cannot handle maximization problems with mixed constraints – only problems in standard form (because quite often the origin is usually not feasible if there are mixed constraints). But this disadvantage vanishes as soon as one learns the Phase I pivoting rules, which can handle such problems (see Chapter 9), so this is not really a disadvantage either. Similarly, what we learned so far cannot handle minimization problems, but we will repair that deficiency in the very next section of this chapter, so that's not really a problem either.

Actually, aside from the boring arithmetic calculations, there really aren't any disadvantages of the algorithm. It gives us a rich amount of information about each problem it solves, and since its inception in 1947, it has saved hundreds of millions, if not billions, of dollars for the companies that have used it. It's no wonder that President Gerald Ford awarded a National Medal of Science in 1975 to George Dantzig for his role in developing the algorithm. Dantzig was awarded several other prizes for his work, and today the Mathematical Programming Society has a prize named after him that they have been awarding periodically since 1982.

Example 5.6 Happy Trails Snacks, Inc., sells three different types of trail mix. A 1-pound bag of Fleetwood Max Mix consists of 3 ounces of chocolate bits, 6 ounces of peanuts, 3 ounces of raisins, and 4 ounces of sunflower seeds. A 1-pound bag of Quicksilver Fresh Air Mix consists of 3 ounces of chocolate bits, 5 ounces of peanuts, and 8 ounces of raisins, and no sunflower seeds. A 1-pound bag of Rodeohead Kid Brand A Mix consists of no chocolate, 5 ounces of peanuts, 6 ounces of raisins, and 5 ounces of sunflower seeds. There are 1,200 ounces of chocolate bits, 3,200 ounces of peanuts, 4,800 ounces of raisins, and 2,000 ounces of sunflower seeds available this week. If each bag of Fleetwood Max Mix generates \$2 profit while each bag of Quicksilver Fresh Air Mix and each bag of Rodeohead Kid Brand A Mix generate \$1 profit, how many bags of each mix should be made from the available resources in order to maximize profit?

We begin by giving a complete, formal setup of the problem. It is as follows (for practice, readers should derive the setup themselves and check it against the following):

Let F be the number of bags of Fleetwood Max Mix.

Let Q be the number of bags of Quicksilver Fresh Air Mix.

Let R be the number of bags of Rodeohead Kid Brand A Mix.

Maximize $P = 2F + Q + R$

subject to:

$3F + 3Q \leq 1{,}200$ (oz. chocolate bits)

$6F + 5Q + 5R \leq 3{,}200$ (oz. peanuts)

$3F + 8Q + 6R \leq 4{,}800$ (oz. raisins)

$4F + 5R \leq 2{,}000$ (oz. sunflower seeds)

$F \geq 0, \; Q \geq 0, \; R \geq 0.$

This is clearly a standard form maximization problem, and the size is 4×3, so we could not readily solve it by either graphical method. Letting S_1, S_2, S_3 and S_4 be the slack variables, in that order, we obtain the following initial tableau, where we have also indicated the steps for the first pivot.

	F	Q	R	S_1	S_2	S_3	S_4	Capacity	Ratios
S_1	3	3	0	1	0	0	0	1,200	400
S_2	6	5	5	0	1	0	0	3,200	$533\frac{1}{3}$
S_3	3	8	6	0	0	1	0	4,800	1,600
S_4	4	0	5	0	0	0	1	2,000	500
P	-2	-1	-1	0	0	0	0	0	

By step 1 of the algorithm, we select the first column as the pivot column, so F is the incoming variable. By step 2, since 400 is the smallest ratio, the first row is the pivot row, and S_1 the outgoing variable. So normalized pivot row is as follows.

NPR = 1	1	0	$\frac{1}{3}$	0	0	0	400.

For the rest of the rows, the pivot operations determined by step 3 are as follows:

Pivot operations

$$r_2 \rightarrow r_2 - 6 \cdot \text{NPR}$$
$$r_3 \rightarrow r_3 - 3 \cdot \text{NPR}$$
$$r_4 \rightarrow r_4 - 4 \cdot \text{NPR}$$
$$r_5 \rightarrow r_5 - (-2)\,\text{NPR} = r_5 + 2 \cdot \text{NPR}.$$

Following these pivot instructions, we obtain the next tableau (which is not optimal since there are still negative entries in the last row, so we have again indicated the steps for the next pivot).

	F	Q	R	S_1	S_2	S_3	S_4	Capacity	Ratios
F	1	1	0	$\frac{1}{3}$	0	0	0	400	Undefined
S_2	0	-1	5	-2	1	0	0	800	160
S_3	0	5	6	-1	0	1	0	3,600	600
S_4	0	-4	5	$-\frac{4}{3}$	0	0	1	400	80
P	0	1	-1	$\frac{2}{3}$	0	0	0	800	

Row r_4 is the pivot row so divide it by 5 to obtain the NPR. Then:

Pivot operations

$$r_1 \rightarrow r_1 - 0 \cdot \text{NPR}$$
$$r_2 \rightarrow r_2 - 5 \cdot \text{NPR}$$
$$r_3 \rightarrow r_3 - 6 \cdot \text{NPR}$$
$$r_5 \rightarrow r_5 + 1 \cdot \text{NPR}.$$

The reader should verify all the steps. Note that we ignore the undefined ratio in the first row (which is caused by the 0 in the pivot column) when selecting the pivot row in step 2. But also notice that the 0 in this column means that in the next tableau, the pivot operation will yield the exact same row $r_1 \rightarrow r_1$. This is because the pivot operations were designed to "clear out" the pivot column and make every entry 0 (except for the 1 in the pivot position). If there is already a 0 in that column for a particular row, there is no need to change the row. Performing the next pivot, we are led to the following tableau.

	F	Q	R	S_1	S_2	S_3	S_4	Capacity
F	1	1	0	$\frac{1}{3}$	0	0	0	400
S_2	0	3	0	$-\frac{2}{3}$	1	0	-1	400
S_3	0	$\frac{49}{5}$	0	$\frac{3}{5}$	0	1	$-\frac{6}{5}$	3,120
R	0	$-\frac{4}{5}$	1	$-\frac{4}{15}$	0	0	$\frac{1}{5}$	80
P	0	$\frac{1}{5}$	0	$\frac{2}{5}$	0	0	$\frac{1}{5}$	880

Since there are no more negative entries in the last row, we have reached an optimal point. We now display the optimal solution, the complete solution including marginal values of all the resources:

$$F = 400 \text{ bags of Fleetwood Max Mix}$$
$$Q = 0 \text{ bags of Quicksilver Fresh Air Mix}$$
$$R = 80 \text{ bags of Rodeohead Kid Brand A Mix}$$
$$P = \$880 \text{ (maximized)}$$
$$S_1 = 0 \text{ leftover oz. chocolate bits}$$
$$S_2 = 400 \text{ oz. leftover peanuts}$$
$$S_3 = 3,120 \text{ leftover oz. raisins}$$
$$S_4 = 0 \text{ leftover oz. sunflower seeds}$$
$$M_1 = \frac{2}{5} = \$0.40 \text{ per ounce (chocolate bits)}$$
$$M_2 = \$0 \text{ per ounce (peanuts)}$$
$$M_3 = \$0 \text{ per ounce (raisins)}$$
$$M_4 = \frac{1}{5} = \$0.20 \text{ per ounce (sunflower seeds).}$$

Despite it being a 4×3 problem whose feasible simplex likely has very many corners, we found an optimal point after only two pivots, so we visited only a small portion of the feasible corner points. Furthermore, the final tableau contains much important information in the last row and last column, including the values of the leftover resources and the marginal values. *From now on, we will include the marginal values in the complete solution for any linear programming problem solved via the simplex algorithm. The reader is expected to do likewise when working the exercises.*

We conclude this section with a brief summary of some minor problems that might occur with the simplex algorithm and how to avoid them. The first issue is rational numbers vs. decimals. Why didn't we change the $\frac{1}{3}$ entry to 0.333 or something similar? The reason is roundoff error. The difference between $\frac{1}{3}$ and 0.333 may be pretty small, but the error is compounded when further row operations are used. After several pivots, your approximate solution may not be so close to the actual solution. You have two choices: either learn how to measure and control roundoff error (and machine error if you use a computer to make the calculations), which can be done by studying a branch of mathematics called *numerical methods* or *numerical analysis*, or avoid decimals altogether and always use rational numbers (fractions) in your calculations. The former might be required for large-scale, real-life problems, but for the problems we'll encounter in this book, we'll stick with the latter solution. Modern computer algebra systems such as Mathematica can perform the row operations using all rational arithmetic and avoid decimals.

The second issue is a bit more subtle, and it is here where we have to confess that we have left out an important fact about the Phase II simplex algorithm. Namely, we haven't addressed the issue of whether it is guaranteed to terminate. We can (and did) make the argument that if a tableau is reached without negative entries in the last row, then this tableau must represent an optimal point. The argument is based on Conclusion 5.3 and is a sound and valid argument. However, we more or less implied without proof that we will always reach such a tableau. Can it happen that we never reach such a tableau and, therefore, the algorithm does not terminate?

Well, one way we can fail to reach such a tableau is if the problem has no solution. For example, in step 2 of the algorithm, what if every ratio is negative; that is, what if there is no positive entry at all in the pivot column? Then we cannot select a pivot row, or, more precisely, each row, were it chosen as the pivot row, would lead to a non-feasible corner. But this is already built into the algorithm. The only way for a maximization problem to fail to have a solution is if the feasible simplex X is either empty or unbounded. Since we are discussing standard form maximization problems, we know that X is not empty, since it contains the origin. Thus, the only possibility is an unbounded feasible set. The algorithm says to stop in this case and conclude (correctly) there is no solution, because the feasible set is unbounded. So it does terminate in this case, however unhappy we may be with the end result.

There is, however, another way in which we can fail to reach an optimal tableau. Consider the case when the problem is *degenerate*. As noted in Section 3.5, a degenerate problem means that one or more of the basic variables have a value of 0 rather than a strictly positive value. Geometrically, it means that the corner point is the intersection of "more than the minimum number of hyperplanes." To give a more concrete example, suppose there are two decision variables, so the problem is $m \times 2$. In this case, the feasible simplex is a convex polygon in a plane. Each corner of a polygon is usually determined by the intersection of exactly two of the constraint lines. But occasionally, it might be the case that three (or more) of the constraint lines all pass through the same corner. For a specific case where this happens, see Exercise 20b, from Section 3.4.

How can we recognize when this happens in a standard form maximization problem, and what effect does it have on pivoting? Suppose there is a tie for the smallest ratio when you are choosing the pivot row in step 2. Recall that since you have already chosen the pivot column, you are bringing the incoming variable into solution (meaning its previous value is zero, but after pivoting, it will be positive.) In step 2, when selecting the pivot row, you are deciding which variable is outgoing (meaning it will have value 0 in the next pivot because it will be a nonbasic variable). Performing the pivot means you are interchanging the incoming variable for the outgoing variable – the incoming switches from nonbasic to basic, and the outgoing switches from basic to nonbasic. None of the other variables change their status. But if there is a tie in the ratios between two rows, you run out of both resources simultaneously – that is, the values of both variables become 0 at the same time. Thus, the outgoing variable becomes 0, as it should, since it is now a nonbasic variable, but the value variable in the other row also becomes 0 in the next tableau, even though it remains a basic variable. So, in the next tableau, there will appear a 0 in the right-hand column. In subsequent pivots, if there are any, the ratio for that row, with basic variable being 0, will be 0, making it the smallest nonnegative ratio (in case the corresponding entry in the pivot column is positive) and, hence, the choice for the next pivot row. But if you pivot on a row with a 0 in the capacity column, the value of the objective function does *not* change. Indeed, the rule for pivoting the objective row is $r \rightarrow r - c \cdot \text{NPR}$ for some (negative) constant c. But since there is a 0 in the right-hand entry for NPR, the right-hand entry for that row r does not change, regardless of the value of c.

This is the problem for the simplex algorithm. Most times a pivot is performed, the ratio of the pivot row is strictly positive, *which will force the value of the objective function to strictly increase*. If the value increases at every pivot, we know the algorithm must terminate since a finite number of corners implies a finite number of possible increases before we hit the optimal point. But if it is possible to perform a pivot where the value of the objective function does not increase but merely stays the same, then we have stalled our march up to the maximum value. If you can stall once, it could happen repeatedly, and you would be "stuck" at a non-optimal corner forever, with each pivot getting you nowhere because all you are doing with each pivot is renaming which variables (with value 0) are going to be the nonbasic ones and which ones will remain basic, despite being value 0. Meanwhile, you are staying at the same corner and not moving on the simplex X to an adjacent corner. After renaming the basic variables several times, you might end up revisiting a tableau you have already seen.

Another way to not have the objective function go up in value when pivoting would be if the adjacent corner represented by the tableau you obtain after pivoting has the same value of the objective function (that is, if the two corners both lie on the same level objective line). But then it's also possible on a subsequent pivot to wind up back on the previous tableau where you started.

When this happens in the simplex algorithm – that after pivoting several times you wind up back at a tableau you have previously visited – you are said to be *cycling*. When cycling happens, it is theoretically possible the algorithm will cycle forever and not converge to a solution. However, this is extremely rare. For an example of this phenomenon, see pages 127–129 of Calvert and Voxman (1989). Even if it does occur, one can often disrupt the cycle, continue to the optimal tableau by violating the rules for choosing the pivot row or column specified by the greedy choice in step 1 or step 2 of the simplex algorithm, and just choose a different row or column (except you may have to use Phase I techniques in case you stray outside the feasible region with your choice). This isn't guaranteed to work – you could slip back into cycling later – but it almost always does work because cycling is so extremely rare. If you are looking for a fix that will guarantee that cycling *never* occurs, one was discovered by R. G. Bland in 1977 (see Bland, 1977), and it is very simple to implement. Bland's rule just specifies what to do in case of a tie when choosing pivot rows and columns:

Bland's Rule: Consider the tableau for an $m \times n$ linear programming problem, with decision variables $x_1, \ldots, x_n$ and slack variables $S_1, \ldots, S_m$. If there is a tie in the most negative entries in the last row when choosing the next pivot column, always choose the column farthest to the left in the tableau. If there is a tie among the smallest ratios when choosing the pivot row, choose the row labeled with the decision variable x_i for which the subscript i is smallest (among those rows that are tied). If there are no decision variables among the tied rows (so that they are all labeled with slack variables), choose the row labeled with the slack variable S_j for which the subscript j is smallest (among the rows that are tied).

Note that the rule for columns can be expressed in exactly the same way as the rule for rows, because the columns are always labeled in order with the decision variables first, so the column farthest to the left of the tableau among those that are tied is either the column corresponding to x_i with i as small as possible or, if there are no decision variable columns among the tied columns, the column corresponding to S_j where j is as small as possible.

Although we do not do so here, it can be shown that if you follow Bland's rule while pivoting, then cycling does not occur, and therefore, the algorithm always converges to the optimal point when the feasible set is bounded.

We conclude this section with an example of a degenerate problem.

Example 5.7 Consider the problem:

$$\text{maximize } P = 3x + 4y$$
$$\text{subject to:}$$
$$x + 10y \le 70$$
$$3x + 5y \le 60$$
$$x + 5y \le 40$$
$$x + y \le 18$$
$$x \ge 0,\ y \ge 0.$$

It is not evident from the setup of the problem that it is a degenerate problem. Notice that it only has two decision variables, so one could solve it by graphing in the decision space, and the degeneracy would be evident in the graph of the feasible region. To solve it via the simplex algorithm, we begin by setting up the initial tableau. As usual, we display the ratios used in step 2 of the algorithm and indicate the pivot row and column, and display the pivoting directions from step 3.

	x	y	S_1	S_2	S_3	S_4	Capacity	Ratios	Pivot operations
S_1	1	10	1	0	0	0	70	7	
S_2	3	5	0	1	0	0	60	12	$r_2 \rightarrow r_2 - 5 \cdot NPR$
S_3	1	5	0	0	1	0	40	8	$r_3 \rightarrow r_3 - 5 \cdot NPR$
S_4	1	1	0	0	0	1	18	18	$r_4 \rightarrow r_4 - NPR$
P	-3	-4	0	0	0	0	0		$r_5 \rightarrow r_5 + 4 \cdot NPR$

Nothing yet indicates degeneracy. We proceed as usual with the *NPR* obtained by dividing the pivot row by the pivot number 10. Performing the pivot will lead to the next tableau.

	x	y	S_1	S_2	S_3	S_4	Capacity	Ratios	Pivot operations
y	$\frac{1}{10}$	1	$\frac{1}{10}$	0	0	0	7	70	$r_1 \rightarrow r_1 - \frac{1}{10} \cdot NPR$
S_2	$\frac{5}{2}$	0	$-\frac{1}{2}$	1	0	0	25	10	
S_3	$\frac{1}{2}$	0	$-\frac{1}{2}$	0	1	0	5	10	$r_3 \rightarrow r_3 - \frac{1}{2} \cdot NPR$
S_4	$\frac{9}{10}$	0	$-\frac{1}{10}$	0	0	1	11	$\frac{110}{9} = 12\frac{2}{9}$	$r_4 \rightarrow r_4 - \frac{9}{10} \cdot NPR$
P	$-\frac{13}{5}$	0	$\frac{2}{5}$	0	0	0	28		$r_5 \rightarrow r_5 + \frac{13}{5} \cdot NPR$

Now the degeneracy is almost evident. As we can see, there is a tie in the smallest ratios for choosing the pivot row. Either row could be chosen, but we use Bland's rule and select S_2 over

S_3 because of the smaller subscript. Thus, the second row becomes the pivot row. Regardless of which of the two rows is chosen, we expect, as discussed earlier, a 0 to appear in the capacity column in the next tableau due to this tie. Performing the next pivot yields the following tableau.

	x	y	S_1	S_2	S_3	S_4	Capacity	Ratios	Pivot operations
y	0	1	$\frac{3}{25}$	$-\frac{1}{25}$	0	0	6	50	$r_1 \to r_1 - \frac{3}{25} \cdot \text{NPR}$
x	1	0	$-\frac{5}{25}$	$\frac{10}{25}$	0	0	10	–	$r_2 \to r_2 + \frac{5}{25} \cdot \text{NPR}$
S_3	0	0	$-\frac{10}{25}$	$-\frac{5}{25}$	1	0	0	0	$r_3 \to r_3 + \frac{10}{25} \cdot \text{NPR}$
S_4	0	0	$\frac{2}{25}$	$-\frac{9}{25}$	0	1	2	2	
P	0	0	$-\frac{3}{25}$	$\frac{26}{25}$	0	0	54		$r_5 \to r_5 + \frac{3}{25} \cdot \text{NPR}$

Note that the degeneracy is now clear as $S_3 = 0$ even though S_3 is a basic variable. It's also worth mentioning that although we could have reduced some of the fractional entries to lowest terms, it's actually better to leave them all over a common denominator of 25, in terms of performing the arithmetic of the row operations. Finally, observe carefully that the pivot row is the fourth row – not the third – despite the fact that $0 \leq 2$. The reason we ignore the ratio of 0 in this problem is because of the negative number in the pivot column and the third row. (Strictly speaking, we should not even have bothered to compute the ratio since the entry in the pivot column is negative.) Had that number in the a_{33} position been positive, we would not ignore that ratio of 0, and we would have chosen the third row as a pivot row instead of the fourth.

We will complete the next pivot momentarily, and we will see that we arrive at the optimal tableau. However, to get an idea of just how robust the simplex algorithm is, we mention in passing that had we failed to notice the negative entry in the a_{33} position and chosen row three as the pivot row, we still would end up at the optimal tableau without any cycling taking place. What would happen (see Exercise 21), is that on the next pivot, we would have stayed at the same corner and just represented it with a different set of basic variables. However, on the very next pivot, we would proceed to the optimal tableau so we would not revisit any previous tableau. Of course, we would have performed an extra pivot, arriving at the optimal tableau in four pivots instead of three, because we did not apply the algorithm exactly according to the rules of choosing the pivot row, which are designed for efficiency.

Performing the final pivot, we arrive at the optimal tableau.

	x	y	S_1	S_2	S_3	S_4	Capacity
y	0	1	0	$\frac{1}{2}$	0	$-\frac{3}{2}$	3
x	1	0	0	$-\frac{1}{2}$	0	$\frac{5}{2}$	15
S_3	0	0	0	-2	1	5	10
S_1	0	0	1	$-\frac{9}{2}$	0	$\frac{25}{2}$	25
P	0	0	0	$\frac{1}{2}$	0	$\frac{3}{2}$	57

The final solution is

$$x = 15$$
$$y = 3$$
$$P = 57 \text{ (maximized)}$$
$$S_1 = 25 \text{ and } M_1 = 0$$
$$S_2 = 0 \text{ and } M_2 = \frac{1}{2}$$
$$S_3 = 10 \text{ and } M_3 = 0$$
$$S_4 = 0 \text{ and } M_4 = \frac{3}{2}.$$

Thus, the only wrinkle was to be careful to choose the correct pivot row in case one or more of the ratios for the rows are 0 and to follow Bland's rule in a tie. Degenerate problems should not cause any serious difficulties for the simplex algorithm in practice.

Variations on Standard Form Problems Involving the Nonnegativity Constraints

In standard form problems, for both maximization and minimization problems, we always assume that the nonnegativity constraints $x_i \geq 0$ are part of the setup. In most problems one would encounter in practice, these inequalities make sense. However, it seems possible that one might be interested in a linear programming problem where the decision variables are not restricted to be nonnegative. For example, what if one of the decision variables – say, x_1 – is required to be nonpositive instead, so we want the solution to satisfy $x_1 \leq 0$? It's easy to convert this to standard form by just making a substitution – invent a new variable that is just the negative of x_1 – call it y_1, for example – so $y_1 = -x_1$. Now, in the objective function and in all the constraints, replace each occurrence of x_1 with $-y_1$, and treat y_1 as the decision variable. Then $y_1 \geq 0$ since $x_1 \leq 0$, and proceed as usual with all nonnegativity constraints in place.

Similarly, what if there is a decision variable with no sign restriction? In this case, use the fact that any real number can be written as the difference between two nonnegative real numbers (in many ways). For example, $\frac{3}{2}$ can be written as $\frac{3}{2} - 0$, or as $2 - \frac{1}{2}$, and similarly, -5 can be written as $0 - 5$ or as $7 - 12$. So if x_1 is an unrestricted decision variable, invent two nonnegative variables y_1 and z_1, and write $x_1 = y_1 - z_1$. Then, in the objective function and in all of the constraints, replace each occurrence of x_1 with $y_1 - z_1$, and then annex the two constraints $y_1 \geq 0$ and $z_1 \geq 0$. Although the new problem has one more decision variable than the original problem, it now has the usual nonnegativity constraint for each decision variable.

In this way, the simplex algorithm can be extended to cover problems where the nonnegativity constraints are modified or absent. (See Exercises 22 and 23.)

Exercises

In Exercises 1–10, simplex tableaux are shown. For each tableau, determine whether or not the tableau corresponds to a feasible point. If the answer is no, explain why not. If the answer is yes, then write down the complete set of data for that corner point by reading off the tableau – that is, the values of all the variables, including the marginal values for each constraint. Finally, determine if the tableau represents an optimal point. If not, determine the next pivot row and pivot column.

1. a.

	x	y	S_1	S_2	S_3	Capacity
S_1	0	-2	1	-1	0	6
x	1	$\frac{1}{2}$	0	2	0	21
S_3	0	3	0	$-\frac{3}{2}$	1	15
P	0	-5	0	-3	0	64

b.   Interpret the meaning of all the entries in the y column.

2.

	x	y	S_1	S_2	S_3	S_4	Capacity
x	1	0	$\frac{4}{3}$	0	0	0	$\frac{100}{3}$
S_2	0	0	$\frac{1}{3}$	1	0	$\frac{1}{3}$	25
S_3	0	0	$-\frac{2}{3}$	0	1	$\frac{11}{3}$	22
y	0	1	$\frac{5}{3}$	0	0	$\frac{7}{3}$	$\frac{77}{3}$
z	0	0	$-\frac{4}{3}$	0	0	$-\frac{7}{3}$	$\frac{400}{3}$

3.

	x	y	S_1	S_2	Capacity
y	-2	1	4	0	125
S_2	5	0	9	1	-66
P	-1	0	-3	0	1875

4.

	x	y	S_1	S_2	Capacity
y	0	1	-4	6	70
x	1	0	8	-1	45
z	0	0	12	9	1,450

5.

	x_1	x_2	x_3	x_4	x_5	S_1	S_2	Capacity
S_2	$\frac{1}{4}$	1	0	0	12	$-\frac{3}{4}$	1	100
x_3	$\frac{3}{4}$	-2	1	3	3	$\frac{11}{4}$	0	23
P	5	-2	0	6	-3	7	0	42

6.

	w	x	y	z	S_1	S_2	S_3	Capacity
w	1	-1	0	0	$\frac{3}{5}$	$\frac{6}{5}$	4	200
y	0	2	1	0	$-\frac{2}{5}$	1	$\frac{4}{5}$	270
z	0	1	0	1	$\frac{1}{5}$	-2	$\frac{3}{5}$	35
P	0	5	0	0	4	7	6	2,018

	x_1	x_2	x_3	x_4	x_5	S_1	S_2	S_3	Capacity
S_1	$-\frac{1}{2}$	8	0	0	1	1	5	10	160
x_4	3	0	0	1	2	0	7	-3	125
x_3	6	-1	1	0	4	0	$\frac{11}{2}$	5	90
P	-5	-9	0	0	-1	0	$\frac{3}{2}$	-9	1,000

7. a.

   b.   Interpret the meaning of all the entries in the x_2 column.

	x	y	z	S_1	S_2	S_3	Capacity
z	0	$\frac{3}{17}$	1	$-\frac{2}{17}$	0	$\frac{31}{17}$	$-\frac{150}{17}$
x	1	$\frac{11}{17}$	0	$\frac{1}{17}$	0	$\frac{5}{17}$	$\frac{203}{17}$
S_2	0	$\frac{20}{17}$	0	$\frac{3}{17}$	1	$\frac{14}{17}$	$\frac{100}{17}$
w	0	$\frac{10}{17}$	0	$-\frac{6}{17}$	0	$-\frac{2}{17}$	$\frac{800}{17}$

8.

	p	q	r	t	w	S_1	S_2	Capacity
w	$\frac{3}{2}$	0	$\frac{1}{2}$	2	1	8	$\frac{7}{2}$	18
q	0	1	$-\frac{1}{2}$	5	0	$\frac{15}{2}$	$\frac{11}{2}$	45
z	-4	0	-1	-8	0	6	5	88

9.

	x_1	x_2	x_3	x_4	x_5	S_1	S_2	S_3	S_4	Capacity
x_3	0	-3	1	12	0	0	1	4	4	77
S_1	0	1	0	-8	0	1	-2	5	3	150
x_5	0	2	0	6	1	0	1	6	-2	225
x_1	1	5	0	6	0	0	2	10	0	101
P	0	60	0	38	0	0	50	65	42	166,280

10.

11. Using the simplex algorithm, solve the problem that was set up in Exercise 11 of Section 3.1, which we reproduce here for convenience: Bruce Jax is responsible for constructing end tables and kitchen tables for the White Room Woodshop. Each end table uses 2 square yards of $\frac{3}{4}$-inch oak boards and takes 2 hours to complete. Each kitchen table uses 4 square yards of oak board and takes 3 hours to complete. This week he has available 36 square yards of oak boards and 32 hours of time. Other resources are unlimited. How many of each item should he make if he is paid $70 for each end table and $100 for kitchen table? Be sure to include the complete data set for the solution.

12. Using the simplex algorithm, solve the following problem (see Exercise 12 of Section 3.1). Joni is putting her design skills to work at the Court and Sparkle Jewelry Emporium. She makes two signature bracelet designs. Each Dawntreader design uses 1 ruby, 6 pearls, and 10 opals. Each Hejira design uses 3 rubies, 3 pearls, and 15 opals. She has 54 rubies, 120 pearls, and 300 opals to work with. If either model yields a profit of $1,800, how many of

each type of bracelet should she make in order to maximize her profit? Be sure to include the complete data set for the solution.

13.   Solve the following problem by the simplex method (see Exercise 15 from Section 3.1). Joanne's Antique Restoration Emporium refurbishes Victorian furniture. Each dining room set requires 16 hours stripping and sanding time, 4 hours refinishing time, and 4 hours in the upholstery shop. Each bedroom set requires 36 hours stripping and sanding time, 4 hours refinishing time, and 2 hours in the upholstery shop. Each month, Joanne's Antique Restoration Emporium has available 720 person-hours in the stripping and sanding shop, 100 person-hours in the refinishing shop, and 80 person-hours in the upholstery shop. Each dining room set generates $600 profit, and each bedroom set generates $900 profit. How many of each type of furniture set should Joanne accept to work on each month in order to maximize her profit?

14.   Solve the following problem (see Exercise 19 from Section 3.1). Green's Heavy Metal Foundry mixes three different alloys composed of copper, zinc, and iron. Each 100-pound unit of Alloy I consists of 50 pounds of copper, 50 pounds of zinc, and no iron. Each 100-pound unit of Alloy II consists of 30 pounds of copper, 30 pounds of zinc, and 40 pounds of iron. Each 100-pound unit of Alloy III consists of 50 pounds of copper, 20 pounds of zinc, and 30 pounds of iron. Each unit of Alloy I generates $100 profit, each unit of Alloy II generates $80 profit, and each unit of Alloy III generates $40 profit. There are 12,000 pounds of copper, 10,000 pounds of zinc, and 12,000 pounds of iron available. The foundry has hired Gary Giante, an outside consultant, to help them maximize its profit. Fortunately, Mr. Giante knows the simplex algorithm.

15.   Solve the following 2×2 problem (see Exercise 3 from Section 3.4) using the simplex method:

$$\text{maximize:} P = x + 3y$$
$$\text{subject to:}$$
$$10x + 16y \leq 240$$
$$15x + 8y \leq 240$$
$$x \geq 0, \ y \geq 0.$$

16.   Solve the following with the simplex method (see Exercise 6 from Section 3.4). The Glass House is a shop that produces three types of specialty drinking glasses. The Runaway is a 12-ounce water tumbler, the Reunion is a 16-ounce beer glass, and the Experience is an elegant stemmed champagne flute glass. A case of Runaway glasses takes 1 hour on the molding machine and 0.1 hour to pack and generates a profit of $40. A case of Reunions glasses takes 1.5 hours on the molding machine and 0.1 hour to pack and generates a profit of $60. A case of Experience glasses takes 1.5 hours on the molding machine and 0.2 hour to pack, and generates a profit of $70. Each week, there are 720 hours of time available on the molding machine and 60 hours available to pack. How many cases of each type should The Glass House manufacture in order to maximize profit?

17.   Solve the following with the simplex method (see Exercise 7 from Section 3.4). The Spooky Boogie Costume Salon makes and sells four different Halloween costumes: the witch, the ghost, the goblin, and the werewolf. Each witch costume uses 3 yards of material and takes 3 hours to sew. Each ghost costume uses 2 yards of material and takes 1 hour to sew. Each

goblin costume uses 2 yards of material and takes 3 hours to sew. Each werewolf costume uses 2 yards of material and takes 4 hours to sew. The profits for each costume are as follows: $10 for the witch, $8 for the ghost, $12 for the goblin, and $16 for the werewolf. If the Spooky Boogie Costume Salon has 600 yards of material and 510 sewing hours available before the holiday, how many of each costume should it make in order to maximize profit, assuming it sells everything they make?

18.   Consider the following variation on Angie and Kiki's lemonade stand (Exercise 10 from Section 3.4). Suppose Kiki's idea was to include a third type of lemonade instead of a third ingredient. Thus, the resources are just lemons and sugar (no limes). They have 60 lemons and 30 tablespoons of sugar available. In addition to the sweet lemonade (three lemons and 2 tablespoons of sugar per glass) and the tart lemonade (four lemons and 1 tablespoon of sugar per glass), they also make a lite lemonade (two lemons and $\frac{4}{3}$ tablespoons of sugar per glass). The prices are $1.25 for a glass of sweet lemonade, $1.50 for a glass of tart lemonade, and $1.00 for a glass of lite lemonade. How many glasses of each should they make in order to maximize their revenue? Use the simplex algorithm.

19.   Solve Problem 11b from Section 3.4 (the Shades of Grey Art Supplies problem) using the simplex algorithm.

20.   Consider the following linear programming problem:

$$\text{maximize } z = 8x + 7y$$

subject to:

$$x + y \le 17$$
$$x + 2y \le 32$$
$$5x \le 12y$$
$$x \ge 0, \ y \ge 0.$$

   a.   Convert the third inequality to standard form, and then set up the initial tableau.
   b.   Explain how you can tell this is a degenerate problem by looking at the initial tableau.
   c.   Graph the feasible region in the decision space. Explain why this is a degenerate problem, based on the graph.
   d.   Solve it via the simplex algorithm.
   e.   Solve by the method of graphing in the decision space. Your answer should agree with what you obtained in Part d.

21.   a.   In Example 5.7, in pivoting from the third to the fourth tableau, we mentioned that we could have still arrived at the optimal tableau (without any cycling occurring), had we mistakenly chosen the third row as the pivot row instead of the fourth. Verify that this is true by performing the pivoting.
   b.   Solve the problem in Example 5.7 by the method of graphing in the resource space, to verify that we obtained the correct answer by the simplex algorithm. Also, explain why, based on the graph, the problem is degenerate.

22.   Consider the linear programming problem:

$$\text{maximize } P = 2x + 2y$$

subject to:

$$5x + 6y \le 120$$

$$5x + 18y \leq 180$$
$$x - 2y \leq 40$$
$$x \geq 0, \ y \text{ unrestricted.}$$

a.   Solve it by the simplex algorithm, using the techniques described in the paragraph at the end of the section.

b.   Solve it by graphing in the constraint space and check that your answer agrees with what you got in Part a.

23.   Consider Problem 13 from Section 3.4. The Jefferson Plastic Fantastic Assembly Corporation manufactures gadgets and widgets for airplanes and starships. Each case of gadgets uses 2 kilograms of steel and 5 kilograms of plastic. Each case of widgets uses 2 kilograms of steel and 3 kilograms of plastic. The profit for a case of gadgets is $360, and the profit for a case of widgets is $200. Suppose the Jefferson Plastic Fantastic Assembly Corporation has 80 kilograms of steel and 150 kilograms of of plastic available on a daily basis and can sell everything it manufactures. How many cases of each should the corporation manufacture if it is obligated to produce at least 10 cases of widgets per day? The objective is to maximize daily profit.

a.   Let x be the number of cases of gadgets and y the number of cases of widgets. Write down the setup to the problem.

b.   Notice that the obligation to produce at least 10 cases of widgets converts to an inequality of the form $y \geq 10$, so this problem has mixed constraints. That is, it is not a standard form maximization problem. As stated in the text, once Phase I pivoting is learned, problems like this can be solved using the full simplex algorithm, with both phases. However, here, taking our cue from the paragraph at the end of the section, we can solve the problem with just Phase II by a substitution. Simply define a new variable $z = y - 10$. In the objective function and in all of the constraints, rewrite y in terms of z. You may also find it convenient to make a substitution for the objective variable P so that the objective function has no constant term. Note that the constraint $y \geq 10$ reduces to the nonnegativity constraint $z \geq 0$. In particular, not only is the new problem now a standard form maximization problem, but it has one fewer structural constraint than the original problem! Solve the resulting problem via Algorithm 5.2, and thereby obtain the optimal solution to the original problem as well.

5.3   Standard Form Minimization Problems

To solve a minimization problem, we make use of a clever trick. We first perform a little "magic spell" to turn it into a maximization problem so we can use what we have already learned. That is not the only possibility. If one insists, one can learn a new set of rules for pivoting minimization problems. Some authors take this approach – for example, see section 6.5 of Calvert and Voxman (1989). In this text, we eschew that in favor of the magic of turning a minimization problem into a maximization problem.

There are actually two different "magic spells" that do this. One invokes the awesome order-reversing power of the number -1; the other invokes the mystery of duality. We illustrate with an example. Consider the following minimization problem:

Example 5.8

$$\text{Minimize: } w = 30x + 24y + 10z$$

subject to:

$$6x + 8y + 2z \geq 30$$
$$12x + 16y + z \geq 30$$
$$12x + 9y + 3z \geq 45$$
$$8x + 4y + 4z \geq 40$$
$$x \geq 0, \ y \geq 0, \ z \geq 0$$

Invent a new variable that is the negative of w: say, $P = -w$. Isn't it true that w reaches its smallest possible value whenever P is as large as possible? Thus, we can view the problem as having the objective:

$$\text{maximize: } P = -30x - 24y - 10z.$$

That is, we simply multiply the objective function by -1 and maximize instead of minimize. Of course, the problem is now the inequalities all point in the wrong direction for a standard form maximization, so we cannot set up an initial tableau. But there's an easy fix for that; multiply each inequality through by -1 as well, which has the effect of reversing the direction of the inequalities. So, after multiplying everything through by -1, the problem reads as follows:

$$\text{maximize: } P = -30x - 24y - 10z$$

subject to:

$$-6x - 8y - 2z \leq -30$$
$$-12x - 16y - z \leq -30$$
$$-12x - 9y - 3z \leq -45$$
$$-8x - 4z - 4z \leq -40$$
$$x \geq 0, \ y \geq 0, \ z \geq 0.$$

This is now in the proper format for a maximization problem, but there are still issues to resolve, as it is just semi-standard form rather than standard form. For example, if you were to set up the initial tableau, you would find that every entry in the last row is already nonnegative, so is the initial tableau the optimal one? Not so fast. Look at the capacity column. Every entry in this column is negative, so the initial tableau, which represents the origin of the decision space as usual, is not even feasible, as all the basic variables are negative! Well, of course, they are – with greater-than-type constraints, the origin typically is not a feasible point, and merely writing the inequality as a $\leq$ type inequality won't change that.

The way out of this is, of course, to use Phase I pivoting instead of the algorithm, since we are beginning at an infeasible point, and that is exactly what Phase I is designed for. So we defer further discussion of this technique to handle minimization problems until we learn Phase I pivoting in Chapter 9. We will make one further point about it before moving on to the second technique. That is, since we will need both phases of the simplex algorithm – I and II – we need not begin with a standard form minimization (as we did in this example). This technique will handle any minimization problem with mixed constraints as well as standard form constraints, as we will see in Section 9.2.

We now move on to the approach using duality. The problem in Example 5.8 is a standard form minimization, and we already know from Chapter 4 that its dual problem will be a standard form maximization. Let's write down the dual problem in this example, using the method discussed in Chapter 4. First, we construct the following condensed tableau.

	x	y	z	Capacity
S_1	6	8	2	30
S_2	12	16	1	30
S_3	12	9	3	45
S_4	8	4	4	40
w	30	24	10	0

Next, we take the transpose to obtain the condensed tableau of the dual problem (and change the row and column labels appropriately).

	u_1	u_2	u_3	u_4	Capacity
T_1	6	12	12	8	30
T_2	8	16	9	4	24
T_3	2	1	3	4	10
Q	30	30	45	40	0

Thus, the statement of the dual problem is the following:

$$\text{maximize: } Q = 30u_1 + 30u_2 + 45u_3 + 40u_4$$
$$\text{subject to:}$$
$$6u_1 + 12u_2 + 12u_3 + 8u_4 \leq 30$$
$$8u_1 + 16u_2 + 9u_3 + 4u_4 \leq 24$$
$$2u_1 + u_2 + 3u_3 + 4u_4 \leq 10$$
$$u_i \geq 0 \text{ for } i = 1, 2, 3, 4.$$

Visibly, and as expected, this is a standard form maximization problem, so can be solved using Phase II pivoting (Algorithm 5.2). Before going further, let's review how to interpret the meaning of the dual variables. Recall that the decision variable u_i of the dual is the same as the marginal value M_i of the primal problem, and similarly, the decision variables x, y, z of the primal are the same as the marginal values (let's call them N_1, N_2, and N_3) of the constraints in the dual. Finally, recall that the maximum value of Q is equal to the minimum value of w by the principle of strong duality.

We can now set up the initial tableau for the maximization problem, as usual, from the statement/setup of the dual problem, by adding the slack variables $T_i, i = 1, 2, 3$ to the constraint inequalities, and then adding the required extra information to the augmented coefficient matrix of the resulting system of equations. Of course, it might be quicker and easier to just insert the missing columns for the slack variables T_i directly into the condensed tableau. Either way, we obtain this following tableau.

	u_1	u_2	u_3	u_4	T_1	T_2	T_3	Capacity
T_1	6	12	12	8	1	0	0	30
T_2	8	16	9	4	0	1	0	24
T_3	2	1	3	4	0	0	0	10
Q	−30	−30	−45	−40	0	0	0	0

The reader should check that after three pivots, we arrive at the final tableau.

	u_1	u_2	u_3	u_4	T_1	T_2	T_3	Capacity
u_3	0	$\frac{5}{4}$	1	0	$\frac{1}{4}$	$-\frac{1}{12}$	$-\frac{5}{12}$	$\frac{4}{3}$
u_1	1	$\frac{5}{4}$	0	0	$-\frac{1}{4}$	$\frac{1}{4}$	$\frac{1}{4}$	1
u_4	0	$-\frac{21}{16}$	0	1	$-\frac{1}{16}$	$-\frac{1}{16}$	$\frac{7}{16}$	1
Q	0	$\frac{45}{4}$	0	0	$\frac{5}{4}$	$\frac{5}{4}$	$\frac{25}{4}$	130

Read off the solution to the problem from the final tableau:

$$u_1 = 1$$
$$u_2 = 0$$
$$u_3 = \frac{4}{3}$$
$$u_4 = 1$$
$$Q = 130 \text{ (maximized)}$$

$$T_1 = 0 \qquad N_1 = \frac{5}{4}$$
$$T_2 = 0 \qquad N_2 = \frac{5}{4}$$
$$T_3 = 0 \qquad N_3 = \frac{25}{4}.$$

But wait a minute – we are not especially interested in the solution to the maximization problem since it is the dual problem. We want to know the solution of the primal, minimization problem. However, the reader can see most of that information immediately, in light of how to interpret the variables. Thus, we can write down the following information:

$$x = \frac{5}{4} \text{ (same as } N_1)$$
$$y = \frac{5}{4} \text{ (same as } N_2)$$
$$z = \frac{25}{4} \text{ (same as } N_3)$$
$$w = 130 \text{ (minimized)}$$

$$M_1 = 1 \ (\text{same as } u_1)$$
$$M_2 = 0 \ (\text{same as } u_2)$$
$$M_3 = \frac{4}{3} \ (\text{same as } u_3)$$
$$M_4 = 1 \ (\text{same as } u_4).$$

The only information missing from this solution is the values of the surplus variables S_i $i = 1, 2, 3, 4$. Of course, since the marginal value M_1 is nonzero, we could deduce that $S_1 = 0$ by complementary slackness (see Section 4.5.3). The same holds for S_3 and S_4. Also, by complementary slackness, since $M_2 = 0$, we expect $S_2 \neq 0$, but we do not know its exact value yet. However, look carefully at the final tableau. We've already said more than once that the important data is in the last column and in the bottom row. But we have not considered *all* of the data in the bottom row. The entries in the bottom row in the columns marked by slack variables T_i are, of course, the marginal values N_i, but what about the entries in the bottom row in the columns marked by decision variables u_i? Notice that the entries are 0 in exactly the columns marked by $u_1, u_3,$ and u_4, and these subscripts match exactly the S_i that are 0. The one entry that is nonzero is in the u_2 column. Could it be that S_2 is equal to that entry, $\frac{45}{4}$? If so, then we would have an interpretation for the dual primal problem of every entry in the last row of the final tableau. We can check as follows. The S_2 slack variable goes with the second constraint in the minimization problem:

$$12x + 16y + z \geq 30.$$

Substituting in the values for x, y, z that we just found yields

$$12\left(\frac{5}{4}\right) + 16\left(\frac{5}{4}\right) + \frac{25}{4} \geq 30$$
$$15 + 20 + \frac{25}{4} \geq 30$$
$$\frac{165}{4} \geq 30.$$

Indeed, we see a surplus of exactly $\frac{45}{4}$ in this inequality at the optimal point. Thus, the complete solution to the primal (minimization) problem is

$$x = \frac{5}{4}$$
$$y = \frac{5}{4}$$
$$z = \frac{25}{4}$$
$$w = 130 \ (\text{minimized})$$

$S_1 = 0$	$M_1 = 1$
$S_2 = \dfrac{45}{4}$	$M_2 = 0$
$S_3 = 0$	$M_3 = \dfrac{4}{3}$
$S_4 = 0$	$M_4 = 1.$

It can be shown (although we will not do so here) that what happened in this problem always occurs. The values of the surplus variables for the minimization (primal) always appear in the bottom row of the final tableau for the dual maximization problem, under the columns labeled with the decision variables of the dual. In particular, whenever one solves a maximization problem using the simplex algorithm, one is actually solving two problems at once, since the solution to the dual problem also appears in the final tableau. This is worth stating as an important result, which we will use in future chapters:

COROLLARY 5.9 *When using the simplex algorithm to solve a linear programming (maximization) problem, one automatically obtains the solution to the dual (minimization) problem with no additional work. The solution to the maximization problem appears in the last column (capacity column) of the final tableau, and the solution to the minimization problem appears in the bottom row of the final tableau. More precisely, in the capacity column, you can read off the values of all the basic variables (decision and slack variables) and the maximum value of the objective function in the bottom-right corner, which is equal to the minimum value of the objective function of the dual minimization. (The nonbasic decision and slack variables are not listed, but, of course, they have value 0.) The marginal values of the maximization are the same as the decision variables of the minimization and appear in the bottom row, under the columns headed by the slack variables for the maximization. The surplus variables of the minimization appear in the bottom row under the columns headed by decision variables for the maximization. Finally, the marginal values of the minimization are the same as the decision variables of the maximization.*

Of course, in this section, we have not yet learned Phase I pivoting, so we can only apply this to standard form maximization problems so far. But later, when we cover Phase I, we can check that the corollary still holds true even for problems not in standard form.

We conclude this section with one more example of solving a minimization problem using the magic of duality. Recall the following example from Chapter 3.

O'Casek's Used Cars. Mr. O'Casek is a used car dealer who just opened a second lot in a nearby town. He needs to ship some of his existing inventory from his first lot to his new one, a drive of 30 miles. He decides that he should transport at least 24 SUVs, at least 30 sedans, and at least 40 compact cars. He calls Moving in Stereo, Inc., who has two types of vehicles to ship the cars – automobile transport vans and flatbed trucks. Each automobile transport van can hold one SUV, two sedans, and four compact cars (simultaneously), while each flatbed truck can hold two SUVs, one sedan, and one compact car (simultaneously). Between fuel and driver expenses, the cost for a transport van amounts to $6 per mile, while the cost for flatbed truck is $4 per mile. How many of each vehicle should O'Casek hire in order to minimize the total cost of the move?

The setup:

Let x be the number of transport vans.
Let y be the number of flatbed trucks.
Minimize: $C = 6x + 4y$
subject to:
$x + 2y \geq 24$ (SUVs)
$2x + y \geq 30$ (sedans)
$4x + y \geq 40$ (compact cars)
$x \geq 0, \ y \geq 0$.

This leads to the condensed tableau.

	x	y	Capacity
S_1	1	2	24
S_2	2	1	30
S_3	4	1	40
C	6	4	0

Take the transpose to obtain the condensed tableau for the dual maximization problem.

	u_1	u_2	u_3	Capacity
T_1	1	2	4	6
T_2	2	1	1	4
P	24	30	40	0

We won't pause here to write out the setup of the maximization problem, which has decision variables $u_i (i = 1, 2, 3)$ and slack variables $T_j (j = 1, 2)$, although it would be good practice for the reader to do so. We proceed right to the initial tableau by inserting the slack variable columns and negating the bottom row.

	u_1	u_2	u_3	T_1	T_2	Capacity
T_1	1	2	4	1	0	6
T_2	2	1	1	0	1	4
P	−24	−30	−40	0	0	0

The reader should check (see Exercise 1) that after three pivots, the following final tableau is reached.

	u_1	u_2	u_3	T_1	T_2	Capacity
u_2	0	1	$\frac{7}{3}$	$\frac{2}{3}$	$-\frac{1}{3}$	$\frac{8}{3}$
u_1	1	0	$-\frac{2}{3}$	$-\frac{1}{3}$	$\frac{2}{3}$	$\frac{2}{3}$
P	0	0	14	12	6	96

We are interested in the solution to the minimization problem only (but see Exercise 2), which appears in the last row:

$$x = 12 \text{ transport vans}$$
$$y = 6 \text{ flatbed trucks}$$
$$C = \$96 \text{ per mile} \times 30 \text{ miles} = \$2,880 \text{ (minimized)}$$
$$S_1 = 0 \text{ (surplus SUVs)} \qquad M_1 = \tfrac{2}{3}$$
$$S_2 = 0 \text{ (surplus sedans)} \qquad M_2 = \tfrac{8}{3}$$
$$S_3 = 14 \text{ (surplus compact cars)} \qquad M_3 = 0.$$

Comparing this to the solution we obtained graphically in Section 3.2, we see agreement. But we now also have information that we did not have with the graphical solution – namely, the marginal values for SUVs, sedans, and compact cars. How should we interpret these? With a greater-than constraint, we consider the possible savings in our objective function by lowering that constraint by one unit. In other words, if we lowered the SUV constraint from a minimum of 24 to a minimum of 23, presumably we'd save $\$\frac{2}{3}$ per mile (times 30 miles or $20 total, lowering the total cost to $2,860). In fact, for each SUV we lower the constraint by, we'd save a total of $20 per SUV (until we lowered it by enough to interfere with the other constraints, leading to an optimum point with a different set of basic variables.) Similarly, in the sedan constraint, for each sedan we lower the constraint below 30, we'd save $\$\frac{8}{3}$ per mile $\times$ 30 miles $=$ $80 in total costs. Finally, why is the marginal value for compact cars 0? Because that constraint is not binding – we are already shipping a surplus of 14 compact cars. If we lowered the minimum to 39 instead of 40, there'd be no savings in cost at all; we'd end up at the same solution point $(x, y) = (12, 6)$, but with 15 surplus cars instead of 14.

In summary, we can now solve any standard form problem (maximum or minimum), by Phase II pivoting (Algorithm 5.2) and obtain the complete solution including the values of slack or surplus variables as well as marginal values.

Exercises

1. Perform the pivoting needed in the O'Casek's used car example, and verify that the final tableau is correct.
2. In the O'Casek used car example, (a) write out the setup of the dual maximization problem in terms of inequalities, and (b) write out the complete solution to this maximization problem by reading it off from the final tableau.
3. Solve the minimization problem using the simplex algorithm (see Exercise 8 from Section 3.1):

$$\text{minimize: } C = 20x + 12y$$
$$\text{subject to:}$$
$$5x + 2y \geq 30$$
$$5x + 7y \geq 70$$
$$x \geq 0, \ y \geq 0.$$

4. Solve the Toys in the Attic problem (Exercise 13 from Section 3.1) using the simplex algorithm.
5. Solve the Poseidon's Wake Petroleum Company problem (Exercise 10 from Section 3.2) using the simplex algorithm.
6. Solve Part a of the Topographic Starship Tour Company problem (Exercise 20 from Section 3.2) using the simplex algorithm.
7. Consider the minimization problem:

$$\text{minimize } C = 60x + 40y$$
$$\text{subject to:}$$
$$10x + 2y \geq 80$$

$$6x + 2y \geq 68$$
$$4x + 5y \geq 40$$
$$x + y \geq 18$$
$$x \geq 0, \ y \geq 0.$$

a. Solve the problem via graphing in the decision space. One of the constraints has an unusual property – what is it?

b. Solve the problem via the simplex algorithm. Your solution should agree with what you obtained in Part a.

8. A dietitian at a hospital must design a meal for a vitamin-deficient patient. Each ounce of mashed potatoes has one unit of protein, six units of vitamins, and two units of fat. Each ounce of broccoli has two units of protein, two units of vitamins, and one unit of fat. Each ounce of chicken breast has eight units of protein, one unit of vitamins, and one unit of fat. Each ounce of corn has three units of protein, five units of vitamins, and two units of fat. The meal must consist of at least 60 units of protein and at least 40 units of vitamins. Also, it should contain at least 12 ounces of food altogether. Design a diet that minimizes the fat content. Use the simplex algorithm.

5.4   Solving Linear Programming Problems with a Computer

There are many specialized software packages today that were developed to solve linear programming problems or more general optimization problems such as integer programming, nonlinear programming, or dynamic programming. It is not our intention to recommend any specific software packages or to even try to provide a list of such packages, which would probably be out of date as soon as this book was published.

Instead, we focus on two mathematical packages that are much more general tools but are widely accessible to beginners and also happen to be able to efficiently solve linear programming problems. Namely, we discuss solving linear programming problems using Wolfram Mathematica and Microsoft Excel.

5.4.1   Using Mathematica

This discussion is based on version 10.4 of Mathematica, and there could well be changes in future versions. When in doubt, consult the Mathematica documentation. The **Linear Programming** command can be found in the Symbolic and Numeric Computation chapter of the documentation, under the section entitled Optimization.

The reader already familiar with Mathematica knows that the correct syntax is very important when typing input. A case in point is how Mathematica uses the different types of certain delimiters, specifically parentheses ( ), brackets [ ], and set braces { }. Parentheses are used in grouping terms for algebraic calculations, so one would type, for example, $3(2x + 7)$ rather than $3[2x + 7]$. Brackets are used in functions and commands, so one would write $Cos[x]$ or $f[x]$ rather than $Cos(x)$ or $f(x)$, and similarly for commands. If A is a square matrix whose determinant is needed, one would type the command **Determinant[A]** as opposed to Determinant(A). Finally, braces are used in making (ordered) lists. For example, a vector or an ordered n-tuple is regarded as a list in Mathematica, so that, for example, the ordered triple $(1,2,5)$ would be written as $\{1,2,5\}$. The user who ignores or forgets these distinctions would quickly become frustrated.

In particular, a matrix is regarded in Mathematica as a list of vectors, or a list of lists. For example, to enter the matrix

$$\begin{bmatrix} 1 & 0 & 0 & 3 \\ 9 & 9 & 9 & 9 \\ 6 & -1 & -2 & -3 \end{bmatrix}$$

as input in Mathematica, one would type

$$\{\{1,0,0,3\},\{9,9,9,9\},\{6,-1,-2,-3\}\}. \tag{5.3}$$

One can force Mathematica to display a matrix A in the usual way, by using the command **MatrixForm[A]**, but it's helpful to know that it is treated as a list of lists internally. In the language of Mathematica, if A and B are matrices, the notation $A.B$ indicates the matrix product AB, or the dot product if A and B are vectors.

DEFINITION 5.10 *Given two n-tuples* $a = (a_1, a_2, \ldots, a_n)$ *and* $b = (b_1, b_2, \ldots, b_n)$ *we write* $a \leq b$ *to mean* $a_i \leq b_i$ *for* $i = 1, 2, \ldots, n$. *We use similar conventions for* $a < b$, $a \geq b$, *and* $a > b$.

For example, if $a = (2,1,4)$ and $b = (3,1,7)$, then $a \leq b$ because $2 \leq 3$, $1 \leq 1$, and $4 \leq 7$. However, since 1 is not strictly less than 1 in the second coordinate, we do not have $a < b$. Unlike the case with single numbers, it is not always possible that vectors can be compared in this way. For example, if $a = (1,2)$ and $b = (2,1)$, then neither $a \leq b$ nor $a \geq b$ holds, so a and b are not comparable.

The most basic linear programming command in Mathematica has the following form (in any Mathematica system command, it is important to duplicate the capitalization for each word, the lack of spaces between words, and the bracket delimiters as shown in this example):

LinearProgramming[c,m,b]

Here, c and b are vectors and m is a matrix. When this command is executed, it finds a vector X (if possible) which minimizes $c.X$ subject to the constraints $m.X \geq b$ and $X \geq 0$. This is a standard form minimization problem. Indeed, if the vector X is the vector of decision variables, then the dot product $c.X$ is the objective function, so the entries of c are the coefficients of the objective function. Also, the condition $X \geq 0$ translates to the standard nonnegativity constraints, while the condition $m.X \geq b$ are the structural constraints. The vector b would correspond to the entries in the capacity column of the system of inequalities, and the matrix m is the coefficient matrix of the system of inequalities.

We illustrate this with a familiar example: the O'Casek's used car problem. The setup of this problem, as noted in the last section, is as follows.

> Let x be the number of transport vans.
> Let y be the number of flatbed trucks.
> Minimize: $C = 6x + 4y$
> subject to:

$$x + 2y \geq 24 \text{ (SUVs)}$$
$$2x + y \geq 30 \text{ (sedans)}$$
$$4x + y \geq 40 \text{ (compact cars)}$$
$$x \geq 0, \ y \geq 0.$$

If we let $c = (6, 4)$, $X = \begin{pmatrix} x \\ y \end{pmatrix}$, $b = \begin{pmatrix} 24 \\ 30 \\ 40 \end{pmatrix}$, and $m = \begin{pmatrix} 1 & 2 \\ 2 & 1 \\ 4 & 1 \end{pmatrix}$, we make the following

observations:

1. $c.X = 6x + 4y$ is the objective function C that we are trying to minimize.

2. the matrix product $m.X$ is $\begin{pmatrix} x + 2y \\ 2x + y \\ 4x + y \end{pmatrix}$, so the structural constraints are precisely the

condition $m.X \geq b$, and

3. $X \geq \begin{bmatrix} 0 \\ 0 \end{bmatrix}$ means $x \geq 0$ and $y \geq 0$, which are precisely the nonnegativity constraints.

Thus, the problem can be stated as the following: Find X in order to minimize $c.X$ subject to $m.X \geq b$ and $X \geq 0$, which is precisely the formulation needed for the Mathematica command. Incidentally, this is called the *matrix formulation* of the problem. Every linear programming problem has a matrix formulation; for example, see Calvert and Voxman (1989).

Thus, to solve this problem using Mathematica, open a Mathematica Notebook, and enter the command:

LinearProgramming[{6,4},{{1,2},{2,1},{4,1}},{24,30,40}]

(We remind the reader that after typing the command, to enter it as input, you press the "Enter" key on your keyboard while holding the "Shift" key down.)

The program will quickly return the output:

{12,6}

and the problem is solved. Indeed, we recognize the solution $x = 12$ transport vans and $y = 6$ flatbed trucks as the optimal solution from our previous work. However, the disadvantage is that a lot of the optimal data is missing and must be found with additional work. Although it is not hard to find the value of $C = 6x + 4y = 6(12) + 4(6) = \$96$ per mile and the values of the surplus variables (much as we did when we solved linear programming problems graphically), it would be somewhat more of an effort to find the marginal values. One of the advantages of Excel over Mathematica is that we can have the program display all the optimal data at once, including many extra bits of information that are useful, for example, for sensitivity analysis.

Of course, for the student who is skilled in writing Mathematica programs, it should be a fairly easy task to write a program that will fill in all the missing information and, indeed, to perhaps display all the intermediate tableaux as the simplex method iterates. We leave that as a challenge for the motivated reader.

However, despite this drawback to the Mathematica approach, there are numerous positive aspects. First of all, Mathematica is extremely powerful, so it will solve your linear programming problems very quickly, including very large problems with many decision variables and many constraints. Secondly, the linear programming command is quite general. There are versions of the command that allow the constraints to be mixed, so it is not required that the problem be in standard form. We'll illustrate an example of this in this section. Also, there are versions of the command that will generalize the nonnegativity constraints so that instead of each decision variable x_i being required to be nonnegative, we can require it to be nonpositive, greater than (or less than) some other number than 0, or restricted to a range of values. In fact, one can put a different range on each decision variable, if desired. Thus, if l is a number, the command

$$\textbf{LinearProgramming[c,m,b,l]}$$

solves the linear programming problem where $x \geq 0$ has been replaced by $x \geq l$. If we replace l by a vector, the command

$$\textbf{LinearProgramming[c,m,b,\{l_1,l_2,\ldots l_n\}]}$$

solves the problem where $x \geq l$ has been replaced with $x_i \geq l_i$ for each $i = 1, 2, \ldots, n$. If each l_i is replaced by a pair l_i, u_i with $l_i \leq u_i$, then the command

$$\textbf{LinearProgramming[c,m,b,\{\{l_1,u_1\},\{l_2,u_2\},\ldots,\{l_n,u_n\}\}]}$$

solves the problem where $x_i \geq l_i$ has been replaced by $l_i \leq x_i \leq u_i$ for all $i = 1, 2, \ldots, n$. One can also specify that some or all of the x_i values be integers instead of real numbers. That is *integer programming*, which we cover in Chapter 9.

To incorporate mixed constraints, in the vector b, replace each b_i with a pair $\{b_i, s_i\}$, where s_i is 1, 0, or -1. If $s_i = 1$, the ith constraint reads, as usual, $m_i.x \geq b_i$ (where m_i is the ith row of m). If $s_i = 0$, the constraint is replaced by an exact equality $m_i.x = b_i$, and if $s_i = -1$, it is replaced by the reverse inequality $m_i.x \leq b_i$.

Finally, to solve a maximization problem, convert it to a minimization problem by multiplying c through by -1 so that we are minimizing $-c.x$ instead of maximizing $c.x$. We illustrate a maximization problem now with mixed constraints:

Example 5.11 The Tillerman Tea Company mixes three blends of loose-leaf tea. Each case of Mild World Blend requires 5 pounds of black tea, 5 pounds of green tea, and 1 pound of an herb/spice blend, and generates $20 profit for the company. Each case of the Peace Strain requires 6 pounds of black tea, 8 pounds of green tea, and 2 pounds of herb/spice blend, and generates $30 profit for the company. Each case of Shadowmoon Chai Mix requires 4 pounds of black tea, 2 pounds of green tea, and 10 pounds of herb/spice blend, and generates $32 profit for the company. Every week, the company has 1,200 pounds of black tea, 1,500 pounds of green tea, and 800 pounds of the herb/spice blend to work with. It can store leftover tea for future use, but the herb/spice blend loses potency, so the company, in its dedication to freshness, decides to completely use up the 800 pounds of spice blend each week. Furthermore, due to a contract with Cat's Bull Natural Foods Store, the company is obligated to produce at least 50 cases of the Shadowmoon Chai each week, and due to a contract with Steven's Grocery, it is obligated to produce at least a total of 130 cases per week of the non-chai blends. How many cases of each tea should be mixed each week in order to maximize profit?

Here is the setup of this problem.

Let x be the number of cases of Mild World Blend.
Let y be the number of cases of Peace Strain.

Let z be the number of cases of Shadowmoom Chai Mix.
Maximize $P = 20x + 30y + 32z$
subject to:
$5x + 6y + 4z \leq 1{,}200$ (lbs. of black tea)
$5x + 8y + 2z \leq 1{,}500$ (lbs. of green tea)
$x + 2y + 10z = 800$ (lbs. of herb/spice blend)
$z \geq 50$ (contract with Cat's Bull Natural Foods)
$x + y \geq 130$ (contract with Steven's Grocery)
$x \geq 0, \ y \geq 0, \ z \geq 0$ (nonnegativity).

According to the directions, we should minimize $-P = -20x - 30y - 32z$, and we should modify the vector b to incorporate the exact equality constraint (the herb/spice blend) and the less-than-type constraints (the black tea and green tea). Also, we can incorporate the contract with Cat's Bull by modifying the nonnegativity constraint for z.

Therefore, the proper Mathematica command would be

LinearProgramming[{–20,–30,–32},

{{5,6,4},{5,8,2},{1,2,10},{1,1,0}},

{{1200,–1},{1500,–1},{800,0},{130,1}},{0,0,50}]

In Mathematica, it should be entered in one line – it's displayed on several lines here as it is too wide to fit in the margins. This yields the output:

{50,125,50}

This corresponds to the solution $x = 50$ cases of Mild World, $y = 125$ cases of Peace Strain, and $z = 50$ cases of Shadowmoon Chai. In this case, the profit is $P = 20(50) + 30(125) + 32(50) = \$6{,}350$ per week, which is maximized. Notice that the amount of black tea used is $5(50) + 6(125) + 4(50) = 1{,}200$ pounds, so there happens to be no leftover black tea. The amount of green tea used is $5(50) + 8(125) + 2(50) = 1{,}350$, so there are 150 pounds of leftover green tea. The amount of herb/spice blend used is $50 + 2(125) + 10(50) = 800$, so there is no leftover herb/space blend, as required. Also, both of the contractual requirements have been met since we are producing 50 cases chai (the required minimum) and $x + y = 50 + 125 = 175$ cases of non-chai, which is 45 cases surplus beyond the required minimum. As mentioned before, we do not have information about the marginal values of the constraints without more work.

On the other hand, there is a command called **DualLinearProgramming[ ]** that does solve both the primal and the dual problem at the same time! It has the exact same arguments (what goes in the brackets) as the **LinearProgramming[ ]** command, and returns a list $\{x, y, z, w\}$ as output. In this output, the x vector is the solution to the primal, and the y vector is the solution to the dual. The interpretation of z and w is not important for our purposes – they arise in the statement of the most general dual problem, when the primal variables are restricted by a range of values other than the natural nonnegativity constraints (and z and w will have values of 0 for standard form problems). We close this subsection by illustrating this phenomenon with a familiar problem, Angie and Kiki's lemonade stand. Recall the setup.

Let x be the number glasses of sweet lemonade.
Let y be the number of glasses of tart lemonade.

$$\text{Maximize } R = 1.25x + 1.5y$$

subject to:

$3x + 4y \le 60$ (lemons)

$x + 2y \le 28$ (limes)

$2x + y \le 30$ (tbsp. sugar)

$x \ge 0, \ y \ge 0.$

When using the **LinearProgramming** or the **DualLinearProgramming** command, we convert the decimals to rational numbers in order to obtain exact solutions. When Mathematica sees a problem with decimals, it automatically switches its algorithm from the simplex algorithm to some interior point method, which only gives an approximate answer, to within some specified tolerance. To avoid that and find exact answers, use rational numbers instead of decimals.

Thus, the appropriate command is

$$\textbf{DualLinearProgramming}[\{-\tfrac{5}{4}, \ -\tfrac{3}{2}\}, \{\{3,4\}, \{1,2\}, \{2,1\}\},$$

$$\{\{60,-1\}, \{28,-1\}, \{30,-1\}\}]$$

The output is

$$\{x,y,z,w\} = \{\{12,6\}, \{-\tfrac{7}{20}, 0, \ -\tfrac{1}{10}\}, \{0,0\}, \{0,0\}\}$$

As expected, the values of z and w are 0 since the problem is standard form. The solution is $x = 12$ glasses sweet lemonade and six glasses tart lemonade. The solution to the dual problem are the marginal values, except we must drop the minus sign in their interpretation because of the factor of -1 by which we multiplied the objective function. Thus, the marginal value of a lemon is $\$\tfrac{7}{20} = \$0.35$, the marginal value of a lime is \$0, and the marginal value of a tablespoon of sugar is $\$\tfrac{1}{10} = \$0.10$. We recall these solutions as correct from our work in earlier chapters. Of course, to find the value of the objective function, we compute $R = 1.25(12) + 1.5(6) = \$24.00$. We will find the **DualLinearProgramming** command useful when we apply linear programming to game theory in Section 7.3.

5.4.2 Using Excel

This discussion is based on Excel 2016, although there would be very little change from several previous versions. Many people are more familiar with Excel than Mathematica, but as we did in Section 5.4.2, we first review some general facts about the use of Excel before solving any linear programming problems.

First, let's describe how to take dot products of vectors, or *arrays*, as they are called in Excel. This is one of the built-in functions, but in Excel, the function is called *sumproduct* rather than dot product. Let's illustrate with an example. Suppose we wish to compute the dot product of the two vectors $A = (\ 2 \quad 3 \quad 5\ )$ and $B = \begin{pmatrix} -1 \\ 4 \\ 2 \end{pmatrix}$. As we know, the answer is $2(-1) + 3(4) + 5(2) = 20$. To compute this as a matrix product, we would need to keep B as a column so the inner dimensions would match. But as vectors, it doesn't matter, and, indeed, in Excel, we'd enter them both as rows. Enter each number in a distinct cell. The screenshot in Figure 5.1 illustrates the result (note that we have enlarged the view of the cells where we entered the data to make it readable in the screenshot).

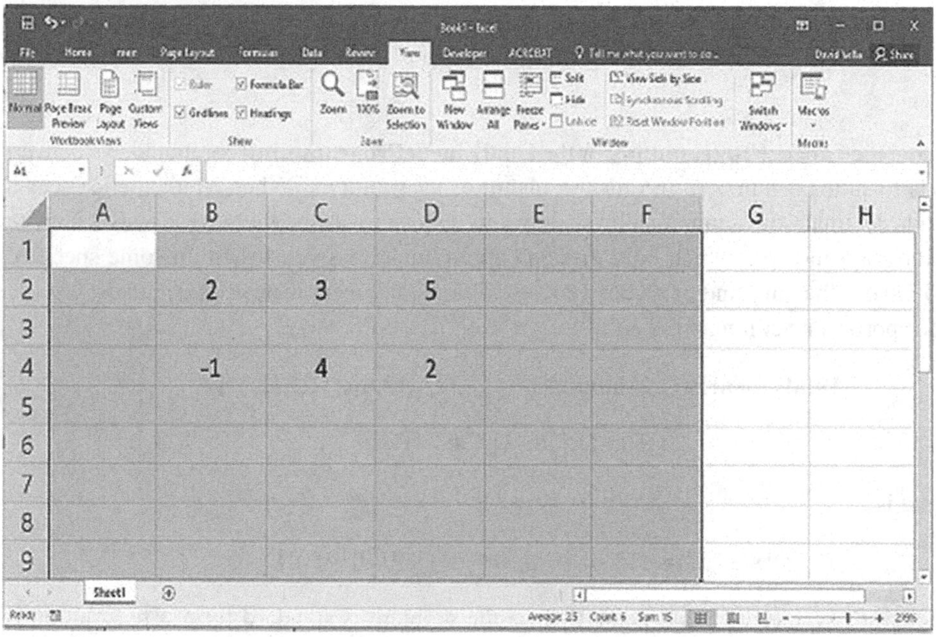

Figure 5.1 Entering data in an Excel spreadsheet.

As you can see, we chose adjacent cells for the entries in each vector, but other than that, we were free to enter the vectors where we wanted. They need not be in adjacent rows (as shown), nor do the cells need to line up. We could have entered the vector B in cells C4, D4, and E4 if we wanted. Notice also that in the screenshot, each number is centered in the cell. This is a cosmetic touch that is not really necessary. When the numbers are entered in the cells, they by default go to the right side of the cell (that is, they are right-justified.) You can center them using the alignment icons in the Home tab on the ribbon (above the spreadsheet). The highlighted icon in the bottom row of the ribbon is for center justification. You can do this cell by cell, or for an entire row by clicking on the row number before clicking the alignment icon, and similarly for an entire column, or even for the entire page, by clicking on the little triangle in the upper left of the spreadsheet (just beneath the cell locater which reads "D9" in the screenshot) before clicking on the alignment icon. There are other cosmetic enhancements that we will illustrate in the following paragraphs. None of these visual enhancements are necessary, but they do make the output organized and easier to read.

Next, select a cell where we will put the result of the computation. For example, if we want to put the output into cell F2, it might help keep things organized if we label it in an adjacent cell such as F1. So, after labeling cell F1 with "dot product" or "solution," then click on cell F2, where the answer will appear. While that cell is highlighted, click on the *function* icon, which is the symbol f_x visible just to the left of the input line above the cells of the spreadsheet. The function you want, as noted before, is called sumproduct, but unless it was recently used, it will not appear in the list of functions in the "Insert Function" dialog box that opens. If you can't find it in the "current" category, click on the "all" category, which lists every built in function in alphabetical order. Here is another screenshot, shown in Figure 5.2.

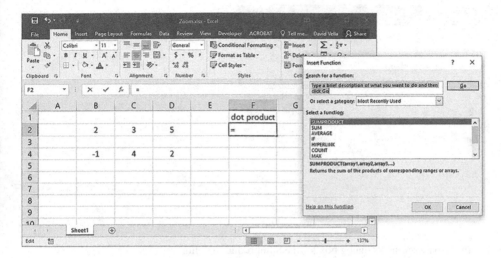

Figure 5.2 Find the Sumproduct function in the Insert Function dialog box.

Once you have the correct function highlighted, click "OK" in the dialog box. A new dialog box called "Function Arguments" opens, as shown in Figure 5.3.

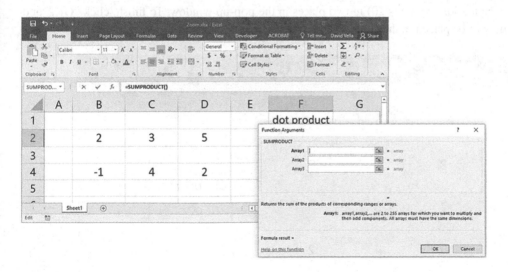

Figure 5.3 The Function Arguments dialog box.

You can see (part of) the name of the function inside the cell F2, as well as in the input field above the spreadsheet. In the dialog box, we need to fill in Array 1 and Array 2 with the vectors A and B. However, you do not do that by typing in the entries of these vectors. Instead you must refer to the cell addresses of the arrays. So, in Array 1, you would fill in B2:D2. The colon indicates the array consists of all the cells between (and including) cells B2 and D2. Similarly, enter B4:D4 in

Array 2. Once this is done, you can see if the arrays are correctly entered right in the dialog box next to the array rows, as shown in Figure 5.4.

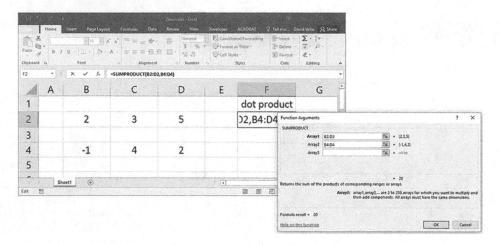

Figure 5.4 The Function Arguments dialog box for the Sumproduct function.

Rather than type in "B2:D2," there is a shortcut to entering this. Click on Array 1 and then move the cursor to cell B2. Click and while holding the mouse button down, drag it to cell D2, and then let go. This will enter the correct cells in Array 1. Then repeat for Array 2. In the dialog box, not only can you check that the arrays have been entered correctly, but you can already see the result of the computation (which is 20) in two places in the pop-up window. To finish, click "OK," and the result will be placed in the chosen cell F2. The result is shown in Figure 5.5.

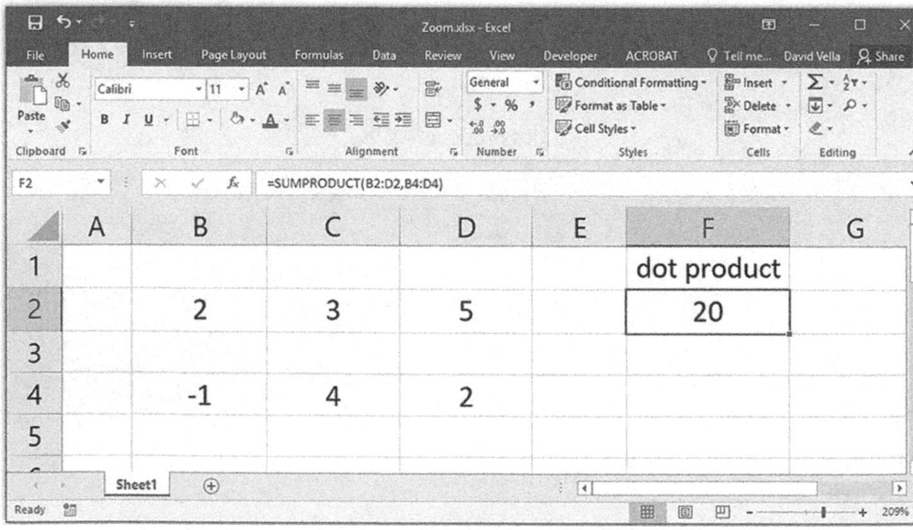

Figure 5.5 Excel puts the output of the function in the highlighted cell F2.

As the reader can see, the solution is now displayed in cell F2, and we have added another cosmetic attribute – the heavy outline around the edge of that cell. This is added using the drop-down menu just to the left of the paint bucket above. Speaking of the paint bucket, we will also use this drop-down menu to put background colors in some of the cells when we solve linear programming problems to make it easier to organize and display the solution.

One last comment before going on to solve linear programming problems. Now that the "sumproduct" function has been attached to cell F2, any change in the entries of the arrays in the cells B2–D2 or B4–D4 will be instantly result in an updated cell F2 with the new correct dot product. For example, change the 3 to a 6 in cell D4, and see the dot product in cell F2 change from 25 to 40. Go ahead and try it!

Now that the reader has reviewed how to use the function icon (and, in particular, how the sumproduct function works), we are ready to solve a linear programming problem using Excel. What is needed is an add-in to Excel called Solver. If nobody has previously used the Solver add-in in your copy of Excel, it may not be installed. To see if it is installed, go to the "Data" tab and look all the way over to the right – it's circled in the screenshot in Figure 5.6.

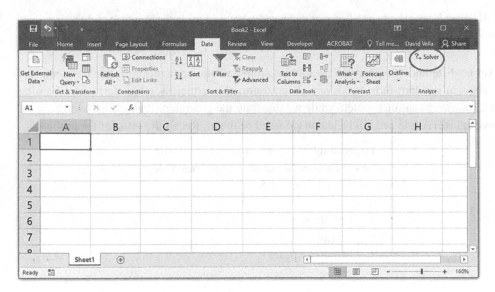

Figure 5.6 If the Solver Add-in is already installed, you will see it in the Data tab of the ribbon, on the extreme right side (it is circled in the screenshot).

If you see the word "Solver" as shown in Figure 5.6, then the add-in is already installed, but don't click on it yet – we have to set up the spreadsheet first. If you do not see the word "Solver" there, then you need to install the add-in. To do that, go to the "Insert" tab. On the ribbon, the third box from the left is entitled "Add-Ins." In that box is a drop-down menu entitled "My Add-ins," circled in the screenshot in Figure 5.7.

Open that drop-down menu and click on the bottom entry, entitled "Manage Other Add-ins." This will take you to a list of add-ins, of which "Solver" will be one. Click on it and then click "Go" to install it. Once you have installed it, it should always appear on the Data tab as indicated earlier; you do not need to install it each time it is used. Finally, be aware that there is an online version of the Solver add-in, but that is not the one you are installing; you want the local one so you can analyze data from your own hard drive, which is the one that appears under "Manage Other Add-ins." Once the Solver add-in is installed, you are ready to go.

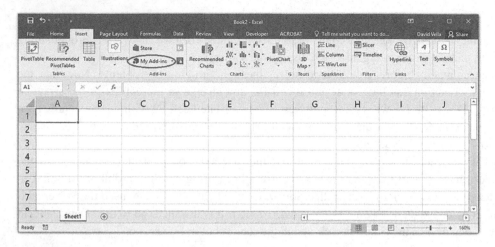

Figure 5.7 If Solver is not already installed, you find it under My Add-ins in the insert tab of the ribbon (circled in the screenshot).

It makes no difference whether you are solving a maximization or a minimization problem, or if the problem is in standard form or has mixed constraints. Nevertheless, let us start with a familiar standard form maximization problem while we illustrate the procedure. Indeed, let's use Excel to solve (once again!) Angie and Kiki's lemonade stand problem. Recall the setup of the problem.

> Let x be the number glasses of sweet lemonade.
> Let y be the number of glasses of tart lemonade.
> Maximize $R = 1.25x + 1.5y$
> subject to:
> $3x + 4y \leq 60$ (lemons)
> $x + 2y \leq 28$ (limes)
> $2x + y \leq 30$ (tbsp. sugar)
> $x \geq 0,\ y \geq 0.$

The first step is to enter the important data on a spreadsheet. We will need two cells to stand for the values of x and y (we begin by entering these values as 0). We also need cells for the two objective coefficients 1.25 and 1.5. We need a cell to stand for the value of R (which we want to maximize), as well as cells for all the coefficients in the structural constraints and for all the numbers in the capacity column. We show two screenshots: the first with just the bare bones data entered (in Figure 5.8) and the second enhanced with all the cosmetic aids including centering the entries of each cell, labeling various important cells, putting in heavy borders and color-coding some of the cells (in Figure 5.9).

The reason we have left a space between column C and column G is that we will need to fill in some things in column E momentarily. But first, let's dress up the display a little bit by adding the cosmetic touches.

Both screenshots contain the same information, but the cosmetically enhanced one is easier to read and remember what every cell stands for. Now, in the second screenshot, there are four colored cells in column E with no data filled in – the purple revenue cell and the three orange cells below it. Since the screen shot is in black and white, you can't see the actual colors we used, and you can make your own color choices anyway. Into these cells we need to put dot product

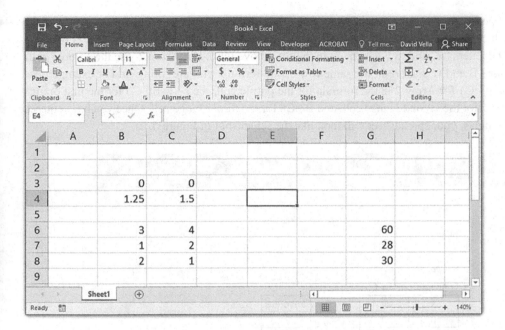

Figure 5.8 Entering the data for the lemonade stand problem.

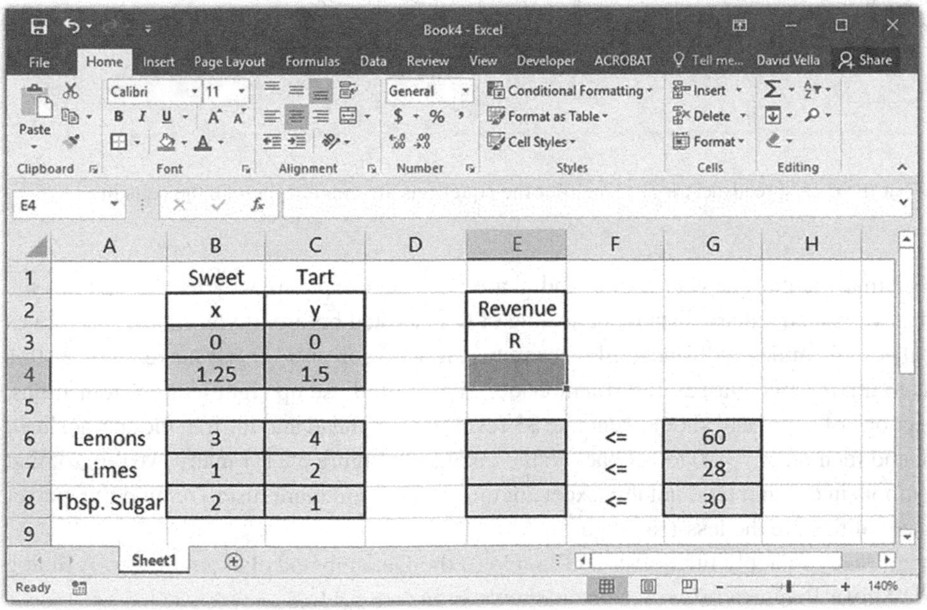

Figure 5.9 The same data with cosmetic additions to help organize the spreadsheet.

formulas. In the purple cell, we would enter the formula for revenue: $1.25x + 1.5y$. In the first orange cell, we put the expression for the number of lemons used, $3x + 4y$; in the second, the expression for the number of limes used, $x + 2y$; and in the third, the formula for the number of tablespoons sugar used, $2x + y$. You enter these formulas using the function icon f_x and the sumproduct function. In each of the four cases, use cells B3:C3 for Array 1 – the *values* of x and y (not the letters "x" and "y," which are in the row above that). For Array 2, use the colored cells in columns B and C that are in the same row as the formula you are filling in. For example, use the cells B4:C4 for Array 2 in the formula for the purple cell, the cells B6:C6 for the first orange cell, B7:C7 for the second orange cell, and B8:C8 for the third orange cell. Once you have entered these formulas, your spreadsheet will look like the screenshot in Figure 5.10.

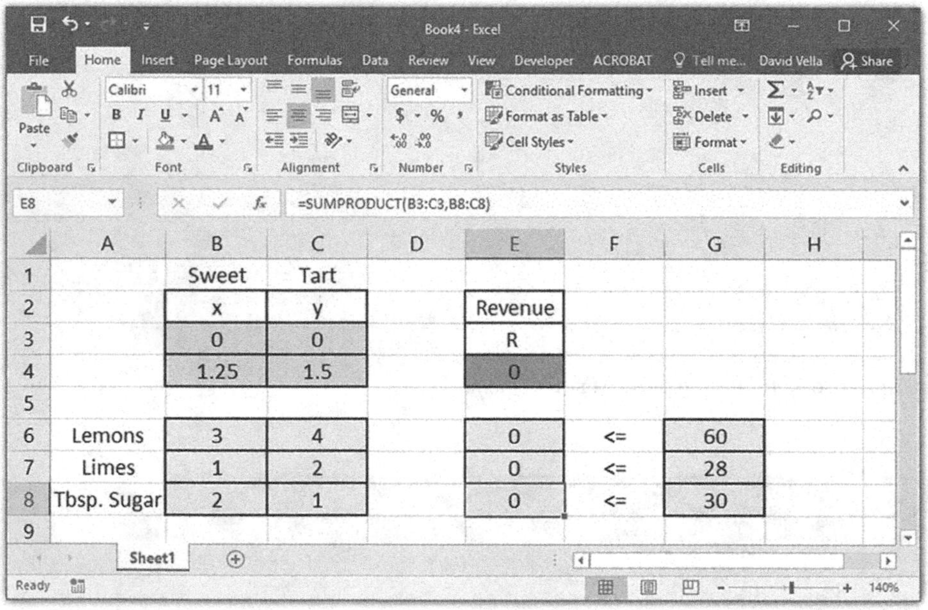

Figure 5.10 What the data spreadsheet looks like after the equations are entered (here in the cells in column E).

Remember that the current values of x and y are 0, which explains why all the cells in the new formulas we just added are displaying a value of 0. As noted before, it you change the values of x or y, all the formulas will update. For example, if we keep $x = 0$ but make $y = 2$, this corresponds to mixing two glasses of tart lemonade. That should use up eight lemons, four limes, and 2 tablespoons of sugar and should generate $3 revenue. Go ahead and try it in the spreadsheet, if you wish, and then reset $y = 0$ to get back to the display in Figure 5.11. Finally, we remark that the "<=" symbols in column F do not tell Excel anything – they are there just to remind the viewer that these inequalities are the less-than type.

Now we are ready to apply the Solver add-in. Go to the Data tab and click on Solver. A dialog box entitled "Solver Parameters" will open, as shown in Figure 5.11.

You should move the dialog box so that you can see the cells behind it. We have not done so in the screenshot because then the picture would be too wide to display here. To fill in the dialog box, proceed as follows: first, click on the empty field next to the words "Set Objective" and then click on the purple cell E4. This tells Excel that the value of the objective function appears in that cell. In the screenshot, this step is already done so that E4 already appears in the field next to "Ste Objective." Then click the radio button for "Max" since this is a maximization problem (it

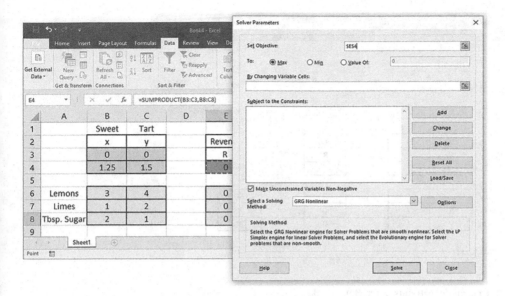

Figure 5.11 The Solver dialog box.

may already be selected as a default setting). Ignore the field entitled "value of," and go down to the field for "By Changing Variable Cells" and click in the blank field. Then go to the cells with the values of x and y – namely B3:C3. Click on cell B3, hold the button down, drag to C3, and then let go. These cell addresses will now be entered in the field in the dialog box. Notice that in this dialog box, there are dollar signs ($) preceding the addresses of all the cells. These need to be there, which is why we recommend clicking and dragging on the cells you want to enter rather than typing them in manually – it is easy to forget the $ symbol when typing.

Next, click on the large empty field under "subject to the constraints." This is where you will enter the structural constraints. To do that, click on the "Add" button to the right of the field. The dialog box will be replaced by a new dialog box entitled "Add Constraint" for entering the constraints, as shown in Figure 5.12.

We will add all three constraints before we click "OK." Here's how. Click on the "Cell Reference" field on the left of the dialog box, and then click cell E6. Remember that the contents of cell E6 represent the total number of lemons used. In the "Constraint" field, you will enter the cell G6 in the same way – don't type in the value (60), but instead the cell address by clicking on the field and then clicking on the cell. Between these two fields in the dialog box, you see a <= sign, indicating that the constraint says that whatever is in cell E6 must be less than or equal to whatever is in cell G6; that is, that $3x + 4y \leq 60$. There is a drop-down menu in that center field if you wish to change the inequality to one of the >= type or even to an exact equality. In any case, once you have entered the correct cells, you would click on "Add" at the bottom to add the second constraint for limes, and then "Add" again to add the third constraint for sugar. When all three have been entered, then click "OK" to return to the previous dialog box, where the added constraints may now be viewed, as shown in Figure 5.13.

Next, just below the large field, there is a check box (already checked as default) that says "Make Unconstrained Variables Non-Negative." This is the set of nonnegativity constraints. When

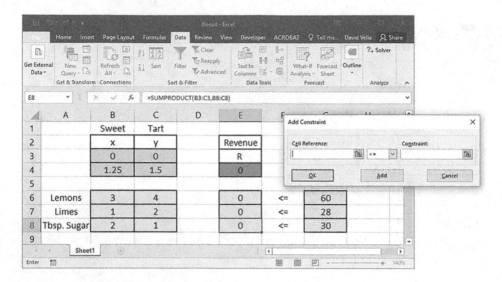

Figure 5.12 Entering constraints in the Solver Add-in.

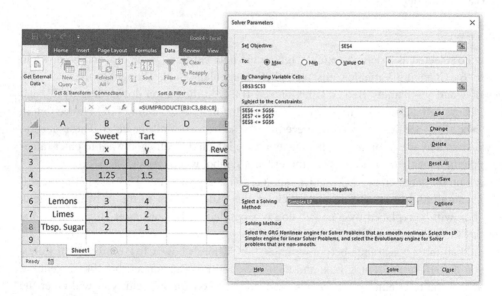

Figure 5.13 The completed Solver dialog box.

that box is checked, you do not need to add those constraints manually. Finally, below that, there is a field labeled "Select a Solving Method." Use the drop-down menu to the right of that field, and select "Simplex LP" (already done in the screenshot in Figure 5.14). Once you have done that, you may click on the "Solve" button at the bottom of the dialog box. A new dialog box will appear, entitled "Solver Results," as shown in Figure 5.14.

The alert reader will notice that the solution already appears on the spreadsheet. The purpose of this new dialog box is to tell Excel what information to save on a separate page. All you need to do with this dialog box is, under the "reports" field, click on both the answer report and the sensitivity

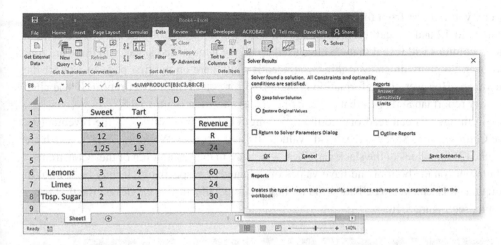

Figure 5.14 Selecting Reports for the solution.

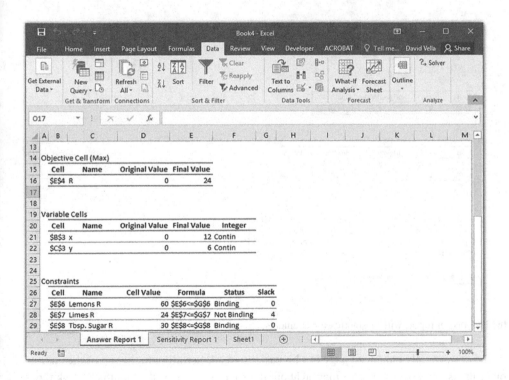

Figure 5.15 The answer report.

report so they are both highlighted (as in Figure 5.15). You do not need the "limits" report. Then click on "OK." At this point, all dialog boxes vanish, and you are left with the original spreadsheet. However, you will notice on the bottom that two new tabs have been added" the answer report and the sensitivity report. These are separate pages that contain all the information you need about the solution of the problem.

What can you tell just from the original spreadsheet? Well, we can see the values of x and y have changed to 12 and 6, and the revenue is now $24. Looking carefully at the orange cells in the E and G columns will reveal that we have used up all the lemons and sugar, and there are four limes left over. So pretty much everything you need is right there on the first page except for the marginal values. However, this information is organized for you in the answer report. If you click on the answer report tab and open that page, it appears as shown in Figure 5.15.

The main things you can read off are the following: (1) the "final value" of the revenue (the Objective Cell), which is $24; (2) the "final" values of x and y (under "Variable Cells"), which are 12 and 6; and (3) the values of the slack variables, which are in the last column of the "Constraints" part of the report on the bottom and have values of $0, 4$, and 0, respectively. It may be helpful to add cosmetic enhancements to call attention to these main details, just as we did on the first page. For example, see Figure 5.16.

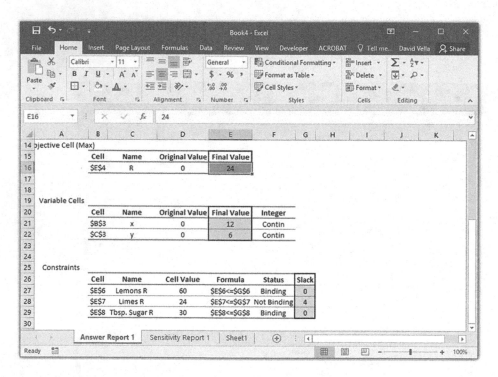

Figure 5.16 The answer report with cosmetic enhancements.

Again, the cosmetically enhanced page is a bit easier to read. Scrolling to the top of the answer report, one can also find out other information about the solution, such as the fact that it took three pivots to solve, and Excel carried out the solution in 0.015 seconds. Figure 5.17 is a similarly cosmetically enhanced shot of the sensitivity report, which we need mainly to find the values of the marginal values (called Shadow Prices by Excel) at the optimal point.

From these two reports, we can easily construct the complete solution:

$$x = 12 \text{ glasses sweet lemonade}$$
$$y = 6 \text{ glasses tart lemonade}$$
$$R = \$24 \text{ (maximized)}$$

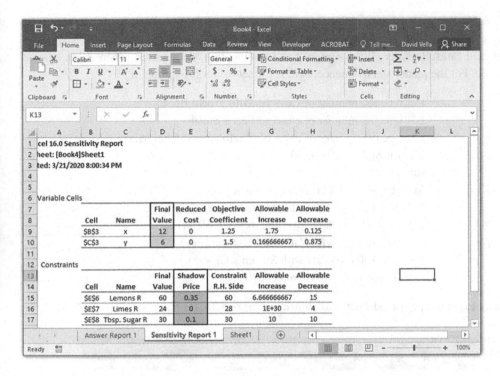

Figure 5.17 The sensitivity report with cosmetic enhancements.

$$S_1 = 0 \text{ leftover lemons}; \ M_1 = \$0.35 \text{ per lemon}$$
$$S_2 = 4 \text{ leftover limes}, \ M_2 = \$0 \text{ per lime}$$
$$S_3 = 0 \text{ leftover tbsp. sugar}, \ M_3 = \$0.10 \text{ per tbsp. sugar}.$$

But as a bonus, the sensitivity report gives us a lot more information. The last two columns entitled "Allowable Increase" and "Allowable Decrease" are precisely what we need to determine the "stable ranges" that were introduced in Chapter 4. For example, look at the objective coefficient 1.25 for the price of a glass of sweet lemonade. The report says the allowable increase is $1.75; that is, increasing the price of sweet lemonade up to a price of $1.25 + $1.75 = $3.00 per glass will keep the current optimal point $(12, 6)$ if no other changes are made. Similarly, a decrease of $0.125 is allowed, so the current optimal point $(12, 6)$ is stable over the range of prices $1.125 to $3.00 for sweet lemonade. Likewise, the stable range for the price of a glass of tart lemonade is $1.50 − 0.875$ to $1.50 + 0.16667$; that is, the stable range is $0.625 to $1.67 per glass. These match the stable ranges were computed in Chapter 4.

Similarly, we find the stable ranges for the constraints: for lemons, it is $60 − 15$ to $60 + 6.67$, or 45 to 66.66667 lemons. For limes, it is 24 to ∞ (The number 1 + E30 is scientific notation. It means 10^{30}, so for all practical purposes, it is ∞.) Finally, for sugar, the stable ranges are 30 ± 10, or 20 to 40. These also agree with the results obtained in Chapter 4.

The upshot is that using Excel to solve linear programming problems requires more steps than using Mathematica, so it takes some getting used to. However the benefit of using Excel is having more important information at the end of the process displayed for you so you do not have to dig it out yourself.

For practice, the reader should try a more complicated example. For example, let's apply Excel to solve the Tillerman Tea problem from the last section. Reader should set up the spreadsheet on their own and carefully enter all the sumproduct formulas that are required. Then check against our example. For the reader's convenience, here is the set up for this problem.

Let x be the number of cases of Mild World Blend.
Let y be the number of cases of Peace Strain.
Let z be the number of cases of Shadowmoon Chai Mix.
Maximize $P = 20x + 30y + 32z$
subject to:
$5x + 6y + 4z \leq 1{,}200$ (lbs. of black tea)
$5x + 8y + 2z \leq 1{,}500$ (lbs. of green tea)
$x + 2y + 10z = 800$ (lbs. of herb/spice blend)
$z \geq 50$ (contract with Cat's Bull Natural Foods)
$x + y \geq 130$ (contract with Steven's Grocery)
$x \geq 0,\ y \geq 0,\ z \geq 0$ (nonnegativity).

One way to set up the spreadsheet is as shown in Figure 5.18.

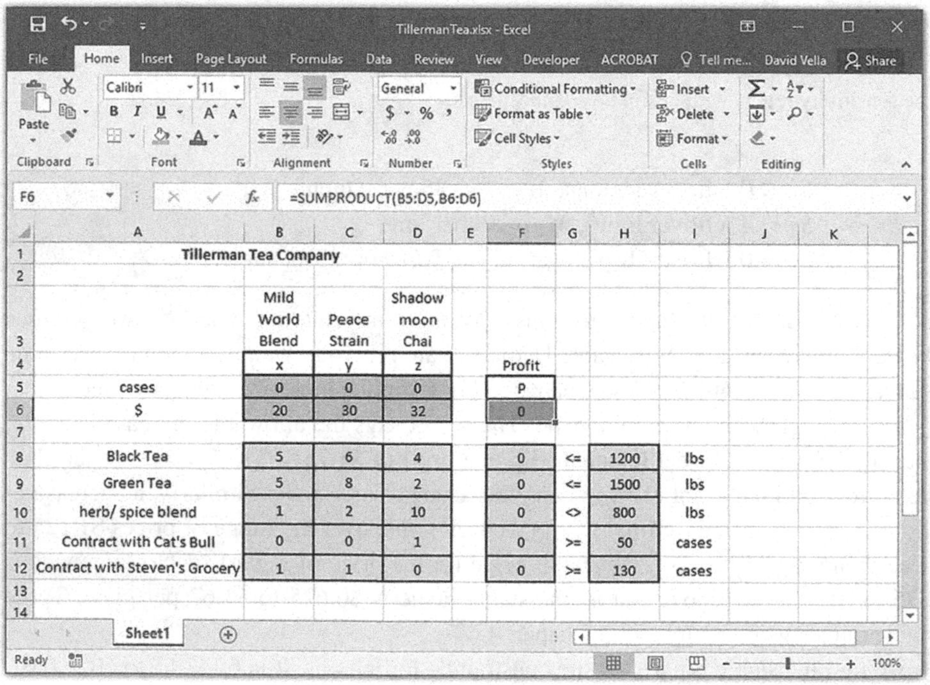

Figure 5.18 The spreadsheet for the Tillerman tea problem.

Notice that the first two inequalities are the less-than type; the third is an exact equality, which we indicate with the symbol "<>" rather than "=" because if you type an equal sign into a cell, Excel thinks you want to enter a function in that cell. The last two are greater-than-type inequalities. The reader should use the Solver add-in to finish the problem. The screenshot in

Figure 5.19 is part of the answer report for this problem. As can be seen, the solution agrees with the solution we found using Mathematica in the last section.

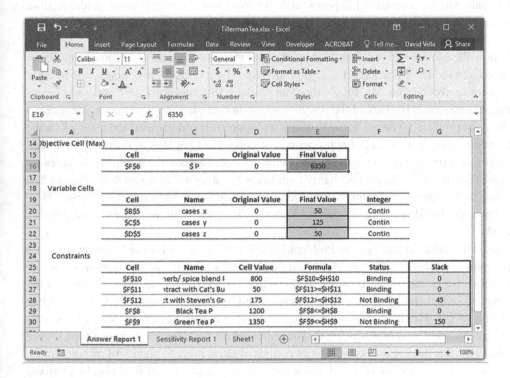

Figure 5.19

This concludes our discussion of computers for the moment. We will have a chance to revisit these techniques in Chapter 7 and again in Chapter 9.

Exercises

1. a. Use Mathematica to solve the problem that was set up in Exercise 11 from Section 3.1, which we reproduce here for convenience: Bruce Jax is responsible for constructing end tables and kitchen tables for the White Room Woodshop. Each end table uses 2 square yards of $\frac{3}{4}$-inch oak boards and takes 2 hours to complete. Each kitchen table uses 4 square yards of oak board and takes 3 hours to complete. This week, he has available 36 square yards of oak boards and 32 hours of time. Other resources are unlimited. How many of each item should he make if he is paid $70 for each end table and $100 for each kitchen table? Then use the **DualLinearProgramming** command in order to determine the marginal values of the resources.

    b. Use *Excel* to solve this problem. Obtain the complete solution, including the marginal values, from the answer and sensitivity reports.

2. Follow the same directions as in Problem 1 (in both Parts a and b) for the following problem (See Exercise 12 of Section 3.1). Joni is putting her design skills to work at the Court and

Sparkle Jewelry Emporium. She makes two signature bracelet designs. Each Dawntreader design uses 1 ruby, 6 pearls, and 10 opals. Each Hejira design uses 3 rubies, 3 pearls, and 15 opals. She has 54 rubies, 120 pearls, and 300 opals to work with. If either model yields a profit of $1,800, how many of each type of bracelet should she make in order to maximize her profit?

3. Follow the same directions as in Problem 1 (both Parts a and b) for the following problem (See Exercise 13 of Section 3.1). Toys in the Attic, Inc., operates two workshops to build toys for needy children. Mr. Tyler's shop can produce 36 Angel dolls, 16 Kings and Queens board games, and 16 Back in the Saddle rocking horses each day it operates. Mr. Perry's shop can produce 10 Angel dolls, 10 Kings and Queens board gamess, and 20 Back in the Saddle rocking horses each day it operates. It costs $144 per day to operate Mr. Tyler's shop and $166 per day to operate Mr. Perry's shop. Suppose the company receives an order from the Kids Dream On charity foundation for at least 720 Angel dolls, at least 520 Kings and Queens board games, and at least 640 Back in the Saddle rocking horses. How many days should the company operate each shop in order to fill the order at least possible total cost?

4. Follow the same directions as in Problem 1 (both Parts a and b) for the following diet mix problem (Problem 9 of Section 3.4). Kerry's Kennel is mixing two commercial brands of dog food for its canine guests. A bag of Dog's Life Canine Cuisine contains 3 pounds of fat, 2 pounds of carbohydrates, 5 pounds of protein, and 3 ounces of vitamin C. A giant-size bag of Way of Life Healthy Mix contains 1 pound of fat, 5 pounds of carbohydrates, 10 pounds of protein, and 7 ounces of vitamin C. The requirements for a week's supply of food for the kennel are that there should be at most 21 pounds of fat, at most 40 pounds of carbohydrates, and at least 21 ounces of vitamin C. How many bags of each type of food should be mixed in order to design a diet that maximizes protein?

5. Follow the same directions as in Problem 1 (both Parts a and b) for the following scheduling problem (Problem 10 of Section 3.4). The Poseidon's Wake Petroleum Company operates two refineries. The Cadence Refinery can produce 40 units of low-grade oil, 10 units of medium-grade oil, and 10 units of high-grade oil in a single day. (Each unit is 1,000 barrels.) The Cascade Refinery can produce 10 units of low-grade oil, 10 units of medium-grade oil, and 30 units of high-grade oil in a single day. Poseidon's Wake receives an order from the Mars Triangle Oil retailers for at least 80 units of low-grade oil, at least 50 units of medium-grade oil, and at least 90 units of high-grade oil. If it costs Poseidon's Wake $1,800 to operate the Cadence Refinery for a day and $2,000 to operate the Cascade Refinery for a day, how many days should Poseidon's Wake operate each refinery to fill the order at least cost?

6. Follow the same directions as in Problem 1 (both Parts a and b) for the following linear programming problem (Problem 13 of Section 3.4). The Jefferson Plastic Fantastic Assembly Corporation manufactures gadgets and widgets for airplanes and starships. Each case of gadgets uses 2 kilograms of steel and 5 kilograms of plastic. Each case of widgets uses 2 kilograms of steel and 3 kilograms of plastic. The profit for a case of gadgets is $360, and the profit for a case of widgets is $200. Suppose Jefferson Plastic Fantastic Assembly Corporation has 80 kilograms of steel available and 150 kilograms of of plastic available on a daily basis and can sell everything it manufactures. How many cases of each should the corporation manufacture if it is are obligated to produce at least 10 cases of widgets per day? The objective is to maximize daily profit.

7. Follow the same directions as in problem 1 (both Parts a and b) for the following problem (Problem 14 of Section 3.4). Mr. Cooder, a farmer in the Purple Valley, has at most 400 acres to devote to two crops: rye and barley. Each acre of rye yields $100 profit per week, while each acre of barley yields $80 profit per week. Due to local demand, Mr. Cooder must plant at least 100 acres of barley. The federal government provides a subsidy to grow these crops in the form of tax credits. The government credits Mr. Cooder four units for each acre of rye and two units for each acre of barley. The exact value of a "unit" of tax credit is not important. Mr. Cooder has decided that he needs at least 600 units of tax credits in order to be able to afford to his loan payments on a new harvester. How many acres of each crop should he plant in order to maximize his profit?

8. Follow the same directions as in problem 1 (both Parts a and b) for the following problem (Problem 7 of Section 3.5). The Spooky Boogie Costume Salon makes and sells four different Halloween costumes: the witch, the ghost, the goblin, and the werewolf. Each witch costume uses 3 yards of material and takes 3 hours to sew. Each ghost costume uses 2 yards of material and takes 1 hour to sew. Each goblin costume uses 2 yards of material and takes 3 hours to sew. Each werewolf costume uses 2 yards of material and takes 4 hours to sew. The profits for each costume are as follows: $10 for the witch, $8 for the ghost, $12 for the goblin, and $16 for the werewolf. If the Sppoky Boogie Costume Shop has 600 yards of material and 510 sewing hours available before the holiday, how many of each costume should it make in order to maximize profit, assuming it sells everything it makes?

9. Solve the problem of the Gigantic ocean liner (Problem 14 of Section 3.1) You may use either Mathematica or Excel. In your solution, find values of all decision and slack variables as well as the objective function, but don't bother with the marginal values.

10. Follow the same directions as in Problem 9 for Derek's diet problem (Problem 16 of Section 3.1).

11. Follow the same Directions as in Problem 9 for Ray's diet problem (Problem 17 of Section 3.1).

12. Follow the same directions as in Problem 9 for the Creative Thought Matters Biological Research problem (Problem 20 of Section 3.1).

13. Consider the Green's Heavy Metal Foundry problem (Problem 19 of Section 3.1).

    a. Use Excel to solve this problem.
    b. From the sensitivity report, find the stable ranges for each constraint – that is, for each resource.

14. Consider the the Court and Sparkle Jewelry Emporium (see Problem 2). From the Excel solution, determine the stable ranges for each of the the resources and for the two objective coefficients.

15. In the Court and Sparkle Jewelry Emporium (see Problem 2), determine the new optimal data:

    a. If the supply of pearls is increased to 150 and no other changes are made.
    b. If the profit for a Hejira model bracelet increases to $2,100 and no other changes are made.

16. In the Jefferson Plastic Fantastic Assembly Corporation (see Problem 6), determine the new optimal data if the supply of plastic is increased to 175 kilograms and no other changes are made.

6   Game Theory

6.1   Introduction

The reader may have heard that game theory is a *mathematical analysis of strategic behavior*; and may know that it has found important applications in many fields of study such as warfare, politics, diplomacy, biology, and especially economics. The theory continues to find new applications, and moreover, it has captured the public imagination in a way that is unusual for mathematics. One of the high-water marks of the theory's public appreciation occurred in 1994, when John F. Nash (later the subject of the biographical book and movie *A Beautiful Mind*), John C. Harsanyi, and Reinhard Selten were awarded the Nobel Prize in Economics for their work in game theory. To quote from the 1994 press release from the Royal Swedish Academy of Sciences:

The foundations for using game theory in economics were introduced in a monumental study by John von Neumann and Oskar Morgenstern entitled *The Theory of Games and Economic Behavior* (1944). Today, 50 years later, game theory has become a dominant tool for analyzing economic issues. In particular, non-cooperative game theory, i.e., the branch of game theory which excluded binding agreements, had had a great impact on economic research. The principal aspect of this theory is the concept of an equilibrium, which is used to make predictions about the outcome of strategic interaction. John F. Nash, Reinhard Selten, and John C. Harsanyi are three researchers who have made eminent contributions to this type of equilibrium analysis.

In fact, the reader may find it instructive to read the entire press release, which can be found at the following website:

www.nobelprize.org/prizes/economic-sciences/1994/press-release/

But what exactly is meant by "analysis of strategic behavior," and what does this have to do with games? The concept of a mathematical game is different from and more general than the usual notion of a parlor game, such as chess, Monopoly, poker, or ticktacktoe. A mathematical game is a special type of optimization problem. In the optimization problems studied so far, such as at the end of Chapter 1 as well as the linear programming problems covered in the previous three chapters, the goal is to make a decision leading to an outcome that best meets an objective. The relevant thing to observe here is that the outcome that results, whatever it is (sales profit, transportation costs, protein content of a diet, etc.) depends only on *your* decision. In a game, the outcome depends not only on your decision but also on the decision(s) of other agents over which you have no control. Those other agents could be other people, but they also could be elements of chance or randomness. So you still attempt to make a decision that will lead to the best outcome for you, but you must do so while taking into account what others might do that could affect your outcome. When you make your decisions in a thoughtful way, taking into account the possible actions of others, you are said to be behaving strategically, and those are the situations of mathematical game theory.

Clearly, this is the sort of situation one finds oneself in a parlor game like chess or poker. Your objective is to "win" without knowing in advance what "moves" your opponent(s) might make. But many other situations from real life also fall into this category that are not related to parlor games at all. For example, in business, if you are deciding between several potential locations for your next franchise store, the eventual success of your new store will at least partly depend on where your competitors open their next stores. In warfare, if you are sending supplies and/or reinforcements to your troops located on a remote island, you must choose which supply route to take. Your enemy can potentially thwart your plans – will they attack this route or that? You do not know in advance, but nevertheless, you must try to anticipate what they might do and what the outcome might be if they do attack. In other words, you must think and act strategically.

In fact, strategic decisions are pervasive in life. Should you apply to that Ivy League college that is a "reach" school for you? The ultimate success of your application being accepted, depends partly on you (your GPA, school rank, your record of community service, how well you impress the admissions officers with your application essay, etc.). But it also depends on how many others are also applying, who they are, and what their records are. Thus, applying to college is a strategic game. In the next presidential election, if there are three candidates, should you vote for your favorite candidate, who is a long shot, or should you vote for your second-favorite candidate who has a much better chance, just to help make sure the candidate you like the least is not elected? The success of either strategy depends partly on how others vote. Thus, voting can be regarded as a strategic game if there are three or more candidates. (Clearly, no such dilemma occurs if there are only two candidates.)

On first reflection, it may seem to be impossible to know what course of action is best for you, given that you have no control over how other players behave. However, your previous experience from parlor games should suggest otherwise. Even if you do not know what moves your opponents will make, it is often still clear that some moves are better than others for you. In chess, you shouldn't make a move that leaves your king unprotected; in poker, some bets are ill advised; and if you repeatedly play "rock" in rock paper scissors, you will surely end up losing in the long run!

So, at least in some strategic situations, we shall see that it is indeed possible to arrive at a "best course of action." On the other hand, as we shall also see, there are some games that do not have a universally accepted "solution." Even so, the concepts and methods of game theory can still be helpful by clarifying the situation and helping to focus on exactly where the strategic analysis comes up short.

Bear in mind, however, that a game, like any other mathematical construct, is only a model of the real-life conflict. As we said in Chapter 1, all models, being deliberately simpler than the real-life problem they are designed to mimic, lead to conclusions that are limited by the assumptions that are built in to the model. Mathematical models in the social sciences, as compared to models in the physical sciences (such as the physics of baseball trajectories mentioned in Chapter 1), are even more susceptible to error, and their conclusions are even more closely tied to the assumptions built into the model. This is because human behavior is notoriously messy and complex. This does not mean that mathematical models in the social sciences are useless; it only means one has to be very clear about what assumptions are being built into the model.

Consequently, we shall begin our study of game theory by an explicit discussion of the assumptions we will be working with in this chapter. In later chapters, after the reader completely understands the theory under these assumptions, we will begin to relax and modify some of those assumptions in order to make the theory more widely applicable.

A mathematical game consists of a set of *players* (players can be individuals or any collection of individuals such as teams, corporations, armies, countries, etc.), each with a collection of actions open to them, called *strategies*. Once the players select their strategies, the play of the game will result in an *outcome*. We stipulate that each outcome of the game results in a *payoff* for each of the players, which we assume can be measured numerically.

For example, the payoff might be binary in nature – a player is either a winner or a loser, like in ticktacktoe or chess. In this case, we might designate the payoff of 1 to indicate a winner and 0 to indicate a loser. In other games, the payoff might be an integer, such as the number of "points" earned. For example in card games such as cribbage, hearts, and certain rummy games, players accumulate points throughout the game, and the game stops as soon as someone reaches a specified number of points agreed upon before the game started. Regarding an election as a game between the candidates, the payoff points might be the number of votes they receive from the voters. Or the payoff might be a real number, positive or negative, such as a monetary amount; for example, your winnings or loss at a poker game, or the yearly profit generated by your newly opened business, or the number of years of a jail sentence you might earn as a result of a poor strategic decision.

Of course, there are certainly strategic situations in which the payoff is is not easily measured numerically. For example, if the game is a battle between two armies, and your army wins, how do you measure the payoff? The number of casualties of the enemy? The difference between the number of casualties between the two sides? The amount you spent on weapons and ammunition and equipment to secure the victory? One way to deal with this ambiguity is to fall back on a binary measure of payoff – 1 for winning the battle and 0 for losing it. But that is a very crude measure. Some victories are decisive; others are so costly so as to be only marginally better than defeat (the classic "Pyrrhic victory").

Or suppose the game consists of you and several of your coworkers competing for a promotion, which is ultimately awarded to you. How do you measure the value of this outcome? Your increase in salary? A promotion also often involves changed status and responsibilities. How are these aspects figured into the payoff? What if, in order to win the promotion, you had to put in so much overtime that it adversely affected other parts of your life, eroding your social life, family life, or health? How do you measure these things numerically?

Utility theory is that part of game theory that deals with how a player values a particular outcome – the *utility* of the outcome. The examples in the last paragraph show that it may be quite difficult to assign a specific utility of an outcome. To make matters worse, the same outcome with the same payoff might be valued differently by different people, so the players in a game may not even agree on the utility of an outcome. To illustrate the difference between payoff and utility, imagine an estate being divided at the reading of a will. One of the heirs might inherit an $80,000 Rolls Royce, while his brother may inherit $60,000 cash. Even though the payoff for the car is $80,000, the person who inherits the car may actually value the $60,000 cash more – perhaps he doesn't drive or doesn't look forward to the bother and expense of maintaining the car. So the utility of the car may actually be less than $60,000 to the person who inherited it.

Techniques have been proposed to attempt to help players objectively assign values to outcomes, but clearly, this aspect of game theory has more to do with psychology than with mathematics and is difficult to model mathematically. (Perhaps a better name for this aspect of the study of games would be "futility theory"?) The reader interested in this aspect of game theory may consult von Neumann and Morgenstern (1944) or Straffin (1993) for more details. In this text, however, until further notice, we will greatly simplify our treatment of games if we make the following assumption:

> Each outcome of a game has a numerical *payoff* (or *value* or *utility*), which is known to all the players before the game starts and upon which all players agree.

Observe that tacit in this statement is the assumption that all the players have the same information about the game. Since "intelligence" is synonymous with "information," this assumption is sometimes expressed by saying that the players are *intelligent*. This does not imply that the players are smart or clever, just that they have information about the payoffs ahead of time.

Our next assumption is that the players will always try to act in a way that will lead to the best possible payoff for themselves. So they will not, for example, choose a bizarre strategy that could lead to a poor payoff for themselves just for the sake of "being unpredictable" or any other reason. The player's goal is always to get the best possible payoff, and the player will always act in a strategic way, not in a way based on wild hopes or attempts to confuse the other players. So this assumption is closer to what one might mean by a "smart" player, and the accepted word for this in game theory is *rational* behavior.

One might argue that in some cases, an attempt to confuse the opponent *is* the rational thing to do. But if your opponent is susceptible to such a ruse, perhaps he or she is not very skillful. Since we want to develop techniques of playing a game that will be successful against any opponent, no matter how skillful, we may as well assume that rational players are not susceptible to such a ploy, that your opponents are skillful enough to see through mere attempts to confuse them and will play optimally, regardless of what you do. Therefore, you should stick to strategies that lead to your best possible payoff and ignore any temptation to engage in another approach based on a whim. In short, we will assume

> Players in a game are skillful, and will act rationally (that is, they will always strive to obtain the best possible payoff for themselves).

In some games, such as chess or ticktacktoe, the play is a long string of sequential moves, with the players alternating turns. In other mathematical games, the entire game consists of just one move to arrive at the outcome. For example, if the game is voting for a candidate, all the voters vote (and since there is no order necessary in voting, we may as well assume that all the votes are cast simultaneously), and that's it – the votes are tallied, and a winner of the election is declared. Similarly, a business deciding on a location for a new store, a football quarterback deciding between a run play or a pass play, and children playing rock paper scissors are all examples of single-move games.

In this chapter, we will focus on these single-move games because they are the easiest to analyze. Later on, we will consider some simple examples of games with sequential moves, and it turns out that with a suitable definition of "strategy" in such a game, one may regard it as a single-move game anyway. So, in theory, there is no loss of generality in restricting attention to single-move games. Thus, we make the following assumption:

> The play of the game consists of each player selecting one of the strategies open to them; then the outcome is reached.

We also, for now, make the additional assumption:

The players will make their selection of strategies *simultaneously* and *independently*.

In particular, independent play means there is no communication allowed between the players before they select their strategies. We may relax some of these assumptions later, but these are the operating assumptions for this chapter.

Games fall into various categories or classes, but before describing this taxonomy of games, it is time for some specific examples.

Example 6.1 *Two-Finger Morra.* Two players, Ann and Bill, face each other with their right hands hidden behind their backs. On the count of three, they move the hidden hand to the space between them, revealing either one or two fingers extended. If the total number of fingers revealed is odd, Ann pays Bill $1. If the total number of fingers revealed is even, Bill pays Ann $1.

Example 6.2 *Rock Paper Scissors.* Ann and Bill play the well-known game of rock paper scissors, following the usual conventions that rock beats scissors, scissors beat paper, and paper beats rock. They decide to follow the payoff scheme that the loser pays the winner $1, but there is no exchange of money of the result is a tie.

Example 6.3 *Saturday Chores at the Anderson Household, Version I.* Mr. and Mrs. Anderson have three children – Ian, Jon, and Laurie – who will help with certain chores. Mr. Anderson rewards them by paying a total of $24: $6 for washing the dishes, $6 for cleaning the basement, and $12 for raking the yard. The children can divide up the tasks any way they want, and they can even share tasks (and split the reward two or three ways). However, past experience shows that if the children discuss their preferences ahead of time, they will just get into an argument. To avoid those arguments, Mr. Anderson has the kids write down their job choice or choices secretly on a piece of paper. The only rule is that all the tasks must be done, so if a job is not claimed by one or more of the children, then Mr. Anderson will assign the task as follows: it will be shared equally by all children who do not choose any task, and if all three have chosen something, then it will be shared equally by all three children.

Here are a few examples: One Saturday, Ian chose cleaning the basement, Jon chose raking and dishes, and Laurie chose dishes. So Ian earns $6 for the basement, Jon earns $12 for the raking, and Jon and Laurie each earn $3 for washing the dishes together. The next Saturday, Ian chose raking, Jon and Laurie both choose dishes, and nobody chose cleaning the basement, which was then split between all three. Thus, each of them earns $2 for their share of the basement in addition to the tasks for which they volunteered. Ian ends up with $14, Jon and Laurie each earn $5 for splitting the dishes and sharing the raking with Ian. The next Saturday, Laurie chooses all three jobs, and the boys choose nothing. So Laurie earns the entire $24 for herself. The following week, nobody chose anything, so all three tasks were assigned to be shared among the three of them, each of them earning $8.

For the purposes of this game, we assume the children are more concerned with earning money than they are in avoiding work.

Example 6.4 *Saturday Chores at the Anderson Household, Version II.* Several years later, the Anderson children are all older. Ian is off at college, leaving Jon and Laurie to share the chores. However, nobody plays in the basement anymore, so there are only two tasks – raking and dishes. Mr. Anderson has modified the rewards as follows: He'll pay Laurie $6 if she chooses the dishes, but he wants to encourage Jon to choose that task more often since he tends to avoid it, so if Jon does the dishes, he earns $8. If they share the job, he pays $7: $4 to Jon and $3 to Laurie. Mr. Anderson does not want to be unfair to Laurie, so he has a similar asymmetry built into the payoffs for raking in Laurie's favor. He'll pay Jon $12 to do the job himself, and he'll pay Laurie $15 to do the job herself. If they share the job, he pays $13: $7 to Laurie and $6 to Jon. The rest of the rules are the same: any unclaimed task is assigned to a child who has claimed nothing, or is split between them if both children or neither have claimed a task.

Example 6.5 *A Marketing Game.* Antares Industries and Bellatrix, Inc., are the only two companies that sell a certain computer chip, the Pentangle 5.1. There is an estimated market for a total of 400,000 of these chips for the year 2021. Each company must set their prices. Production costs dictate that the price should be at least $80 per chip, while industry regulations require that the price be no more than $120 per chip. Let x be the price set by Antares and y the price set by Bellatrix. Based on reliable market studies of which both companies are aware, it has been determined that the percentage of the total market that goes to Antares depends on x and y and is reasonably approximated by the function:

$$E(x, y) = 330 - 3x - 2.7y + 0.03xy. \tag{6.1}$$

What prices x and y should be set by the companies if their goal is to receive the largest-possible share of the market?

The preceding examples illustrate single-move games of various types. Before we embark on a study of how to play these games strategically, we discuss some ways of classifying games. The first way to distinguish types of games from one another is based on how many players there are. In Example 6.3, the players are the three children, since they are the ones selecting strategies (that is, writing their job choices on a piece of paper.) Mr. Anderson is not considered a player – all he does is pay the rewards and enforce the rules. So this is a three-player game.

In the remaining examples, there are just two players – Ann and Bill in the first two games, Jon and Laurie in Example 6.4, and the two companies Antares and Bellatrix in Example 6.5. Games with more than two players have additional complications that require different mathematical tools than do two-person games. Thus, games are classified into two-player games vs. n-player games (where n is 3 or more). While we will address some aspects of n-player game theory, the main focus in this text is two-player games.

The next way to classify games is related to the total payoffs. Observe that in the first two examples, there is a direct exchange of money between the players, so that whatever one player wins, the other loses. Thus, the sum of the payoffs for both players is always 0, no matter what the outcome of the game is. Such games are called *zero-sum* games. Two person, zero-sum games are the easiest to analyze, so we will devote a large portion of our study to games of this type. In fact, it has been known since 1928 that these types of games always have a solution, a vital theorem first proved by John von Neumann (discussed in the following sections).

A minor variation of a zero-sum game is a *constant-sum* game, which means that the sum of the payoffs to the players is a constant; that is, it does not depend on the outcome of the game. Constant-sum games can be solved by the same techniques that apply to zero-sum games. In our examples, Examples 6.3 and 6.5 are constant-sum. In Example 6.5, we stated that the goal of each company was to have the largest-possible market share, and we also stated that these were the only two companies producing the chip in question. That means that whatever prices are set by the companies, the two market shares add up to 100% (if we regard the payoffs as percentages of the market) or add up to 1 (if we regard the payoffs as fractions of the market). In Example 6.3, it is constant-sum even though there are three players; the payoffs always add up to the $24 total that Mr. Anderson is paying for the chores to be done.

A game that is not constant-sum (in particular, it is not zero-sum), is said to be *variable-sum*, because the sum of the payoffs to all the players varies from outcome to outcome. Some other authors use the term *general sum game* for this type of game. Example 6.4 is variable-sum. For example, if Jon chooses dishes and Laurie raking, then the payoffs are $8 for Jon and $15 for Laurie for a total of $23. If Jon chooses raking and Laurie dishes, the payoffs are $12 for Jon and $6 for Laurie for a total of $18. If they both share both jobs, the payoff is $10 each for a total of $20. Thus, the sum of the payoffs depends on the outcome of the game. We also note that in Example 6.5, had the goals of the companies been to maximize their revenues instead of market shares, then the game would no longer be constant-sum but variable-sum.

In a two-person game that is zero-sum or constant-sum, whatever gains are made by one player are balanced by an equal loss to the other. Since both players want to maximize their own gain, it is clear that in such a game, the players have diametrically opposed interests. In variable-sum games, it may be possible to move to a different outcome and have *both* players gain something, since the sum of their payoffs could increase, the proverbial "win-win situation." Thus, although each player is still interested in optimizing his or her own payoff, it is no longer true that the players' interests are diametrically opposed; the players may be at least partly interested in cooperating so that both of their payoffs can improve. Therefore, games that are variable-sum are sometimes called *games of partial conflict*.

Yet a third way to classify games is by their size, by which we mean the number of strategies open to each player. If the game just involves two players, we will see in the next section that there is a certain matrix associated with the game, called the *payoff matrix*. If one of the players has m strategies and the other has n, this matrix will be of size $m \times n$, and we will say the game itself is an $m \times n$ game. In Example 6.1, both Ann and Bill have two strategies (put out one finger or two), so this is a 2×2 game. Similarly, Example 6.2 (rock paper scissors) is 3×3. Example 6.3 is a three-person game, so we won't worry about the size right now. Moving on to Example 6.4, each of the children has four strategies (volunteer for one of the two chores, or both, or neither), so this is a 4×4 game. Although all these examples have square payoff matrices, that is not necessary – in some games, one of the two players might have a different number of strategies available than the other.

In Example 6.5, each of the two companies chooses to set a price to be any real number in the interval $[80, 120]$, so each player has an infinite number of strategies. Such a game is called an *infinite game*, or even a *continuous game*, since the strategies range over a continuous interval. One might argue that, in reality, since prices can only change in increments of $0.01, there are not really an infinite number of prices available, just a very large finite number. Technically, that is correct, but when we examine this game in more detail later, we will ignore that and actually assume the prices can range continuously over the interval [80, 120]. In this text, we mostly focus on finite games, but we will take a small detour to look at some infinite games, including the complete solution to the marketing game in Example 6.5.

Other ways of classifying games exist, but they depend on the game's solution, so we will defer further discussion of the taxonomy of games until the next section. In preparation for learning to solve games, let us now narrow our attention to two-person games, and consider two ways to describe and display the important information of the games in a convenient format. The first way is via *tree diagrams*.

For example, consider the two-finger morra game in Example 6.1. Select a player (it does not matter which one), and label a node of the tree with his or her name. Draw branches down from this node corresponding to and labeled with the strategies for that player. At the end of each of those branches, put a node labeled with the name of the other player. From each of those nodes, draw further branches down corresponding to and labeled with the second player's strategies. Since the game ends after each of the two players has selected a strategy (that is, a branch of the tree diagram), the ends of these branches are terminal nodes corresponding to the outcomes of the game. We can then label each outcome with its payoff.

In our example, we'll keep things alphabetical so we choose Ann for the first node and Bill second. The payoffs are labeled by ordered pairs, where the first coordinate is Ann's payoff and the second is Bill's. The result is shown in Figure 6.1.

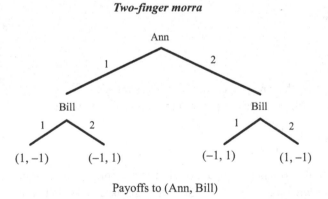

Two-finger morra

Payoffs to (Ann, Bill)

Figure 6.1 Game tree for two-finger morra.

Clearly, all the important information about the game is displayed in this diagram, which is called the *game tree*. Since we are assuming that the players make their selections simultaneously, we could just as well have chosen to put Bill at the top node instead of Ann. Being at the root of the tree (top node) does not imply that player moves first. However, later, when we drop the assumption of simultaneous play and look at games where the players do indeed move in sequence, then it will be important to draw the game tree with the first player to move at the top, etc. When a two-person game is depicted by using the game tree, the game is said to be in *extensive form*. Notice that since this game is zero-sum, as soon as we know one payoff, we can determine the other, so it is not necessary to use ordered pairs for the payoffs. For zero-sum (and, more generally, constant-sum) games, we will often just put one payoff – for example, just Ann's payoffs – on the terminal nodes without losing any information.

If one makes a table of the payoffs so that the rows of the table correspond to Ann's strategies and the columns to Bill's strategies, the following 2×2 matrix A, called the *payoff matrix* of the game, results.

		Bill	
		1 finger	2 fingers
Ann	1 finger	$(1, -1)$	$(-1, 1)$
	2 fingers	$(-1, 1)$	$(1, -1)$

Ann, who is called the *row player*, chooses a row (her strategies), and Bill, the *column player,* chooses a column (his strategies). The position of the matrix where the row and column intersect is the outcome of the game (more precisely, it is the payoff associated with the outcome.) For example, if Ann puts out one finger (first row), and Bill puts out two fingers (second column), then the payoff is the $(-1, 1)$ in the first row and second column. This means Ann pays Bill \$1 (because the total number of fingers revealed is odd).

As noted earlier, since this is a zero-sum game, it suffices to know one player's payoffs. *By convention, we always display the row-player's payoffs.* Thus, for two-person zero-sum (or constant-sum) games, the payoff matrix is an ordinary matrix with numerical entries.

		Bill	
		1 finger	2 fingers
Ann	1 finger	1	-1
	2 fingers	-1	1

In fact, when it is understood which player is the row player and what the strategies are, we omit all the labels on the rows and columns and simply write the payoff matrix as the following:

$$A = \begin{bmatrix} 1 & -1 \\ -1 & 1 \end{bmatrix}$$

Of course, if a game is not constant-sum, the row player's payoffs do not determine the column player's payoffs, so we must retain the ordered pair payoffs as the entries in the matrix. When a two-person game is described by giving its payoff matrix, the game is said to be in *normal form*. Clearly, the normal form has the exact same information as the game tree, so it is easy to convert the extensive form to the normal form and vice versa. For simultaneous-play games, they are equivalent (and we'll see later that we can make them equivalent for sequential-play games as well.)

In Example 6.2, rock paper scissors, each player has three choices for a strategy, so the payoff matrix is 3×3. We leave it as an exercise for the reader to construct the payoff matrix and the game tree for this game. Example 6.3 is a three-person game, and Example 6.5, while it is a two-person, it is infinite, so neither of these games have payoff matrices. Example 6.4 can be expressed in normal form, and we conclude this section by doing so.

Let Jon be the row player and Laurie the column player. We know each player has four strategies, so the payoff matrix is 4×4. Since the game is variable-sum, we must use ordered pair payoffs, with Jon's payoff in the first coordinate as he is the row player. The result is as follows.

| | **Laurie** | | | |
	No chore	Dishes	Raking	Both
No chore	$(10, 10)$	$(12, 6)$	$(8, 15)$	$(0, 21)$
Dishes	$(8, 15)$	$(10, 10)$	$(8, 15)$	$(4, 18)$
Jon Raking	$(12, 6)$	$(12, 6)$	$(10, 10)$	$(6, 13)$
Both	$(20, 0)$	$(16, 3)$	$(14, 7)$	$(10, 10)$

Let's check some of these entries. Suppose the children both choose the same strategy, so the outcome is on the main diagonal. Then they split whatever chore they chose (if any) and split any chore they did not choose as well, since it is assigned to them. Hence, they always end up splitting both chores. So Jon earns $4 for the dishes and $6 for the raking for a total of $10, while Laurie earns $3 for the dishes and $7 for the raking, also for a total of $10, and this explains the entries $(10, 10)$ on the main diagonal. What if Jon chooses raking, and Laurie chooses both? Then Jon earns $6 for the raking, Laurie earns $7 for the raking and $6 for the dishes, for a total of $13, so the payoff here is $(6, 13)$. The other entries can be verified in the same manner and are left for the reader to check.

Exercises

1. Express the rock paper scissors game from Example 6.2 in both normal form and extensive form. That is, draw the game tree and find the payoff matrix as well.
2. Draw the game tree for the game in Example 6.4, Saturday chores at the Anderson household, version II.
3. The game of two-finger morra (Example 6.1) can be extended to three-finger morra. The same rules apply: if the total number of fingers revealed is even, the column player pays the row player $1, while if the total is odd, the row player pays the column player $1. Draw the game tree, and find the payoff matrix for three finger Morra.
4. The game of two-finger morra (Example 6.1) can be extended to four-finger Morra. The same rules apply: if the total number of fingers revealed is even, the column player pays the row player $1, while if the total is odd, the row player pays the column player $1. Find the payoff matrix for four-finger morra.
5. In the game of Example 6.3, Saturday chores at the Anderson household, version I, it was mentioned that each player has eight strategies open to them. List the strategies.
6. Consider the marketing game Antares vs. Bellatrix of Example 6.5. Construct a finite version of this game as follows: Instead of allowing the set prices to be any number in the interval $[80, 120]$, assume the companies are debating between just a few specific choices (all of which lie within this interval). For example, suppose Antares is choosing between the two prices $80 and $110, while Bellatrix is choosing between the two prices $90 and $110. Under these assumptions, the game is reduced to a 2×2 game. Let Antares be the row player and Bellatrix the column player. Find the payoff matrix, assuming each company still wants to obtain the largest-possible market share. [**Hint:** Use Eq. (6.1) for $E(x, y)$ from the example to determine the payoffs.]
7. Repeat Exercise 6 if Antarcs is deciding between the two prices $80 and $110 while Bellatrix is deciding between the three prices $80, $100, and $110. (This exercise illustrates that payoff matrices need not be square matrices.)

8.   Repeat Exercise 6 if Antares and Bellatrix are both deciding between the five prices $80, $90, $100, $110, and $120.

9.   This is a variation on four-finger morra. Each player simultaneously puts out one, two, three, or four fingers. Again, if the total number of fingers is even, the column player pays the row player, while if the total number is odd, the row player pays the column player. However, instead of the payment being $1, the number of dollars will be the (nonnegative) difference between the number of fingers they show. Find the payoff matrix.

10. A guessing game. Bill secretly puts x in his hand, where $x = 0, 1, 3,$ or 5. Ann must guess how many dollars Bill is holding. If she guesses correctly, Bill pays her twice the number of dollars he is holding, plus one dollar more; that is, he pays her $\$(2x + 1)$. If she guesses incorrectly, she must pay Bill the (positive) difference between her guess and the actual number, plus a $2 penalty. Assume Ann is the row player and find the payoff matrix.

6.2   Dominant Strategies and Nash Equilibrium Points

In this section, we begin our study of how to solve games. To illustrate the two main ideas introduced in this section, we draw upon an actual episode from World War II, the Battle of the Bismarck Sea. The episode lends itself well to game theoretic analysis, as was shown originally by O. G. Haywood, Jr., who published an article about this battle in the *Journal of Operations Research* in 1954. Since then, it has been discussed in a number of texts, including Luce and Raiffa (1957), Dixit and Skeath (1999), and Casti (1996).

The situation was this: In 1943, the island of New Guinea was divided between Japanese forces (who controlled the northern half) and the Allied forces (who controlled the southern half). New Guinea is the world's second-largest island (after Greenland) and is located roughly between Australia and the island of New Britain, with the Bismarck Sea to the north of New Britain and the Solomon Sea to the south of New Britain and to the east of New Guinea.

A Japanese convoy was being assembled at their base in Rabaul on New Britain. Between February 7 and February 14, Allied forces began to detect the preparations of the Japanese forces at Rabaul, but they did not know the destination of the convoy. On February 16, codebreakers in Melbourne, Australia, and Washington, DC, decrypted a Japanese message that indicated that convoys were being planned for three different destinations, with the Japanese base at Lae, New Guinea, being one of them. Subsequently, a message from the Japanese 11th Air Fleet was intercepted, indicating that destroyers and six transports would land at Lae around March 5. (Later, another message was decrypted giving a landing date of March 12, but the Japanese chose the earlier date.)

Since Rabaul is at the northeast tip of New Britain, the convoy would have to sail around the entire island of New Britain to land in New Guinea, which is near the southwest end of New Britain. They could sail north of New Britain, in the Bismarck Sea, or south of New Britain, in the Solomon Sea. New Britain itself is fairly large – the largest island of the Bismarck Archipelago, and either route would entail a trip of three days. The Allied forces, led by Lieutenant General George C. Kenney, under the direction of Supreme Allied Commander of the Southwest Pacific, General Douglas MacArthur, decided to attack the convoy as it sailed.

The plan was, once the convoy was located, to shadow the convoy for the remainder of the journey and inflict as much damage as possible to the convoy before it reached Lae. But the Allied forces did not know if the Japanese would take the northern route or the southern route, and they did not have enough reconnaissance aircraft to search both routes simultaneously. To complicate matters, weather was predicted to be rainy with poor visibility in the Bismarck Sea (the northern route), while the Solomon Sea (the southern route) was expected to be sunny and clear. This adverse weather would not affect the length of the trip – the convoy would take three days on either route regardless of the weather, but it could affect the Allies' ability to locate the convoy for bombing.

To model this as a game, the Japanese commanders have two options open to them: to sail the northern route (Bismarck Sea) or the southern route (Solomon Sea), so those are their strategies. Similarly, the Allies have similar choices where to send the reconnaissance planes: search the northern route or search the southern route (a decision that, unlike that of the Japanese fleet, can be changed on a daily basis, if necessary). But what are the numerical payoffs we can use? Since enemies at war are the paramount example of players with diametrically opposed interests, this should surely be a zero-sum game or a constant-sum game. What are the players goals? The Allies want to destroy as much of the convoy as they can, but there is no way they can know ahead of time how much damage they might inflict, so they cannot measure their payoff by the number of ships sunk or the number of Japanese troops killed. However, it seems reasonable to say that the longer they can shadow the convoy, the more damage they can inflict. So it makes sense from the Allied point of view to say that their payoff is the number of days they have available to bomb the convoy.

What about the Japanese point of view? Well, they obviously want to get as far as they can along their route before they are located by the Allies so they can minimize the number of days the Allies have to bomb them. So what is good for the Japanese from this perspective is bad for the Allies and vice versa. Some authors, by decree, take the payoff to the Japanese fleet to be just the negative of what the payoff to the Allies is. This forces the game to be zero-sum, of course. Thus, if the Allies bomb the convoy for two days, we simply think of this as a payoff of -2 for the Japanese.

This gets the job done, but it might be more natural to think of it another way. What the Japanese want is to maximize the number of days they are able to sail uninhibited by Allied warplanes. So this number is their payoff. Thus, if the Allies bomb the convoy for two days, then the payoff to the Japanese is one day of clear sailing, so to speak. If the Allies bomb for one day, then the Japanese payoff is two days clear sailing. It is clear that the sum of the payoffs is three, the number of days of the entire trip, so we have a constant-sum game here. Recall that in constant-sum games, the players' interests are also diametrically opposed – they are just minor variations of zero-sum games. The example at hand illustrates this point quite vividly.

So, carefully reviewing the assumptions we have built into our model so far, we have agreed that

- The total time for the convoy's journey is three days on either route.
- The Allies will decide on which route to send their reconnaissance planes on a daily basis, until the convoy is located.
- Once the Allies locate the convoy, they will bomb it continuously (at least in the daylight hours), for the remainder of the trip. (In particular, they will not lose track of the convoy and require a second search once they've found it, even if the Japanese take the northern route with poor visibility.)

Now, to determine the payoffs, we need to make some assumptions about how many days the Allies will have to bomb the convoy under various conditions determined by how the players choose their strategies. In a sense, it seems possible that there is an entire range of days between 0 and 3, including fractional amounts, which might result. However, to make the game easier to analyze, we will make some simplifying assumptions based on reasonable guesses. So, to the preceding assumptions, we add the further assumptions:

- If the Japanese sail south and the Allies search south, they will locate the convoy virtually immediately, leaving the Allies three full days to bomb the convoy.
- If the Allies search the wrong route first, they will spend the entire first day looking in the wrong place but will switch their search the next day and will then be searching along the correct route. Thus, an incorrect guess the first day costs the Allies one full day.
- If the Japanese sail north, the poor weather will delay the Allies locating them by one full day.

The assumptions imply that if the Allies search north but the Japanese sail south, the Allies will waste the first day but switch to searching south the second day and find the convoy immediately, resulting in two days of bombing. If both players go north, the Allies will be searching in the correct place, but the poor weather and visibility will delay them in locating the convoy by a day, again resulting in two days of bombing. If they both go south, as noted before, the Allies find them immediately and have three days to bomb the convoy. Finally, if the Japanese sail north but the Allies search south, the Allies lose one day while looking in the wrong place and a second day due to the poor weather once they start looking in the right place, resulting in a single day of bombing. We now have enough information to draw the game tree and find the payoff matrix.

The game tree is shown in Figure 6.2.

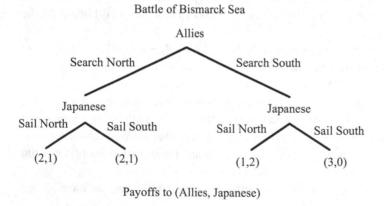

Battle of Bismarck Sea

Allies

Search North   Search South

Japanese   Japanese
Sail North   Sail South   Sail North   Sail South

(2,1)   (2,1)   (1,2)   (3,0)

Payoffs to (Allies, Japanese)

Figure 6.2 Extensive form of the game.

Again, since this is a constant-sum game, knowing the payoffs for the Allies will determine the payoffs to the Japanese, so we really do not need the ordered pair payoffs. In the payoff matrix that follows, we therefore only list the payoffs to the Allies, the row player. In interpreting these payoffs, it is helpful to remember that the row player wants to maximize this number, while the column player prefers the smaller numbers, since, because the game is constant-sum, a poor payoff to the row player means a better payoff for the column player. So the payoff matrix is as follows.

		Japanese	
		North	South
Allies	North	2	2
	South	1	3

Now, we finally address the question of what the players do and why. Although we could draw the same conclusions by looking at the game tree, we will focus on the payoff matrix A instead. First of all, since the name of this episode is the "Battle of the Bismarck Sea" and not the "Battle of the Solomon Sea," the astute reader will no doubt have already concluded that the Japanese (at least) decided to take the northern route. Perhaps this is true, but the point is to understand *why* they chose the northern route. When this example is presented to my classes, there are typically several distinct reasons given by the students as to why the Japanese should choose the northern route. Here are some frequent responses the students offer:

- The Japanese should go north so they can take advantage of the poor weather conditions.
- The Japanese should go north because if they do, the average payoff in the "north" column of the payoff matrix is 1.5, whereas if they go south, the average payoff in the "south" column is 2.5. Since the Japanese would prefer a smaller number, the column with the smaller average is better for them.
- The Japanese should go north because if the Allies search south, then they are bombed for only one day compared to the three days they would be bombed if they went south. On the other hand, if the Allies search north, then they will be bombed two days either way, and so they are no worse off if the Allies go north than the would be if they go south.
- The Japanese should go north because the worst that could happen is two days of bombing, whereas if they go south, they risk a possible three days of bombing.

Let's consider each of these responses carefully. As for the first response, in one sense, it is a good reason for the Japanese to go north. However, the person giving that response is missing the point! They made no appeal whatsoever to the payoff matrix to come to this conclusion; that is, they did not learn anything at all about the situation by viewing it through the eyes of game theory. They completely ignored the game model and could have offered that suggestion as soon as they heard the weather forecast for the region, with no more than a vague conviction that the bad weather would help the Japanese. We will not consider this response further.

The second response is a better reason for going north from our perspective, since the students who give this reason are at least trying to use the information in the payoff matrix. However, there is a flaw in their argument. Have you detected it? Since the suggestion is to look at, for each of the Japanese strategies, the *average* payoffs to the Japanese of what happens if the Allies go north vs. if the Allies go south, there is a tacit assumption that it is equally likely that the Allies will go north as it is that they go south. But, of course, it is not equally likely. The Allies are not indifferent to the Japanese; they are acting strategically to try to damage the Japanese convoy as much as possible. The Allies will not just choose a route randomly, but in a deliberate way to give themselves the best possible payoff.

In other words, students who give this response have forgotten one of the assumptions we built in to game theory – namely, that your opponent is skillful and will act rationally (not randomly). Therefore, students giving this response, even though they have come to the correct conclusion that going north is better for the Japanese, have not given a valid justification for the Japanese decision.

Before considering the other two responses, we should at least mention in passing that later, when we relax the assumptions of game theory, we may very well encounter situations where we may be playing an indifferent opponent rather than one who is "out to get you." Such situations are called *games against nature*. In that case, the person who argues by average payoffs, such as in the second response, is using one of several distinct acceptable approaches to solving such a game.

Nevertheless, in a game against a rational opponent, the justification by average payoffs is incorrect, so we move on to the third response. We can summarize this response as follows. The strategy of going north for the Japanese has the property that *regardless of what the Allies do, the Japanese are always at least as well off, if not better, when they choose north over south.*

Let's carefully check this. If the Allies go south, the row player (the Allies) get a payoff of 1 if the Japanese choose north, as opposed to 3 if the Japanese go south, so in this case, going north is strictly better for the Japanese (since $1 < 3$). On the other hand, if the Allies go north, then either choice by the Japanese results in a payoff of 2 for the Allies, so the Japanese are certainly no worse off by going north than south in this case as well. The conclusion is that the Japanese are always better off (or at least as well off) by going north, regardless of the Allies strategy.

When one strategy S has this property compared to another strategy T – that regardless of what the opponent does, S always leads to a better payoff than T (or at least a tie in the payoffs) – we say S *dominates* T, and we call it a *dominant strategy*. The poorer strategy T is said to be *dominated* by S. Notice it is easy to check if one strategy dominates another by looking at the payoff matrix (remembering that the lower numbers are better for Japan): Every payoff in the "north" column for the Japanese is less than (or equal to) the corresponding payoff in the "south" column, so the strategy of going north is dominant for the Japanese. (Clearly, for the row player, dominance would mean the payoff in row S is greater than (or equal to) the corresponding payoff in the row T since the row player prefers the larger numbers in the payoff matrix). Let's make a formal definition.

DEFINITION 6.6   *In a two-person game, a strategy S dominates* a strategy T *if each payoff in S is at least as good as the corresponding payoff in T.*

By the remarks before the definition, we have the following corollary.

COROLLARY 6.7   *In a two-person constant-sum game, with payoff matrix A consisting of payoffs to the row player, then Column S dominates Column T if every entry in S is less than or equal to the corresponding entry in T – that is, if $a_{is} \leq a_{it}$ for all i. On the other hand, row S dominates row T if every entry in S is greater than or equal to the corresponding entry in T – that is, if $a_{sj} \geq a_{tj}$ for all j. We say the dominance is* strict *if at least one of the inequalities is strict.*

If you are a player in a two-person game with just two strategies, S and T, and S dominates T, you should clearly play S, as in the case of the Japanese in the Bismarck Sea example. If there are more than two strategies and S dominates T, it is not necessarily the case you should play S – you would need to know how S compares to these other strategies besides T. However, one thing is clear: you should not play T. You would always be at least as well off, if not better, by playing S. Therefore, in any two-person game, *rationality dictates that you should never play a dominated strategy.*

In the Battle of the Bismarck Sea, do the Allies have a dominant strategy? The answer is no, they do not. If the Japanese go north, the Allies are better off going north also, with a payoff of 2 instead of the 1 they get if they went south. On the other hand, if the Japanese go south, the Allies are better off going south as well, with a payoff of 3 instead of the 2 they would get if they went north. Thus, the best strategy for the Allies depends on what the Japanese do. By definition, a dominant strategy is one that is independent of what your opponent does, since it will produce a better payoff regardless of what your opponent does. Alternatively, just look at the payoff matrix and observe, in contradiction to Corollary 6.7, that neither row has the property that all of its entries are greater than or equal to the corresponding entries in the other row. We will continue to analyze the Allied position momentarily. First, let's consider the fourth student response for justifying the Japanese action.

The catchwords in this response are the phrase "worst that can happen." If you are trying to maximize your payoff, it seems odd to be focused on the worst that can happen, rather than the best possible outcome. A moment's reflection, however, will reveal why this may be a good approach. If a player wants the best payoff, he or she must choose the row or column containing it as his or her strategy. However, that row or column may also contain much worse payoffs, and which payoff in that row or column you actually receive depends on your opponent's choice, not yours (once you have selected a strategy, your job is done). In fact, no matter what row or column you choose, the same is true – you open yourself up to risk the worst outcome in that row or column, and you are not in a position to do anything about it once your strategy is selected. Also, don't forget that in a constant-sum game, your opponent is actively trying to give you that worst payoff, and we have postulated that the opponent are skillful and will act rationally. Given that, it seems like a prudent way to play, then, is to consider the worst thing that can happen for each of your potential choices and choose the row or column that contains the *best of the worst-case scenarios* in order for you to minimize the damage done if your opponent is skillful enough to force the worst case.

In a payoff matrix for a constant-sum game, the worst that can happen from the column player's perspective is the largest entry in each column (since a large payoff to the row player means a small payoff for herself). That is, the column maximum is the worst-case scenario for the column player. So, if the column player wanted to adopt this point of view of minimizing the damage done, he or she should choose the column with the smallest maximum element – the minimum of the column maxima. This approach for the column player can be called the *minimax* approach. On the other hand, the row player must do the opposite and follow a *maximin* approach. That is because for the row player, the worst-case scenario for each row is the minimum entry in that row – the row minimum – and the best of the worst-case scenarios is then the row with the largest of these row minima.

The students who suggest response 4 are following this minimax approach. In the first column, sail north, the maximum is the 2 in the a_{11} position. In the second column, sail south, the maximum is the 3 in the a_{22} position. Therefore, the column with the best of these worst-case scenarios is the first column, since 2 is better than 3 for the Japanese. Therefore, the conclusion is to sail north.

As we shall see, there are some constant-sum games where this minimax approach will not necessarily work. What is required to make it work is the happy accident that if *both* players follow this approach, it leads to the same outcome for the game; in other words, the same outcome is simultaneously the best of the worst-case scenarios for both players. As we shall see momentarily, this actually does happen in the Battle of the Bismarck Sea. Before we go on to the Allies, let's just summarize what we have learned. We have two poor reasons for the Japanese choosing north (one based on average payoffs and one not based on payoffs at all) and two good reasons for the Japanese to go north (one based on a dominant strategy for the Japanese – sail north – and the other based on the minimax approach, once we see that the same reasoning leads to the same outcome for the Allies).

Turning to the Allies, we have already observed that neither strategy is dominant for the Allies. So, instead, let's look at this approach of taking the best of the worst-case scenarios. The minimum in the first row (search north) is a 2; both of the entries are tied for this dubious honor. The minimum in the second row (search south) is the 1 in the a_{21} position. Since a 2 is better than a 1, the maximin strategy for the Allies is the first row – search north. So, if both players follow this approach, they both lead to the same outcome, which is that the Japanese sail north and the Allies search north. That outcome, with its payoff of 2 for the Allies, is simultaneously the best of the worst-case scenarios for both players.

In terms of the payoff, the 2 located in the a_{11} position is characterized by being simultaneously the maximum of the row minima and also the minimum of the column maxima; that is, it is simultaneously the smallest in its row (ties are not a problem – it's still the smallest in the row) and the largest in its column. Such points in the payoff matrix of a constant-sum game are special and will be discussed in the following paragraphs. But first, we want to observe that whenever the best of the worst-case scenarios occurs at the same outcome of a game (whether it is constant-sum or not), this outcome has an important stability property. Namely, *neither player stands to gain by a unilateral change of strategy*.

Let's verify this for the Battle of the Bismarck Sea. The minimax point and the maximin point are both the same – the 2 associated with the outcome where both players choose north. Can either player gain from switching to the southern route? No, if the Allies switch unilaterally (which means that only the Allies are changing – the Japanese stick to the northern route), their payoff goes down from 2 to 1, so there is no gain there. On the other hand, if the Japanese switch unilaterally, the payoff stays the same at 2 (in the a_{12} position). But there is no gain for the Japanese in a unilateral switch. This leads us directly to one of the fundamental concepts of game theory, given in its most general form by John Nash in the 1950s, and named in his honor.

DEFINITION 6.8   *In any game (two-person, n-person, constant-sum or variable-sum), an outcome of the game is said to be a* Nash equilibrium *if it has the property that no player stands to gain by a unilateral change in strategy.*

Since neither player can gain by a unilateral switch, there is no temptation for a rational player to switch. Therefore, a game that is in a Nash equilibrium outcome is in a very stable state, explaining why Nash called it an "equilibrium point." We have just seen that both players choosing north leads to a Nash equilibrium in the Battle of the Bismarck Sea. Thus, our "solution" to this game is that both players should choose the northern route for their strategies, leading to a payoff of two days, bombing for the Allies and one day of unimpeded sailing for the Japanese.

Finally, we note that although the Allies did not have a dominant strategy, it is still possible to determine their best strategy without resorting to the minimax approach. Since we have assumed that all the players have the same information about the game, it should be clear to both players that the Japanese have a dominant strategy. The rationality assumption implies that dominated strategies should not be played, so both players can conclude that the best thing for the Japanese to do is go north. Since both players will ignore the dominated strategy, why not cross it off the matrix? In the new, smaller matrix with the dominated column deleted, some new dominant strategies for the row player might appear that were not present in the full matrix. These are called *higher-order dominant strategies*. In some games, one can iterate this process, crossing off higher-order dominated rows and columns until the entire matrix is reduced to a 1×1 matrix, which must be the solution point (and in fact, this point will be a Nash equilibrium).

This is what happens in the Battle of the Bismarck Sea. The Allies do not have a dominant strategy, but they can see the Japanese do and conclude the Japanese will ignore the "sail south" strategy. Crossing off this dominated column from the payoff matrix reduces it to a 2×1 matrix.

		Japanese
		North
	North	2
Allies	South	1

Now the Allies do indeed have a (higher-order) dominant strategy on this reduced matrix – to search north – because the 2 payoff is greater than 1, the payoff for going south. From this matrix, deleting the dominated second row, we are left with the 1×1 matrix [2] corresponding to both players choosing north, which is the predicted optimal solution, and, as mentioned, it is the Nash equilibrium of this game.

In summary, we have derived the optimal solution to the Battle of the Bismarck Sea – both players should go north, which results in two days of bombing the fleet after one day unimpeded sailing. We have two distinct valid lines of argument to draw these conclusions. On the one hand, we have seen that the outcome (north, north) is a Nash equilibrium. Secondly, we can argue that the Japanese must go north by dominance, and knowing that, the Allies must go north by higher-order dominance. Either way, we arrive at the same conclusion.

Before we continue to explore properties of dominant strategies and Nash equilibria, we pause to compare our solution to what really happened in the Battle of the Bismarck Sea. In the real battle, both players did indeed choose the northern route. Even though the payoff of 2 that we filled into the matrix was a guess, that is pretty much what actually happened – the Allies bombed the Japanese convoy for two days, March 2 and 3, resulting in terrible losses for the Japanese. (In addition to the bombing raids, there were also follow-up attacks by PT Boats through March 4.)

While the actual figures are not relevant to our discussion, the following estimates will indicate the scope of the victory for the Allies: The convoy had 16 ships – eight destroyers and eight troop transports – and more than 100 Japanese fighter planes to guard the convoy. All the transports and half the destroyers were destroyed, and more than two-thirds of the Japanese planes were shot down. By contrast, the Allies lost 13 aircrew – 10 in battle and 3 in an accident. Of the nearly 7,000 Japanese troops that were headed for New Guinea, only about 1,200 made it there. About 2,700 more were saved by destroyers and submarines and returned to Rabaul, their point of departure in New Britain. Nearly 3,000 Japanese troops were killed in the battle, which became a turning point in the war in the South Pacific.

The heavy casualties for the Japanese were partly due to the Allies using new, improved tactics in the bombing raid, including attacking the Japanese ships from multiple directions at once, and low-altitude attacks such as mast-height bombing and skipbombing. Of course, we did not attempt to model any of these tactical decisions in our game, which was only about the decision where to send reconnaissance planes.

If there is one thing that is clear from the previous example, it is that being able to spot dominant strategies and Nash equilibria is important in game theory. While it is possible to do this with a game in extensive form, it is easier to work with the payoff matrix, so we will tend to look at most of our games from now on in normal form. Let A be the payoff matrix of a two-person game. There is a graphical device that is useful for locating Nash equilibria.

Suppose you are the row player and have tentatively decided on a strategy (that is, you have tentatively chosen a row). In order to determine if an outcome is a Nash equilibria or not, you would have to analyze what would happen if you made a unilateral change in strategy. That means you would have to consider whether or not you would be better off if you switched rows while the column player stayed in the same column. To answer that question, you should know which entry in each column is the best outcome for you. Similarly, the column player should know which entry in each row is the best for him. We can indicate this by drawing a diagram of the positions of the payoff matrix and within each column, drawing vertical arrows pointing to the position of the best outcome in that column for the row player, and within each row, drawing horizontal arrows pointing to the best outcome in that row for the column player. Following Straffin (1993), we call this the *movement diagram* of the game.

In the Battle of the Bismarck Sea, or in any constant-sum game where the payoff matrix only displays the row player's payoffs, that means the vertical arrows point to the largest entry in the column, while the horizontal arrows point to the smallest element of the row. So the movement diagram form the Battle of the Bismarck Sea is shown in Figure 6.3.

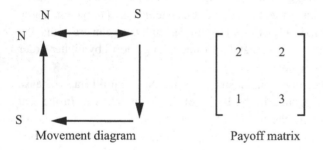

Movement diagram   Payoff matrix

Figure 6.3 Comparing the movement diagram to the payoff matrix.

We have placed the diagram next to the payoff matrix to illustrate the construction of the diagram. Visibly, vertical arrows (which represent potential unilateral changes by the row player) point to the largest payoff in a column, while horizontal arrows (indicating potential unilateral changes by the column player) point to the smallest entry in a row. A tie between different payoffs that are equally desired by the relevant player are indicted with a double-headed arrow, such as that in the top row of Figure 6.3.

Notice that an outcome of a game that has both a vertical and a horizontal arrow pointing to it (such as the NN outcome in this game) must be a Nash equilibrium since, by definition, neither player stands to gain by switching away from that outcome unilaterally, as indicated by the arrowheads. Furthermore, the diagram also graphically shows that the Japanese have a dominant strategy – all of the entries in the north column have arrowheads pointing to them, indicating that for the Japanese, north is better than (or at least as good as) south, regardless of what row is chosen by the Allies. That the Allies do not have a dominant strategy is also easy to read from the movement diagram – one of the vertical arrows points up and the other points down, indicating that the row that is best for the Allies depends on which column the Japanese choose.

We have already mentioned that the concepts of dominant strategy and Nash equilibrium also apply to variable-sum games. We now illustrate that two-person variable-sum games also have movement diagrams. They are constructed in the same manner; the only difference being that since the payoffs are ordered pairs, we decide which outcome in a column is best for the row player by looking at the first coordinates of these ordered pairs (and choosing the highest possible) and which outcome on a row is best for the column player by looking at the second coordinates of the ordered pairs (and choosing the highest possible). Let's illustrate this with the example of the game of Saturday chores at the Anderson household, version II. Recall from the last section that the payoff matrix is given by the following.

		Laurie			
		No chore	Dishes	Raking	Both
	No chore	$(10, 10)$	$(12, 6)$	$(8, 15)$	$(0, 21)$
Jon	Dishes	$(8, 15)$	$(10, 10)$	$(8, 15)$	$(4, 18)$
	Raking	$(12, 6)$	$(12, 6)$	$(10, 10)$	$(6, 13)$
	Both	$(20, 0)$	$(16, 3)$	$(14, 7)$	$(10, 10)$

From this, we draw the movement diagram in Figure 6.4.

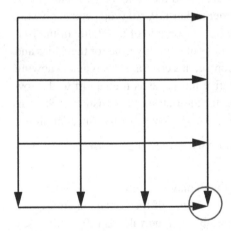

Figure 6.4 The movement diagram for the game in Example 6.4.

Indeed, the largest entries in the first coordinates are all in the bottom row, and the largest entries in the second coordinates are all in the rightmost column. Thus, if it was not clear from the matrix itself, the movement diagram shows that for each of the children, the strategy of volunteering for both chores dominates all three of the other strategies. So in this game, both players have a dominant strategy. Also, the outcome where these dominant strategies intersect is a Nash equilibrium (circled on the movement diagram).

Does this constitute a solution to this game? After all, a Nash equilibrium that results from playing two dominant strategies seems especially stable. Variable-sum games can have some subtle dynamics, so we need to be careful. It was mentioned in the last section that variable-sum games are games of partial conflict and that it may be possible by partial cooperation to find outcomes in which both players improve their payoffs (the win-win situations that do not happen with constant-sum games).

In this case, there is no such improvement that is possible (under our assumptions of simultaneous, rational play with no communication between the players and in which the players truly want to maximize the money they make), because by inspecting the payoff matrix, we find no payoffs with both coordinates larger than or equal to 10 simultaneously, except the other payoffs of $(10, 10)$ that all lie on the main diagonal. Thus, each player can see that no further improvements over the $(10, 10)$ payoff are possible for both players at once; any improvement for one player results in the loss for the other player (similar to what happens in a constant-sum game).

Therefore, in this game, it may seem to be safe to say that we have an acceptable solution to the game (although we will not formally discuss variable-sum games until a later chapter). Rational play dictates that both players should volunteer for both jobs, resulting in a payoff of $10 for each of the children. Perhaps this example can be regarded as an application of game theory to behavioral psychology, which might be useful to parents who want to coax their children into volunteering to do their fair share of household chores without directions from the parents and without arguments amongst themselves. Actually, we will see later (Exercise 30 in Section 6.4) that we are being a little too naive here – it may be possible to find other outcomes where the payoff to both players is more than $10 if we use *mixed strategies,* which are defined in Section 6.3.

Incidentally, when a game is solved, the optimal payoffs to the players are an important part of the solution, so it's helpful to give this bit of data a name. Its called the *value* of the game. Thus, the value of the Anderson household chores game is the ordered pair $(10, 10)$ (you need to give the value for both players because it is a variable-sum game). In the example of the Battle of the Bismarck Sea, the value is the ordered pair $(2, 1)$ meaning two days of bombing time for the Allies and one day of unimpeded travel for the Japanese convoy. Of course, in a constant-sum game, knowing one payoff is sufficient, so for such games, we typically give the value as the payoff to the row player instead of an ordered pair. So the value of the game for the Battle of the Bismarck Sea can be regarded as simply the number 2 instead of the ordered pair $(2, 1)$. We make a formal definition:

DEFINITION 6.9   *The* Solution *of a two-person game consists of the optimal strategy for each player, as well as the value of the game.*

So the solution of the Battle of the Bismarck Sea is that both players choose the north strategy, and the value is 2 (or the ordered pair $(2, 1)$). The apparent solution of the Anderson Household Chores game is that Jon and Laurie each volunteer for both jobs, and the value is $(10, 10)$. We say "apparent solution" because we have not yet formally studied the added complications of variable-sum games in general. *For the remainder of this chapter, we consider only constant-sum games.*

Since the games on which we now focus are constant-sum, and we follow the usual convention of writing the payoff to the row player only in the payoff matrix, the concept of a Nash equilibrium reduces to something simpler, a *saddle point* of the payoff matrix:

DEFINITION 6.10   *Let A be the payoff matrix of a constant-sum game. A* saddle point *of A is an entry of A that is simultaneously the smallest entry in its row and the largest entry in its column.*

For example, consider the payoff matrix of the Battle of the Bismarck Sea:

$$A = \begin{bmatrix} 2 & 2 \\ 1 & 3 \end{bmatrix}.$$

There is a saddle point in a_{11} position. The entry 2 there is the largest entry in its column and is (tied for) the smallest in its row. This position corresponds to the outcome of the game in which both players choose north, which we already know is a Nash equilibrium. This is no accident:

PROPOSITION 6.11   *In the payoff matrix A for a constant-sum game, an entry corresponds to a Nash equilibrium if and only if it is a saddle point.*

Proof   Let a be an entry that is a saddle point. Since a is the smallest entry in its row, it represents the best-possible outcome in that row for the column player, so in the movement diagram, the horizontal arrow for that row would point to that entry. Similarly, since it is the largest entry in tis column, it represents the best-possible outcome in that column for the row player, so in the movement diagram, the vertical arrow for that column would point to that entry a. Thus, neither player stands to gain by a unilateral move away from that outcome, which is exactly the definition of a Nash equilibrium. We have shown that if a is a saddle point, then it corresponds to a Nash equilibrium.

The phrase "if and only if" means the implication goes both ways, so we must also show that the converse holds; that is, we must show that if we have a Nash equilibrium, there is a saddle point at that outcome. So suppose a is the payoff located at the Nash equilibrium outcome. Suppose there is a (strictly) larger entry b in the column containing a. In that case, the row player can benefit by switching (unilaterally) from a's row to b's row. This contradicts the fact that a is a Nash equilibrium, so no such b exists. Thus, a is the largest entry in its column. Similarly, if there is an entry c in the row containing a that is (strictly) smaller than a, then the column player could benefit from a (unilateral) switch to the column containing c, yielding the same contradiction to a being a Nash equilibrium. Thus, no such c exists and a is therefore the smallest entry in its row. But then a is a saddle point by definition.   ∎

DEFINITION 6.12   *Any constant-sum game (in particular, any zero-sum game) in which the payoff matrix has a saddle point is said to be* strictly determined.

In a strictly determined game, both players should choose the strategy that leads to the saddle point outcome. Thus, the solution to a strictly determined game is easy to find – choose a strategy containing the saddle point, and the value of the game (for the row player) is the value of the saddle point.

REMARK 6.13   *In a strictly determined (constant-sum) game, we can regard the value v of the game as a sort of guaranteed minimum payoff for each player. If the players play optimally, the result will be the saddle point outcome and each player will obtain the value guaranteed there (v for the row player and c − v for the column player, if the payoffs sum to c.) However, if either player deviates from his or her optimal strategy, it may be possible for the other player to have an even better payoff.*

REMARK 6.14   *Since the solution to a strictly determined game is obvious to both players, nothing that the players might say to each other before or during the game should affect their strategy choices. Thus, secrecy is not important in such a game.*

The definition of a strictly determined game is more complicated in the variable-sum case and is given in the next chapter. As we shall see, a strictly determined game might have more than one saddle point. However, not all games are strictly determined.

The next order of business is to describe an alternate method of locating saddle points. Since a saddle point is the smallest entry in its row, the first step is to record the minimum of each row. This can be done by placing this row minimum to the right of the matrix (in the same row), or it can simply be indicated by underlining the entry right in the matrix. Also, since a saddle point is the highest entry in its column, the next step is to record the maximum in each column. Either indicate this by placing this maximum just below the matrix (in the same column), or indicate it in the matrix itself (not by underlining, which indicates a row minimum – perhaps by placing a bar above the entry). Then find the maximum of the row minima (which is the best of the worst-case scenarios for the row player) and the minimum of the column maxima (the best of the worst-case scenarios for the column player). If these values are the same, then the outcome(s) where this occurs will be a saddle point (or saddle points). Here's an example. Suppose the payoff matrix is the following:

$$A = \begin{bmatrix} 1 & -2 & 0 & -1 \\ 3 & 4 & 3 & 6 \\ 5 & 1 & -1 & -4 \end{bmatrix}$$

We record the row minima and column maxima:

$$
\begin{array}{cc}
 & \text{Row minima} \\
A = \begin{bmatrix} 1 & -2 & 0 & -1 \\ 3 & 4 & 3 & 6 \\ 5 & 1 & -1 & -4 \end{bmatrix} & \begin{matrix} -2 \\ 3 \\ -4 \end{matrix} \\
\text{Column maxima} \quad \begin{matrix} 5 & 4 & 3 & 6 \end{matrix} &
\end{array}
$$

The largest row minima is 3 (second row), and the smallest column maxima is also 3 (third column), which indicates that the entry $a_{23} = 3$ is a saddle point, which is easily verified.

On the other hand, if you use the underline/overline method, you would obtain the following:

$$A = \begin{bmatrix} 1 & \overline{-2} & 0 & -1 \\ \underline{3} & \overline{4} & \overline{\underline{3}} & \overline{6} \\ \underline{5} & 1 & -1 & \underline{-4} \end{bmatrix}$$

With this method, the saddle point at $a_{23} = 3$ is indicated by being an entry with both an underline and an overline simultaneously. Observe that both of the entries in the second row that have value 3 are underlined, since they are tied for being the row minimum, but only the one in the third column is a saddle point. In this game, we haven't specified whether it is zero-sum or constant-sum, but it does not matter as far as locating a saddle point is concerned.

Consider the following 4×6 payoff matrix, which we may as well assume represents the payoffs in a zero-sum game:

$$A = \begin{bmatrix} -2 & 0 & -4 & 5 & -3 & 0 \\ 1 & \boxed{1} & 2 & 3 & \boxed{1} & 4 \\ 0 & -2 & 1 & -7 & -1 & -3 \\ 2 & \boxed{1} & 2 & 2 & \boxed{1} & 1 \end{bmatrix}$$

By one of the preceding methods or by drawing a movement diagram, you will find that all four of the entries that are boxed are saddle points. This illustrates that a strictly determined game can have more than one saddle point (i.e., Nash equilibrium). So the solution to this game is that the row player should choose either the second or the fourth row, the column player should choose either the second or fifth column, and regardless of these choices, the outcome is at one of the four saddle points. The value v of the game is the common value 1 of all the saddle points.

This does raise the question of how typical this example is. Following Straffin (1993), let's call two Nash equilibria *equivalent* if they have the same payoffs (i.e., same value) for each player. This definition applies to Nash equilibria even if the game is variable-sum or if there are more than two players, when the concept of a saddle point no longer applies. In the preceding game, all four saddle points are equivalent because they all have the same value 1. Furthermore, we call two Nash equilibria *interchangeable* if any (or all) of the players change strategies from a strategy containing one of the equilibria to a strategy containing the other and the outcome of the game is still a Nash equilibrium.

In the preceding example, all four of the equilibria are interchangeable because if either player (or both) switches from the second row to the fourth row or from the second column to the fifth column, the outcome is still one of the four saddle points. Clearly, that is because the saddle points are arranged as corners of a rectangle within the payoff matrix. The next theorem shows that in a zero-sum game (or a constant-sum game), the Nash equilibria are always equivalent and interchangeable.

THEOREM 6.15 *Any two Nash equilibria in a two-person constant-sum game are equivalent and interchangeable.*

Proof In a constant-sum game, a Nash equilibrium is the same thing as a saddle point. So choose two saddle points from the payoff matrix $A = (a_{ij})$. If they both lie in the same row, they both have a value equal to the row minimum, whence they must have the same value and so are equivalent. Similarly, they are equivalent if they both lie in the same column, as they both are equal to the column maximum. In either case, they are clearly interchangeable as well, since only one of the players changes his or her strategy, and the resulting outcome is the other saddle point.

It remains to handle the case when the two saddle points are neither in the same row nor in the same column. For concreteness, let's say one of the saddle points is a_{pq} and the other is a_{rs} with $p \neq r$ and $q \neq s$ by assumption. Consider the following two entries of A: a_{ps} and a_{rq}. These two entries and the two given saddle points are at four distinct locations in A since $p \neq r$ and $q \neq s$. These four locations are positioned at four corners of a rectangle in A, similar to the example, with the given saddle points at opposite corners of this rectangle (again, because $p \neq r$ and $q \neq s$).

Now, since a_{pq} is a saddle point, it is the minimum value in its row; therefore,

$$a_{pq} \leq a_{ps}.$$

On the other hand, a_{ps} is in the same column as a_{rs}, the other saddle point, which must be the maximum value in its column. Therefore,

$$a_{ps} \leq a_{rs}.$$

Combining these two in equalities yields

$$a_{pq} \leq a_{ps} \leq a_{rs}.$$

Similarly, we can compare the values of the two saddle points by going through a_{rq} instead of u_{ps}. Since a_{pq} is the largest entry in its column, which contains a_{rq}, and a_{rs} is the smallest entry in its row, which also contains a_{rq}, we obtain the reverse inequalities:

$$a_{pq} \geq a_{rq} \geq a_{rs}.$$

It follows that

$$a_{pq} = a_{rs},$$

which means the two saddle points are equivalent.

Furthermore, looking back over all of these inequalities, we see all four entries must then have the same value, since a_{rq} and a_{ps} were between the other two. In particular, a_{rq} is (tied for) the smallest value in its row, since $a_{rq} = a_{rs}$, and also (tied for) the largest value in its column since $a_{rq} = a_{pq}$; therefore, a_{rq} is also a saddle point. Similarly, a_{ps} is also a saddle point. Thus, all four Nash equilibria are equivalent. Furthermore, they are interchangeable since, regardless of whether the row player chooses row p or row r and whether the column player chooses column q or column s, the intersection of the row and column chosen is one of the four saddle points. ∎

In the next section, we will consider games that are not strictly determined – that is, without saddle points.

Exercises

1. Draw the movement diagram for the game rock paper scissors. Conclude that in this game, there are no dominant strategies and that none of the outcomes are a Nash equilibrium.

    The matrices in Exercises 2–10 are assumed to be payoff matrices for zero-sum games. For each one do the following: (a) Determine if there are any dominant and dominated strategies for either player, and list them. (2) Locate all saddle points, if any. You may use your favorite method to locate them (i.e., movement diagrams, comparing row minima to column maxima, or the overline/underline method). (3) If the game is strictly determined, give the complete solution (optimal strategies for each player and value of the game).

2.

$$\begin{bmatrix} 1 & 2 \\ 3 & 4 \end{bmatrix}$$

3.

$$\begin{bmatrix} 0 & 1 & -3 \\ 3 & 2 & 4 \end{bmatrix}$$

4.

$$\begin{bmatrix} 0 & 1 & -3 \\ -2 & 3 & 4 \end{bmatrix}$$

5.

$$\begin{bmatrix} -4 & -1 & 7 \\ 2 & -1 & 3 \\ 8 & -3 & 11 \end{bmatrix}$$

6.

$$\begin{bmatrix} 1 & 0 & -4 \\ -3 & 1 & -6 \\ 0 & 6 & -3 \end{bmatrix}$$

7.

$$\begin{bmatrix} 1 & 0 & 1 & -3 \\ -2 & 0 & 1 & 1 \\ 3 & 2 & -5 & 0 \end{bmatrix}$$

8.

$$\begin{bmatrix} -1 & -3 & 1 & -9 \\ 5 & -3 & 7 & -4 \\ 4 & -2 & 0 & -2 \end{bmatrix}$$

9.

$$\begin{bmatrix} 1 & 0 & 2 & 0 & 3 \\ -3 & -1 & 1 & -3 & 0 \\ 3 & 0 & 4 & 0 & 5 \\ 0 & -1 & 0 & -4 & 1 \end{bmatrix}$$

10.

$$\begin{bmatrix} -4 & 1 & 0 & 0 & -2 \\ -3 & 3 & 3 & 0 & -1 \\ 1 & 1 & 1 & 1 & 0 \\ 5 & -3 & -2 & -4 & 5 \end{bmatrix}$$

11. In Exercise 3 of Section 6.1, we extended two-finger morra to a version for three fingers. Draw the movement diagram, and determine whether or not this is a strictly determined game.

12. In Exercise 4 of Section 6.1, we extended two-finger morra to a version for four fingers. Draw the movement diagram and determine whether or not this is a strictly determined game.

13. In Exercise 7 of Section 6.1, we considered a finite version of the Antares vs. Bellatrix marketing game. Determine whether or not this is a strictly determined game, and if so, give the solution.

14. In Exercise 8 of Section 6.1, we considered a finite version of the Antares vs. Bellatrix marketing game. Determine whether or not this is a strictly determined game, and if so, give the solution.

15. In Exercise 9 of Section 6.1, we considered a variation on four-finger morra. Determine whether or not this is a strictly determined game, and if so, give the solution.

16. Consider the guessing game of Exercise 10 from Section 6.1. Determine whether or not this is a strictly determined game, and if so, give the solution.

17. Consider the following variation of four-finger morra. Ann and Bill simultaneously put out one, two, three, or four fingers. If the total sum of the fingers shown is odd, Ann pays Bill the a number of dollars equal to the smaller of the two numbers shown. If the total number

of fingers revealed is even, then Bill pays Ann a number of dollars equal to the smaller of the two numbers shown. Write down the payoff matrix, draw the movement diagram, and list any dominant strategies and saddle points, if any. If the game is strictly determined, give the solution.

18. a.   Consider the following variation of four-finger morra. Ann and Bill simultaneously put out one, two, three, or four fingers. If the total number of fingers shown is a number that has a remainder of 1 when it is divided by 3, then Ann pays Bill $1. If the total number of fingers shown is a number that gives a remainder of 2 when divided by 3, then Bill pays Ann $1. However, if the total number of fingers shown is exactly divisible by 3 with no remainder, then no payment is made. Write down the payoff matrix, draw the movement diagram, and list any dominant strategies and saddle points, if any. If the game is strictly determined, give the solution.

     b.   Repeat this exercise with the same rules except using up to three fingers only.

19. This exercise is about *translations* of games. Let A be a matrix consisting of the payoffs for a constant-sum game, which is strictly determined, and suppose that the payoffs to the row player and the column player sum to the constant c. To clarify things, think of the entries of A as being ordered pairs of payoffs to both players rather than just to the row player. Suppose we modify the matrix A by adding a real constant b to each entry of the matrix A to obtain a new matrix B. That is, we add b to both coordinates, so if the ij entry of A is (a_{ij}, a'_{ij}), then the ij entry of B is $(a_{ij}, a'_{ij}) + (b, b) = (a_{ij} + b, a'_{ij} + b)$ for all i and j. The game with payoff matrix B is said to be a *translation* of the game with payoff matrix A. Since we are assuming A is constant-sum, we typically just write the payoffs to the row player rather than the complete ordered pair. In that case, we can think of the matrix B as being obtained from A by $b_{ij} = a_{ij} + b$ for all i, j, provided we remember the convention that b is also added to the column player's payoffs.

     a.   Explain why the resulting matrix B still represents the matrix of payoffs for the row player in a constant-sum game.

     b.   What constant do the payoffs sum to in this modified game?

     c.   Explain why this game is still strictly determined and why the saddle points in B occur in exactly the same locations as in A.

     d.   If the value of the original game is $v = v_A$, what is the value v_B of the modified game?

20. This exercise is about *scalings* of games. Let A be a matrix consisting of the payoffs for a constant-sum game, which is strictly determined, and suppose that the payoffs to the row player and the column player sum to the constant c. To clarify things, think of the entries of A as being ordered pairs of payoffs to both players, rather than just to the row player. Suppose we modify the matrix A by multiplying each entry by a positive real constant $m > 0$ to obtain a new matrix B. That is, we multiply both coordinates by m, so if the ij entry of A is (a_{ij}, a'_{ij}), then the ij entry of B is $m(a_{ij}, a'_{ij}) = (ma_{ij}, ma'_{ij})$ for all i and j. The game with payoff matrix B is said to be a *scaling* of the game with payoff matrix A. Since we are assuming A is constant-sum, we typically just write the payoffs to the row player rather than the complete ordered pair. In this case, we can think of the matrix B as being obtained from A by $b_{ij} = ma_{ij}$ for all i, j provided we remember the convention that the column players payoffs are scaled by the same factor m.

a. Explain why the resulting matrix B still represents the payoff matrix of a constant-sum game. What do the entries sum to in B?

b. Explain why this game is still strictly determine and why the saddle points in B occur in exactly the same locations as in A.

c. If the value of the original game is $v = v_A$, what is the value v_B of the modified game?

6.3 Mixed-Strategy Constant-Sum Games

In this section, we will begin our study of two-person constant-sum games that are not strictly determined. However, we begin by solving just one more strictly determined game: it is the Antares vs. Bellatrix game that was the infinite game in Example 6.5 of Section 6.1. It will turn out that, despite being strictly determined and despite being infinite, the ideas behind its solution will be useful for solving 2×2 games that are not strictly determined. Once we understand how to solve 2×2 games, we will build up to larger games by various techniques until we are finally able to solve any two-person constant-sum game of any size using linear programming.

6.3.1 Antares vs. Bellatrix

Recall the following example from Section 6.1: Antares Industries and Bellatrix, Inc., are the only two companies that sell a certain computer chip, the Pentangle 5.1. There is an estimated market for 400,000 of these chips in the year 2021. Each company must set its prices. Production costs dictate that the price must be at least \$80 per chip, while industry regulations dictate that the price should be no more than \$120 per chip. Let x be the price set by Antares and y the price set by Bellatrix. Based on reliable market studies of which both companies are aware, it has been determined that the percentage of the total market which goes to Antares depends on x and y and is reasonably approximated by the function given by

$$E(x, y) = 330 - 3x - 2.7y + 0.03xy.$$

What are the prices x and y that each company should choose, if their goal is to obtain the largest-possible market share?

In our earlier discussion of this game, we observed that this can be regarded as a two-person constant-sum game. The strategies for each company are the prices they set (x and y). The values of these variables must be in the real interval $[80, 120]$, so each player has an infinite number of strategies. Once x and y are chosen, then Antares receives a payoff of $E(x, y)$ while Bellatrix receives a payoff of $100 - E(x, y)$. These numbers are percentages of the total market, so they sum to the constant 100, as expected.

If the game were finite, we would construct the payoff matrix by letting the row player (Antares) choose a row (that is, the company chooses a price x); the column player (Bellatrix) choose a column (that is, Bellatrix chooses a price y); and entering $E(x, y)$ at the point in the matrix where the row x and the column y meet (We encourage the reader to review his or her solutions to Exercises 6, 7, and 8 in Section 6.1, which were finite versions of this game.) Then, we search for a saddle point in this matrix (see Exercises 13 and 14 in Section 6.2).

In our case, since we have infinitely many choices for x and y, it does not make sense to construct a matrix, but it should be clear that the function $E(x, y)$ *serves the exact same purpose as the payoff matrix* – it gives, for each choice (x, y), the payoff $E(x, y)$ to the row player Antares. The attentive reader will recognize that the graph of $E(x, y)$ is a surface. So for such infinite, continuous games, this surface replaces the payoff matrix. Of course, we are only concerned with that part of the surface which is defined for $x \in [80, 120]$ and similarly for y.

Now, the argument that a saddle point in the payoff matrix is the same thing as a Nash equilibrium, which we gave in Proposition 6.11 of Section 6.2, is actually still valid when applied to this surface. A choice of (x, y) for which $E(x, y)$ is simultaneously a minimum in its "row" (being in a row means holding y fixed) and a maximum in its ''column" (which means holding x fixed) will represent an outcome of the game where neither player stands to gain by a unilateral change in strategy, which is a Nash equilibrium. So we should search for such a saddle point on the surface $E(x, y)$ in the relevant ranges for x and y.

Finally, the reader should recall Chapter 1, where we studied surfaces of the form $F(x, y) = A - Bx - Cy + Dxy$. We saw that any function of this form was called a *saddle surface* precisely because it could always be converted to standard form $F(x, y) = D(x - p)(y - q) + v$, which is a surface with a saddle point at the location $(x, y) = (p, q)$ and where $F(p, q) = v$. Thus, to solve the Antares vs. Bellatrix infinite game, we must put $E(x, y)$ into standard form. By the techniques we discussed in Chapter 1, the standard form of $E(x, y)$ is

$$E(x, y) = .03(x - 90)(y - 100) + 60. \tag{6.2}$$

Therefore, there is a saddle point at $(x, y) = (90, 100)$ and $E(90, 100) = 60$. Notice that the saddle point is located in the required range of prices $80 \le x, y \le 120$. Therefore, the Antares vs. Bellatrix game is strictly determined (even though it is infinite). Antares should set its price at $90; Bellatrix should set its price at $100. If they both follow this recommendation, the value of the game is $v = 60$. Thus, Antares will have 60% of the market, and Bellatrix will have 40%.

Since this is a Nash equilibrium, if either player were to unilaterally deviate from this strategy, we might expect the payoff in market shares to the other player to possibly improve. But actually, in this game, that does not happen. For example, if Antares sets its price at the optimum choice of $90 per chip, then by evaluating (6.2), it is clear that $E(90, y) = 60$ for all y. So in this case, if Bellatrix changes its price unilaterally, it won't affect the payoff to Antares at all. Similarly, if Bellatrix sets its price at the optimum of $100 per chip, a unilateral change in price by Antares will not affect its payoff since $E(x, 100) = 60$ for any x. (Not all saddle points behave this way, but if the game is one where the payoffs are given by a saddle surface function of crossed type, then the behavior is like this one – no unilateral change of strategies affects the payoffs at all.)

We remark in passing that we can also find the revenue for each company. Assuming a total market of 400,000 chips, the total revenue for Antares is $400,000\left(\frac{E}{100}\right)x = $400,000(0.6)(90) = $21,600,000; and the revenue for Bellatrix is $400,000\left(\frac{100-E}{100}\right)y = $400,000(0.4)(100) = $16,000,000. Remember, however, that the goal of each company is to maximize market shares, not revenue, since we observed previously a game with revenues as payoffs would be variable-sum.

6.3.2 Mixed Strategies, Probabilities, and Expected Payoffs

We have remarked earlier that in a two-person constant-sum strictly determined game, such as the Battle of the Bismarck Sea, secrecy is not important. Both players are aware that the best outcome for them is the Nash equilibrium, and knowing that your opponent will try to obtain this outcome will not sway you from trying to do the same thing. By contrast, in a game without a saddle point, the opposite is true – knowing ahead of time what your opponent will do would definitely affect your strategic decision. For example, in rock paper scissors, if you knew before the play that your opponent will choose rock, it is clear that you would then choose paper in order to win, while if you knew your opponent will choose paper, you would choose scissors, etc. In a game like this, secrecy is important. To be successful in rock paper scissors (and other games without saddle points), one way to keep your opponent guessing about your intentions would be to play different strategies in different plays of the game. Sometimes play rock, other times paper, and other times scissors. This is what is meant by a *mixed strategy*, as opposed to a *pure strategy*, which means you would always play that strategy no matter how often the game is played, such as the sail north strategy in the Battle of the Bismarck Sea.

The first task we face is to devise some notation for these mixed strategies. We do so for the row player by writing a row vector $p = (p_i)$, whose ith component p_i is the relative frequency with which the row player plays the ith row. Similarly, for the column player we write a column vector $q = (q_j)$, whose jth component q_j is the relative frequency with which the column player plays the jth column in repeated play of the game. For example, in rock paper scissors (where the strategies are listed in that order), we might have

$$p = \begin{bmatrix} 0.3 & 0.2 & 0.5 \end{bmatrix}, \qquad q = \begin{bmatrix} 0.4 \\ 0.4 \\ 0.2 \end{bmatrix},$$

which we interpret as the mixed strategies where the row player plays rock 30% of the time, paper 20% of the time, and scissors 50% of the time, and the column player would play rock and paper 40% of the time each and scissors the remaining 20% of the time. Of course, geometrically, there is no difference between a row vector and a column vector, so it might seem that we write one as a row and one as a column just as an artifice to remind ourselves which one belongs to the row player and which to the column. However, we shall see soon that there is another good reason for writing them this way. For the moment, we should just note that the entries in a mixed strategy are nonnegative fractions whose sum is 1, since the percentages must all sum to 100%. Also note that a pure strategy can also be written in this form – for example, the unit vector $p = (1, 0, 0)$ is the pure strategy of "always play rock."

The reader may object that the concept of a mixed strategy depends on repeated play of the same game. For some games, repeated play makes sense – for example, rock paper scissors is often played repeatedly for several rounds. But what if a game is only to be played once, such as the Battle of the Bismarck Sea? We can get around this difficulty by viewing the entries of the mixed strategies as *probabilities*. That is, on any one play of the game, p_i represents the probability that the ith row is played, and q_j represents the probability that the jth column is played. Students who have studied probability theory may recognize that one way to interpret the meaning of a probability is to think of it as relative frequency when an experiment is repeated many times, and this relative frequency is exactly how we defined the entries of p and q in the first place. Thus, the concept of a mixed strategy as a vector of probabilities makes sense whether the game is played once or repeated several times.

We hasten to point out that if the game is repeated and a particular mixed strategy is used, the entries of the strategy only tell you the relative frequency of each strategy *in the long run*;

they do not tell you what to do exactly on any one play of the game. For example, suppose the row player in two-finger morra (Example 6.1 in Section 6.1) decides to use the mixed strategy $p = (0.5, 0.5)$. That is, half the time they put out one finger, and half the time they put out two fingers.

There are many different sequences of play that will lead to an overall percentage of 50/50. For example, one way to achieve that mix is to simply alternate – play one finger every other play and two fingers every other play. Do you see why that is not a good idea? As soon as a skillful opponent notices that pattern, he or she can predict what you will do next and will use that information to his or her advantage. Following a predicable pattern is tantamount to telling your opponent what your strategy is, and as we noted before, in a game without saddle points, it is essential to keep that information secret. What is necessary in repeated play is that each play of the game the actual strategy you use is randomly chosen, just as long as it averages out to the 50/50 mix in the long run. For example, you might want to toss a fair coin before each play and use the number of fingers dictated by the coin toss. That way each play is random, but if the coin is fair, you will average out to a 50/50 mix in the long run.

We claim that by using a combination of pure and mixed strategies, we can give a solution to any two-person constant-sum game. Proving this claim is something we will work toward over the next several sections; it will be a bit of a journey before we get there. The first step is to assess how much we'll need to know about probability theory in order to work with mixed strategies. Probability theory is a fascinating branch of mathematics, but to learn it properly would be a detour from our main goals. Fortunately, one can get by in game theory with just a little bit of knowledge of probability theory.

A *probability experiment* is any activity that has an observable outcome. For example, you could toss a coin and observe whether it lands showing heads or tails. Another experiment is to roll a pair of dice and note the ordered pair of outcomes such as $(2, 5)$ or $(6, 6)$. Another could be taking a poll of families and recording the number of children under 18 living at home and their genders. The concept is quite general. Let S be the set of all possible outcomes in a probability experiment. S is called the *sample space* of the experiment. Any subset E of S is called an *event* (so an event is just a collection of some of the possible outcomes). The main points we need to know about probability are the following:

- The probability of E, written $P(E)$, is a numerical measure of how likely it is for the event E to happen (that is, how likely it is that the outcome of the experiment belongs to E). $P(E)$ is a fraction between 0 and 1; the closer it is to 1, the more likely it is that E happens. If S is finite, we interpret $P(E) = 1$ to mean that the event E is inevitable and $P(E) = 0$ to mean that the event E is impossible. For example, $P(\varnothing) = 0$ and $P(S) = 1$. If you sum up all the probabilities of all the outcomes of an experiment, you obtain 1. More generally, if you sum up the probabilities of all the outcomes in any event E, you obtain $P(E)$.

- Two events E and F are said to be *independent* if knowing that one of them occurs does not affect the probability of the other occurring. (Examples will be given in the following paragraphs.)

- If E and F are *independent* events, then the probability that they both happen is obtained by multiplying the individual probabilities; that is,

$$P(E \cap F) = P(E) \cdot P(F). \tag{6.3}$$

• If every outcome of a probability experiment with n outcomes has a certain value v_i assigned to it, then the weighted average of these values, using the probabilities p_i as weights, is called the *expected value* of the experiment, written

$$E = p_1 v_1 + p_2 v_2 + \cdots + p_n v_n = \sum_{i=1}^{n} p_i v_i. \qquad (6.4)$$

We will illustrate all four of these points momentarily. Hopefully, the examples will be sufficient to be able to quickly get back to game theory. Should the reader require more depth in their understanding of probability, we have included more details in the Appendix at the end of the text.

Let the experiment be to toss two coins, a nickel and a dime. We can record the outcome of the experiment as an ordered pair such as HT (we won't bother with the parentheses or the comma), which means heads on the nickel and tails on the dime (we will always list the outcome of the nickel first). Suppose first that both coins are perfectly fair, so the probability of H or T on either coin is $\frac{1}{2}$. There are just four possible outcomes, so $S = \{HH, HT, TH, TT\}$. Let E be the event that both coins come up tails, let F be the event that both coins have the same result, and let G be the event that the dime shows heads. We can list the outcomes of each of these events as follows, which the reader should check:

$$E = \{TT\}, \qquad F = \{HH, TT\}, \qquad G = \{HH, TH\}.$$

We can see that E consists of just one outcome (such events are called *simple events*), while F and G are made up of several outcomes. To understand the intersection of two events, you can describe it either verbally or by appealing to the sets that make up the events. For example, in words, $F \cap G$ means the event that F and G both happen; that is, $F \cap G$ means the outcomes on both coins are the same, while the dime must show H. The only outcome with those properties is HH, so $F \cap G = \{HH\}$. But this is clearly the result when you simply find the intersection of the two sets F and G.

Now, since we are assuming that the coins are fair, it stands to reason that all four of the outcomes in S are equally likely (such sample spaces are said to be *uniform*). Since, by the first bullet point, the probabilities add up to 1 and they are equal, it follows that the probability of each outcome is $\frac{1}{4}$: $P(HH) = P(HT) = P(TH) = P(TT) = \frac{1}{4}$. Then we have, also by the same bullet point, that $P(E) = \frac{1}{4}$, $P(F) = \frac{1}{4} + \frac{1}{4} = \frac{1}{2}$, and similarly $P(G) = \frac{1}{2}$.

Next, let A be the event of obtaining heads on the nickel, so $A = \{HH, HT\}$. Recall that G is the event of obtaining heads on the dime. We claim that A and G are independent events. Intuitively, this makes sense – knowing that A occurs does not affect the probability of G, because knowing the outcome of the nickel has no effect whatsoever on the outcome of the dime. For an example of two events that are not independent, consider the events E and G. If you know that E occurs, then there is no way that the dime is showing heads, so G is now impossible. In other words, without knowing anything, $P(G) = \frac{1}{2}$, but given that E has occurred, the probability of G does change – it drops to 0 in this case. This illustrates the definition of independent events. (We remark that there is a more mathematically precise definition of independent events using something called *conditional probability*, which the interested reader can find in the Appendix. For our purposes, this intuitive approach is sufficient.)

To illustrate Eq. (6.3), consider the independent events A and G. Observe that $A \cap G = \{HH\}$, so $P(A \cap G) = \frac{1}{4}$. Thus, we have

$$P(A \cap G) = \frac{1}{4} = \frac{1}{2} \cdot \frac{1}{2} = P(A) \cdot P(G),$$

so the formula holds in this case. However, the formula will fail if the events are not independent. For example, consider E and G, which we know are not independent. Here $E \cap G = \varnothing$, whence $P(E \cap G) = 0$ by the first bullet point. On the other hand, $P(E) \cdot P(G) = \frac{1}{8} \neq 0$. None of these illustrations we have done depend on the assumption that the coins were fair. For example, suppose that the coins are weighted so that for the nickel, $p(H) = 0.7$ (so $P(T) = 0.3$ since the probabilities sum to 1), while for the dime, $p(H) = 0.6$ (and $P(T) = 0.4$). It is still true that events A and G are independent events since, even with weighted coins, the outcome of one coin has no effect on the other. Thus, $P(HH) = P(A \cap G) = P(A) \cdot P(G) = (0.7)(0.6) = 0.42$. In a similar way, we can determine the probabilities of each of the four outcomes in S.

Outcome	Probability
HH	$(0.7)(0.6) = 0.42$
HT	$(0.7)(0.4) = 0.28$
TH	$(0.3)(0.6) = 0.18$
TT	$(0.3)(0.4) = 0.12$

This is a perfectly valid probability function, since the probabilities still sum to $1 : 0.42 + 0.28 + 0.18 + 0.12 = 1$. The reader should compute the probabilities of E, G and $E \cap G$ to see whether or not E and G are still not independent in this case.

It remains to illustrate the notion of expected value (6.4), for which we use the following example. A robbery has occurred at the Gold Tooth Jewelry Shop. The thieves made off with four emeralds worth \$8,000 each, five rubies worth \$7,200 each, 10 sapphires worth \$3,500 each, and 80 garnets worth \$400 each. Also, the inexperienced thieves were surprised to find in their haul one diamond worth \$35,000, which they didn't even know they were holding in their hands! Danny Steele, the newspaper reporter for the *Boston Rag* who is covering the robbery, wants to know the average value of the stolen gems. As usual, we compute average values by adding up the values and dividing by the total number of stones:

$$E = \frac{35,000 + (8,000 + \cdots) + (7,200 + \cdots) + \cdots + (400 + \cdots + 400)}{100}$$

$$= \frac{35,000 + 4 \cdot 8,000 + 5 \cdot 7,200 + 10 \cdot 3,500 + 80 \cdot 400}{100} = \$1,700.$$

Thus, the stolen gems had an average value of \$1,700. Note that we could write this in the following form:

$$E = \left(\frac{1}{100}\right) 35,000 + \left(\frac{4}{100}\right) 8,000 + \left(\frac{5}{100}\right) 7,200$$

$$+ \left(\frac{10}{1,000}\right) 3,500 + \left(\frac{80}{100}\right) 400 = \$1,700.$$

It's the same sum as the first one, but in a slightly less cumbersome form. The second sum has only five terms, one for each of the different values (or different types of gem), whereas the numerator in the first sum is a sum of 100 terms, one for each gem stolen (although we wrote the numerator in a somewhat abbreviated form in the second line). Notice also that in the second sum, we can interpret the meaning of the fractional coefficients of each of the five values. Indeed, they are the probabilities of choosing that value if one selected a gem from the stolen gems at random. That is, there is only one diamond, so the probability of selecting it randomly is $\frac{1}{100}$. By contrast, there are 80 garnets, so the probability of selecting a garnet at random is $\frac{80}{100}$, the coefficient of the value of a garnet in the sum, and similarly for the other values. Thus, when written in the second form, we can interpret the sum as a weighted average of the values, where each value is weighted according to the probability that it occurs. But, of course, the form of this sum is exactly as in Eq. (6.4), the definition of expected value.

We make three further remarks. First, like any average, the expected value does not say anything about the value of any individual item. For example, none of the stones in the robbery were worth exactly \$1,700. So expected values are most useful for analyzing aggregate behavior or long-term behavior. Second, the value attached to the outcome of the experiment does not have to be monetary value; it could measure anything in which we might be interested. For example, if the experiment is to poll a community to determine how many children under 18 live at home, for each household, the value is the number of children. In this case, the probabilities would be the probability that a random household in this community has zero children, one child, two children, etc. Then the expected value in this case would be the average number of children per household, as in statements like "the average household in Putnam County has 2.08 children." Finally, we mention that we have seen in Chapter 2 that we can write the expected value using matrix multiplication as a dot product of the vector of probabilities p_i with a vector of the values v_i.

To compute expected value, it is sometimes convenient to make a table. In the first column, list the outcomes; in a second column, list the probabilities p_i of the outcomes; in a third column, list the values v_i; and in a fourth column, list the products $p_i v_i$. Then the expected value E is the sum of the entries in the last column, by Eq. (6.4). For example, suppose that the poll of households in Putnam County yielded the following data: 1,000 households with zero children, 1,200 households with one child, 4,500 households with two children, 2,800 households with three children, 400 households with four children, zero households with five children, and 100 households with six children. Note that 10,000 households were polled altogether. Thus, the probability of a household chosen at random having no children is $\frac{1,000}{10,000} = 0.1$, etc. We can compute the expected number of children per household using Table 6.1.

Table 6.1 Computing expected value.

Outcome	p_i	v_i	$p_i v_i$
No children	0.1	0	0
1 child	0.12	1	0.12
2 children	0.45	2	0.9
3 children	0.28	3	0.84
4 children	0.04	4	0.16
5 children	0	5	0
6 children	0.01	6	0.06

Summing the last column yields an expected value of $E = 0 + 0.12 + 0.9 + 0.84 + 0.16 + 0 + 0.06 = 2.08$.

This concludes our detour into introductory probability theory, and we now return to the study of mixed-strategy games.

Consider the following payoff matrix for a zero-sum game. It does not matter what the actual game is; if you like, you can think of it as a version of two-finger morra, where we have modified the payoffs:

$$A = \begin{array}{c} \\ S \\ T \end{array} \begin{array}{cc} S & T \\ \left[\begin{array}{cc} 2 & -2 \\ -4 & 7 \end{array} \right]. \end{array}$$

It is easy to check that this game is not strictly determined – there is no saddle point. Under repeated play, using mixed strategies, we'd like to compute the expected payoffs for each player. Suppose that the row player decides to use the mixed strategy $p = [0.2, 0.8]$, and the column player decides to use the mixed strategy $q = \begin{bmatrix} 0.6 \\ 0.4 \end{bmatrix}$. If we think of each play of the game as a probability experiment, we can compute the expected payoff for the row player as follows. First, observe that there are four possible outcomes – SS, ST, TS, and TT – depending on what strategies each player chooses in any given play. We already know that the row player chooses row S with probability 0.2 (based on the mixed strategy p), and the column player chooses column S with probability 0.6 (based on the mixed strategy q.) So what is the probability of the outcome SS? In order for SS to occur, both players must choose S: In other words, the outcome SS is the intersection of the two events "row chooses S" and "column chooses S." Can we compute the probability of this intersection? According to Eq. (6.3), we can, as long as the events are independent. But that is precisely one of the assumptions we have built into our game models – that the players choose their strategies simultaneously and independently. Therefore, we are justified in multiplying the individual probabilities to obtain the probability of the intersection: $P(SS) = (0.2)(0.6) = 0.12$. We can compute the probability of each of the four outcomes in the same manner. Also, the values v_i that go with the outcomes are precisely the entries of the payoff matrix. We are now prepared to compute the expected payoff to the row player. We construct a table in Table 6.2.

Table 6.2 Expected payoffs in a mixed-strategy game.

Outcome	p_i	v_i	$p_i v_i$
SS	$(0.2)(0.6) = 0.12$	2	$(0.2)(0.6)(2) = 0.24$
TS	$(0.8)(0.6) = 0.48$	-4	$(0.8)(0.6)(-4) = -1.92$
ST	$(0.2)(0.4) = 0.08$	-2	$(0.2)(0.4)(-2) = -0.16$
TT	$(0.8)(0.4) = 0.32$	7	$(0.8)(0.4)(7) = 2.24$

Thus, in the long run, the expected payoff to the row player is

$$E = 0.24 + (-1.92) + (-0.16) + 2.24 = 0.4$$

This means that, on average, the row player will win 0.4 per game, if the players stick to these strategies. Since the game is zero-sum, this means that, on average, the column player will lose 0.4

per game. Of course, if the players change their strategy to a different mixed strategy, this expected payoff will also change. This payoff E depends on p and q, so we denote it $E(p,q)$. We would like to be able to compute $E(p,q)$ quickly for various p and q, and it would be rather tedious to construct a table like Table 6.2 for each calculation, especially if the payoff matrix is larger. So we seek a shortcut. Let's look again at the sum of the entries in the last column, but written out in full:

$$E = (0.2)(0.6)(2) + (0.8)(0.6)(-4) + (0.2)(0.4)(-2) + (0.8)(0.4)(7) = 0.4$$

Since this is a sum of products, it suggests a matrix multiplication. Moreover, each summand is a product of three terms, so perhaps it is a product of three matrices? Indeed, the three factors in each summand come from the entries in $p = [0.2, 0.8]$, the entries of $q = \begin{bmatrix} 0.6 \\ 0.4 \end{bmatrix}$, and the entries in the payoff matrix $A = \begin{bmatrix} 2 & -2 \\ -4 & 7 \end{bmatrix}$. By considering the sizes of the matrices, we can guess in which order they should be multiplied. After a little trial and error, we see that the calculation for E is exactly the same as the matrix product:

$$E = E(p,q) = pAq$$
$$= \begin{bmatrix} 0.2 & 0.8 \end{bmatrix} \begin{bmatrix} 2 & -2 \\ -4 & 7 \end{bmatrix} \begin{bmatrix} 0.6 \\ 0.4 \end{bmatrix}$$
$$= (0.2)(0.6)(2) + (0.8)(0.6)(-4) + (0.2)(0.4)(-2) + (0.8)(0.4)(7) = 0.4$$

Thus, once again, the itemized calculation (this time, of expected payoff in a game) is made simple and automatic via matrix multiplication. If one repeats this exercise for any payoff matrix, regardless of size, we find, in general, that the same result holds. We state the result as Proposition 6.16.

PROPOSITION 6.16 *Let A be the payoff matrix of any two-person, constant-sum game. Suppose the row player decides on the (pure or mixed) strategy p and the column player decides on the (pure or mixed) strategy q. Then, in repeated play, the expected payoff to the row player is the matrix product:*

$$E(p,q) = pAq. \tag{6.5}$$

Note that the size of the matrices always works out correctly. If A is $m \times n$, then p is $1 \times m$ and q is $n \times 1$. Thus, it makes sense to form the matrix product pAq, and the size of the result is 1×1, a single number (which is the expected payoff.) In fact, this is the real reason we write p as a row matrix and q as a column matrix.

So, for example, in the preceding game, the row player changes to $p = [0.5, 0.5]$ and the column player changes to $q = \begin{bmatrix} 0.3 \\ 0.7 \end{bmatrix}$, then the expected payoff for the row player is now

$$E(p,q) = pAq$$
$$= \begin{bmatrix} 0.5 & 0.5 \end{bmatrix} \begin{bmatrix} 2 & -2 \\ -4 & 7 \end{bmatrix} \begin{bmatrix} 0.3 \\ 0.7 \end{bmatrix} = 1.45.$$

Clearly, the row player is better off in the long run with these choices of mixed strategies than with the original choice. This raises several questions: If you know what mixed strategy your opponent is using, is there a best-possible responding strategy for you? If so, how do we find it? We ask the same questions if you have no idea what strategy your opponent is using: is there an optimal strategy for you, and if so, how is it found? In other words, what is the solution to the game? We pursue these topics in the next section.

Exercises

1.　In the marketing game Antares vs. Bellatrix, suppose that, instead of Eq. (6.1), the market share for Antares is given by this equation:

$$E(x, y) = .04xy - 3.6x - 4.4y + 441.$$

Solve the game under this hypothesis. Give the optimal prices for both companies, the value of the game in terms of market shares, and also the revenue that each company will enjoy, assuming a total market of 400,000 chips.

2.　Consider a probability experiment that consists of rolling an ordinary, six-sided die which is labeled with the integers $1, 2, 3, 4, 5$, and 6 on the sides.

　　a.　List the outcomes in the sample space S. If the die is fair, what are the probabilities associated with each outcome?

　　b.　Let A be the event of rolling an even number, and let B be the event of rolling a number (strictly) greater than 2. Determine the probabilities of A and B.

　　c.　Find $P(A \cap B)$ and determine whether or not A and B are independent events.

3.　Suppose the die in Exercise 2 is not fair, but weighted so that the probabilities of each outcome are as follows, for some unknown x.

Outcome	p_i
1	0.1
2	0.1
3	0.1
4	0.2
5	0.2 + x
6	x

　　a.　Find x.

　　b.　Let A be the event of rolling an even number, and let B be the event of rolling a number (strictly) greater than 2. Determine the probabilities of A and B.

　　c.　Find $P(A \cap B)$ and determine whether or not A and B are independent events.

4.　Consider a probability experiment that consists of rolling two fair dice and recording the outcomes showing on the top face of each as an ordered pair. (Assume the two dice are distinguishable. For example, they could be different colors, so the first entry in the ordered pair always corresponds to the same die – for example, the red die if you are rolling a red one and a white one.)

a. List the entire sample space S of ordered pairs.

b. Since the dice are fair, each outcome is equally likely. What is the probability of any one outcome?

c. Let A be the event of rolling a sum of 10 or greater. Let B be the event of rolling doubles (i.e., the same number on each die.) Let C be the event of rolling a 5 or a 6 on the red die. Find $P(A)$, $P(B)$, and $P(C)$.

d. Find $P(A \cap B)$, $P(A \cap C)$, and $P(B \cap C)$. Are any two of A, B, or C independent events?

5. The Saratoga Spitfires Soccer Club is selling raffle tickets as a fund-raiser for its upcoming trip to a tournament. The club prints a total of 300 tickets to sell. There is one grand-prize ticket worth \$250, four second-prize tickets worth \$50 each, and ten third-prize tickets worth \$25 each. The remaining 285 tickets have no value. Assume that all 300 tickets will be sold.

a. Considering the purchase of a ticket as a probability experiment, there are four possible outcomes (grand-prize ticket, second-prize ticket, third-prize ticket, and worthless ticket). Compute the expected value of a ticket (construct a table as we did in the text for the household polling example). What is the interpretation of the result?

b. If the price of a raffle ticket is \$10, how much money will the team make toward its trip?

6. Past records at the local Bob & Jerry's Grateful Scoop Ice Cream Shop indicate that they get 180 customers on a hot and sunny day, 120 customers on a hot and cloudy day, 100 customers on a mild and sunny day, 80 customers on a mild and cloudy day, and 30 customers on a cool day. Suppose that for the next month, the probability that it is hot and sunny is 0.2, the probability that it is hot and cloudy is 0.4, the probability that it is mild and sunny is 0.1, the probability that it is mild and cloudy is 0.2, and the probability that it is cool is 0.1. Find the expected number of customers per day to visit the store.

7. Mick and Keith are playing, and Mick (the row player) says he will not be satisfied unless he wins an average of \$0.50 per game. Meanwhile, Keith (the column player) wants to win at least an average of \$0.40 per game. Mick decides never to play paper, but to play rock half the time and scissors half the time. Keith decides to play rock a quarter of the time, paper half the time, and scissors a quarter of the time. If both players stick to these mixed strategies in repeated play, what is the expected payoff for both players? (Use Eq. (6.5).) Does Mick have satisfaction? Does Keith get what he wants?

In Exercises 8–15, you are given the payoff matrix A of a zero-sum game. Compute the expected payoff to the row player with using the indicated mixed strategies. (Use Eq. (6.5).)

8. $A = \begin{bmatrix} 1 & 0 \\ -3 & 2 \end{bmatrix}$, $\quad p = \begin{bmatrix} 0.5 & 0.5 \end{bmatrix}$, $\quad q = \begin{bmatrix} 0.6 \\ 0.4 \end{bmatrix}$

9. $A = \begin{bmatrix} -2 & 5 \\ 4 & -1 \end{bmatrix}$, $\quad p = \begin{bmatrix} \frac{1}{3} & \frac{2}{3} \end{bmatrix}$, $\quad q = \begin{bmatrix} \frac{2}{3} \\ \frac{1}{3} \end{bmatrix}$

10. $A = \begin{bmatrix} 1 & 0 \\ 0 & -3 \\ -2 & 1 \end{bmatrix}$, $\quad p = \begin{bmatrix} \frac{1}{4} & \frac{1}{2} & \frac{1}{4} \end{bmatrix}$, $\quad q = \begin{bmatrix} \frac{1}{4} \\ \frac{3}{4} \end{bmatrix}$

11. $A = \begin{bmatrix} 2 & -1 & 3 & -2 \\ -3 & 0 & 1 & 2 \end{bmatrix}$,

$p = \begin{bmatrix} 0.7 & 0.3 \end{bmatrix}$, $\qquad q = \begin{bmatrix} 0.3 \\ 0.4 \\ 0.1 \\ 0.2 \end{bmatrix}$

12. $A = \begin{bmatrix} 1 & -3 & 5 \\ -2 & 4 & -6 \\ 0 & -1 & 2 \end{bmatrix}$,

$p = \begin{bmatrix} 0.4 & 0.3 & 0.3 \end{bmatrix}$, $\qquad q = \begin{bmatrix} 0.2 \\ 0.6 \\ 0.2 \end{bmatrix}$

13. Use the same matrix as Exercise 12, but with the following mixed strategies:

$p = \begin{bmatrix} 0.5 & 0.2 & 0.3 \end{bmatrix}$, $\qquad q = \begin{bmatrix} 0.5 \\ 0.5 \\ 0 \end{bmatrix}$

14. $A = \begin{bmatrix} 1 & -4 & 2 & 0 \\ 1 & 1 & 1 & -5 \\ -3 & -1 & 8 & 4 \\ 0 & 2 & -9 & -1 \end{bmatrix}$,

$p = \begin{bmatrix} \frac{1}{5} & \frac{2}{5} & \frac{2}{5} & 0 \end{bmatrix}$, $\qquad q = \begin{bmatrix} \frac{1}{5} \\ \frac{1}{5} \\ 0 \\ \frac{3}{5} \end{bmatrix}$

15. $A = \begin{bmatrix} 1.4 & -2.1 & 3.9 & -1.3 & 0.1 \\ -2.2 & 1.7 & 0 & 3.3 & -0.8 \\ -0.9 & 1.2 & -4.1 & -2.6 & 4.9 \\ 2.0 & -4.5 & 2.8 & 1.9 & -5.6 \end{bmatrix}$,

$p = \begin{bmatrix} 0.1 & 0.2 & 0.3 & 0.4 \end{bmatrix}$, $\qquad q = \begin{bmatrix} 0.1 \\ 0 \\ 0.2 \\ 0.5 \\ 0.2 \end{bmatrix}$

16. a. In Exercise 9, what is the expected payoff for the column player?

    b. Suppose the given matrix A in Exercise 9 is the matrix of payoffs for the row player in a constant-sum game, rather than a zero-sum game, where the payoffs sum to the constant 2. What are the expected payoffs for the row player and the column player now?

17. Jimmy and Robert are sitting at the Four Sticks Tavern, playing the variation on four-finger from Exercise 9 in Section 6.1. Recall that each player simultaneously puts out one, two, three, or four fingers. If the total number of revealed fingers is even, the column player pays the row player, while if it is odd, the row player pays the column player. The difference is, instead of the payoff being $1, the number of dollars is equal to the (nonnegative) difference

between the number of fingers shown. Suppose both players use the same mixed strategy (so $p = q^T$) of playing one finger 10% of the time, two fingers 20% of the time, three fingers 30% of the time, and four fingers 40% of the time. What is the expected payoff to Robert if he is the row player?

18.   Recall the guessing game from Exercise 10 in Section 6.1. Bill secretly puts $x in his hand, where $x = 0, 1, 3$, or 5. Ann must guess how many dollars Bill is holding. If she guesses correctly, Bill pays her twice the number of dollars he is holding, plus one dollar more; that is, he pays her $(2x + 1). If she guesses incorrectly, she must pay Bill the (positive) difference between her guess and x, plus an extra $2. Assume Ann is the row player. Ann decides to guess according to the mixed strategy of guessing $0 with probability $\frac{1}{2}$, guessing $1 with probability $\frac{1}{4}$, guessing $3 with probability $\frac{1}{8}$, and guessing $5 with probability $\frac{1}{8}$. Bill, on the other hand, decides never to put $5 in his hand, but to use the remaining three strategies with equal probability, so with probability $\frac{1}{3}$ each. Under repeated play with these assumptions, what is the expected payoff to Ann?

6.4   Solving Mixed-Strategy Games: The Minimax Theorem in the 2 × 2 Case

In this section, we begin our study of how to answer the questions raised at the end of the last section. Namely, given a constant-sum game that is not strictly determined, is there an optimal way to play it, and if so, how do we find the optimal strategy? We consider these questions in two cases – when we know what strategy our opponent is using and when we don't. Since we will be in the latter situation most often, we begin with that case.

We begin with the simplest case: that of 2 × 2 constant-sum games. We will discuss several different approaches to solving these games, and we will find that all such games do indeed have an optimal solution. In Section 6.4.1, we take an approach that leads naturally to a fundamental theorem about zero-sum games, called the minimax theorem, first proved by John Von Neumann in 1928. Our natural proof in the 2 × 2 case does not easily extend to larger games, however, so in Section 6.4.2, we will consider an alternate approach to solving a game that is more useful for larger games.

As a bonus, we'll be able to completely answer the question about how to play in the case when you know your opponent's strategy for matrices of any size in Section 6.4.2. In the next chapter, we extend the results of this chapter to larger games in the case when you do not know your opponent's strategy.

6.4.1   The Derived Game

In this subsection, we only consider 2 × 2 games in the case when we have no idea of what strategy our opponent might use. We begin with an example. Suppose a zero-sum game has the payoff matrix

$$A = \begin{bmatrix} 4 & -2 \\ -3 & 1 \end{bmatrix}.$$

Clearly, this matrix has no saddle point, so the game it models is not strictly determined. We have already agreed that in playing such a game we redefine what me mean by a strategy. The row player, instead of simply considering the two rows as her strategies, thinks of a strategy as an ordered pair of fractions like $p = (0.2, 0.8)$, where we interpret this to mean she plays the first row with probability 0.2 and the second row with probability 0.8 on any one play of the game, and similarly for the column player. We call these ordered pairs mixed strategies. Furthermore, we have seen that once the row player has settled on a mixed strategy to use – say, $p = (0.2, 0.8)$ – and once the column player has also selected his strategy – say, $q = \begin{bmatrix} 0.5 \\ 0.5 \end{bmatrix}$ – then we can find the expected payoff to the row player using Eq. (6.5):

$$E(p, q) = pAq = \begin{bmatrix} 0.2 & 0.8 \end{bmatrix} \begin{bmatrix} 4 & -2 \\ -3 & 1 \end{bmatrix} \begin{bmatrix} 0.5 \\ 0.5 \end{bmatrix} = -0.6$$

We have observed that different choices of p and q lead to different expected payoffs to the row player. We are now interested in the question of how each player should choose his or her mixed strategies, p and q, which would lead to the best-possible expected payoff for himself or herself. But making a rational decision to choose the best-possible option when the (expected) payoffs are known in advance is exactly what a game is. It appears that we are replacing one game with another. Indeed, that is exactly what we are doing. We make the following formal definition.

DEFINITION 6.17   *Let A be the payoff matrix for a two-person, zero-sum game. We consider a new game, called the* derived game *of A, to be the game where the (pure) strategies are precisely the mixed strategies p and q in the original game, and where the payoff $E(p, q)$ to the row player for a particular outcome of the derived game is given by the formula (6.5):*

$$E(p, q) = pAq.$$

REMARK 6.18   *In this section, we only consider derived games for a 2×2 matrix. However, it should be clear that the definition of the derived game applies to matrices of any size. Moreover, although we have only stated the definition for zero-sum games, the concept extends to constant-sum or even variable-sum games without change.*

REMARK 6.19   *Since the expected payoff to the column player is just the negative of $E(p, q)$, it follows that the derived game is zero-sum if A is. Similarly, if A represents a constant-sum game, then the derived game is also constant-sum (with the same constant), and finally, if A represents a variable-sum game, the derived game will also turn out to be variable-sum.*

REMARK 6.20   *Since there are an infinite number of mixed strategies (that is, infinitely many different possible choices for p and also for q), the derived game is infinite, even though A is finite.*

Since the derived game is infinite, we cannot write down a payoff matrix. However, we have seen that sometimes an infinite game can be solved (the Antares vs. Bellatrix game from Section 6.3.1, for example). Let's look more carefully at a mixed strategy $p = (p_1, p_2)$. Since the p_i values are probabilities whose sum is 1, we can rewrite such a mixed strategy in a slightly simplified form, taking $p_1 = x$ and $p_2 = 1 - x$ as $p = (x, 1 - x)$, where $0 \le x \le 1$. Similarly, the mixed strategy $q = \begin{bmatrix} q_1 \\ q_2 \end{bmatrix}$ can be rewritten as $q = \begin{bmatrix} y \\ 1 - y \end{bmatrix}$, where $0 \le y \le 1$. So the expected payoff $E(p, q)$, in the case of a 2×2 matrix A, really only depends on the two parameters x and y. Let us abuse the notation a bit and call it $E(x, y)$ in this case. Do we have an explicit formula for $E(x, y)$ as a function of x and y? Indeed we do, as

$$E(x, y) = E(p, q) = pAq$$

$$= \begin{bmatrix} x & 1-x \end{bmatrix} \begin{bmatrix} 4 & -2 \\ -3 & 1 \end{bmatrix} \begin{bmatrix} y \\ 1-y \end{bmatrix}$$

$$= \begin{bmatrix} 7x-3 & 1-3x \end{bmatrix} \begin{bmatrix} y \\ 1-y \end{bmatrix}$$

$$= 10xy - 4y - 3x + 1.$$

Following what we did in the solution of the Antares vs. Bellatrix game, we note that (1) the function $E(x, y)$ is what replaces the payoff matrix, since the players choose x and y so as to make $E(x, y)$ as beneficial to themselves as they can; and (2) just like the Antares vs. Bellatrix example, this function is a saddle surface, so the derived game is strictly determined. Indeed, we rewrite $E(x, y)$ in standard form:

$$E(x, y) = 10xy - 4y - 3x + 1$$
$$= 10(xy - 0.3x - 0.4y) + 1$$
$$= 10(xy - 0.3x - 0.4y + 0.12) + 1 - 1.2$$
$$= 10(x - 0.4)(y - 0.3) - 0.2.$$

We see that the saddle point is located at $(x, y) = (0.4, 0.3)$, and E has the value -0.2 there. So $x = 0.4$ means $p = \begin{bmatrix} 0.4 & 0.6 \end{bmatrix}$, and $y = 0.3$ means $q = \begin{bmatrix} 0.3 \\ 0.7 \end{bmatrix}$. Since this represents a saddle point, it is a Nash equilibrium in the derived game. Thus, as long as both players play these strategies, the payoff to the row player is $v = -0.2$. As with any Nash equilibrium, if a player unilaterally deviates from the optimal strategy, it is possible for the payoff for the other player to improve (and, therefore, the first player's payoff would decrease since it is a constant-sum game). However, like the Antares vs. Bellatrix example, it turns out in this case that as long as one of the players plays the optimal mix, the payoff will stay at the value $v = -0.2$, regardless of the choice of the other player (indeed, consider the factored form of the expected payoff).

Since the expected payoff $E(x, y)$ gives the long-term average payoff in the original, finite game, played with mixed strategies instead of pure strategies, these statements also hold for the original game. To be precise, the choice of mixed strategies $\widehat{p} = \begin{bmatrix} 0.4 & 0.6 \end{bmatrix}$ and $\widehat{q} = \begin{bmatrix} 0.3 \\ 0.7 \end{bmatrix}$ will result in an average payoff of $v = -0.2$ per game to the row player in the long run. If either player deviates from this unilaterally, he or she cannot improve his or her payoff, so we think of $v = -0.2$ as the value of the game, and this solution represents a Nash equilibrium in the original game under repeated play. (The carat on p and q is our way of indicating the optimal strategies.) It is one of life's little ironies that we were able to solve the smallest game possible (where each player has only two strategies) by converting it to an infinite game! What we did for this game will work for any 2×2 game.

THEOREM 6.21 (Minimax Theorem for 2×2 Constant-Sum Games.) *Let A be the payoff matrix for any 2×2 constant-sum game. If the game is strictly determined, then we already know the solution corresponds to a saddle point. If the game is not strictly determined, then the derived* *game of A is strictly determined. In particular, in either case, there are optimal strategies $\widehat{p}$ and $\widehat{q}$ and a real number v (called the value of the game) such that*

1. $E(\widehat{p},q) \geq v$ *for any q.*
2. $E(p,\widehat{q}) \leq v$ *for any p.*
3. $E(\widehat{p},\widehat{q}) = v.$

REMARK 6.22   *The inequalities (1) and (2) may be interpreted as saying that v is a "guaranteed minimum payoff" for each player, analogous to the saddle point value in a strictly determined game.*

Proof Let $A = \begin{bmatrix} a & b \\ c & d \end{bmatrix}$ be the payoff matrix. If the matrix has a saddle point, then $\widehat{p}$ and $\widehat{q}$ are the pure strategies that yield a saddle point solution, and v is the value of that saddle point. We already know (1)–(3) hold in that case. Suppose A has no saddle point, so the game is not strictly determined. We claim that for 2×2 constant-sum games, the presence of a saddle point is equivalent to at least one player having a dominant strategy. Indeed, if one player has a dominant strategy, then we should not play the dominated strategy, so we delete it from the matrix A, leaving a matrix that has either just one row or just one column. In either case, one of the two entries will be a higher-order dominant strategy for the other player, so the payoff matrix reduces to a 1×1 matrix. That entry must be a saddle point in the original matrix A (this is what happened in the matrix for the Battle of the Bismarck Sea).

Conversely, suppose one of the entries is a saddle point. We will show that one of the players has a dominant strategy. We assume that the saddle point occurs at the entry $a_{11} = a$. (The argument is similar if it is located elsewhere, so there is no loss of generality in making this assumption. Alternately, interchange the rows and/or columns, if necessary, to assure the saddle point is located in the a_{11} position.) Since a is a saddle point, we have $a \leq b$ and $a \geq c$. We now consider the options for d. If $d \leq b$, then the first row is dominant for the row player. On the other hand, if $d \geq c$, then the first column is dominant for the column player. To avoid either of these cases, we must simultaneously have $d < c$ and $d > b$. But this is impossible, since when combined with the other inequalities we obtain $d < c \leq a \leq b < d$, or $d < d$, a contradiction. Thus, either the row player or the column player has a dominant strategy.

Now, since A has no saddle point, we seek mixed-strategy solutions. Let $\widehat{p} = \begin{bmatrix} x & 1 - x \end{bmatrix}$ and $\widehat{q} = \begin{bmatrix} y \\ 1 - y \end{bmatrix}$. Then we have

$$E(x, y) = E(\widehat{p},\widehat{q}) = \widehat{p}A\widehat{q}$$

$$= \begin{bmatrix} x & 1 - x \end{bmatrix} \begin{bmatrix} a & b \\ c & d \end{bmatrix} \begin{bmatrix} y \\ 1 - y \end{bmatrix}$$

$$= (a - b - c + d)\,xy - (d - b)x - (d - c)y + d.$$

(The reader should verify the matrix multiplication!) As we did in the example, we now put this in standard form. Since there is no saddle point in the original matrix, it follows from the previous paragraph that neither player has a dominant strategy. It follows that (after interchanging rows and/or columns if necessary) we may assume a and d are the two largest entries of A. Indeed, if the two largest entries were in the same row or same column, there would be a dominant strategy.

Thus, they are in different rows and columns, hence at opposite corners. Then after interchanging rows and/or columns if necessary (see Exercise 8 at the end of this section), we can arrange them so that the largest entry is at the corner containing a and the second-largest entry is at the opposite corner d. It follows that $a - b > 0$ (otherwise, there would be a dominant column) and $d - c > 0$ (for the same reason). Similarly, $a - c > 0$ and $d - b > 0$ (otherwise, there would be a dominant row). Write $D = a - b - c + d$ and note $D = (a - c) + (d - b) > 0$ since it is the sum of two positive numbers. It follows we can rewrite $E(x, y)$ as

$$
\begin{aligned}
E(x, y) &= Dxy - (d - b)x - (d - c)y + d \\
&= D\left(xy - \frac{d - b}{D}x - \frac{d - c}{D}y\right) + d \\
&= D\left(xy - \frac{d - b}{D}x - \frac{d - c}{D}y + \frac{d - b}{D} \cdot \frac{d - c}{D}\right) + d - D\frac{d - b}{D} \cdot \frac{d - c}{D} \\
&= D\left(\left(x - \frac{d - c}{D}\right)\left(y - \frac{d - b}{D}\right)\right) + \frac{dD - (d - b)(d - c)}{D} \\
&= D\left(\left(x - \frac{d - c}{D}\right)\left(y - \frac{d - b}{D}\right)\right) + \frac{d(a - b - c + d) - (d - b)(d - c)}{D} \\
&= D\left(\left(x - \frac{d - c}{D}\right)\left(y - \frac{d - b}{D}\right)\right) \\
&\quad + \frac{ad - db - dc + d^2 - d^2 + db + dc - bc}{D} \\
&= D\left(\left(x - \frac{d - c}{D}\right)\left(y - \frac{d - b}{D}\right)\right) + \frac{ad - bc}{D}.
\end{aligned}
$$

Now observe that this is the standard form of a saddle surface with saddle point located at $x = \frac{d - c}{D}$ and $y = \frac{d - b}{D}$. Furthermore, since $d - c$, $d - b$, and D are all positive, we see the saddle point is located in the first quadrant; that is, $x > 0$ and $y > 0$. Furthermore, $D = (a - c) + (d - b)$ and $a - c > 0$, so $d - b < D$. It follows that $y = \frac{d - b}{D} < 1$. Similarly, $D = (a - b) + (d - c)$ and $a - b > 0$, so $d - c < D$ and whence $x = \frac{d - c}{D} < 1$. Thus, both x and y are positive fractions less than 1, so they can be interpreted as probabilities.

We have shown that if A is a 2×2 payoff matrix for a constant-sum game that is not strictly determined, then the derived game is strictly determined, and the saddle point of the derived game occurs at a point that corresponds to two mixed strategies for the original game A. Since (1)–(3) hold for the derived game, they also hold for the original game. This completes the proof. ∎

REMARK 6.23 *It can be shown that if A is strictly determined, then E(x, y) is still a saddle surface. However, the saddle point of this surface will be outside the domain $0 \leq x \leq 1$ and $0 \leq y \leq 1$, so the global saddle point of E(x, y) does not correspond to a mixed strategy since x and y are not probabilities outside this domain. Nevertheless, one of the corners of the patch of E(x, y) when $0 \leq x, y \leq 1$ still has the property that (1)–(3) hold. Since the corners correspond to $x = 0$ or 1 and $y = 0$ or 1, and these correspond to pure strategies in A, we recover the saddle point solution to A in this case. Thus, the statement that the derived game of A is strictly determined always holds, whether or not A is strictly determined. We explore some aspects of this in the exercises.*

REMARK 6.24 *Note that when A is 2×2 and not strictly determined, the inequalities in (1) and (2) will turn out to be equalities as in the preceding example. This is clear by examining the* *standard form of E(x, y) – if either player plays the recommended mix, the expected outcome is the constant v regardless of what the other player does. For larger matrices, we'll see that this theorem also holds but not necessarily with equalities in (1) and (2).*

COROLLARY 6.25   *(See Taylor and Pacelli (2008)) Let $A = \begin{bmatrix} a & b \\ c & d \end{bmatrix}$ be a 2×2 payoff matrix for a constant-sum game without a saddle point. Assume that a and d are the two largest entries in A. Then there is a formula for the optimal solution. Namely, if $\widehat{p}$ and $\widehat{q}$ represent the optimal strategies and v represents the value of the game, then*

$$\widehat{p} = \begin{bmatrix} \frac{d-c}{D} & \frac{a-b}{D} \end{bmatrix}$$

$$\widehat{q} = \begin{bmatrix} \frac{d-b}{D} \\ \frac{a-c}{D} \end{bmatrix}$$

$$v = \frac{ad - bc}{D},$$

where $D = a - b - c + d$ as in the preceding proof.

Proof   This follows immediately from the standard form of $E(x, y)$ in the previous proof. We know $\widehat{p} = (x, 1 - x)$, and $x = \frac{d-c}{D}$. It follows that

$$1 - x = 1 - \frac{d-c}{D} = \frac{D - (d - c)}{D} = \frac{a - b}{D}.$$

Similarly,

$$1 - y = 1 - \frac{d-b}{D} = \frac{D - (d - b)}{D} = \frac{a - c}{D}.$$

This gives the first two formulas. The value of v we simply read off from directly from the standard form of $E(x, y)$. The corollary is proved.   ∎

REMARK 6.26   *Note that the numerator for v is equal to* det *A. This holds only if a and d are the two largest entries in A.*

Let's look at the example we went through before the proof of the minimax theorem, where we have $A = \begin{bmatrix} 4 & -2 \\ -3 & 1 \end{bmatrix}$. We solved this game using the expected payoff function $E(x, y)$ and obtained the standard form:

$$E(x, y) = 10(x - 0.4)(y - 0.3) - 0.2,$$

from which we deduced the solution to the game:

$$\widehat{p} = \begin{bmatrix} 0.4 & 0.6 \end{bmatrix}$$

$$\widehat{q} = \begin{bmatrix} 0.3 \\ 0.7 \end{bmatrix}$$

$$v = -0.2.$$

Notice that the two largest entries of A are indeed in the corners corresponding to a and d, in that order. Therefore, instead of going through the process of putting $E(x, y)$ into standard form, we could just use the corollary. We have $a = 4$, $b = -2$, $c - -3$, and $d = 1$. Therefore, $D = 4 - (-2) - (-3) + 1 = 10$, and the corollary yields

$$\widehat{p} = \left[\begin{array}{cc} \frac{d-c}{D} & \frac{a-b}{D} \end{array} \right] = \left[\begin{array}{cc} \frac{1-(-3)}{10} & \frac{4-(-2)}{10} \end{array} \right] = \left[\begin{array}{cc} \frac{4}{10} & \frac{6}{10} \end{array} \right]$$

$$\widehat{q} = \left[\begin{array}{c} \frac{d-b}{D} \\ \frac{a-c}{D} \end{array} \right] = \left[\begin{array}{c} \frac{1-(-2)}{10} \\ \frac{4-(-3)}{10} \end{array} \right] = \left[\begin{array}{c} \frac{3}{10} \\ \frac{7}{10} \end{array} \right]$$

$$v = \frac{ad - bc}{D} = \frac{(4)(1) - (-2)(-3)}{10} = -\frac{2}{10},$$

which agrees with our previous solution. In the exercises, we explore how to use this corollary when the two largest entries in A are not in the a and d positions.

When the payoff matrix is larger than 2×2, the expected payoff $E(p,q)$ in the derived game depends on more than two variables, so its graph is no longer a surface but something of higher dimension. Consequently, our proof of the minimax theorem does not easily generalize. Furthermore, there is no simple formula for the optimal solution as in the preceding corollary. To handle larger games, we need a different approach. One such approach is explored in the next subsection.

6.4.2 Equalizing Expectation

Since secrecy is important in a game that is not strictly determined, you normally would not expect to have any information about how your opponent will play when it is time to select your strategy. However, suppose you happen to know what mix your opponent is using, and you have reason to believe that he will continue to use that mix no matter what you do. Then how should you play?

We claim that instead of playing your optimal mix, you should respond by using a pure strategy. To see this, imagine you are the row player, and you know what mix q the column player will use. Use the formula $E_i = p_i A q$ to compute the expected payoff E_i for each pure strategy p_i. If you were to use a mixed strategy p, your expected payoff $E = pAq$ would be a (weighted) average of the various E_i values. But any weighted average of the numbers E_i will always be between the smallest E_i and the largest. In particular, it will be less than the largest E_i, so you can obtain a better expected payoff by simply using the pure strategy p_i that produced the largest E_i than you could by using any mixed strategy. The argument works the same way if you are the column player, except that you want E to be small, not large, since the column player wants to minimize the payoff to the row player. This proves a useful principle. Following Straffin (1993), we call it the expected value principle.

THEOREM 6.27 *The Expected Value Principle. Suppose you are playing a two-person constant-sum game, and you know what mix your opponent is using, and you have reason to believe that she will continue to use that mix no matter what you do. The best response to this is to play that pure strategy that gives you the best expected payoff.*

Before we illustrate this with examples, let's be clear about which of our basic assumptions we are dropping. If you were to use a pure strategy in a constant-sum game with no saddle point and your opponent was rational and skillful, he would notice that you were using a pure strategy, and he would retaliate by changing his own strategy to take advantage of your play. So when we assume you have reason to believe that your opponent will not change his mix regardless of what you do, we are no longer assuming your opponent is rational or skillful. In most situations, your opponent will be rational and skillful, so the expected value principle will not apply directly in these situations. However, we can imagine cases where it would apply (for example, in a game against nature or an indifferent opponent).

Now for some examples. Suppose that the payoff matrix is 2×2 with strategies we call S and T:

$$A = \begin{array}{c} \\ S \\ T \end{array} \begin{array}{cc} S & T \\ \left[\begin{array}{cc} 3 & -1 \\ -4 & 6 \end{array} \right] \end{array}.$$

Suppose the column player will use $q = \left[\begin{array}{c} 0.7 \\ 0.3 \end{array} \right]$ no matter what the row player does. How should the row player respond? Playing the first row S means $p = p_1 = \left[\begin{array}{cc} 1 & 0 \end{array} \right]$, so we compute

$$E_1 = p_1 A q = \left[\begin{array}{cc} 1 & 0 \end{array} \right] \left[\begin{array}{cc} 3 & -1 \\ -4 & 6 \end{array} \right] \left[\begin{array}{c} 0.7 \\ 0.3 \end{array} \right] = 1.8.$$

If we play the second row T, then $p = p_2 = \left[\begin{array}{cc} 0 & 1 \end{array} \right]$, whence

$$E_2 = p_2 A q = \left[\begin{array}{cc} 0 & 1 \end{array} \right] \left[\begin{array}{cc} 3 & -1 \\ -4 & 6 \end{array} \right] \left[\begin{array}{c} 0.7 \\ 0.3 \end{array} \right] = -1.$$

Since $E_1 = 1.8$ is the better expected payoff for the row player, the expected value principle recommends playing row 1, the pure strategy S. Actually, we can cut down on the calculation involved. We made two computations, $p_1 A q$ and $p_2 A q$ where p_1 and p_2 are row matrices. We can combine them into a single calculation by constructing a single matrix P that has p_1 as its first row and p_2 as its second row. Then, similar to examples we have seen in Chapter 2 where we have promoted matrix multiplication as a good way to handle itemized computations, the single matrix product $P A q$ will be a (column) matrix that contains both the E_i values at once as entries. However, notice that $P = I$, the identity matrix! This means that $P A q = I A q = A q$. In other words, the product $A q$ is a matrix whose entries are the E_i values (which are the expected payoffs to the row player for using the pure strategies p_i).

We illustrate this in the example at hand:

$$A q = \left[\begin{array}{cc} 3 & -1 \\ -4 & 6 \end{array} \right] \left[\begin{array}{c} 0.7 \\ 0.3 \end{array} \right] = \left[\begin{array}{c} 1.8 \\ -1 \end{array} \right] = \left[\begin{array}{c} E_1 \\ E_2 \end{array} \right].$$

Thus, this single computation achieves the same end as both of the individual computations. More generally, if A has m rows, the single matrix product $A q$ will replace m distinct matrix products of three matrices each. A similar result holds if it is the row player whose mix p is known and the column player is trying to use the expected value principle to determine the best pure strategy response. We simply compute $p A$ and choose the column that yields the smallest entry (which is the best expected payoff for the column player). For example, if the row player decides to use the mix $p = \left[\begin{array}{cc} 0.2 & 0.8 \end{array} \right]$ and the column player knows this, then

$$p A = \left[\begin{array}{cc} 0.2 & 0.8 \end{array} \right] \left[\begin{array}{cc} 3 & -1 \\ -4 & 6 \end{array} \right] = \left[\begin{array}{cc} -2.6 & 4.6 \end{array} \right].$$

Since -2.6 represents the better payoff for the column player, the expected value principle recommends that the first column S should be played. Summarizing, we have the following corollary.

COROLLARY 6.28   *To use the expected value principle in practice, if the column player's mix q is known, the row player should respond by choosing the pure strategy corresponding to the row that yields the highest entry in Aq. If the row player's mix p is known, the column player should respond by choosing the pure strategy corresponding to the column that yields the lowest entry in pA.*

The expected value principle and its corollary are valid for any size matrices. For example, suppose A is the 3×4 payoff matrix:

$$A = \begin{bmatrix} -2 & 4 & 1 & 0 \\ 2 & 3 & -5 & -1 \\ 3 & -6 & 2 & 2 \end{bmatrix}.$$

Suppose that the row player knows that the column player will always use the mix $q = \begin{bmatrix} 0.1 \\ 0.3 \\ 0.4 \\ 0.2 \end{bmatrix}$. How should the row player respond? According to the corollary, we compute Aq:

$$Aq = \begin{bmatrix} -2 & 4 & 1 & 0 \\ 2 & 3 & -5 & -1 \\ 3 & -6 & 2 & 2 \end{bmatrix} \begin{bmatrix} 0.1 \\ 0.3 \\ 0.4 \\ 0.2 \end{bmatrix} = \begin{bmatrix} 1.4 \\ -1.1 \\ -0.3 \end{bmatrix}.$$

The largest expected payoff is 1.4, which is obtained by choosing the first row. On the other hand, suppose the column player knows that the row player will always use the mix $p = \begin{bmatrix} 0.5 & 0.2 & 0.3 \end{bmatrix}$. What is the best response for the column player? We compute pA:

$$pA = \begin{bmatrix} 0.5 & 0.2 & 0.3 \end{bmatrix} \begin{bmatrix} -2 & 4 & 1 & 0 \\ 2 & 3 & -5 & -1 \\ 3 & -6 & 2 & 2 \end{bmatrix} = \begin{bmatrix} 0.3 & 0.8 & 0.1 & 0.4 \end{bmatrix}$$

The best expected payoff for the column player is 0.1, which is obtained by choosing the third column.

Now let's return to our usual assumptions about games, including that all the players are rational and skillful. For the rest of this section, we also return exclusively to the 2×2 case. Even though the expected value principle is not used in this situation to decide on a strategy, it still can be helpful in thinking about the derived game and its solution.

For concreteness, we continue with the payoff matrix $A = \begin{bmatrix} 4 & -2 \\ -3 & 1 \end{bmatrix}$, which we already solved using the derived game approach. Suppose the column player tentatively decides to use the mixed strategy $q = \begin{bmatrix} 0.5 \\ 0.5 \end{bmatrix}$. With that mix, the expected payoffs to the column player if she chose a pure strategy are determined using Corollary (6.28):

$$\begin{bmatrix} E_1 \\ E_2 \end{bmatrix} = Aq = \begin{bmatrix} 4 & -2 \\ -3 & 1 \end{bmatrix} \begin{bmatrix} .5 \\ .5 \end{bmatrix} = \begin{bmatrix} 1 \\ -1 \end{bmatrix}.$$

If the row player knew that the column player would stick with this particular q, then she would clearly choose the first row are her pure strategy response because it has the best payoff among all of her options, pure or mixed, as indicated in the expected value principle (Theorem 6.27). If the row player somehow started with playing the second row, she'd have a pretty strong incentive to switch (unilaterally) to the first, as there is a significant improvement possible in her payoffs.

However, there is no reason to assume the column player would stick with this mixed strategy, and a different choice might lead to a situation with less of a difference between E_1 and E_2. For example, if we repeat this exercise with the tentative choice of $q = \begin{bmatrix} 0.4 \\ 0.6 \end{bmatrix}$ instead of $q = \begin{bmatrix} 0.5 \\ 0.5 \end{bmatrix}$, we obtain

$$\begin{bmatrix} E_1 \\ E_2 \end{bmatrix} = Aq = \begin{bmatrix} 4 & -2 \\ -3 & 1 \end{bmatrix} \begin{bmatrix} 0.4 \\ 0.6 \end{bmatrix} = \begin{bmatrix} 0.4 \\ -0.6 \end{bmatrix}$$

Again, there is incentive for the row player to switch from the second row to the first, but perhaps this time, it is a bit less compelling because what she stands to gain is only half of what it was previously. Going further in this direction, if $q = \begin{bmatrix} \frac{1}{3} \\ \frac{2}{3} \end{bmatrix}$, we obtain

$$\begin{bmatrix} E_1 \\ E_2 \end{bmatrix} = Aq = \begin{bmatrix} 4 & -2 \\ -3 & 1 \end{bmatrix} \begin{bmatrix} \frac{1}{3} \\ \frac{2}{3} \end{bmatrix} = \begin{bmatrix} 0 \\ -\frac{1}{3} \end{bmatrix}.$$

This time, a unilateral switch on the row player's part will only be a gain of $\frac{1}{3}$. As a final example, suppose $q = \begin{bmatrix} 0.301 \\ 0.699 \end{bmatrix}$, which is very close to what we already know is the optimal mix $\widehat{q}$. We obtain

$$\begin{bmatrix} E_1 \\ E_2 \end{bmatrix} = Aq = \begin{bmatrix} 4 & -2 \\ -3 & 1 \end{bmatrix} \begin{bmatrix} 0.301 \\ 0.699 \end{bmatrix} = \begin{bmatrix} -0.194 \\ -0.204 \end{bmatrix}.$$

In this case, the difference between E_1 and E_2 is minuscule. Furthermore, no matter what mixed strategy the row player uses, the expected payoff will be somewhere between E_1 and E_2. So the incentive to switch to $p = (1,0)$ is now very weak.

Pushing this argument to its logical conclusion, suppose we can find a mix q for the column player such that E_1 and E_2 turn out to be equal and call the common value v. In this case, the expected value for the row player would be the same number v no matter what strategy p the row player chooses. This effectively removes any incentive to switch strategies, since switching has no effect whatsoever on the expected payoff! Since the very definition of a Nash equilibrium is an outcome where neither player has any incentive to make a unilateral change of strategies, it is clear that, if we can find an outcome of the game where each player finds a mix that has the property that using it forces her opponent's expected payoff to be a constant value, independent of her mix, then that outcome must be a Nash equilibrium.

We can see that is what happens in our example. If we use $q = \begin{bmatrix} 0.3 \\ 0.7 \end{bmatrix}$ for the column player, then

$$\begin{bmatrix} E_1 \\ E_2 \end{bmatrix} = Aq = \begin{bmatrix} 4 & -2 \\ -3 & 1 \end{bmatrix} \begin{bmatrix} 0.3 \\ 0.7 \end{bmatrix} = \begin{bmatrix} -0.2 \\ -0.2 \end{bmatrix}.$$

Since $E_1 = E_2 = v = -0.2$, there is no particular incentive for the row player to prefer one mix over another, so she may as well choose the mix that equalizes the column player's payoff as well. Let $p = \begin{bmatrix} 0.4 & 0.6 \end{bmatrix}$, so we have

$$pA = \begin{bmatrix} 0.4 & 0.6 \end{bmatrix} \begin{bmatrix} 4 & -2 \\ -3 & 1 \end{bmatrix} = \begin{bmatrix} -0.2 & -0.2 \end{bmatrix}.$$

Again, $E_1 = E_2 = v = -0.2$, so nothing can be gained by the column player by changing his mix either. Thus, we arrive at the same solution to this game that we found when we used the derived game approach – a Nash equilibrium with mixed strategies (even though there is no Nash equilibrium in pure strategies) at the point where

$$\widehat{p} = \begin{bmatrix} 0.4 & 0.6 \end{bmatrix}$$

$$\widehat{q} = \begin{bmatrix} 0.3 \\ 0.7 \end{bmatrix}$$

$$v = -0.2.$$

A moment's reflection should convince the reader that this always happens for a 2×2 matrix. Indeed, we already pointed out that if A is 2×2, then the inequalities (1) and (2) in the minimax theorem are actually equalities. This means that if either player plays the optimal mix, then E is constant and independent of the opponent's mix. Summarizing, we have shown the following.

PROPOSITION 6.29 *Let A be a 2×2 payoff matrix for a two-person constant-sum game. If the game is not strictly determined, then the Nash equilibrium in mixed strategies occurs at the point where each player chooses a mix that will equalize E_1 and E_2, the expected payoffs for the opponent that results from the two pure strategies.*

We need not find these optimal mixes $\widehat{p} = (x, 1 - x)$ and $\widehat{q} = \begin{bmatrix} y \\ 1 - y \end{bmatrix}$ by trial and error, nor by resorting to the derived game approach. This proposition gives an alternate approach. Simply express E_1 and E_2 as functions of x or y, set them equal to each other, and solve for x or y. In the preceding example, the procedure looks like this. To find $\widehat{q}$, set the components of $A\widehat{q}$ equal to each other and solve for y:

$$A\widehat{q} = \begin{bmatrix} 4 & -2 \\ -3 & 1 \end{bmatrix} \begin{bmatrix} y \\ 1 - y \end{bmatrix} = \begin{bmatrix} 6y - 2 \\ 1 - 4y \end{bmatrix}.$$

Since $A\widehat{q} = \begin{bmatrix} E_1 \\ E_2 \end{bmatrix}$, and by the proposition, we seek the point where $E_1 = E_2$, we have

$$6y - 2 = 1 - 4y$$
$$10y = 3$$
$$y = \frac{3}{10}.$$

Thus, $\widehat{q} = \begin{bmatrix} y \\ 1 - y \end{bmatrix} = \begin{bmatrix} \frac{3}{10} \\ \frac{7}{10} \end{bmatrix}$, which agrees with our previous answer. The value v of the game is the common value of E_1 or E_2, so we find it by substituting the value of y we found into either equation:

$$E_1 = 6y - 2 = 6\left(\frac{3}{10}\right) - 2 = -0.2.$$

Similarly, to find $\widehat{p} = \begin{bmatrix} x & 1-x \end{bmatrix}$, we equate the two entries in $\widehat{p}A$:

$$\widehat{p}A = \begin{bmatrix} x & 1-x \end{bmatrix} \begin{bmatrix} 4 & -2 \\ -3 & 1 \end{bmatrix} = \begin{bmatrix} 7x - 3 & 1 - 3x \end{bmatrix}.$$

Setting the E_i equal:

$$7x - 3 = 1 - 3x$$
$$10x = 4$$
$$x = \frac{4}{10}.$$

Thus, $\widehat{p} = \begin{bmatrix} x & 1-x \end{bmatrix} = \begin{bmatrix} \frac{4}{10} & \frac{6}{10} \end{bmatrix}$, as expected. We can also verify that our earlier computation for v is correct, since, again, v is the common value of the E_i:

$$E_1 = 7\left(\frac{4}{10}\right) - 3 = -\frac{2}{10}.$$

We mention in passing that once we have found $\widehat{p}$ and $\widehat{q}$, another way to determine v is to use Eq. (6.5), as indicated by (3) in the minimax theorem:

$$v = E(\widehat{p}, \widehat{q}) = \begin{bmatrix} \frac{4}{10} & \frac{6}{10} \end{bmatrix} \begin{bmatrix} 4 & -2 \\ -3 & 1 \end{bmatrix} \begin{bmatrix} \frac{3}{10} \\ \frac{7}{10} \end{bmatrix} = -\frac{2}{10}.$$

We have thus obtained the optimal solution to this game, and it agreed with that obtained via the derived game approach.

To close this chapter, we illustrate this proposition graphically. The relations

$$E_1 = 7x - 3$$
$$E_2 = 1 - 3x$$

have graphs that are straight lines. We plot them in a common coordinate system, and since we seek where they are equal, that is the point where the lines meet. Of course, since x is a probability, we only care about the domain $0 \le x \le 1$, so that is the only part we plot in Figure 6.5.

Recall that E_1 represents the expected payoff to the row player assuming the column player plays the first column, and E_2 is the expected payoff assuming the column player plays the second column. Of course, the column player, being rational, will not in reality play either pure strategy but some mix of the two columns. In that case, the expected payoff for the row player will be a weighted average of the values given by E_1 and E_2 and, therefore, will lie somewhere between these values. In other words, the actual expected payoff to the row player will be somewhere between the two plotted lines, in the shaded region shown in Figure 6.6.

Assuming the column player is skillful and acts rationally, he will attempt to push this expected payoff as low as possible, to the worst case possible for the row player, which would be somewhere along the bottom edge of the shaded region. The maximin point would be the highest point on this bottom edge – the best of the worst-case scenarios for the row player. But clearly, this maximin point is exactly the point where the two lines cross. This gives an alternate explanation

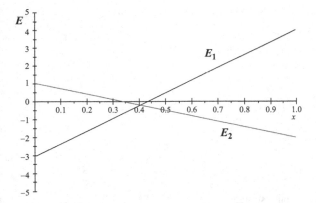

Figure 6.5 Expected payoffs for the row player if the column player uses a pure strategy.

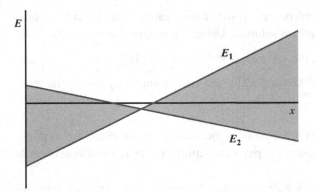

Figure 6.6 Expected payoff to the row player assuming the column player uses a mixed strategy.

for why the Nash equilibrium occurs at the point where the expected values E_1 and E_2 are equal.

There is a similar graph depicting the situation for the column player. This time, we plot the lines for the expected payoffs as functions of y:

$$E_1 = 6y - 2$$
$$E_2 = 1 - 4y.$$

The resulting graph is shown in Figure 6.7.

Again, the actual payoffs, given that the row player will most likely use some mixed strategy, would lie in the region between the lines. Furthermore, assuming the row player is rational and skillful, she will try hard to push her expected payoff as close as possible to the top edge of the region, which represents the worst-case scenarios for the column player. The minimax point would then be the lowest point on the top edge, which is again the point where the lines cross. This would be the best of the worst-case scenarios for the column player, and again, it occurs at the point where the expected values E_1 and E_2 are equalized.

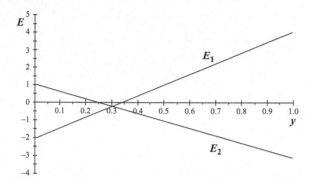

Figure 6.7 Expected payoffs if the row player uses a pure strategy.

To summarize, in case A is a 2×2 matrix for a constant-sum game without saddle points, we have several options for finding the optimal solution. Under the usual assumptions that your opponent is skillful and rational, we seek the solution $\widehat{p} = \begin{bmatrix} x & 1-x \end{bmatrix}, \widehat{q} = \begin{bmatrix} y \\ 1-y \end{bmatrix}$ and $v = \widehat{p}A\widehat{q}$, as described by the minimax (Theorem 6.21), and we can find it by one of the three following techniques.

1. Using the *derived game* approach, as in the proof of the minimax theorem and the example just prior to the proof, where we express the payoff function $E(x, y)$ as a saddle surface in standard form.

2. Using the formulas given in Corollary 6.25, assuming the two largest entries of A are on the main diagonal. (In the exercises, when we explore what happens when the two largest entries of A are not on the main diagonal, we'll see that the formulas in this corollary can be generalized to a technique known as the *method of oddments*, first discovered by J. Williams (1954).)

3. *Equalizing the expectation*, which amounts to setting the two entries of $\widehat{p}A$ equal and solving for x to find $\widehat{p}$ and setting the two entries of $A\widehat{q}$ equal and solving for y to find $\widehat{q}$. (In fact, the graph of the shaded region in Figure 6.6 is nothing more than a "side view" of the saddle surface $E(x, y)$.) While the graphs of the expectation lines are helpful in thinking about why this method works, in the 2×2 case, it is not necessary to draw the graph to find the solution of the game, since we know that we seek the point of intersection of the two lines and can do so algebraically.

Finally, under the assumption that your opponent is indifferent rather than rational, such as in a game against nature, if you know the strategy your opponent is using, you should respond with a pure strategy, as described by the expected value principle. In the next chapter, we look at constant-sum games that are larger than 2×2 and see which of these techniques can be generalized to handle larger games, aside from the expected value principle, which we already know is valid for any payoff matrix. Ultimately, we show that the minimax theorem (Theorem 6.21) holds for any size payoff matrix. However, we will need to introduce yet another method for solving games before we can prove it.

Exercises

1. One hazy day in winter, Paul and Art are playing the game two-finger morra, with zero-sum payoff matrix, as given in Section 6.2:

$$A = \begin{array}{c} \\ 1 \\ 2 \end{array} \begin{array}{cc} 1 & 2 \\ \begin{bmatrix} 1 & -1 \\ -1 & 1 \end{bmatrix} \end{array}.$$

   Neither of them has ever studied game theory, but because of the symmetry of the payoff matrix, they both intuitively believe that the optimal solution is for them to play each strategy half the time. And doing so, they believe that in the long run, they will break even. (Such zero-sum games, where $v = 0$, are called *fair games*.) Find the optimal solution using the derived game approach, and see if their intuition is correct.

2. Recall the zero-sum game from Exercise 8 in Section 6.3, with

$$A = \begin{bmatrix} 1 & 0 \\ -3 & 2 \end{bmatrix}.$$

   Find the optimal solution using the derived game approach.

3. Recall the zero-sum game from Exercise 9 in Section 6.3, with

$$A = \begin{bmatrix} -2 & 5 \\ 4 & -1 \end{bmatrix}.$$

   Find the optimal solution using the derived game approach.

4. When Ann and Nancy tire of playing the card game hearts, they dream up other games. One game they called *little queen* is a zero-sum game with the following payoff matrix, where Ann is the row player:

$$A = \begin{bmatrix} 6 & -1 \\ -3 & 2 \end{bmatrix}.$$

   Find the optimal solution using the derived game approach.

5. Notice that in the payoff matrix A from Exercise 4, the largest entry is a_{11}, and the second largest entry is a_{22}. Thus, the corollary to the minimax theorem applies. Solve the game using the corollary, and verify that the answer agrees with that in Exercise 4.

6. Consider the matrix $A = \begin{bmatrix} -2 & 5 \\ 4 & -1 \end{bmatrix}$ from Exercise 3. The largest two entries are not on the main diagonal, so we cannot use the corollary to the minimax theorem directly. However, if we list the column player's strategies in the reverse order, the effect is to switch the two columns of A so we arrive at a matrix new matrix $B = \begin{bmatrix} 5 & -2 \\ -1 & 4 \end{bmatrix}$, which represents essentially the same game as A.

   a. The corollary does apply to B, so use it to find the optimal solution to B.
   b. Since the difference between A and B only involved transposing the columns, it follows that the optimal mix $\widehat{p}$ for the row player should be the same in matrix A as in B. That is, $\widehat{p}_A = \widehat{p}_B$. On the other hand, since the columns where interchanged, it follows that to obtain $\widehat{q}_A$ from $\widehat{q}_B$, we merely switch the coordinates. Explain why these statements are true, and find the optimal mixes for A using your solution to

B from Part a. [**Hint:** To explain why the statements are true, think of the method of equalizing expectations. How do the calculations change when the columns are interchanged?]

c. Explain why $v_A = v_B$, and rewrite v_A in terms of the determinant of A. Find v_A and check that your answer agrees with your solution from Exercise 3.

7. Consider the matrix $A = \begin{bmatrix} -2 & 1 \\ 6 & -3 \end{bmatrix}$ The corollary to the minimax theorem does not apply directly to A. However, if you list the row player's strategies in the reverse order, the effect is to interchange the rows of A to obtain the matrix $C = \begin{bmatrix} 6 & -3 \\ -2 & 1 \end{bmatrix}$, and the corollary does apply to C. Repeat the steps in Exercise 6 to solve the game with payoff matrix A by comparing it to the solution of C.

8. Consider the zero-sum payoff matrix $A = \begin{bmatrix} 1 & 0 \\ -3 & 2 \end{bmatrix}$ from Exercise 2. We continue to study the effect of interchanging rows and columns on the game. Exercises 6 and 7 explored what happens when you interchanging just the rows or just the columns. Interchanging just rows or just columns to this matrix will not yield a matrix to which the corollary applies. However, if you switch both the rows and the columns (that is, list the strategies for both players in reverse order), you obtain the matrix $B = \begin{bmatrix} 2 & -3 \\ 0 & 1 \end{bmatrix}$, to which the corollary does apply. By combining the results of Exercises 6 and 7, show how to obtain the solution of A from the solution of B. In particular, pay close attention to the formula for v, and observe that $\det A = \det B$. Similarly, the denominators D are the same for both matrices. (Why?) This explains why the formulas in Corollary 6.25 still hold when d is the largest entry instead of a.

9. Let $A = \begin{bmatrix} a & b \\ c & d \end{bmatrix}$ be any 2×2 payoff matrix for a zero-sum game without a saddle point. There is a method for finding the optimal solution, called the *method of oddments*, discovered by J. Williams (1954). The algorithm is as follows:

    i.   To find $\widehat{p}$, the optimal strategy for the row player, take the absolute value of the differences of the entries across the rows. The resulting positive numbers (called *oddments* by Williams) are the odds of playing the rows in $\widehat{p}$, except that you must switch the coordinates so that the oddment in the first row gives the probability of playing the second row, and vice versa. To obtain the probabilities from the odds, simply put them over the sum of the oddments (since "odds" are, by definition, just the ratio of the probabilities).

    ii. To find $\widehat{q}$, the optimal strategy for the column player, take the absolute value of the differences of the entries across the columns. The resulting oddments are the odds of playing the columns in $\widehat{q}$, except that you must switch the coordinates so that the first oddment corresponds to the second row, and vice versa. To obtain the probabilities from the odds, simply put them over the sum of the oddments (since "odds" are, by definition, just the ratio of the probabilities).

    iii. To find v, the value of the game, you can just use $v = \widehat{p}A\widehat{q}$ (or you can derive a formula similar to that in Corollary 6.25, where v is a fraction with the sum of the oddments in the denominator, and $\pm \det A$ in the numerator, where the positive sign is taken if the two largest elements of A are on the main diagonal and the negative sign if otherwise).

Here is an example illustrating the method of oddments. Suppose $A = \begin{bmatrix} 2 & -1 \\ -4 & 3 \end{bmatrix}$.

We find $\widehat{p}$ as follows:

		Absolute value of differences	Switch positions	Oddments
2	-1	$\|2-(-1)\| = \|3\| = 3$	↘	7
-4	3	$\|-4-3\| = \|-7\| = 7$	↗	3

Thus, the row player should play the rows with odds 7 to 3. This means the probabilities are $\frac{7}{7+3} = \frac{7}{10}$ and $\frac{3}{7+3} = \frac{3}{10}$. Therefore $\widehat{p} = \begin{bmatrix} 0.7 & 0.3 \end{bmatrix}$ Similarly, we find $\widehat{q}$.

	2 -4	-1 3
Absolute value of differences	$\|2-(-4)\| = \|6\| = 6$	$\|-1-3\| = \|-4\| = 4$
Switch positions	↘	↙
Oddments	4	6

Thus, the column player should play the columns with odds 4 to 6, so the probabilities are $\frac{4}{4+6} = \frac{4}{10}$ for the first column and $\frac{6}{4+6} = \frac{6}{10}$ for the second column. Therefore, $\widehat{q} = \begin{bmatrix} 0.4 \\ 0.6 \end{bmatrix}$. The value of the game is given by

$$v = \widehat{p}A\widehat{q} = \begin{bmatrix} 0.7 & 0.3 \end{bmatrix} \begin{bmatrix} 2 & -1 \\ -4 & 3 \end{bmatrix} \begin{bmatrix} 0.4 \\ 0.6 \end{bmatrix} = 0.2.$$

This is the optimal solution. Note that the two largest entries of A are on the main diagonal, so an alternate calculation for the value is

$$v = \frac{\det A}{10} = \frac{2}{10} = 0.2.$$

a. Solve this game by either the method of derived games, or by Corollary 6.25, and verify that the preceding solution is correct.

b. By combining Exercises 6, 7, and 8, Corollary 6.25 to the minimax theorem extends to any non-strictly determined 2×2 payoff matrix. Show that this extended corollary is the same thing as the method of oddments. [**Hint:** Consider cases according to where the largest two entries of A are located.]

10. Resolve the game with payoff matrix, as in Exercise 2, $A = \begin{bmatrix} 1 & 0 \\ -3 & 2 \end{bmatrix}$ via the method of oddments. Check that your answer agrees with Exercise 2.

11. Repeat Exercise 10 for the matrix $A = \begin{bmatrix} -2 & 5 \\ -4 & 1 \end{bmatrix}$ of Exercise 3.

12. Repeat Exercise 10 for the matrix $A = \begin{bmatrix} -2 & 1 \\ 6 & -3 \end{bmatrix}$ of Exercise 7.

In Exercises 13 – 20, solve the zero-sum game with the given payoff matrices using the method of equalizing expectation.

13. $A = \begin{bmatrix} 1 & 0 \\ -3 & 2 \end{bmatrix}$ as in Exercise 2.

14. $A = \begin{bmatrix} -2 & 5 \\ 4 & -1 \end{bmatrix}$ as in Exercise 3.

15. $A = \begin{bmatrix} 6 & -1 \\ -3 & 2 \end{bmatrix}$ as in Exercise 4.

16. $A = \begin{bmatrix} -2 & 1 \\ 6 & -3 \end{bmatrix}$ as in Exercise 7.

17. $A = \begin{bmatrix} 12 & -18 \\ -10 & 0 \end{bmatrix}$.

18. $A = \begin{bmatrix} 1 & 8 \\ 7 & 2 \end{bmatrix}$. Compare your solution to that of Exercise 14. Explain.

19. $A = \begin{bmatrix} 2 & 9 \\ 8 & 3 \end{bmatrix}$. Compare your solution to that of Exercises 14 and 18. Explain.

20. $A = \begin{bmatrix} 12 & -2 \\ -6 & 4 \end{bmatrix}$. Compare your answer to that of Exercise 15. Explain.

21.   Based on Exercises 18–20, formulate a conjecture about solutions to translated and scaled games that are not strictly determined. Compare your conjecture to the remark at the end of the exercises in Section 6.2, which explored the same question for strictly determined games. Can you prove your conjecture?

22.   Consider the matrix $A = \begin{bmatrix} -2 & 1 \\ 6 & -3 \end{bmatrix}$ from Exercise 7. Suppose that the column player decides to ignore the optimal mix and instead play the mix $q = \begin{bmatrix} 0.1 \\ 0.9 \end{bmatrix}$, regardless of what the row player does. Use the expected value principle to decide how the row player should respond to this.

23.   Consider the matrix $A = \begin{bmatrix} -2 & 1 \\ 6 & -3 \end{bmatrix}$ from Exercise 7. Suppose that the row player decides to ignore the optimal mix and instead play the mix $p = \begin{bmatrix} \frac{1}{4} & \frac{3}{4} \end{bmatrix}$, regardless of what the column player does. Use the expected value principle to decide how the column player should respond to this.

24.   Consider the matrix $A = \begin{bmatrix} 1 & -4 & 2 & 0 \\ 1 & 1 & 1 & -5 \\ -3 & -1 & 8 & 4 \\ 0 & 2 & -9 & -1 \end{bmatrix}$ from Exercise 14 of the last Section.
Suppose that the column player decides to ignore the optimal strategy mix and instead play the mix $q = \begin{bmatrix} .2 \\ .6 \\ 0 \\ .2 \end{bmatrix}$, regardless of what the row player does. How should the row player respond?

25. Consider the variation on four-finger morra in Exercise 9 of Section 6.1. If the column player

    decided to always use the mix $q = \begin{bmatrix} 0.4 \\ 0 \\ 0.4 \\ 0.2 \end{bmatrix}$, how should the row player respond?

26. Consider the guessing game from Exercise 10 of Section 6.1. If Ann decided to use the following mix for her guesses, $p = \begin{bmatrix} 0.1 & 0.2 & 0.3 & 0.4 \end{bmatrix}$, and Bill knows this, how should Bill respond?

27. Answer the same question as 25, except use the variation of four-finger morra in Exercise 18 of Section 6.2.

28. Consider the variation on four-finger morra in Exercise 17 of Section 6.2. If the row player decided to use the mix $p = \begin{bmatrix} 0 & 0.1 & 0.4 & 0.5 \end{bmatrix}$, how should the column player respond?

29. Consider the finite version of the market game Antares vs. Bellatrix given in Exercise 7 of Section 6.1. Suppose the market for 400,000 chips per year will extend into the foreseeable future, and both companies will reset their prices annually. Suppose that $\frac{1}{3}$ of the time, Antares sets its price at $80, and the remaining $\frac{2}{3}$ of the time, it sets their price at $110. Bellatrix is aware of this proportion and has reason to believe that Antares will continue this practice into the future. How should Bellatrix set its price?

30. Consider the variable-sum game in Example 6.4 of Section 6.1, the Saturday chores at the Anderson household, version II (Example 6.4). In the text, we determined that the payoff matrix is as follows.

		Laurie			
		No chore	Dishes	Raking	Both
	No chore	(10, 10)	(12, 6)	(8, 15)	(0, 21)
Jon	Dishes	(8, 15)	(10, 10)	(8, 15)	(4, 18)
	Raking	(12, 6)	(12, 6)	(10, 10)	(6, 13)
	Both	(20, 0)	(16, 3)	(14, 7)	(10, 10)

   a. Suppose that Jon decides to use the mix $p = \begin{bmatrix} 0.1 & 0.3 & 0.4 & 0.2 \end{bmatrix}$ and Laurie

   decides to use the mix $q = \begin{bmatrix} 0 \\ 0.3 \\ 0.6 \\ 0.1 \end{bmatrix}$. With these mixes, what is the expected payoff to

   both children? [**Hint:** We have not formally worked with variable-sum games yet, but you can determine this. Just remember that the payoffs to Jon are the first coordinates of the ordered pair entries in A, and the payoffs to Laurie are the second coordinates. The formula $E = pAq$ still applies, but to each player's payoffs separately.]

   b. Suppose that Jon becomes aware that Laurie is using the mix given in Part a and has reason to believe she will continue to do so, regardless of what he does. How should he respond? [**Hint:** Apply the expected value principle, just using Jon's payoffs.]

31. In this exercise, we consider the derived game of A in the case when A is itself strictly determined. Consider a zero-sum game with payoff matrix $A = \begin{bmatrix} 0 & 1 \\ -1 & d \end{bmatrix}$, where d is not specified.

a.   Show that no matter what value is assigned for d, the matrix A has a saddle point, so it represents a strictly determined game. What is the optimal solution of the game, as predicted by the minimax theorem?

b.   Consider the derived game of A. Find $E(x, y)$ in standard form, and show that it is a saddle surface if $d \neq 0$. However, show that the saddle point is outside of the range $0 \leq x \leq 1$ and $0 \leq y \leq 1$, so it does not represent an optimal solution to the game.

c.   What happens if $d = 0$?

32.   Consider the same matrix A as in Exercise 31. Instead of using the derived game, try to solve it using the method of equalizing expectations. Show that you never obtain the correct solution, no matter what d is.

   [**Hint**: It will be helpful to consider the graph of the expected value lines E_1 and E_2 for the column player's payoffs, and to consider various ranges for d separately.] (The same thing would happen if you used the method of oddments instead of the method of equalizing expectations.) In summary, while this is not a general proof, it illustrates the fact that if a constant-sum game A has a saddle point, you must find the saddle point to obtain the optimal solution (as stated in the first sentence of the minimax theorem). The method of equalizing expectations, the method of oddments, and the location of the global saddle point of the derived game will all fail to give the correct solution, so these methods should only be used for games that are not strictly determined.

7   More Game Theory

This chapter is devoted to expanding our repertoire of techniques to solve constant-sum games, in particular so that we can handle games of any size $m \times n$. In the first section, we generalize our ad hoc methods, especially the method of equalizing expectation. In Section 7.2, we show how to use linear programming to solve constant-sum games, and as a corollary, we finally are able to prove the minimax theorem in full generality. In Section 7.3, we show how to use Mathematica or Excel to solve constant-sum games. In the last section, we focus on variable-sum games.

7.1   Solving Larger Constant-Sum Games

7.1.1   Games with Dominated Strategies

Consider a zero-sum game with the following payoff matrix:

$$A = \begin{bmatrix} 1 & 2 & -3 \\ -1 & 0 & 1 \end{bmatrix}.$$

None of the techniques we learned in the last chapter apply to this game. There is no saddle point, but the matrix is not 2×2, so the derived game is not helpful, and neither the methods of equalizing expectation nor oddments apply to a 2×3 matrix. However, we have already observed that in any game, you should never play a dominated strategy. In this game, the first column dominates the second, so the column player would never play the second column. Like we did in the previous chapter, we can even cross that column off, resulting in a smaller matrix:

$$B = \begin{bmatrix} 1 & -3 \\ -1 & 1 \end{bmatrix}.$$

Now the reduced game is 2×2, so we can solve this game by using equalizing expectation or oddments. For example, if we use oddments, subtracting across the rows, we obtain 4 and $|-2| = 2$, and switching positions gives oddments of $(2, 4)$, whence $\widehat{p} = \left(\frac{1}{3}, \frac{2}{3} \right)$. Similarly, we obtain $\binom{4}{2}$ for the oddments for the column player, so $\widehat{q} = \begin{pmatrix} \frac{2}{3} \\ \frac{1}{3} \end{pmatrix}$. The value of the game is

$$v = \begin{bmatrix} \frac{1}{3} & \frac{2}{3} \end{bmatrix} \begin{bmatrix} 1 & -3 \\ -1 & 1 \end{bmatrix} \begin{bmatrix} \frac{2}{3} \\ \frac{1}{3} \end{bmatrix} = \begin{bmatrix} -\frac{1}{3} \end{bmatrix}.$$

However, this is the solution to the game with matrix B, not A. The only difference, however, is that A has an extra column that is never played. Therefore, we need to record that information in $\widehat{q}$. The optimal solution for the row player, as well as the value, is the same for A and B. Thus, the optimal solution for the original game A is

$$\widehat{p} = (\ \tfrac{1}{3} \ \ \tfrac{2}{3} \ )$$

$$\widehat{q} = \begin{pmatrix} \tfrac{2}{3} \\ 0 \\ \tfrac{1}{3} \end{pmatrix}$$

$$v = -\frac{1}{3}.$$

This technique will help whenever you have dominated strategies. One deletes the rows and columns that are dominated, including higher-order domination, and the result is a reduced game with a smaller matrix. If you happen to end up with a 2×2 matrix, you can solve it by the techniques of Chapter 6, and then just remember to include a 0 for each row and column that are dominated in the optimal strategies to solve the original game.

Here is another example:

$$A = \begin{bmatrix} 3 & -4 & -1 \\ 4 & -3 & 0 \\ -5 & 6 & 2 \\ -5 & 7 & 8 \end{bmatrix}.$$

In this game, there is no saddle point. One checks that none of the columns dominate any of the others. However, the second row dominates the first, and the fourth row dominates the third. Eliminating the first and third rows leaves

$$\begin{bmatrix} 4 & -3 & 0 \\ -5 & 7 & 8 \end{bmatrix}.$$

Now, the second column dominates the third, so by higher-order dominance, we can eliminate the third column, leaving

$$B = \begin{bmatrix} 4 & -3 \\ -5 & 7 \end{bmatrix}.$$

As there is no further dominance, the reduced game is the 2×2 matrix B. We solve this game using equalizing expectation this time for practice:

$$\widehat{p}B = \begin{bmatrix} x & 1-x \end{bmatrix} \begin{bmatrix} 4 & -3 \\ -5 & 7 \end{bmatrix} = \begin{bmatrix} 9x - 5 & 7 - 10x \end{bmatrix}.$$

Setting the expected values equal, we obtain

$$9x - 5 = 7 - 10x$$

$$19x = 12$$

$$x = \frac{12}{19}$$

$$v = 9 \left(\frac{12}{19} \right) - 5 = \frac{13}{19}.$$

So $\hat{p} = \left[\begin{array}{cc} \frac{12}{19} & \frac{7}{19} \end{array}\right]$. For the column player:

$$B\hat{q} = \left[\begin{array}{cc} 4 & -3 \\ -5 & 7 \end{array}\right] \left[\begin{array}{c} y \\ 1-y \end{array}\right] = \left[\begin{array}{c} 7y - 3 \\ 7 - 12y \end{array}\right].$$

Equalizing expectation:

$$7y - 3 = 7 - 12y$$
$$19y = 10$$
$$y = \frac{10}{19}.$$

So $\hat{q} = \left[\begin{array}{c} \frac{10}{19} \\ \frac{9}{19} \end{array}\right]$. Again, these solutions are for the matrix B. To obtain the complete solution to the original game A, we adjust the optimal strategies to indicate which rows and colums of A are not used because they were dominated. The solution we seek is

$$\hat{p} = \left[\begin{array}{cccc} 0 & \frac{12}{19} & 0 & \frac{7}{19} \end{array}\right]$$

$$\hat{q} = \left[\begin{array}{c} \frac{10}{19} \\ \frac{9}{19} \\ 0 \end{array}\right]$$

$$v = \frac{13}{19}.$$

A strategy that is not used at all (indicated by a 0 in the optimal solution mix) is said to be *inactive*. The point of the example is that, obviously, a dominated strategy is inactive. Surprisingly, we'll see in the next section that sometimes, a strategy will be inactive even if it is not dominated.

7.1.2 The 2 × n and the m × 2 Case

After eliminating all the dominated strategies, the reduced game matrix may still not be 2 × 2. The next best case is that the reduced matrix has just two rows or just two columns. That is, it will be 2 × n or m × 2. In this section, we show how to handle these cases by using the method of equalizing expectation. We begin with the following zero-sum example:

$$A = \left[\begin{array}{cc} -2 & 3 \\ 4 & -2 \\ 1 & 2 \end{array}\right].$$

The reader can check that there are no saddle points in this matrix, and no dominance either, so the matrix cannot be reduced to a smaller size. With three rows, the optimal strategy for the row player has three coordinates and cannot be written with just one unknown x. However, the column player can. The optimal strategy for the column player is still written $\hat{q} = \left[\begin{array}{c} y \\ 1 - y \end{array}\right]$. So let's focus on the column player for the moment. Calculating $A\hat{q}$ as usual, we obtain

$$\left[\begin{array}{cc} -2 & 3 \\ 4 & -2 \\ 1 & 2 \end{array}\right] \left[\begin{array}{c} y \\ 1 - y \end{array}\right] = \left[\begin{array}{c} 3 - 5y \\ 6y - 2 \\ 2 - y \end{array}\right] = \left[\begin{array}{c} E_1 \\ E_2 \\ E_3 \end{array}\right].$$

So this time, we obtain three equations, E_i, for the expected payoff. But recall what each one means – it is the payoff obtained assuming the row player uses one of the three pure strategies. Let's plot the lines on a single coordinate system. We did not have to do this for a 2×2 matrix because we knew with just two lines, we were looking for the point where they intersected. The result is shown in Figure 7.1.

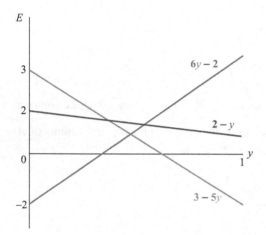

Figure 7.1 Expected payoffs corresponding to the row player's three pure strategies.

It is still true that if the row player mixes the strategies she uses, then the expected payoff will be a weighted average of the payoffs given by the E coordinates of these three lines and, consequently, will lie in the region between these lines. It is also still true that assuming the row player is rational and skilled, that she will try to reach the top portion of this graph, which represents the worst-case scenarios for the column player. So, we should still seek the best of the worst-case scenarios, which is again the lowest point on the top edge of the graph. Visibly, that is the point of intersection of the lines with equations $E_2 = 6y - 2$ and $E_3 = 2 - y$. To find the point, we set these two lines equal:

$$2 - y = 6y - 2$$
$$4 = 7y$$
$$y = \frac{4}{7}$$
$$v = 2 - y = 2 - \frac{4}{7} = \frac{10}{7}.$$

Thus, we have the optimal strategy for the column player and the value of the game:

$$\widehat{q} = \begin{bmatrix} \frac{4}{7} \\ \frac{3}{7} \end{bmatrix}, v = \frac{10}{7}.$$

In summary, the optimal solution (for the player with just two strategies) is still at an intersection of two of the lines we obtain in $A\widehat{q}$ (in the $m \times 2$ case) or in $\widehat{p}A$ (in the $2 \times n$ case), but we need to plot all the lines in order to determine which two lines to equalize.

What about the player with more than two strategies – in this example, the row player? Since $\widehat{p} = (x, y, 1 - x - y)$, the calculation of $\widehat{p}A$ will lead to expressions for E in terms of two variables x and y, so they are no longer lines but surfaces (in fact, planes). Of course, there are only two such planes. Here are the details:

$$\widehat{p}A = \begin{bmatrix} x & y & 1 - x - y \end{bmatrix} \begin{bmatrix} -2 & 3 \\ 4 & -2 \\ 1 & 2 \end{bmatrix}$$

$$= \begin{bmatrix} 3y - 3x + 1 & x - 4y + 2 \end{bmatrix} = \begin{bmatrix} E_1 & E_2 \end{bmatrix}.$$

Again, we note that E_i represents the expected payoff to the row player when we assume the column player is using the ith pure strategy. Thus, again, when the column player mixes his strategies, the actual payoff will be a weighted average of the E_i, so it will lie in the region in space between the planes. Assuming the column player is rational and skillful, he will act to push the payoff down to the bottom part of the region, which represents the worst-case scenarios for the row player. So we again seek the best of the worst-case scenarios, which is the highest point on the bottom part of the region. As in the case for lines, we can argue that this must be on the intersection of the two planes. However, this is not enough information to determine the solution because the intersection of two planes is a line, not a unique point. Let's find the equation of the line of intersection, which we do by equalizing the values of E_i:

$$-3x + 3y + 1 = x - 4y + 2$$
$$7y = 4x + 1$$
$$y = \frac{4}{7}x + \frac{1}{7}.$$

This is the relationship between x and y, but we also need the values of E on this line, which we obtain by choosing either equation for E and substituting in $\frac{4}{7}x + \frac{1}{7}$ for y. For example, choosing E_1, we have

$$E = -3x + 3\left(\frac{4}{7}x + \frac{1}{7}\right) + 1 = \frac{10}{7} - \frac{9}{7}x.$$

Remember that the row player is free to choose x, and she is trying to find the highest point on this line (inside the region where $0 \leq x, y \leq 1$), because that is the best of the worst-case scenarios for her. Clearly, since $0 \leq x \leq 1$ and the slope is negative, the highest point must occur when $x = 0$. Since $y = \frac{4}{7}x + \frac{1}{7}$, this forces $y = \frac{1}{7}$, and so $\widehat{p} = \begin{bmatrix} 0 & \frac{1}{7} & \frac{6}{7} \end{bmatrix}$, and $E = \frac{10}{7}$, of course, agreeing with the value we found earlier when we worked with the column player.

We have solved this 3×2 game using the method of equalizing expectation for both players (more precisely, the method of finding the best of the worst-case scenarios for each player). The complete solution is

$$\widehat{p} = \begin{bmatrix} 0 & \frac{1}{7} & \frac{6}{7} \end{bmatrix}$$

$$\widehat{q} = \begin{bmatrix} \frac{4}{7} \\ \frac{3}{7} \end{bmatrix}$$

$$v = \frac{10}{7}.$$

Notice that the strategy in the first row is inactive, even though it was not dominated by either of the other two rows. This is not a coincidence! One way to explain why is to notice that once we determined the optimal strategy for the column player and obtained $y = \frac{4}{7}$, we knew that if we substituted $y = \frac{4}{7}$ into either the second or the third equation, we would get the value of the game $v = \frac{10}{7}$. But notice if we substitute it into the first equation, we obtain $E_1 = 3-5y = 3-5\left(\frac{4}{7}\right) = \frac{1}{7}$, which is less than the value of the game. In other words, we can calculate the expected payoffs from just knowing $\widehat{q}$ for all three strategies for the row player. We obtain

$$\frac{4}{7}\begin{bmatrix} -2 \\ 4 \\ 1 \end{bmatrix} + \frac{3}{7}\begin{bmatrix} 3 \\ -2 \\ 2 \end{bmatrix} = \begin{bmatrix} \frac{1}{7} \\ \frac{10}{7} \\ \frac{10}{7} \end{bmatrix}.$$

This says that any mix of the second and third strategies for the row player will yield $\frac{10}{7}$ payoff, while the first strategy yields $\frac{1}{7}$ if played as a pure strategy. But any mix that included the first strategy part of the time and any mix of the second and third the rest of the time, would result in a weighted average of $\frac{1}{7}$ and $\frac{10}{7}$, which is less than $\frac{10}{7}$. Therefore, by an argument that is essentially a generalization of the expected value principle, you should respond with just the mix of the second and third and never use the first to actually obtain the $\frac{10}{7}$ expected payoff.

Graphically, you can see this in the first graph we made for the column player. When $y = \frac{4}{7}$, the coordinate on the line with equation $E = 3 - 5y$ from the first row that we did not use to find the minimax point, has E coordinate $\frac{1}{7}$, which is below the minimax point with E coordinate $\frac{10}{7}$.

So this will generalize to any $m \times 2$ game. The graph for the column player will consist of m lines. The worst cases for the column player will always be the top edge. The minimax point, the best of the worst cases, will always be the lowest point on the top edge, which is the intersection of (usually) just two of the lines. Furthermore, this means that if $\widehat{q} = (t, 1 - t)$ has been determined using those two lines, and t is substituted into one of those two lines, you will obtain the value of the game v, but if you substitute t into any of the other $m - 2$ lines, you will obtain an expected payoff of less than v because these lines are all below the minimax point. Therefore, the row player should not use any of the strategies these lines represent – they are all inactive!

We illustrate with another example, this time 5×2. Consider a zero-sum game with payoff matrix

$$A = \begin{bmatrix} -5 & 6 \\ -3 & 2 \\ -1 & 1 \\ 2 & 0 \\ 4 & -3 \end{bmatrix}.$$

The reader can check there is no saddle point and no dominance. Letting $\widehat{q} = \begin{bmatrix} y \\ 1 - y \end{bmatrix}$, we obtain

$$A\widehat{q} = \begin{bmatrix} -5 & 6 \\ -3 & 2 \\ -1 & 1 \\ 2 & 0 \\ 4 & -3 \end{bmatrix}\begin{bmatrix} y \\ 1 - y \end{bmatrix} = A\widehat{q} = \begin{bmatrix} 6 - 11y \\ 2 - 5y \\ 1 - 2y \\ 2y \\ 7y - 3 \end{bmatrix}.$$

Plotting the lines representing expected payoff to the row player under the pure strategy assumptions E_i, $i = 1, 2, 3, 4, 5$, we obtain the graph in Figure 7.2.

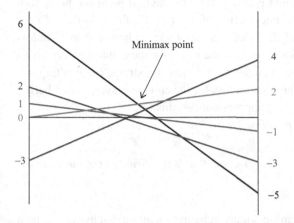

Figure 7.2 At the mix that produces the minimax point, the expected payoff to the row player is lower than v on every other line, so those strategies are inactive for the row player.

The intersection of the E_1 and E_4 lines is best the of the worst-case scenarios for column. Equalizing E_1 and E_4 we obtain

$$6 - 11y = 2y$$
$$6 - 13y$$
$$y = \frac{6}{13}$$
$$v = 2y = \frac{12}{13}.$$

Thus, $\widehat{q} = \begin{bmatrix} \frac{6}{13} \\ \frac{7}{13} \end{bmatrix}$, and for the row player, the second, third, and fifth rows are inactive. Thus, deleting the inactive strategies, the game reduces to solving this 2×2 subgame:

$$\begin{bmatrix} -5 & 6 \\ 2 & 0 \end{bmatrix}.$$

Using oddments (or equalizing expectation) for the row player now gives $\left(\frac{2}{13} \quad \frac{11}{13} \right)$. Therefore, the solution to the original game is

$$\widehat{p} = \begin{bmatrix} \frac{2}{13} & 0 & 0 & \frac{11}{13} & 0 \end{bmatrix}$$
$$\widehat{q} = \begin{bmatrix} \frac{6}{13} \\ \frac{7}{13} \end{bmatrix}$$
$$v = \frac{12}{13}.$$

The $2 \times n$ case is similar. With two rows, you begin by solving for $\widehat{p} = \begin{bmatrix} x & 1 - x \end{bmatrix}$. There are n lines to plot, corresponding to the expected payoff if the column player uses one of the n pure strategies. Since the column player will mix strategies, the actual payoff is between all the lines. Therefore, the worst-case scenarios for the row player will be the bottom edge of the graph, and the best of the worst-case scenarios (the maximin point) will be the highest point on the bottom edge of the graph, which will correspond to the intersection of two of the n lines. At that mix for the row player, the expected payoff for any of the other pure strategies is higher than v (unless more than two lines intersect at this point). Therefore, they are worse for the column player, so all these other lines should remain inactive. Then we can delete these inactive strategies, resulting in a 2×2 submatrix and solve that to determine the optimal mix for the column player, remembering to record the inactive strategies in $\widehat{q}$ with a 0 in the appropriate coordinates.

In this way, we can solve any game that is $m \times 2$ or $2 \times n$.

COROLLARY 7.1   *The solution to any $m \times 2$ or $2 \times n$ game that is not strictly determined is the same as the solution to a 2×2 subgame.*

This corollary is a special case of a theorem on square subgames that we discuss in the next subsection. We close this section with a final example. Consider the zero-sum game with the 3×5 payoff matrix:

$$A = \begin{bmatrix} -5 & -3 & 0 & 1 & 4 \\ 6 & 5 & 3 & -2 & -3 \\ 7 & 5 & 4 & -2 & -1 \end{bmatrix}.$$

First, observe there is no saddle point, so the game is not strictly determined. Next, we check for dominance. There is no dominance among any of the columns; however, the third row dominates the second row. Deleting the second row yields the reduced matrix:

$$\begin{bmatrix} -5 & -3 & 0 & 1 & 4 \\ 7 & 5 & 4 & -2 & -1 \end{bmatrix}.$$

Now the fourth column dominates the fifth, so delete the fifth by higher-order dominanace, leaving this matrix:

$$B = \begin{bmatrix} -5 & -3 & 0 & 1 \\ 7 & 5 & 4 & -2 \end{bmatrix}.$$

At this point, there is no more dominance, so we must solve this 2×4 game. Let $\widehat{p} = \begin{bmatrix} x & 1-x \end{bmatrix}$ and observe

$$\widehat{p}B = \begin{bmatrix} x & 1-x \end{bmatrix} \begin{bmatrix} -5 & -3 & 0 & 1 \\ 7 & 5 & 4 & -2 \end{bmatrix}$$

$$= \begin{bmatrix} 7-12x & 5-8x & 4-4x & 3x-2 \end{bmatrix}.$$

Plotting the four lines yields the graph in Figure 7.3.

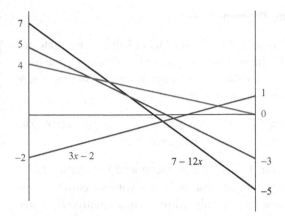

Figure 7.3 Expected payoff corresponding to the four pure strategies for the column player.

The maximin point is the intersection of the first line and the fourth. Equalizing these:

$$7 - 12x = 3x - 2$$
$$9 = 15x$$
$$x = \frac{9}{15} = \frac{3}{5}$$
$$v = 3x - 2 = -\frac{1}{5}.$$

Thus, $\widehat{p} = \begin{bmatrix} \frac{3}{5} & \frac{2}{5} \end{bmatrix}$. Since the second and third strategies are inactive, we delete these columns and arrive at the following 2×2 subgame:

$$\begin{bmatrix} -5 & 1 \\ 7 & -2 \end{bmatrix}.$$

Using oddments, we obtain $\widehat{q} = \begin{bmatrix} \frac{3}{15} \\ \frac{12}{15} \end{bmatrix} = \begin{bmatrix} \frac{1}{5} \\ \frac{4}{5} \end{bmatrix}$. Now, we need to put back in zeros for all the inactive strategies, including the first ones we deleted because they were dominated. We arrive at the solution to the original 3×5 game:

$$\widehat{p} = \begin{bmatrix} \frac{3}{5} & 0 & \frac{2}{5} \end{bmatrix}$$

$$q = \begin{bmatrix} \frac{1}{5} \\ 0 \\ 0 \\ \frac{4}{5} \\ 0 \end{bmatrix}$$

$$v = -\frac{1}{5}.$$

Exercises on this appear at the end of the section.

7.1.3 A 3 × 3 Example: The Square Subgame Theorem

Given a game with payoff matrix A, a *subgame* is the game obtained by deleting some strategies from the original game (so the payoff matrix B of the subgame is obtained by deleting some rows and/or columns of A). If B is a square matrix, we say that the subgame it represents is a *square subgame*. The following theorem is a well-known result in game theory:

THEOREM 7.2   Square Subgame Theorem. *Consider a zero-sum game with payoff matrix A. The solution to the game is the same as the solution to some square subgame.*

This theorem is exploited frequently in Williams (1954), and is mentioned in Straffin (1993) immediately following the statement of the minimax theorem, although without attribution. It says that every payoff matrix for a zero-sum game can be reduced to a square matrix B of size $k \times k$. The case $k = 1$ is exactly the case of a strictly determined game – the submatrix B is just the saddle point.

When A is $2 \times n$ or $m \times 2$, the largest-possible square subgame is 2×2 – the case when $k = 2$ – so if we assume the game is not strictly determined, the solution must be the same as one of its 2×2 subgames. This is precisely the corollary from the last subsection.

Suppose a zero-sum game is 3×3. If all three strategies are active, then the subgame is the whole game; that is, $B = A$. In that case, you can use the method of equalizing expectation to solve it. Here is an example:

Example 7.3 *Rock Paper Scissors. (Example 6.2 from Chapter 6).* Ann and Bill play the well-known game of rock paper scissors, following the usual conventions that rock beats scissors, scissors beats paper, and paper beats rock. They decide to follow the payoff scheme that the loser pays the winner $1, but there is no exchange of money of the result is a tie.

Here is the payoff matrix:

$$A = \begin{array}{c} \\ \text{Rock} \\ \text{Paper} \\ \text{Scissors} \end{array} \begin{array}{ccc} \text{Rock} & \text{Paper} & \text{Scissors} \\ \left[\begin{array}{ccc} 0 & -1 & 1 \\ 1 & 0 & -1 \\ -1 & 1 & 0 \end{array} \right] \end{array}.$$

Although we don't know in advance that all three strategies are active, let's assume they are and see what happens. The first thing to notice is that the game has symmetry – it doesn't matter who is the row player. This is apparent in the matrix because $A^T = -A$. (Such a matrix is said to be *skew-symmetric*. The negative is necessary for a symmetric game because of our convention that the payoffs are written for the row player.) It follows that $\widehat{p}^T = \widehat{q}$; that is, Ann and Bill have the same optimal strategy. Furthermore, you can probably guess that this is a fair game, so we expect that $v = 0$, and you might even guess what the optimal strategy is. But let's see how to calculate it. So let $\widehat{p} = (x, y, 1 - x - y)$, and compute $\widehat{p}A$:

$$\begin{bmatrix} x & y & 1 - x - y \end{bmatrix} \begin{bmatrix} 0 & -1 & 1 \\ 1 & 0 & -1 \\ -1 & 1 & 0 \end{bmatrix}$$

$$= \begin{bmatrix} x + 2y - 1 & 1 - y - 2x & x - y \end{bmatrix} = \begin{bmatrix} E_1 & E_2 & E_3 \end{bmatrix}.$$

Each of the E_i is a plane in the (x, y, E) three-dimensional space. Just as before, the actual payoff is in the region between the three planes, and the worst-case scenarios for the row player are at the bottom of this region. The maximin point – the best of the worst-case scenarios is the highest point on the bottom and will be the unique point that is the intersection of the three planes. (If you cannot visualize the three planes and the fact that there is a unique intersection point, just wait until we solve the equations algebraically.) The solution is when all three of the E_i agree:

$$x + 2y - 1 = 1 - y - 2x = x - y.$$

We could set up the augmented coefficient matrix and use elimination, but it is easy in this case to just solve by substitution. Equating the first and the third equation leads to

$$x + 2y - 1 = x - y$$
$$3y = 1$$
$$y = \frac{1}{3}.$$

Equating the second and the third leads to

$$1 - y - 2x = x - y$$
$$1 = 3x$$
$$x = \frac{1}{3}.$$

To find v, plug in x and y into any of the equations, such as the third:

$$v = x - y = \frac{1}{3} - \frac{1}{3} = 0.$$

Thus, the solution is

$$\widehat{p} = \begin{bmatrix} \frac{1}{3} & \frac{1}{3} & \frac{1}{3} \end{bmatrix}$$
$$\widehat{q} = \widehat{p}^T$$
$$v = 0.$$

This is, no doubt, the solution you predicted. As you can see, the solution is unique, and does involve all three strategies. Incidentally, one could have predicted that all strategies had to be active. If any player never used one strategy – say, Bill never used rock – then Ann can respond by heavily favoring scissors, which will result in either a tie or a win for her, depending on what Bill does.

Unfortunately, this example is deceptively easy. Suppose that you had to solve a 3×3 game where there was at least one inactive strategy (something that may not be obvious ahead of time). Then by the square subgame theorem, if it was not a strictly determined game, then the solution would be the same as a 2×2 subgame - but there are 9 different such subgames that must be checked.

It becomes more tedious as the size increases. For a 4×4 game, if it were not strictly determined, the solution could be one of 36 different 2×2 subgames or 16 different 3×3 subgames or the entire 4×4 game itself if all strategies are active. You would have to check them all to find the optimal solution (actually, you will know when you have the solution, so you might be lucky and only have to check a few – see Williams (1954) for details and examples).

Clearly, we need a more efficient way of solving larger games. In the next section, we show how to do that using linear programming.

Exercises

Solve the games with the following payoff matrices. Assume the game is zero-sum.

1.

$$A = \begin{bmatrix} 1 & -2 & 3 \\ -3 & 1 & -1 \end{bmatrix}$$

2.

$$A = \begin{bmatrix} 1 & 2 \\ 3 & 4 \\ 5 & 6 \end{bmatrix}$$

3.

$$A = \begin{bmatrix} 1 & 2 \\ -3 & 4 \\ 5 & -6 \end{bmatrix}$$

4.

$$A = \begin{bmatrix} -2 & 1 & -1 \\ 3 & -4 & 0 \end{bmatrix}$$

5.

$$A = \begin{bmatrix} 1 & 4 & -2 \\ 0 & 3 & -3 \\ -3 & -2 & 2 \end{bmatrix}$$

6. a.

$$A = \begin{bmatrix} 6 & -7 & -7 \\ -3 & 5 & 5 \\ 1 & 1 & 1 \end{bmatrix}$$

   b.   This game has more than one optimal solution. Find all of the solutions.

7.

$$A = \begin{bmatrix} 1 & 2 & -1 & -2 \\ -3 & 0 & 1 & 3 \\ -6 & -5 & 0 & 2 \end{bmatrix}$$

8.   The following game has more than one solution given by two distinct square subgames. Find the solutions and determine which 2×2 subgames give the correct solution. Why doesn't the third 2×2 subgame give a correct solution?

$$A = \begin{bmatrix} 4 & -6 \\ -2 & 6 \\ 1 & 0 \end{bmatrix}$$

9. Consider a variation on three-finger morra. Each player simultaneously puts out one, two, or three fingers. If the total number of fingers is even, the column player pays the row player, while if the total number is odd, the row player pays the column player. However, instead of the payment being \$1, the number of dollars will be the (nonnegative) difference between the number of fingers they show. Find the optimal solution by equalizing expectation.

10. Consider the usual version of three-finger morra. Each player simultaneously puts out one, two, or three fingers. If the total number of fingers is even, the column player pays the row player \$1, while if the total number is odd, the row player pays the column player \$1. Find the optimal solution by equalizing expectation. What is unusual about this problem?

7.2 Solving Constant Sum-Games with Linear Programming

Given that the method of equalizing expectation becomes more and more tedious with the size of the game, we seek an alternate method to solve constant-sum games. In this section, we show how to do so using linear programming. Since solving a game is a type of optimization problem, it is not such a stretch to believe the linear programming can be applied. In order to motivate the process, we begin with a 2×2 example.

Suppose we have a constant-sum game with payoff matrix

$$A = \begin{bmatrix} a & b \\ c & d \end{bmatrix}.$$

We are going to make the additional assumption that all the entries in A are strictly positive. This is really no loss of generality, because if we start with an arbitrary payoff matrix, we can always translate the game by adding a positive constant M to all the payoffs. As we know, the solution to the translated game is the same optimal strategies as the original, with the only difference being that the value of the translated game is the value of the original plus M.

Let's consider the row player. We have been writing the unknown optimal strategy $\hat{p} = (x, 1-x)$, but in this section, we prefer to write it as $\hat{p} = (p_1, p_2)$, where

$$p_1 + p_2 = 1. \tag{7.1}$$

They key to converting this to a linear programming problem is to look at what Theorem 6.21 says about $\hat{p}$ – namely that it provides a guaranteed minimum payoff no matter what the column player does. In symbols

$$E(\hat{p}, q) \geq v_{\text{row}}$$

for all strategies q, for some fixed real number v_{row}. But we know how to compute E, so this becomes

$$\hat{p} A q \geq v_{\text{row}}$$

for all strategies q. Similar to what we did when we considered equalizing expecation, we apply this in particular to the two pure strategies $q = U_1 = \begin{bmatrix} 1 \\ 0 \end{bmatrix}$ and $q = U_2 = \begin{bmatrix} 0 \\ 1 \end{bmatrix}$. Of course, we can do them both at once by making the computation $\widehat{p}AI = \widehat{p}A$, since the matrix with the columns U_1 and U_2 is the identity matrix. Thus:

$$\begin{bmatrix} p_1 & p_2 \end{bmatrix} \begin{bmatrix} a & b \\ c & d \end{bmatrix} = \begin{bmatrix} ap_1 + cp_2 & bp_1 + dp_2 \end{bmatrix}$$

and each of these must be $\geq v_{\text{row}}$. Thus, we obtain a pair of inequalities:

$$ap_1 + cp_2 \geq v_{\text{row}}$$
$$bp_1 + dp_2 \geq v_{\text{row}}.$$

Now this is beginning to look like a set of constraints. However, it is not yet a linear programming problem for several reasons. First of all, what is the objective function? Clearly, the row player would like v_{row} to be as large as possible. However, in a linear programming problem, the objective function must be expressed in terms of the unknowns p_1 and p_2, and it is not clear exactly how to do so here. Furthermore, making v_{row} as large as possible would be a maximization problem, and the inequalities seem to be going the wrong way for a maximization problem. Finally, in a linear programming problem, the right side of the inequalities must be known quantities, and in this system, the right side is the unknown v_{row}. Also, although we do have the inequalities $p_1 \geq 0$ and $p_2 \geq 0$ since they are probabilities, we also haven't used yet the additional constraint in Eq. (7.1) that the unknowns sum to 1.

Fortunately, we can fix all these problems at once! We simply observe that the v_{row} is some sort of weighted average of the payoffs for the individual outcomes, $a, b, c,$ and d, and since we have assumed that all of these are positive, it follows that $v_{\text{row}} > 0$ also. So we simply define $w = \frac{1}{v_{\text{row}}}$, (and note $w > 0$ as well) and multiply through the inequalities by w. Because $w > 0$, we do not have to worry about reversing the inequalities:

$$ap_1 w + cp_2 w \geq 1$$
$$bp_1 w + dp_2 w \geq 1.$$

Furthermore, we think of $wp_i = \frac{p_i}{v_{\text{row}}}$ as our variables instead of the p_i:

$$t_1 = p_1 w$$
$$t_2 = p_2 w.$$

Observe that

$$t_1 + t_2 = p_1 w + p_2 w = (p_1 + p_2)w = w$$

since the probabilities sum to 1. Also, $w = \frac{1}{v_{\text{row}}}$ is minimized whenever v_{row} is maximized, and the previous equation expresses w in terms of our unknowns t_i. Finally, note that, again, since $w > 0$, we have $t_i \geq 0$ because $p_i \geq 0$. So our transformation into a linear programming problem is complete. The row player must

$$\text{minimize } w = t_1 + t_2$$

subject to:

$$at_1 + ct_2 \geq 1$$
$$bt_1 + dt_2 \geq 1$$
$$t_1 \geq 0, \ t_2 \geq 0.$$

The reader may object that at this point, we haven't verified that $E(\widehat{p}, q) \geq v_{\text{row}}$ for all q yet – just for the pure strategies q. But this is automatic now. Let's illustrate with a particular mixed strategy $q = \begin{bmatrix} 0.6 \\ 0.4 \end{bmatrix}$. We have

$$E(\widehat{p}, q) = \widehat{p} A q$$

$$= \begin{bmatrix} ap_1 + cp_2 & bp_1 + dp_2 \end{bmatrix} \begin{bmatrix} 0.6 \\ 0.4 \end{bmatrix}$$

$$= 0.6 \, (ap_1 + cp_2) + 0.4(bp_1 + dp_2)$$

$$= [0.6 \, (ap_1 + cp_2) + 0.4(bp_1 + dp_2)] w v_{\text{row}}$$

the last equality because $w v_{\text{row}} = 1$. But now

$$E(\widehat{p}, q) = \left[0.6(ap_1 w + cp_2 w) + 0.4(bp_1 w + dp_2 w) \right] v_{\text{row}}$$
$$= [0.6(at_1 + ct_2) + 0.4(bt_1 + dt_2)] v_{\text{row}}.$$

But if we assume the constraints are satisfied for the two pure strategies, then the two terms in parentheses are each ≥ 1. Therefore:

$$E(\widehat{p}, q) \geq (0.6 + 0.4) v_{\text{row}} = v_{\text{row}}$$

as desired. Clearly, the same argument works for any mix $q = \begin{bmatrix} f \\ g \end{bmatrix}$ because $f + g = 1$. It follows that if we solve the linear programming problem and convert back to the original variables $v_{\text{row}} = \frac{1}{w}$, $p_1 = \frac{t_1}{w} = t_1 v_{\text{row}}$, and $p_2 = \frac{t_2}{w} = t_2 v_{\text{row}}$, we will have solved for the row player's optimal strategy since we found a solution that satisfies the minimax theorem.

Before we do an example, consider the column player. We do a similar analysis. We let $\widehat{q} = \begin{bmatrix} q_1 \\ q_2 \end{bmatrix}$, where

$$q_1 + q_2 = 1. \tag{7.2}$$

Then we seek a solution satisfying the minimax theorem, which says there is a $\widehat{q}$ that gives a guaranteed minimum payoff to him (that is, a maximum payoff to the row player), which we call v_{col}. Thus,

$$E(p, \widehat{q}) \leq v_{\text{col}}$$

for all p. We again start by applying this to the two pure strategies $p = U_1 = \begin{bmatrix} 1 & 0 \end{bmatrix}$ and $p = U_2 = \begin{bmatrix} 0 & 1 \end{bmatrix}$. This leads us to find $IA\widehat{q} = A\widehat{q}$ and set each component $\leq v_{\text{col}}$. Thus:

$$A\widehat{q} = \begin{bmatrix} a & b \\ c & d \end{bmatrix} \begin{bmatrix} q_1 \\ q_2 \end{bmatrix} = \begin{bmatrix} aq_1 + bq_2 \\ cq_1 + dq_2 \end{bmatrix},$$

which leads to the system of inequalities:

$$aq_1 + bq_2 \leq v_{\text{col}}$$
$$cq_1 + dq_2 \leq v_{\text{col}}.$$

As in the previous case, we note that $v_{\text{col}} > 0$ because it is a weighted average of the entries of A. So let $z = \frac{1}{v_{\text{col}}}$, and multiply through by z:

$$aq_1 z + bq_2 z \leq 1$$
$$cq_1 z + dq_2 z \leq 1.$$

Define new variables

$$u_1 = q_1 z$$
$$u_2 = q_2 z.$$

Just as for the row player, note that

$$u_1 + u_2 = (q_1 + q_2) z = z.$$

Also, note that the column player wants v_{col} to be as small as possible, which means that its reciprocal z is as large as possible. Thus, the column player must solve this problem:

$$\text{maximize } z = u_1 + u_2$$
$$\text{subject to:}$$
$$au_1 + bu_2 \leq 1$$
$$cu_1 + du_2 \leq 1$$
$$u_1 \geq 0,\ u_2 \geq 0.$$

Finally, just as before, if we find a solution to this, it automatically yields $E(p, \widehat{q}) \leq v_{\text{col}}$ for all p, not just the pure strategies. To find the column player's optimal solution, we note that $q_1 = \frac{u_1}{z} = u_1 v_{\text{col}}$, $q_2 = \frac{u_2}{z} = u_2 v_{\text{col}}$, where $v_{\text{col}} = \frac{1}{z}$.

Now comes the interesting part. Note that the linear programming problem for the column player is a standard form maximization, and the linear programming problem for the row player is a standard form minimization. But look more closely. The condensed tableau for the column player is

$$\begin{bmatrix} \begin{bmatrix} a & b \\ c & d \end{bmatrix} & \begin{matrix} 1 \\ 1 \end{matrix} \\ \begin{matrix} 1 & 1 \end{matrix} & 0 \end{bmatrix}$$

and the condensed tableau for the row player is exactly the transpose of this! It follows that these linear programming problems are dual problems. Furthermore, by the strong duality principle (see Section 4.5), it follows that the maximum value of z is the same as the minimum value of w.

Therefore, $v_{\text{row}} = \frac{1}{w} = \frac{1}{z} = v_{\text{col}}$. So we can drop the subscripts and just call them both v, the value of the game.

Let's illustrate this with an example. Suppose the payoff matrix is

$$B = \begin{bmatrix} -3 & 2 \\ 5 & -4 \end{bmatrix}.$$

We must begin by translating this game to a game with all positive entries. All we have to do is add a constant $M > 4$ to all the entries. It does not matter what the exact value of M is. We could use $M = 4.001$, $M = 5$, or $M = 100$ if we wish. We will use $M = 5$ to avoid fractional entries in the matrix. Adding 5 to each entry, we obtain

$$A = \begin{bmatrix} 2 & 7 \\ 10 & 1 \end{bmatrix}.$$

Let's write down the linear programming problem for each player. For the column player, it is

$$\text{maximize } z = u_1 + u_2$$
$$\text{subject to:}$$
$$2u_1 + 7u_2 \leq 1$$
$$10u_1 + u_2 \leq 1$$
$$u_1 \geq 0, \ u_2 \geq 0.$$

Remember that once we solve this, we convert back to the game by

$$\hat{q} = \begin{bmatrix} q_1 \\ q_2 \end{bmatrix} = \begin{bmatrix} u_1 v \\ u_2 v \end{bmatrix},$$

where $v = \frac{1}{z}$ is the value of the game A. Then the last step is to subtract 5 from v to get the value of the original game B.

Similarly, the row player must solve the dual problem:

$$\text{minimize } w = t_1 + t_2$$
$$\text{subject to}$$
$$2t_1 + 10t_2 \geq 1$$
$$7t_1 + t_2 \geq 1$$
$$t_1 \geq 0, \ t_2 \geq 0.$$

Once solved, we convert back to the game by

$$\hat{p} = \begin{bmatrix} p_1 & p_2 \end{bmatrix} = \begin{bmatrix} t_1 v & t_2 v \end{bmatrix},$$

where $v = \frac{1}{w}$ is again the value of the game A. Then we subtract 5 to obtain the value of the original game B.

Let's solve these problems. We can solve each one by graphing in the decision space because there are only two variables. For the row player, we obtain the feasible region shown in Figure 7.4.

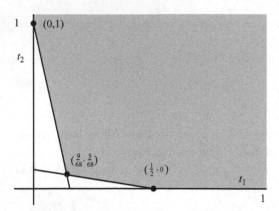

Figure 7.4 The feasible set for the row player's problem.

The reader should check to see the coordinates of the three corners are as in the following table.

(t_1, t_2)	$w = t_1 + t_2$
$(1, 0)$	$1 + 0 = 1$
$\left(\frac{9}{68}, \frac{5}{68}\right)$	$\frac{9}{68} + \frac{5}{68} = \frac{14}{68} = \frac{7}{34}$
$\left(\frac{1}{2}, 0\right)$	$\frac{1}{2} + 0 = \frac{1}{2}$

Clearly, the minimum value of w is $\frac{7}{34}$. This gives $v_A = \frac{1}{w} = \frac{34}{7}$, and

$$\widehat{p} = \left[\begin{array}{cc} t_1 v & t_2 v \end{array} \right] = \left[\begin{array}{cc} \frac{9}{14} & \frac{5}{14} \end{array} \right].$$

Now for the column player, we obtain the feasible region, as shown in Figure 7.5.

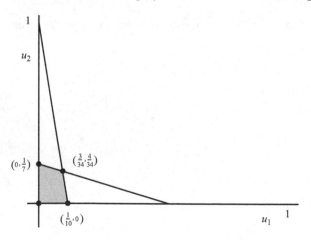

Figure 7.5 The feasible set for the column player's problem.

The corresponding corners are shown in the following table.

(u_1, u_2)	$z = u_1 + u_2$
$(0,0)$	$0 + 0 = 0$
$\left(\frac{1}{10}, 0\right)$	$\frac{1}{10} + 0 = \frac{1}{10}$
$\left(\frac{3}{34}, \frac{4}{34}\right)$	$\frac{3}{34} + \frac{4}{34} = \frac{7}{34}$
$\left(0, \frac{1}{7}\right)$	$0 + \frac{1}{7} = \frac{1}{7}$

The maximum point is clearly $\left(\frac{3}{34}, \frac{4}{34}\right)$, so $z = \frac{7}{34}$. Converting back to the game notation we have $v_A = \frac{1}{z} = \frac{34}{7}$, and

$$\widehat{q} = \begin{bmatrix} u_1 v \\ u_2 v \end{bmatrix} = \begin{bmatrix} \frac{3}{7} \\ \frac{4}{7} \end{bmatrix}.$$

All that remains to do is subtract 5 from the value, so the solution to the original game is

$$\widehat{p} = \begin{bmatrix} \frac{9}{14} & \frac{5}{14} \end{bmatrix}$$

$$\widehat{q} = \begin{bmatrix} \frac{3}{7} \\ \frac{4}{7} \end{bmatrix}$$

$$v_B = v_A - 5 = \frac{34}{7} - 5 = -\frac{1}{7}.$$

The reader can easily verify the solution using oddments or equalizing expectation.

The work involved using this method is, of course, much more than the previous methods we have studied. However, unlike any of the previous methods, this technique generalizes to any size payoff matrix with no conceptual change. Furthermore, don't forget that if you solve the linear programming problems using the simplex algorithm, you can solve both of the problems at once because they are dual problems. We close this section by illustrating this on the previous example.

With the simplex algorithm, we focus on the maximization problem, which is the problem that the column player must solve. The initial tableau is the following.

	u_1	u_2	S_1	S_2	Capacity
S_1	2	7	1	0	1
S_2	10	1	0	1	1
z	-1	-1	0	0	0

According to our phase two pivoting algorithm, including Bland's rule, we pivot on $a_{21} = 10$. Dividing R_2 by 10 to obtain the *NPR*, we obtain

$$R_1 \to R_1 - 2NPR$$

$$R_3 \to R_3 + NPR.$$

Performing the indicated row operations, we obtain the next tableau.

	u_1	u_2	S_1	S_2	Capacity
S_1	0	$\frac{68}{10}$	1	$-\frac{2}{10}$	$\frac{8}{10}$
u_1	1	$\frac{1}{10}$	0	$\frac{1}{10}$	$\frac{1}{10}$
z	0	$-\frac{9}{10}$	0	$\frac{1}{10}$	$\frac{1}{10}$

We then pivot on the u_2 column. The ratios are $\frac{8}{68}$ and 1, so we choose the first row as the pivot row. Dividing by $\frac{68}{10}$ to obtain the NPR, we obtain the following.

	u_1	u_2	S_1	S_2	Capacity
u_2	0	1	$\frac{10}{68}$	$-\frac{2}{68}$	$\frac{8}{68}$
u_1	1	$\frac{1}{10}$	0	$\frac{1}{10}$	$\frac{1}{10}$
z	0	$-\frac{9}{10}$	0	$\frac{1}{10}$	$\frac{1}{10}$

$$R_2 \rightarrow R_2 - \frac{1}{10}NPR$$
$$R_3 \rightarrow R_3 + \frac{9}{10}NPR$$

Completing the pivot, we obtain the next tableau.

	u_1	u_2	S_1	S_2	Capacity
u_2	0	1	$\frac{10}{68}$	$-\frac{2}{68}$	$\frac{8}{68}$
u_1	1	0	$-\frac{1}{68}$	$\frac{7}{68}$	$\frac{6}{68}$
z	0	0	$\frac{9}{68}$	$\frac{5}{68}$	$\frac{14}{68}$

This is the optimal tableau. The solution to the maximization problem is

$$u_1 = \frac{6}{68}, \; u_2 = \frac{8}{68}, \; z = \frac{14}{68}.$$

The slack variables have value 0 here, but we aren't really interested in those. The marginal values, however, are the solution to the dual minimization problem:

$$t_1 = \frac{9}{68}, \; t_2 = \frac{5}{68}, \; w = \frac{14}{68}.$$

In each case, $v_A = \frac{68}{14} = \frac{34}{7}$, so converting back to the game notation, we have

$$\hat{p} = \begin{bmatrix} t_1 v & t_2 v \end{bmatrix} = \begin{bmatrix} \frac{9}{14} & \frac{5}{14} \end{bmatrix}$$

$$\hat{q} = \begin{bmatrix} u_1 v \\ u_2 v \end{bmatrix} = \begin{bmatrix} \frac{6}{14} \\ \frac{8}{14} \end{bmatrix} = \begin{bmatrix} \frac{3}{7} \\ \frac{4}{7} \end{bmatrix}$$

$$v_B = v_A - 5 = \frac{34}{7} - 5 = -\frac{1}{7}.$$

This agrees with our previous solution.

7.2.1   The $m \times n$ Case and the General Minimax Theorem

Having explored the method for a 2×2 game, we now consider the general $m \times n$ game. The analysis goes through without change in the general case. First, translate the game, if necessary, to obtain a payoff matrix A with all positive entries. For the column player, we define new variables

$z = \frac{1}{v}$ and $u_j = q_j z = \frac{q_j}{v}$; $j = 1, 2, \ldots, n$, where $\widehat{q} = \begin{bmatrix} q_1 \\ q_2 \\ \vdots \\ q_n \end{bmatrix}$. We find, by considering the m

pure strategies the row player can play, that the column player must solve this linear programming problem:

$$\text{maximize } z = u_1 + u_2 + \cdots + u_n$$

subject to:

$$a_{11}u_1 + a_{12}u_2 + \cdots + a_{1n}u_n \leq 1$$
$$a_{21}u_1 + a_{22}u_2 + \cdots + a_{2n}u_n \leq 1$$

$$\vdots$$

$$a_{m1}u_1 + a_{m2}u_2 + \cdots + a_{mn}u_n \leq 1$$
$$u_j \geq 0, \ j = 1, 2, \ldots, n,$$

where we note that the coefficient matrix of the system of structural inequalities is just the payoff matrix A. Just as in the preceding example, if these inequalities hold, then the solution satisfies $E = pA\widehat{q} \geq v$ for all strategies p, not just the pure strategies.

Similarly, for the row player, we define new variables $w = \frac{1}{z}$ and $t_i = p_i w = \frac{p_i}{v}$; $i = 1, 2, \ldots, m$, where $\widehat{p} = \begin{bmatrix} p_1 & p_2 & \cdots & p_m \end{bmatrix}$. We find, by considering the n pure strategies that the column player could play, that the row player must solve this linear programming problem:

$$\text{minimize } w = t_1 + t_2 + \cdots + t_m$$

subject to:

$$a_{11}t_1 + a_{21}t_2 + \cdots + a_{m1}t_m \geq 1$$
$$a_{12}t_1 + a_{22}t_2 + \cdots + a_{m2}t_m \geq 1$$

$$\vdots$$

$$a_{1n}t_1 + a_{2n}t_2 + \cdots + a_{mn}t_m \geq 1$$
$$t_i \geq 0, \ i = 1, 2, \ldots, m$$

where we note that the coefficient matrix of the system of structural inequalities is exactly A^T, the transpose of the coefficient matrix. Also, one checks just as in the preceding example that if these inequalities hold, then $E = \widehat{p}Aq \leq v$ for all strategies q, not just the pure strategies.

Furthermore, these are dual problems because the condensed tableau for the problem the column player must solve is

$$\begin{bmatrix} & & & 1 \\ & A & & \vdots \\ & & & 1 \\ 1 & \dots & 1 & 0 \end{bmatrix}$$

and the condensed tableau for the problem that the row player must solve is exactly the transpose of this:

$$\begin{bmatrix} & & & 1 \\ & A^T & & \vdots \\ & & & 1 \\ 1 & \dots & 1 & 0 \end{bmatrix}.$$

Before doing examples, we pause to note that we can now prove the minimax theorem (Theorem 6.21):

THEOREM 7.4   *Let B be the payoff matrix for any $m \times n$ constant-sum game. There are optimal strategies $\widehat{p}$ and $\widehat{q}$ and a real number v (called the value of the game) such that*

1. $E(\widehat{p}, q) \geq v$ *for any q.*
2. $E(p, \widehat{q}) \leq v$ *for any p.*
3. $E(\widehat{p}, \widehat{q}) = v$.

Proof   First, translate B so that the payoff matrix A for the translated game has all positive entries (add a constant M to all the entries in B to obtain A). Then the linear programming problem is a standard form maximization problem with a feasible set that is nonempty and bounded. (It's nonempty because the origin is a feasible point when every $a_{ij} > 0$.) Thus, by our rules for phase two pivoting, we know there is an optimal solution. Furthermore, then the dual problem (a standard form minimization problem) also has an optimal solution. It follows that the game has a solution satisfying the three conditions of the theorem by our previous remarks, for the matrix A. Then the same results hold for the original game B with $v_B = v_A - M$.   ■

REMARK 7.5   *It is possible that the solution may occur at a saddle point of the matrix A. However, unlike the previous methods, the linear programming technique will solve the game correctly even if it is strictly determined. (Of course, it is easier to find a saddle point by inspection if there is one.)*

Example 7.6 Solve the zero-sum game with payoff matrix

$$B = \begin{bmatrix} -2 & 3 & -1 \\ 4 & -5 & 2 \end{bmatrix}.$$

We begin by translating the game, adding 6 to each entry to obtain

$$A = \begin{bmatrix} 4 & 9 & 5 \\ 10 & 1 & 8 \end{bmatrix}.$$

The column player must solve the problem

$$\text{maximize } z = u_1 + u_2 + u_3$$
$$\text{subject to}$$
$$4u_1 + 9u_2 + 5u_3 \leq 1$$
$$10u_1 + u_2 + 8u_3 \leq 1$$
$$u_j \geq 0, \ j = 1, 2, 3.$$

Setting up the initial tableau (which is easy to do directly from A – we don't even really need to write down the problem first), we obtain the following tableau.

	u_1	u_2	u_3	S_1	S_2	Capacity
S_1	4	9	5	1	0	1
S_2	10	1	9	0	1	1
z	−1	−1	−1	0	0	0

We pivot in the first column and the second row, dividing R_2 by 10 to obtain the NPR, and then we perform the row operations.

	u_1	u_2	u_3	S_1	S_2	Capacity
S_1	4	9	5	1	0	1
u_1	1	$\frac{1}{10}$	$\frac{9}{10}$	0	$\frac{1}{10}$	$\frac{1}{10}$
z	−1	−1	−1	0	0	0

$$R_1 \rightarrow R_1 - 4NPR$$
$$R_3 \rightarrow R_3 + NPR,$$

The leads to the next tableau.

	u_1	u_2	u_3	S_1	S_2	Capacity
S_1	0	$\frac{86}{10}$	$\frac{14}{10}$	1	$-\frac{4}{10}$	$\frac{6}{10}$
u_1	1	$\frac{1}{10}$	$\frac{9}{10}$	0	$\frac{1}{10}$	$\frac{1}{10}$
z	0	$-\frac{9}{10}$	$-\frac{1}{10}$	0	$\frac{1}{10}$	$\frac{1}{10}$

Next, we pivot in the second column and first row, dividing R_1 by $\frac{86}{10}$ to obtain the NPR, and then we perform the row operations.

	u_1	u_2	u_3	S_1	S_2	Capacity
u_2	0	1	$\frac{14}{86}$	$\frac{10}{86}$	$-\frac{4}{86}$	$\frac{6}{86}$
u_1	1	$\frac{1}{10}$	$\frac{9}{10}$	0	$\frac{1}{10}$	$\frac{1}{10}$
z	0	$-\frac{9}{10}$	$-\frac{1}{10}$	0	$\frac{1}{10}$	$\frac{1}{10}$

$$R_2 \to R_2 - \frac{1}{10}NPR$$
$$R_3 \to R_3 + \frac{9}{10}NPR,$$

This leads to the following tableau.

	u_1	u_2	u_3	S_1	S_2	Capacity
u_2	0	1	$\frac{14}{86}$	$\frac{10}{86}$	$-\frac{4}{86}$	$\frac{6}{86}$
u_1	1	0	$\frac{76}{86}$	$-\frac{1}{86}$	$\frac{9}{86}$	$\frac{8}{86}$
z	0	0	$\frac{4}{86}$	$\frac{9}{86}$	$\frac{5}{86}$	$\frac{14}{86}$

This tableau is optimal. The solution is

$$z = \frac{14}{86}, \ u_1 = \frac{8}{86}, \ u_2 = \frac{6}{86}, \ u_3 = 0.$$

We also can read off the solution to the row player's problem from the bottom row:

$$t_1 = \frac{9}{86}, \ t_2 = \frac{5}{86}.$$

Converting back to game notation, $v = \frac{86}{14} = \frac{43}{7}$, so

$$\widehat{q} = \begin{bmatrix} u_1 v \\ u_2 v \\ u_3 v \end{bmatrix} = \begin{bmatrix} \frac{8}{14} \\ \frac{6}{14} \\ 0 \end{bmatrix}$$

$$\widehat{p} = \begin{bmatrix} t_1 v & t_2 v \end{bmatrix} = \begin{bmatrix} \frac{9}{14} & \frac{5}{14} \end{bmatrix}$$

$$v_B = v_A - 6 = \frac{43}{7} - 6 = \frac{1}{7}.$$

We leave it as an exercise to verify that this is the correct solution.

7.2.2 The Square Subgame Theorem Revisited

In Section 7.1.3, we stated the square subgames theorem without proof. We can now prove this theorem. Suppose that A is a payoff matrix for a zero-sum game of size $m \times n$. Imagine solving the game using linear programming via the simplex algorithm. In the initial tableau, there are $m + 1$ rows, the first m of them labeled with slack variables S_i. Suppose that in the final tableau, there are k slack variables that are basic (so they are still listed in the left column). This means that there are $m - k$ decision variables that are basic. The basic decision variables correspond to the strategies for the column player that are active. Thus, the column player has $m - k$ active strategies.

Now, by complementary slackness Proposition 4.8, we know $S_i \cdot M_i = 0$ for all i. If S_i is a basic variable, its value is nonzero, whence $M_i = 0$. So if there are k basic slack variables, there must be k marginal values equal to 0. But the marginal values are the decision variables for the dual problem, which is the problem that the row player must solve. If k of the M_i are 0, that leaves $m - k$ of them to be nonzero. But these nonzero marginal values correspond to the active strategies for the row player.

The assumption that $S_i \neq 0$ if S_i is a basic variable is almost always true, but it can fail if the problem is degenerate. Nevertheless, degeneracy is marked by the fact that not only is $S_i \cdot M_i = 0$, but they are both zero simultaneously. So either way, if S_i is basic, then $M_i = 0$, and the argument holds that there are $m - k$ nonzero marginal values.

We have shown that if there are exactly $l = m - k$ active strategies for the column player, then there must also be (exactly) l active strategies for the row player. Thus, the solution to the game is the same as an $l \times l$ subgame, which is a square subgame. This proves the theorem.

Exercises

1. Consider a zero-sum game with payoff matrix

$$A = \begin{bmatrix} 5 & -3 \\ -3 & 1 \end{bmatrix}.$$

   a. Write down the linear programming problems for both the row player and the column player.
   b. Solve the problems from Part a using graphing in the decision space, and use you answers to solve the game.
   c. Instead, solve the column player's game using the simplex algorithm, and use your answer to solve the game for both players.
   d. Solve the game by another method (for example, by equalizing expectation), and verify that your solutions in Parts b and c are correct.

2. Consider a zero-sum game with payoff matrix

$$A = \begin{bmatrix} -2 & -1 \\ -3 & 2 \\ -5 & 5 \end{bmatrix}.$$

   a. Write down the linear programming problems for both the row player and the column player.
   b. Your problem for the column player should have only two decision variables. Solve it by graphing in the decision space. Did you obtain an optimal pure strategy? Explain.
   c. Your problem for the row player should have three decision variables, so you can't solve it by graphing in the decision space. However, based on your answer from Part a, you should be able to tell which square subgame is equivalent to the original game. Based on that, solve the game for the row player as well.
   d. Solve the game for the column player via the simplex algorithm and thereby completely solve the original game for both players. Your answer should agree with what you got in Parts b and c.

3. Consider a zero-sum game with the payoff matrix

$$A = \begin{bmatrix} -20 & 19 \\ 4 & -6 \end{bmatrix}.$$

Write down the linear programming problems that both players must solve, and solve the game using the simplex algorithm.

4. Consider a zero-sum game with the following payoff matrix:

$$A = \begin{bmatrix} 1 & -2 & 3 & -4 \\ -5 & 1 & -6 & 2 \end{bmatrix}.$$

a. Write down the linear programming problems that both players must solve.

b. One of your two problems has just two decision variables. Solve that linear programming problem by graphing in the decision space.

c. Use your answer to determine which 2×2 subgame is equivalent to the original game, and solve the game for both players (you may use any method you like on the 2×2 subgame).

d. Solve the game using the simplex algorithm instead.

5. Recall that if a zero-sum game has a payoff matrix such that $A^T = -A$, the game is said to be *symmetric*. (This is a bit of a misnomer because the matrix itself is said to be *antisymmetric*, but don't worry about that.) An example is the game rock paper scissors. Prove that a symmetric game is always a fair game; that is, it has value $v = 0$. [**Hint:** Just as we did in rock paper scissors, argue that in any symmetric game, we must have $\widehat{p} = \widehat{q}^T$, that is, both players have the same optimal strategy. Then apply Part 3 of the minimax theorem, and take the transpose of both sides.]

6. We mentioned in the text that it does not matter what you translate the payoff matrix by, as long as the result has all positive entries. To illustrate this, solve the game in Exercise 1 again, but use a different constant than you used in Exercise 1 to translate the matrix. (For example, if you added 4 in Exercise 1, try adding 5 or 6 instead.)

7. In Section 6.1, in Exercise 6, we considered a finite version of the Antares vs. Bellatrix game, where each company was deciding between two prices. Solve that game using linear programming.

8. In Section 6.1, in Exercise 7, we considered a finite version of the Antares vs. Bellatrix game, where one of the companies was deciding between two prices and the other between three prices. Solve that game using linear programming.

7.3   Using a Computer to Solve Constant-Sum Games

In Section 5.4, we showed how to solve linear programming problems using either Mathematica or Excel. We now apply this to solving constant-sum games. We begin by solving with Mathematica.

We illustrate the process by solving rock paper scissors. Here is the payoff matrix from Example 7.3:

$$A = \begin{matrix} & \begin{matrix} \text{Rock} & \text{Paper} & \text{Scissors} \end{matrix} \\ \begin{matrix} \text{Rock} \\ \text{Paper} \\ \text{Scissors} \end{matrix} & \begin{bmatrix} 0 & -1 & 1 \\ 1 & 0 & -1 \\ -1 & 1 & 0 \end{bmatrix} \end{matrix}.$$

We begin by translating the game so that the payoff matrix has all positive entries. Adding 2 to each entry, we obtain

$$B = \begin{bmatrix} 2 & 1 & 3 \\ 3 & 2 & 1 \\ 1 & 3 & 2 \end{bmatrix}.$$

The column player's linear programming problem is then

maximize $z = u_1 + u_2 + u_3$

subject to

$2u_1 + u_2 + 3u_3 \leq 1$

$3u_1 + 2u_2 + u_3 \leq 1$

$u_1 + 3u_2 + 2u_3 \leq 1$

$u_i \geq 0,\ i = 1,2,3.$

The row player's linear programming problem is the dual of this:

minimize $w = t_1 + t_2 + t_3$

subject to

$2t_1 + 3t_2 + t_3 \geq 1$

$t_1 + 2t_2 + 3t_3 \geq 1$

$3t_1 + t_2 + 2t_3 \geq 1$

$t_i \geq 0,\ i = 1,2,3.$

In Section 5.4, we observed that Mathematica had two different commands, the LinearProgramming command and the DualLinearProgramming command. Since we want the solution to both problems, clearly the DualLinearProgramming command is the one to use for game theory. When we solve a game using the simplex algorithm by hand, we focus our attention on the problem for the column player, which is a standard form maximization. Mathematica uses a minimization as a standard format, so if we were solving a maximization using Mathematica, we would convert it to a minimization by multiplying through (both the objective function and the constraints) by a factor of -1.

Such a technique would certainly work here, but there is a more direct way to solve the game. The row player's problem is already a standard form minimization. So if we apply the DualLinearProgramming command directly to the row player's problem, we do not have to convert anything. Following the proper syntax for the command that we learned in Section 5.4, we would enter the following:

DualLinear Programming[{1,1,1},{{2,3,1},{1,2,3},{3,1,2}},{1,1,1}]

Entering this, Mathematica returns the output:

$$\{\tfrac{1}{6}, \tfrac{1}{6}, \tfrac{1}{6}\}, \{\tfrac{1}{6}, \tfrac{1}{6}, \tfrac{1}{6}\}, \{0,0,0\}, \{0,0,0\}$$

Recall that the first ordered list of numbers is the solution to the primal problem, and the second ordered list is the solution to the dual problem, and we can ignore that last two sets that record slack variables and other information not important to the game. Thus,

$$t_1 = t_2 = t_3 = \frac{1}{6}$$
$$u_1 = u_2 = u_3 = \frac{1}{6}.$$

To find the solution to the game, we need the value of the objective function, which is

$$t_1 + t_2 + t_3 = \frac{1}{2} = u_1 + u_2 + u_3.$$

It follows that the value of the game $v_B = 2$ and that we obtain $\widehat{p}$ and $\widehat{q}$ by multiplying all these decision variables by v_B. Thus,

$$\widehat{p} = \begin{bmatrix} \frac{1}{3} & \frac{1}{3} & \frac{1}{3} \end{bmatrix}$$

$$\widehat{q} = \begin{bmatrix} \frac{1}{3} \\ \frac{1}{3} \\ \frac{1}{3} \end{bmatrix}$$

$$v_A = v_B - 2 = 2 - 2 = 0.$$

This agrees with the solution we obtained by equalizing expectations. Each player should play each strategy with equal probability $\frac{1}{3}$, and it is a fair game.

Of course, this is a symmetric game, so the solution had a particularly easy form. Let's try an example that is not symmetric. Suppose the payoff matrix is

$$A = \begin{bmatrix} -5 & -3 & 0 & 1 & 4 \\ 6 & 5 & 3 & -2 & -3 \\ 7 & 5 & 4 & -2 & -1 \end{bmatrix}.$$

We previously solved this game immediately following Corollary 7.1 by reducing it to a 2×2 subgame. Here, we don't need to check for dominance or saddle points. We begin by translating the game by adding 6 to all the payoffs, obtaining

$$B = \begin{bmatrix} 1 & 3 & 6 & 7 & 10 \\ 12 & 11 & 9 & 4 & 3 \\ 13 & 11 & 10 & 4 & 5 \end{bmatrix}.$$

For practice, like we did above for rock paper scissors, the reader should write down both linear programming problems. We will skip that step and go right to the Mathematica command. The key is to remember that the payoff matrix is (part of) the condensed tableau for the column player, so you read across the rows when writing the constraints, while the transpose of the payoff matrix is (part of) the condensed tableau for the row player, so you read down the columns when writing the constraints. Thus, the command to enter is as follows:

DualLinearProgramming[{1,1,1},

{{1,12,13},{3,11,11},{6,9,10},{7,4,4},{10,3,5}},{1,1,1,1,1}]

(Enter it on one line in Mathematica.) Mathematica returns the following:

$$\left\{\left\{\frac{3}{29},0,\frac{2}{29}\right\},\left\{\frac{1}{29},0,0,\frac{4}{29},0\right\},\left\{0,\frac{1}{29},0\right\},\{0,0,0\}\right\}$$

So the value is the reicprocal of the objective function $\frac{3}{29}+0+\frac{2}{29}=\frac{5}{29}$:

$$v_B=\frac{29}{5}.$$

Multiplying through by v_B, it follows that

$$\widehat{p}=\left[\begin{array}{ccc}\frac{3}{5}&0&\frac{4}{5}\end{array}\right]$$

$$\widehat{q}=\left[\begin{array}{c}\frac{1}{5}\\0\\0\\\frac{4}{5}\\0\end{array}\right]$$

$$v_A=v_B-6=\frac{29}{5}-6=-\frac{1}{5}.$$

This agrees with the previous solution from Section 7.1.2.

Next, let's use Excel to solve constant-sum games. Again, we use what we learned in Section 5.4. We can illustrate with the same two games we solved earlier. We again begin by translating the payoff matrix until it has all positive entries. For rock paper scissors, the translated matrix we used earlier is

$$B=\left[\begin{array}{ccc}2&1&3\\3&2&1\\1&3&2\end{array}\right].$$

The Solver feature in Excel can solve either the maximization or the minimization, so it doesn't matter which one we solve. Just be sure to include the sensitivity report in the solution, which will contain the solution to the dual problem under the shadow prices column. We will work with the maximization problem (the column player's problem) so that we don't have to bother taking the transpose of the condensed tableau.

So, entering the data on the spreadsheet as we did in Section 5.4, we obtain what is shown in the screenshot in Figure 7.6.

The reader will recall that the cells we have colored in column F contain formulas. We are using the variables x,y,z instead of u_1,u_2,u_3 because it is easier to enter variables in Excel without subscripts. The formula in the objective cell is just $x+y+z$, which we must maximize. Recall that it is entered in the cell using the sumproduct function, using the arrays in the cells B4–D4 and B5–D5. The other formulas are entered similarly, multiplying the appropriate arrays.

Once this data is entered, we open the Solver and enter the required information. The set objective cell is F5 in our example; make sure the "Max" radio button is checked. Changing variable cells are the cells containing the values of x,y,z, which are B4–D4 in our example. Add the three constraints, be sure the checkbox "make unconstrained variables non-negative" is checked, and the method of solution is "Simplex LP." The result appears as shown in Figure 7.7.

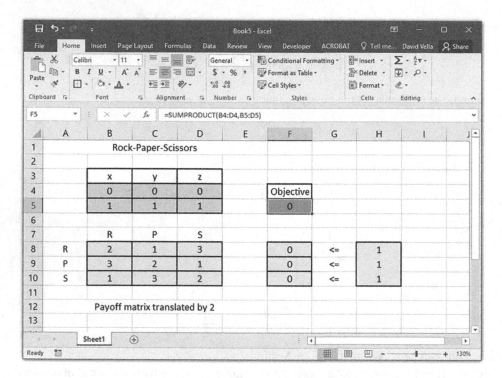

Figure 7.6 The spreadsheet for the translated rock paper scissors game.

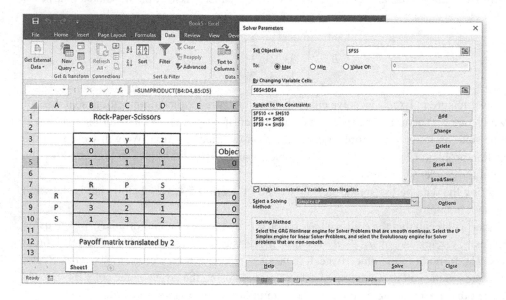

Figure 7.7 Fill in the Solver pop-up window as usual.

Then click the Solve button, and highlight the answer report and the sensitivity report in the new dialog box, as shown in Figure 7.8.

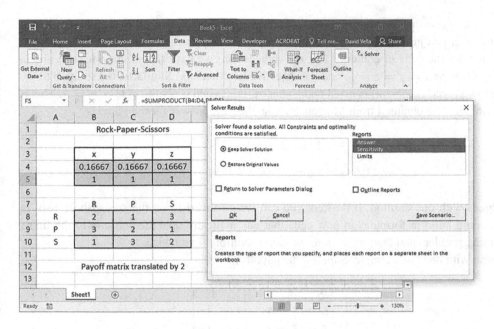

Figure 7.8 Select both the answer report and the sensitivity report.

Then click OK. As we did in Section 5.4, we've converted the answers to fraction format instead of decimals, and we've copied some of the answer report data and some of the sensitivity report data on Sheet 1 so we can view everything at once, as shown in Figure 7.9.

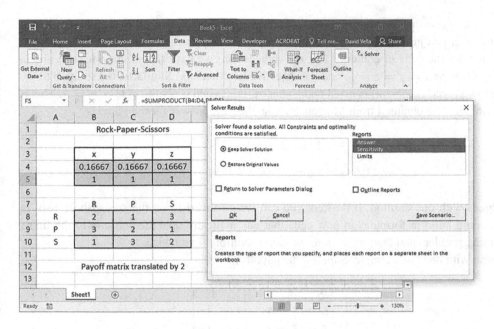

Figure 7.9 The relevant parts of both reports are copied to the spreadsheet displaying the solution.

Because we do not need the values of the slack or surplus variables, we did not copy the entire answer report or sensitivity report. One can see both the final values of x, y, z and the values of the shadow prices on the sensitivity report, so for solving games, you may opt to skip the answer report altogether and just highlight the sensitivity report when you click OK.

Reading from the cells in column M of the spreadsheet, we have the solution to the column player's problem as $x = y = z = \frac{1}{6}$, and the objective function is $\frac{1}{2}$. Similarly, the solution to the dual (the row player's problem) is listed under shadow prices and is also $t_1 = t_2 = t_3 = \frac{1}{6}$. This is the same solution we obtained from Mathematica, so the final answer for the game is

$$\hat{p} = \begin{bmatrix} \frac{1}{3} & \frac{1}{3} & \frac{1}{3} \end{bmatrix} = \hat{q}^T$$
$$v_A = v_B - 2 = 2 - 2 = 0,$$

which agrees with our solution. If desired, you can record this data on the spreadsheet also. In fact, you could build in formulas in new cells that automatically take the solution to the linear programming problem and convert them to the game solution, if you wish.

We finish with one last example. Suppose the payoff matrix is

$$A = \begin{bmatrix} -5 & -3 & 0 & 1 & 4 \\ 6 & 5 & 3 & -2 & -3 \\ 7 & 5 & 4 & -2 & -1 \end{bmatrix},$$

the other game we solved above using Mathematica. We translated the matrix by adding 6 to everything. Just to illustrate that it does not matter what we translate by, as long as we obtain positive numbers, let's add 10 this time, to obtain the matrix:

$$B = \begin{bmatrix} 5 & 7 & 10 & 11 & 14 \\ 16 & 15 & 13 & 8 & 7 \\ 17 & 15 & 14 & 8 & 9 \end{bmatrix}.$$

We enter this on a new spreadsheet as part of the condensed tableau for the column players problem, as shown in Figure 7.10.

Figure 7.11 shows the final spreadsheet after using Solver - and after converting certain cells from decimals to fractions.

Reading off the solution to the linear programming problems, we have for the column player:

$$u = \frac{1}{49}, \, v = 0, \, w = 0, \, x = \frac{4}{49}, \, y = 0$$
$$z = \frac{5}{49}.$$

We have for the row player:

$$t_1 = \frac{3}{49}, \, t_2 = 0, \, t_3 = \frac{2}{49}.$$

To convert to the solution of the game, we multiply through by $v_B = \frac{1}{z} = \frac{49}{5}$ and then subtract 10 to obtain v_A. The result is

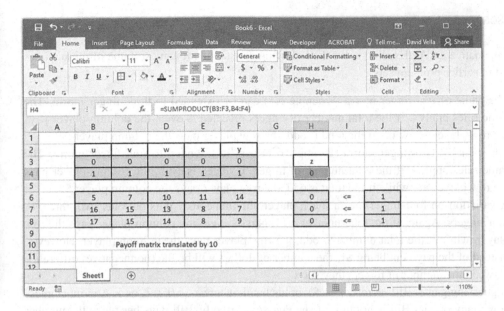

Figure 7.10 The spreadsheet for the translated game.

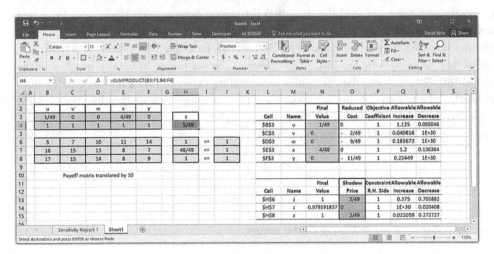

Figure 7.11 The solution to the translated game.

$$\widehat{p} = \begin{bmatrix} \frac{3}{5} & 0 & \frac{4}{5} \end{bmatrix}$$

$$\widehat{q} = \begin{bmatrix} \frac{1}{5} \\ 0 \\ 0 \\ \frac{4}{5} \\ 0 \end{bmatrix}$$

$$v_A = \frac{49}{5} - 10 = -\frac{1}{5}.$$

This soluton agrees with our previous one.

Exercises

Use Mathematica and/or Excel to solve the following constant-sum games. Some of these games appeared in earlier problem sets.

1. The game of two-finger morra (Example 6.1) can be extended to three-finger morra. The same rules apply – if the total number of fingers revealed is even, the column player pays the row player $1, while if the total is odd, the row player pays the column player $1.

2. The game of two-finger morra (Example 6.1) can be extended to four-finger morra. The same rules apply – if the total number of fingers revealed is even, the column player pays the row player $1, while if the total is odd, the row player pays the column player $1.

3. A variation on four-finger morra. Each player simultaneously puts out one, two, three, or four fingers. Again, if the total number of fingers is even, the column player pays the row player, while if the total number is odd, the row player pays the column player. However, instead of the payment being $1, the number of dollars will be the (nonnegative) difference between the number of fingers they show.

4. A guessing game. Bill secretly puts x in his hand, where $x = 0, 1, 3$, or 5. Ann must guess how many dollars Bill is holding. If she guesses correctly, Bill pays her twice the number of dollars he is holding, plus one dollar more; that is, he pays her $(2x + 1)$. If she guesses incorrectly, she must pay Bill the (positive) difference between her guess and the actual number, plus a $2 penalty. Assume Ann is the row player.

5. Consider the marketing game Antares vs. Bellatrix of Example 6.5. Construct a finite version of this game as follows: Instead of allowing the set prices to be any number in the interval $[80, 120]$, assume the companies are debating between just a few specific choices (all of which lie within this interval). Assume that both companies are deciding between the prices $80, $86, $96, $106, and $120.

6. Solve the game with the following payoff matrix:

$$A = \begin{bmatrix} 1 & 2 & -1 & -2 \\ -3 & 0 & 1 & 3 \\ -6 & -5 & 0 & 2 \end{bmatrix}.$$

7. Solve the game with the following payoff matrix:

$$A = \begin{bmatrix} -2 & -1 \\ -3 & 2 \\ -5 & 5 \end{bmatrix}.$$

8. Solve the game with the following payoff matrix:

$$A = \begin{bmatrix} 0 & -1 & 3 & -4 & 7 \\ -2 & 1 & 0 & -1 & 3 \\ 1 & -3 & -4 & 5 & -1 \\ 3 & 3 & -2 & 0 & -5 \\ -4 & -3 & 6 & -2 & 8 \end{bmatrix}.$$

(Answers in decimal form are acceptable.)

7.4 Variable-Sum Games

Now that we have shown how to solve any constant-sum game, we take a look at variable-sum games. That is, we no longer assume that the payoff for the column player plus the payoff for the row player is a constant c, which means that knowing the row player's payoff is not sufficient to determine the column player's payoff. Thus, we will need to write all payoffs as ordered pairs (x, y), where x is the row player's payoff and y is the column player's payoff. We will see that some concepts carry over from the constant-sum theory, and some do not. In particular, for some games, there is no universally accepted notion of a solution to the game.

7.4.1 Dominance and Nash Equilibrium Points

First, we observe that some concepts from constant-sum theory carry over with very little change. We urge the reader to go back and review the definition of one strategy dominating another, which we derived during our discussion of the Battle of the Bismarck Sea. This definition does not change, so the concept of a dominant strategy applies equally well to variable-sum games.

What does change, though, is that the corollary immediately following this definition is no longer valid. How you tell if a strategy is dominant requires comparing payoffs in a row or in a column, but these payoffs are now ordered pairs, not single numbers. Since the first coordinates are the payoffs for the row players, in order to tell if one row dominates another, you just look at the first coordinates. If the first coordinate of every entry in row A is at least as large as the corresponding first coordinate in row B, then we say row A dominates row B. Similarly, if the second coordinate of every entry in column A is at least as large as the corresponding second coordinate in column B, we say column A dominates column B. The following payoff matrix illustrates this:

$$A = \begin{bmatrix} (5,5) & (2,11) \\ (10,2) & (3,4) \end{bmatrix}.$$

Looking at the first coordinates, we see $10 > 5$ and $3 > 2$, so the second row dominates the first. Similarly, looking at the second coordinates, we see $11 > 5$ and $4 > 2$, so the second column dominates the first. In fact, we have already seen an example of dominance in a variable-sum game when we considered the second version of the Anderson Family Saturday chores game in Section 6.2.

Furthermore, in that example, we also illustrated that the notion of a movement diagram is valid for variable-sum games. You still draw the vertical arrows pointing to the best payoff for the row player in a given column and the horizontal arrow pointing to the best payoff for the column player in a given row. The best payoff for the row player is now the ordered pair with the largest first coordinate, and the best payoff for the column player is the ordered pair with the largest second coordinate. So the movement diagram for the payoff matrix A above is as shown in Figure 7.12.

As usual, the fact that the arrowheads point in the same direction for each row indicates the second column is dominant, and the fact that the arrowheads both point down in each column indicates the second row is dominant.

Finally, we observe that the Definition 6.8 of a Nash equilibrium is also valid and unchanged for variable-sum games. It is still an outcome of the game where neither player stands to gain by a

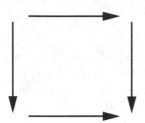

Figure 7.12 Movement diagram.

unilateral change in strategy. What is different now, however, is that there is no such thing as a saddle point, because the definition of a saddle point depended on the column player wanting smaller payoffs (for the row player). Now, rationality dictates that the column player wants larger payoffs for himself, which is determined by the second coordinate of the ordered pair. In a constant-sum game, having a larger second coordinate is equivalent to having a smaller first coordinate, so we could dispense with the second coordinate altogether and define a saddle point. For a variable-sum game, the two are not necessarily equivalent. So, while the notion of a Nash equilibrium still makes sense, the notion of a saddle point does not.

Also, we observe that an outcome of the game that is the intersection of dominant strategies is still a Nash equilibrium, just as in the case of a constant-sum game. The preceding game with payoff matrix A is an example. The outcome in the bottom-right corner (where both arrowheads point in the movement diagram), with payoffs $(3, 4)$, is a Nash equilibrium. Indeed, if either player unilaterally changes his or her strategy, his or her own payoffs decrease, so neither player can gain by a unilateral change of strategy. So does this point with payoffs $(3, 4)$ make a good "solution" to the game?

There is something unusual about this game and its alleged "solution." Observe that the outcome in the upper-left corner, with payoffs $(5, 5)$, is actually better for both players simultaneously! However, you can't arrive at this outcome from the Nash equilibrium by a *unilateral* change of strategy – it would require *both* players to change their strategy. Another problem is that the $(5, 5)$ outcome is not stable; as can be seen by looking at the movement diagram, each player can benefit by changing his or her strategy (unilaterally). But if one player does so, he or she will be disappointed if the other player does as well, which would more than cancel the added benefit he or she would get by a unilateral change.

This unusual situation comes up fairly often in variable-sum games, but it never happens in a constant-sum game. We will have more to say about this in the next subsection. Of course, it doesn't always come up. In the Anderson Family Saturday chores game we considered in Section 6.2, the outcome of both children choosing both chores was a Nash equilibrium (which was, in fact, the intersection of dominant strategies), with a payoff of $(10, 10)$. In that game, there was no other outcome that was better than that for both players. That is why, in Section 6.2, we said it looks as if we solved the Anderson chore game. There is no added benefit coming from a unilateral change of strategy at the $(10, 10)$ outcome, nor is it possible to make a joint change of strategies that can benefit both players simultaneously, at least not with pure strategies. However, before we can say whether or not we have solved that game, we need to consider what happens if we allow mixed strategies, the topic of the next subsection. Exercises on movement diagrams, dominant strategies, and (pure-strategy) Nash equilibria appear at the end of the section.

7.4.2 Mixed Strategies, Payoff Polygons, and Pareto Efficiency

With constant-sum games, we introduced mixed strategies as a way to handle games that were not strictly determined; that is, they had no saddle points, which were the same things as pure strategy Nash equilibria. The same idea works for variable-sum games. It is quite possible that a variable-sum game has no pure-strategy Nash equilibrium. Yet Nash proved that every game has at least one equilibrium point, so there must be a way to find one using mixed strategies. We'll show an example presently, but before embarking on the example, we caution the reader not to get his or her hopes up too much. In the example of the last subsection, we did find a Nash equilibrium (even one using pure strategies), yet we had reason to possibly reject that outcome as an alleged "solution" to the game.

Consider the game with the following payoff matrix:

$$A = \begin{bmatrix} (3,5) & (6,1) \\ (7,2) & (5,7) \end{bmatrix}.$$

The reader should check that none of the four outcomes is a Nash equilibrium (for example, draw a movement diagram). The concept of a mixed strategy is the same as before, as is the notation. Thus, playing $p = [0.4, 0.6]$ can be interpreted as playing the first row 40% of the time in the long run with repeated play or, equally well, as saying that on any one play of the game, the first row is played with probability 0.4. So, for example, if $p = [0.4, 0.6]$ and $q = \begin{bmatrix} 0.7 \\ 0.3 \end{bmatrix}$, how do we determine the expected payoff to each player? The formula $E(p,q) = pAq$ is still valid (assuming that the players choose simultaneously and independently, as usual), only this time, we must be careful about what we use for A. If you want the row players payoffs, you use just the first coordinates in the payoffs in A. To obtain the column players payoffs, in the constant-sum case, all you had to do was subtract from the constant. However, here you repeat the calculation pAq, this time using the second coordinates in A. So for the given example:

$$E_R = \begin{bmatrix} 0.4 & 0.6 \end{bmatrix} \begin{bmatrix} 3 & 6 \\ 7 & 5 \end{bmatrix} \begin{bmatrix} 0.7 \\ 0.3 \end{bmatrix} = \begin{bmatrix} 5.4 \end{bmatrix}$$

$$E_C = \begin{bmatrix} 0.4 & 0.6 \end{bmatrix} \begin{bmatrix} 5 & 1 \\ 2 & 7 \end{bmatrix} \begin{bmatrix} 0.7 \\ 0.3 \end{bmatrix} = \begin{bmatrix} 3.62 \end{bmatrix},$$

where the subscripts indicate the player (R for the row player and C for the column player.) Or, writing the expected payoffs as an ordered pair, we have $E = (E_R, E_C) = (5.4, 3.62)$. The reader is asked to check (Exercise 7) that if $p = q^T = (0.5, 0.5)$, we obtain the payoffs $(5.25, 3.75)$ instead.

Recall the definition of the derived game. It is a new game where the pure strategies in the derived game are the same as the mixed strategies in the original game. Of course, the derived game is infinite, but we exploited this idea in Chapter 6 to prove a special case of the minimax theorem. The proof was based on the idea that for a constant-sum game, the derived game is also constant-sum and is strictly determined. In our example, note that in the first case, the sum of the payoffs is 9.02, and in the second case, the sum is 9.0. This shows, if it wasn't already obvious by considering pure strategies, that the derived game is also variable-sum. We haven't defined what we mean by a strictly determined game in the variable case yet, so we cannot yet see if the same approach will help us in the variable-sum case.

However, we can try to guess an approach. In the case when a constant-sum game was not strictly determined, we found the Nash equilibrium, at least in the 2×2 case, by equalizing the expected payoffs. Will the same approach help here? Well, there are two sets of payoffs now – which ones do we want to equalize in order to find the Nash equilibrium? Actually, we have already answered this question. Go back and read carefully the argument leading to Proposition 6.29. We must equalize our *opponent's* payoffs. If the game was constant-sum, equalizing one player's payoffs meant also equalizing the other's, so we didn't need to distinguish between them. But now it is important to distinguish between them. Thus, we reiterate Proposition 6.29 as follows to emphasize the variable-sum case:

PROPOSITION 7.7   *In a 2×2 variable-sum game, if each player plays a mix that equalizes his or her opponent's payoffs, the resulting outcome is a Nash equiilibrium.*

We illustrate with the preceding game. The row player must equalize the column player's payoffs, so we set $p = (x, 1 - x)$; compute pA, using the second coordinates in the ordered pairs for A; and set the two equal. Thus:

$$\begin{bmatrix} x & 1-x \end{bmatrix} \begin{bmatrix} 5 & 1 \\ 2 & 7 \end{bmatrix} = \begin{bmatrix} 3x+2 & 7-6x \end{bmatrix}.$$

So

$$3x + 2 = 7 - 6x$$
$$9x = 5$$
$$x = \frac{5}{9}.$$

Thus, the row player's mixed strategy is $p = \begin{bmatrix} \frac{5}{9} & \frac{4}{9} \end{bmatrix}$. The expected payoff to the column player is $3x + 2 = 3\left(\frac{5}{9}\right) + 2 = \frac{11}{3} \approx 3.6667$._

Similarly, the column player must equalize the row player's payoffs:

$$\begin{bmatrix} 3 & 6 \\ 7 & 5 \end{bmatrix} \begin{bmatrix} y \\ 1-y \end{bmatrix} = \begin{bmatrix} 6-3y \\ 2y+5 \end{bmatrix}$$

So

$$6 - 3y = 2y + 5$$
$$1 = 5y$$
$$y = \frac{1}{5}.$$

Thus, the column player's mixed strategy is $q = \begin{bmatrix} \frac{1}{5} \\ \frac{4}{5} \end{bmatrix}$, and the expected payoff to the row player is $2y + 5 = 2\left(\frac{1}{5}\right) + 5 = \frac{27}{5} = 5.4$.

So, if both players play these mixed strategies, then the expected payoff to both is $(E_R, E_C) = (5.4, 3.6667)$. And here is the key part: because of Remark 6.24 (or by checking the solution we just found), if either player deviates unilaterally from this mix, his or her own payoff does not change. Thus, there is nothing to gain by a unilateral change of strategies – which is exactly the definition of a Nash equilibrium!

Even though we successfully found a Nash equilibrium for this game, there are two problems with proposing this as the "solution" to the game. First, just like the game in the last subsection, there are other solutions that yield better payoffs for both players. For example, the reader is asked to check (Exercise 7) that when $p = \begin{bmatrix} 0.3 & 0.7 \end{bmatrix}$ and $q = \begin{bmatrix} 0.6 \\ 0.4 \end{bmatrix}$, the expected payoff to both is $(E_R, E_C) = (5.6, 3.82)$. Like the earlier example, we cannot reach this outcome from the Nash equilibrium by a unilateral change of strategies; it would require both players to change. There may be other mixed strategies that lead to even better payoffs.

It would be nice if we had a way to "see" all the possible payoffs at once. Fortunately, there is such a way. Following Straffin (1993), we call this the *payoff polygon*. To construct it, begin by plotting all the ordered pairs that represent the payoffs for the outcomes of the game. Then take the smallest convex set (see Section 3.3 for a discussion of convex sets) that contains these points, also called the *convex hull* of these points. The convex hull of a finite number of points is always a simple closed polygon. The points in the matrix represent the payoffs to each player when pure strategies are used, and the points in their convex hull represent all weighted averages of these points and consequently represent all the possible payoffs to each player when mixed strategies are used. The payoff polygon for this game is shown in Figure 7.13.

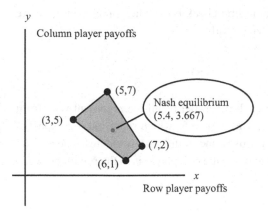

Figure 7.13 Payoff polygon.

The interior dot indicates the location of the Nash equilibrium. There are points in the payoff polygon that are to the right and above the dot, which correspond to expected payoffs that are better for both players. This leads to the following definition.

DEFINITION 7.8 *If a point in the payoff polygon is located so that there are other points that are better for both players, or better for one player and the same for the other player, we say the point is* Pareto inefficient. *If there are no such points, we say the point is* Pareto efficient.

The concept is named after Vilfredo Pareto (1848–1923), an Italian economist and social scientist. The reader should convince himself or herself that the only points on the payoff polygon that are Pareto efficient are those on an edge line with a negative slope, farthest away from the origin in the "northeast" direction. Indeed, if a point is any interior point, there are other points to the right and above it that are still within the payoff polygon. The same is true if the point is located on any edge with a zero slope, undefined slope, or positive slope, and it is also true of any point located on an edge with a negative slope that is not the farthest away from the origin in a northeast direction. So, in the preceding example, any point on the edge connecting $(5, 7)$ to $(7, 2)$ is Pareto efficient, and these are the only such points.

The Nash equilibrium in this game is not Pareto efficient, which is one of the reasons we hesitate to suggest it as a solution to the game. There is another reason, which is related to the assumptions we built into our game theory models. Perhaps it has already occurred to the reader. In order to find the Nash equilibrium, each player had to find the mix that equalized the opponent's payoff. But that violates our assumption of rationality. We stipulated that each player acts in a way to give himself or herself the best possible payoff, yet to find the equilibrium point, each player totally ignored his or her own payoffs!

Is there a way to analyze this game by having each player look at his or her own payoffs instead? Suppose a player ignores his opponent's payoffs and just plays the game as if it were zero-sum with that player's payoffs in the matrix. Let's illustrate this in the game with the row player. The row player's payoffs are

$$\begin{bmatrix} 3 & 6 \\ 7 & 5 \end{bmatrix}.$$

Playing as if this were a zero-sum game, we would first check for saddle points, and there are none. So now the row player tries to equalize her own payoffs:

$$pA = \begin{bmatrix} x & 1-x \end{bmatrix} \begin{bmatrix} 3 & 6 \\ 7 & 5 \end{bmatrix} = \begin{bmatrix} 7 - 4x & x + 5 \end{bmatrix}.$$

We know we are supposed to set these equal to each other, but recall the argument why from Section 6.4.2. We plot the two lines $E = 7 - 4x$ and $E = x + 5$, which represent the expected payoffs to the row player assuming the column player uses one of the pure strategies. Then, if the column player uses a mixed strategy, the expected payoff to the row player is in the region between the lines. In this game, the graph is shown in Figure 7.14.

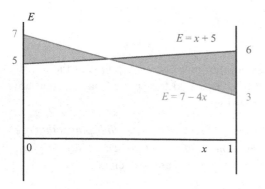

Figure 7.14 Expected payoffs for the row player.

In a zero-sum game against a skilled, rational opponent as the column player, the row player assumed the column player would force her payoffs to worst case, which is the bottom edges of the shaded region. Thus, it made sense to choose the maximin point because it was the best of the worst-case scenarios. However, the argument breaks down in the variable-sum case. If we assume the column player is rational, then he is not actively trying to force the row player's payoffs to the bottom edge. Instead, he is trying to force his own payoffs to be large and doesn't care at all about the row player's payoffs. That removes part of the reason for choosing the maximin point. Nevertheless, there is still a good reason to choose it. It is because it represents a guaranteed minimum payoff. Even if the column player is not trying to hurt the row player, his choices could, by coincidence, hurt her by forcing her payoffs toward the bottom. For this reason, we would still choose the maximin point because no matter what the column player does, the maximin is the minimum the row player would earn. In fact, it is more likely that the row player would actually earn a higher amount than that minimum than she would when actually playing a zero-sum game, precisely because the column player is not trying to deliberately hurt her.

Continuing along this line of reasoning, we solve for x:

$$7 - 4x = x + 5$$
$$2 = 5x$$
$$x = \frac{2}{5}.$$

To summarize, the row player who reasons as we did would play $p = \begin{bmatrix} 0.4 & 0.6 \end{bmatrix}$, and her guaranteed minimum payoff would be $v = x + 5 = 5.4$, regardless of what the column player did, and in general, she could actually earn higher than this minimum since the column player is not actively trying to minimize her expected payoff.

So, on the one hand, the row player can play $p = \begin{bmatrix} \frac{4}{9} & \frac{5}{9} \end{bmatrix}$ to equalize the column player's expected payoff (in the hopes that the column player approaches the game similarly, so that the outcome will be a Nash equilibrium). On the other hand, she could play $p = \begin{bmatrix} 0.4 & 0.6 \end{bmatrix}$ to guarantee herself a minimum payoff of 5.4 regardless of what the column player does (and may even end up with a higher payoff; however, in this case, we should not expect that the outcome is a Nash eqyilibrium).

It seems like we can use different mixed (or pure) strategies, depending on what our goal is. In Straffin (1993), special names are given to these various approaches. The strategy of equalizing the opponent's payoffs (which leads to a Nash equlibrium if both players do it) is called the *equalizing strategy*. The strategy where you just pretend you are playing a zero-sum game and ignore your opponent's payoffs is called the *prudential strategy*, and the guaranteed minimum payoff for that player is called the *security level* for that player.

The equalizing and the prudential strategies aren't the end of the story either. After all, the derived game has infinitely many possible strategies, and all we are doing is singling out a few for special attention. Since neither player knows what his or her opponent will do, it is not always clear what the best course of action is. For example, as pointed out in Straffin (1993), if one player suspects he or she knows what mix the opponent is using (for example, the prudential mix), he or she may very well counter that by using a pure strategy according to the expected value principle, called the *counter-prudential strategy* there. Table 11.1 on page 70 of Straffin (1993) is in fact just a small part of the payoffs in the derived game.

We conclude this subsection by finding the prudential strategy for the column player in this game. The column player's payoffs are

$$A = \begin{bmatrix} 5 & 1 \\ 2 & 7 \end{bmatrix}.$$

If the column player plays this as if it were a zero-sum game, we would first check for saddle points. However, the definition of a saddle point was based on the convention that we list the payoffs to the row player in a matrix. We have two choices here. One approach is to reverse the defintion of a saddle point so that it is an entry that is simultaneously the smallest in its column and largest in its row. The other approach may be less confusing since we do not have to change the definition of a saddle point: simply take the transpose of the matrix so that the column player *becomes* the row player. In this case:

$$A^T = \begin{bmatrix} 5 & 2 \\ 1 & 7 \end{bmatrix}.$$

Now this matrix has no saddle points under the usual definition. So we seek to equalize the payoffs. We can work with A^T or A for this step – it doesn't matter. Using A^T, we would write $q^T A^T = \begin{bmatrix} x & 1-x \end{bmatrix} A^T$ and set the two entries equal, or using A, we set the two entries of Aq equal. Note that $(Aq)^T = q^T A^T$, so we obtain the same result. Thus,

$$Aq = \begin{bmatrix} 5 & 1 \\ 2 & 7 \end{bmatrix} \begin{bmatrix} x \\ 1-x \end{bmatrix} = \begin{bmatrix} 4x+1 \\ 7-5x \end{bmatrix}.$$

So the minimax point occurs when

$$4x + 1 = 7 - 5x$$
$$9x = 6$$
$$x = \frac{2}{3}.$$

Thus, the column player would play $q = \begin{bmatrix} \frac{2}{3} \\ \frac{1}{3} \end{bmatrix}$, and his security level is $v = 4x + 1 = 4\left(\frac{2}{3}\right) + 1 = \frac{11}{3} \approx 3.6667$. What happens if both players play their prudential strategies? In Exercise 7, the reader is asked to verify that the payoffs will be $(E_R, E_C) = (5.4, 3.6667)$. These payoffs happen to be exactly the same as at the Nash equilibrium, but that is something one would not expect to happen in general. This does illustrate that different outcomes (i.e., different strategy choices) can have the exact same payoffs. Of course, we already knew this for pure strategies – for example, the Saturday chores at the Anderson household game had four different outcomes with the same payoff $(10, 10)$.

Since this outcome has the same payoffs as the Nash equilibrium, it also illustrates another fact. Playing the game prudentially can also lead to an outcome that is not Pareto efficient. So if we reject the Nash equilibrium as a "solution" because of Pareto inefficiency, we should also reject the prudential solution for the same reason, even though the latter approach does not violate the rational play assumption. Some other combinations of strategies in the derived game are considered in the exercises.

7.4.3 Goals of Play, the Relative Zero-Sum Game, and Strictly Determined Games

We have seen several approaches to analyzing variable-sum games, and even in some very simple cases, we were not sure what outcome to call the "solution" to the game. The reason for this is that in a variable-sum game, what's good for the individual player isn't necessarily good for the group (of players) together. If you are at an outcome of a constant-sum game and move to a different outcome, which benefits one of the players, then the other player necessarily loses something, precisely because the sum of the payoffs is constant. In a variable-sum game, there may be moves that benefit both players simultaneously (a "win-win" situation). Thus, such games are not strictly competitive, as we have already noted, and it may be partially in the players' interests to cooperate with one another. Notice that in a constant-sum game, the lack of moves to an outcome that simultaneously benefits both players means exactly that every outcome of a constant-sum game is Pareto efficient.

The fact that the individual's interests do not necessarily align with the interests of the group makes it difficult to say what rational behavior is, exactly, in a variable-sum game. One way to try to resolve this difficulty is to consider varying what the goal of a game is. We are free to change the objective function in a linear programming problem, so why not change what the objective of a game is? We can break down our possibilities into goals that favor the individual vs. goals that favor the group. Within each category, there are a couple of different possibilities.

Goals that would benefit the group include the following:

1. Trying to get an outcome with the highest total sum of the payoffs.
2. Trying to get an outcome which is Pareto efficient.

Goals that would benefit the individual include the following:

1. Trying to obtain the highest-possible minimum payoff
2. Trying to obtain an outcome in which your payoff most exceeds your opponent's payoff.
3. Trying to obtain a Nash equilibrium.

Let's look at these options more carefully. Trying to get an outcome that has the highest sum of the payoffs has two problems. The first is that it may require the two players to coordinate their play, which implies that the players communicate with one another. This violates our assumption that players act simultaneously and independently. We will address this in the next chapter when we change the assumptions under which we operate. The other problem is that the payoffs $(1, 19)$ and $(10, 10)$ both have a sum of 20, but one of them is not so good for the row player unless it is possible for the players to combine their winnings and agree how to split them. That also requires communication between the players and assumes that the payoffs can be combined and split however the players want, which is also an additional assumption. Since we don't seem able to implement this goal without additional and/or changed assumptions, we abandon it for now.

On the other hand, it may be possible to obtain an outcome that is a Pareto-efficient outcome without communicating or coordinating actions, so we keep this in mind as a possible goal. Note that if we do obtain an outcome that has the highest-possible sum of payoffs, then it necessarily is a Pareto-efficient outcome, so the first condition is a stronger condition than the second.

Turning now to the goals that benefit the individual player, we recognize the first one as what happens when the prudential strategy is played. Furthermore, the technique fits in what what we have previously called rational behavior. The only problem, as we have seen by example, is that it may not lead to a Pareto-efficient outcome. (It also may not lead to a Nash equilibrium, even if the payoffs are the same in a Nash equilibrium as in the example in the previous subsection.)

What about the second goal, of trying to have your payoffs exceed your opponent's by the widest margin? This also has some potential problems. For example, is a payoff of $(10, 0)$ really preferred by the row player to a payoff of $(50, 48)$, just because you beat the column player by 10 as opposed to beating him by 2? Our rationality assumption seems to dictate that the $(50, 48)$ would be preferred because 50 is better than 10. And yet it seems possible to imagine games for which the $(10, 0)$ payoff would be preferred. It's all a matter of how you measure the payoff. If the game is an economic competition between two companies and the outcomes are in millions of dollars, if you earned \$50 million, you probably wouldn't care if your competitor earned \$48 million. As far as you are concerned, \$50 million is a lot better than \$10 million. On the other hand, if these payoffs represented the outcomes of a wartime battle between two armies, even though 10 is less than 50, it seems that $(10, 0)$ is a pretty decisive victory, while $(50, 48)$ seems to almost be a tie. (It's difficult to say because we haven't specified what these numbers are measuring anyway.)

The reader may object that a wartime battle isn't a good example because that seems to be a game that should be zero-sum, as the players have opposed interests. But that is the beauty and advantage of using the second goal. It converts the variable-sum game into a zero-sum game! Indeed, if you are interested in how much your payoff exceeds the opponent's payoff, then if (r, s) is the payoff at a particular outcome, as far as the row player is concerned, the gain is $r - s$, and as far as the column player is concerned, the gain is $s - r$, and these two numbers sum to zero. We illustrate using the game from the last subsection. The payoff matrix is

$$A = \begin{bmatrix} (3,5) & (6,1) \\ (7,2) & (5,7) \end{bmatrix}.$$

If we adopt the second goal (for individual gain) that each player most wants to beat the opponent by the widest margin, at the $(6, 1)$ payoff, for example, the row player gains $6 - 1 = 5$ (which is the amount by which the column player loses). Replacing each ordered pair (r, s) with $r - s$, the resulting zero-sum game has payoff matrix:

$$B = \begin{bmatrix} -2 & 5 \\ 5 & -2 \end{bmatrix}.$$

In Exercise 7d, the reader is asked to verify that the solution to this zero-sum game is

$$p = \begin{bmatrix} 0.5 & 0.5 \end{bmatrix}$$

$$q = \begin{bmatrix} 0.5 \\ 0.5 \end{bmatrix}$$

$$v = 1.5.$$

Since this game is based on measuring your payoff relative to your opponent's payoff, let's call it the *relative zero-sum game* associated to the given variable-sum game.

If the players play these strategies p and q in the original variable-sum game, the expected payoffs are

$$\begin{bmatrix} 0.5 & 0.5 \end{bmatrix} \begin{bmatrix} 3 & 6 \\ 7 & 5 \end{bmatrix} \begin{bmatrix} 0.5 \\ 0.5 \end{bmatrix} = \begin{bmatrix} 5.25 \end{bmatrix}$$

and

$$[\ 0.5 \quad 0.5\ ] \begin{bmatrix} 5 & 1 \\ 2 & 7 \end{bmatrix} \begin{bmatrix} 0.5 \\ 0.5 \end{bmatrix} = [\ 3.75\ ].$$

That is, $(E_R, E_C) = (5.25, 3.75)$. We make the following observations. First, indeed, the row player's expected payoff exceeds the column player's expected payoff by 1.5, as suggested by the relative zero-sum game. Second, the sum of the payoffs is 9, which is not the largest sum; in fact, it is not even Pareto efficient, as can be seen if you plot this point in the payoff polygon in the last subsection. Thus, while playing the associated zero-sum game removes any ambiguity in what rational behavior means, it still leads to an outcome, like most of the methods that benefit the individual, which is not necessarily good for the group.

Finally, what about the third goal of trying to obtain a Nash equilibrium? We recognize how to do that, namely by playing the equalizing strategy (but both players must do so to reach the Nash equilibrium). We have already noted examples where the Nash equilibrium can fail to be Pareto efficient so that it would fail to satisfy either of the goals that benefit the group. But there is even another problem with Nash equilibrium points in variable-sum games. Consider a game with payoff matrix:

$$A = \begin{bmatrix} (3,2) & (4,6) \\ (5,3) & (0,1) \end{bmatrix}.$$

The reader should observe that there is no dominance in this game, but there are two pure-strategy Nash equilibria – namely the outcomes with payoffs $(5,3)$ and $(4,6)$. For example, draw the movement diagram to see this. Furthermore, the payoff polygon looks like the graph shown in Figure 7.15.

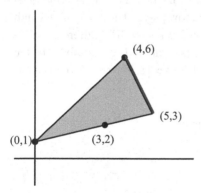

Figure 7.15 Two different Pareto-efficient Nash equilibriums.

The Pareto-efficient points are marked along the thick edge in the upper right. They include the two Nash equilibria. However, even though the Nash equilibria are Pareto efficient, there is still a problem. Notice the two Nash equilibria have different payoffs as $(4,6) \neq (5,3)$. That is, the Nash equilibria are not *equivalent* in the sense of Theorem 6.15, nor are they *interchangeable*. Indeed, the $(5,3)$ outcome is preferred by the row player, and the $(4,6)$ outcome is preferred by the column player. In order to obtain his or her preferred Nash equilibrium, each player chooses the strategy containing it, so the second row and second column are chosen. But then, the outcome is

$(0, 1)$, which is not a Nash equilibrium. (In fact, looking at the payoff polygon, one might argue that the $(0, 1)$ outcome is the worst possible, both from the individual and the group perspective!) So the problem is that in a variable-sum game, Theorem 6.15 can fail. Nash equilibria need not be equivalent and interchangeable.

In summary, there are all kinds of subtleties in variable-sum games, most of which arise from the conflict between outcomes that benefit the individual vs. outcomes that benefit the group. Nash may have proved that every game has an equilibrium point, but such points can fail to be Pareto efficient, and they can cause problems if there are more than one that are not equivalent and interchangeable.

However, by chance, if a game has the property that the outcome(s) that benefit the individual just happen to also be the one(s) that benefit the group, and if there is no confusion caused by by non-interchangeable and inequivalent equilibrium points, then it would be clear how to play such a game. It is these games that we will refer to as strictly determined in the variable-sum case.

DEFINITION 7.9   *A variable-sum game is said to be* strictly determined *if*

1.   *It has a unique Pareto-efficient Nash equilibrium, or*
2.   *If it has more than one Pareto-efficient Nash equilibrium, then they are all equivalent and interchangeable.*

In Straffin (1993), such games are called *solvable in the strict sense*.

Consider the following example. Robby and Jim are two botany students assigned to do a joint experimental lab growing hyacinths in a greenhouse. They each are to grow a different strain of hyacinth but must grow them together using the same combination of soil nutrients and pesticides. Robby can choose between two different soil nutrients, Densmore's A mix or Densmore's B mix. Jim can choose between two different pesticides, Rayman's Aphid-B-Gone and Zarek's Unknown Soldier Ants. We can model this as a game with Robby being the row player. His strain of hyacinth is known as *Spanish caravan*, and Jim's strain is known as *Indian summer*. The different strains are affected differently by various combinations of nutrients and pesticides. The payoff will be the number of biomass units of hyacinths that are harvested after five weeks. Based on trial runs of the experiment, the expected payoffs are given by the matrix:

$$
A = \begin{array}{c} \\ A\text{-mix} \\ B\text{-mix} \end{array} \begin{array}{c} \begin{array}{cc} \text{Aphid-} & \text{Unknown} \\ \text{B-Gone} & \text{Soldier} \\ & \text{Ants} \end{array} \\ \left[\begin{array}{cc} (8,7) & (9,4) \\ (6,5) & (1,9) \end{array} \right] \end{array}.
$$

The reader can check that the first row is dominant, so by higher-order dominance, the outcome with payoff $(8, 7)$ is a Nash equilibrium. The payoff polygon looks like the graph in Figure 7.16.

The Pareto-efficient points are on two edges of the polygon connecting $(1, 9)$ to (8.7) and connecting $(8, 7)$ to $(9, 4)$. Clearly, $(8, 7)$ is the only Nash equilibrium in pure strategies, and it happens to be Pareto efficient. If there were a Nash equilibrium using mixed strategies as well, it would be located at an interior point, so it would not be Pareto efficient. Therefore, this game is strictly determined. Robby should choose the A-mix, and Jim should choose the Aphid-B-Gone pesticide.

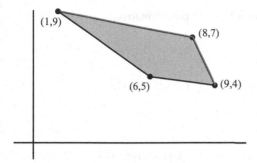

Figure 7.16 The payoff polygon for the botany students.

So, this game is strictly determined because individual gain aligns with collective gain. In fact, notice that the outcome with payoff $(8, 7)$ is actually the highest sum of the payoffs possible, 15, so this solution is the best for the group according to the stronger first condition as well. What about the relative zero-sum game? Since the two students are supposed to cooperate in the experiment, the relative zero-sum game, which is based on strict competition, is probably not how one would expect to analyze this game. Nevertheless (maybe each student is secretly trying to impress the professor with how much better a botanist they are than the other student), if we analyze that game, we obtain the payoff matrix:

$$B = \begin{bmatrix} 1 & 5 \\ 1 & -8 \end{bmatrix}.$$

It is interesting to note that this game also has first row dominant and a Nash equilibrium (saddle point) in the upper-left corner. Thus, we arrive at the same solution as using the "strictly determined" approach for the variable-sum game. That raises an interesting question to close out this section: if a variable-sum game is strictly determined, must it always be the case that the relative zero-sum game is also strictly determined with the same solution?

Exercises

1. Consider the following payoff matrix:

$$A = \begin{bmatrix} (0, 5) & (4, 2) \\ (3, -1) & (5, 4) \end{bmatrix}.$$

   Draw the movement diagram. Determine all instances of dominance, and determine if there is a pure-strategy Nash equilibrium.

2. Follow the same directions as Exercise 1 for the following payoff matrix:

$$B = \begin{bmatrix} (1, 3) & (4, 2) \\ (2, 1) & (3, 4) \end{bmatrix}.$$

3. Follow the same directions as Exercise 1 for the following payoff matrix:

$$C = \begin{bmatrix} (1,3) & (3,4) \\ (2,1) & (4,2) \end{bmatrix}.$$

4. Follow the same directions as Exercise 1 for the following payoff matrix:

$$D = \begin{bmatrix} (1,3) & (2,1) \\ (4,2) & (3,4) \end{bmatrix}.$$

5. Draw the payoff polygons for the payoff matrices $A-D$ of Exercises 1–4. In each case, determine if any Nash equilibrium your found is Pareto efficient or not. Determine if the game is strictly determined. Notice that different payoff matrices can have the same payoff polygon!

6. Among those games in Exercises 1–4 that are strictly determined, does the solution satisfy the first condition to benefit the group (that the outcome has the highest-possible sum of payoffs)?

7. Consider the game from Section 6.4.2 with payoff matrix

$$A = \begin{bmatrix} (3,5) & (6,1) \\ (7,2) & (5,7) \end{bmatrix}.$$

a. Verify that if both players play using a 50/50 mix, so $p = q^T$, then the expected payoff is $(E_R, E_C) = (5.25, 3.75)$, as claimed in the text.

b. Verify that if $p = \begin{bmatrix} 0.3 & 0.7 \end{bmatrix}$ and $q = \begin{bmatrix} 0.6 \\ 0.4 \end{bmatrix}$, then the expected payoff is $(E_R, E_C) = (5.6, 3.82)$, as claimed in the text.

c. Verify that if both players play their prudential strategies (determined in the text), then the payoffs will be $(E_R, E_C) = (5.4, 3.6667)$.

d. Verify the solution to the relative zero-sum game is what is claimed in the text.

8. Continuing with the game of Exercise 7, we intend to see what happens in the derived game when players try various strategies. We have already determined in the text what the payoffs are if both players use the equalizing strategy (and we know this outcome is a Nash equilibrium). Also, in Exercise 7c, we determined the payoffs if they both use their prudential strategy.

a. Determine the payoffs if the row player uses the equalizing strategy and the column player plays prudentially.

b. Determine the payoffs if the row player uses the prudential strategy and the column player uses the equalizing strategy.

c. Determine the payoffs if the row player uses the prudential strategy and column player (guessing correctly that the row player is playing prudentially) uses the expected value principle to respond with a pure strategy.

d. Determine the payoffs if the row player uses the equalizing strategy and the column player (guessing incorrectly that the row player is playing prudentially) uses the expected value principle to respond to the prudential strategy with a pure strategy.

e. Determine the payoffs if each player thinks the other is playing prudentially and uses the expected value principle to respond.

f. Suppose the row player is using $p = (x, 1 - x)$, and the column player is using $q = \begin{pmatrix} y \\ 1 - y \end{pmatrix}$. Find a formula for the expected payoffs to each player (E_R, E_C). Your solution should be an ordered pair with a saddle surface equation of crossed type in each coordinate.

9. Consider the four games in Exercises 1–4, with payoff matrices A, B, C, and D. For each game, find the relative zero-sum game, and solve it. Find an example of a strictly determined variable-sum game where the relative zero-sum game is also strictly determined, but with a different solution than predicted by the original variable-sum game. Also, find an example of a strictly determined variable-sum game whose relative zero-sum game is not strictly determined. These examples show (in two different ways) that the question posed at the end of the section has a negative answer.

10. For any of the four matrices A, B, C, and D from Exercises 1–4 that do not have a pure strategy Nash equilibrium, find the equalizing strategies that yield a mixed strategy Nash equilibrium and the corresponding expected payoffs.

11. a. Find a Nash equilibrium in the game with payoff matrix:

$$\begin{bmatrix} (4, -1) & (1, 6) \\ (-2, 3) & (2, 2) \end{bmatrix}.$$

b. Solve the relative zero-sum game. If both players use the optimal strategies suggested by the relative zero-sum game, what are their expected payoffs?

c. Find the expected payoffs if each player plays prudentially.

12. a. Show that the following game is strictly determined:

$$\begin{bmatrix} (1, 1) & (-1, 3) \\ (2, -2) & (4, 0) \end{bmatrix}.$$

b. Solve the relative zero-sum game, and show it is also strictly determined.

13. Consider the following payoff matrix:

$$\begin{bmatrix} (3, -2) & (1, 4) & (3, -2) \\ (-2, 4) & (-1, -2) & (4, -3) \end{bmatrix}.$$

Draw the payoff polygon, and show that the game is strictly determined.

14. Consider the following payoff matrix:

$$\begin{bmatrix} (4, -1) & (1, 2) & (-2, 1) \\ (0, 3) & (-2, 2) & (4, -3) \\ (-3, 1) & (-2, 5) & (1, 1) \end{bmatrix}.$$

Draw the payoff polygon, and show that the game is strictly determined. Show also that, in this game, every Pareto-efficient point has the same sum of the payoffs for both players. In particular, the solution also satisfies the first condition that benefits the group.

15. Consider the following payoff matrix:

$$\begin{bmatrix} (5, 1) & (-3, 2) & (6, 3) \\ (-2, 0) & (0, 4) & (1, -2) \end{bmatrix}.$$

a. Draw the movement diagram and the payoff polygon, and show that the game is not strictly determined.

b. If each player tries for his best Nash equilibrium point, what is the outcome of the game?

c. With only a minor change to the matrix, the game can have quite different dynamics. Suppose that instead of $(0,4)$, the payoff to the column player changes to c, so that $(0,c)$ becomes the payoff in the $2,2$ position of the matrix. Show that if $c > 0$, then there is no change to the movement diagram at all, so there are still two Nash equilibria (even if $c = 0$, there are still two Nash equilibria in pure strategies), but if $0 \le c \le \frac{7}{3}$, then the game becomes strictly determined.

d. Notice also that if $c \ge 0$, the second column dominates the first column. Reanalyze the game by deleting the dominated column in the original case, when $c = 4$, and in the case, when $0 \le c \le \frac{7}{3}$. You should notice that despite the fact that the payoff polygon changes, the predictions for the outcomes remain the same. (However, this is *not* always true! See chapter 11, exercise 4 of Straffin, 1993 for an example where deleting dominated strategies does affect the predicted outcome.)

16. Consider the variable-sum game with payoff matrix:

$$A = \begin{bmatrix} (3,2) & (-1,0) & (4,-3) \\ (4,1) & (-2,2) & (2,5) \end{bmatrix}.$$

a. Show there is no dominance in this game. Also show there is no pure-strategy Nash equilibrium. Draw the payoff polygon.

b. In the remainder of this exercise, we will find a mixed-strategy Nash equilibrium. Suppose the row player plays $p = (x, 1 - x)$. Using the column player's payoffs in A, compute $pA = (E_1, E_2, E_3)$. Each of the E_i represents the expected payoff to the column player if he uses the ith pure strategy. Plot all three lines in the E, x plane as we did in Section 7.1.2 for constant-sum games. We will show that we seek a point that is the intersection of two of the lines, just as in the constant-sum case. However, remember now that the column player wants E to be as large as possible since these are his payoffs, not the row player's. The top edge of the graph represents the best-case scenarios for the column player. Find the point that is the lowest point on this top edge (the minimax point).

c. Compute $pA = (E_1, E_2, E_3)$ using this specific p that you found in Part b. Observe that two of the coordinates are equal (because we equalized two of the expected payoffs in Part b), while the third is smaller (because we chose the minimax point in Part b. Now suppose the column player uses the mix $q = \begin{bmatrix} x \\ y \\ 1-x-y \end{bmatrix}$. Compute pAq, the expected payoff to the column player. You should obtain the expression $pAq = \frac{13}{9} - \frac{1}{3}y$. Since the column player wants to maximize his expected payoff, it is clear that the maximum occurs when $y = 0$. Conclude that the column player should never play the second column. It becomes an inactive strategy, just as in the constant-sum case.

d.   Now, since $y = 0$, the column player should play $q = \begin{bmatrix} x \\ 0 \\ 1-x \end{bmatrix}$. Compute

$Aq = \begin{bmatrix} E_1 \\ E_2 \end{bmatrix}$, this time using the row player's payoffs in A, to find the value of x that equalizes the expected payoffs $E_1 = E_2$ for the row player.

e.   Explain why the p you found in Part b together with the q you found in Part d yield a Nash equilibrium. Find the corresponding payoff (E_R, E_C), and plot it on the payoff polygon your found in Part a. Is the game strictly determined?

REMARK 7.10   *The techniques of Exercise 16 should work on any $m \times 2$ or $2 \times n$ variable-sum game and yield a Nash equilibrium that is the solution of a 2×2 square subgame, just as in the constant-sum case. This suggests that the square subgame theorem should hold for variable-sum games in order to find a Nash equilibrium. That is, we avoid the mention of a "solution" to the game and state the theorem as "a Nash equilibrium in a variable-sum game agrees with the Nash equilibrium of some square subgame." However, this Nash equilibrium is not generally Pareto efficient, so it may not be worth the effort to compute.*

8   Sensitivity Analysis, Ordinal Games, and *n*-Person Games

In this chapter, we begin to look at what happens if we change the assumptions under which we work. For example, what if play is not simultaneous, or what if communication is allowed? These changes in the assumptions can be regarded as a type of sensitivity analysis for the game theory models. We explore some of this in Section 8.1.

Another assumption that we can abandon is that we have exact numerical payoffs for each outcome. We may know we prefer one outcome to another without knowing exactly what the payoff is. If we can rank all the outcomes in order of preference, we are playing what is known as an *ordinal game* (the payoffs are ordinal numbers instead of real numbers). These games are explored in Sections 8.2 and 8.3.

Finally, we drop the assumption that there are only two players and look at *n*-person games in Sections 8.4 and 8.5. We only give a brief treatment of such games. A lot more could be said about them, but the techniques used to analyze them are quite different from what we have used to analyze two-person games, and they are better left for another book.

8.1   Sensitivity Analysis in Game Theory

For any model, a sensitivity analysis is a study of how sensitive the solutions are to a small change in the setup of the problem. We take a slightly wider point of view to include a small change in the *assumptions* under which a model is constructed and the resulting change to the solution *method* in addition to the solutions themselves.

The former type of sensitivity analysis can certainly be done. For example, suppose we have a payoff matrix for a constant-sum game that is not strictly determined. We might ask, if all the payoffs but one remain fixed and one changes slightly, how does that affect the optimal solution? We could even observe that, for small changes, the game remains not strictly determined, but for a larger change past a certain threshold, the game might suddenly become strictly determined, and we could define a stable range for the changing payoff based on this. For constant-sum games, one way to solve them is via linear programming, and we already have discussed sensitivity analysis in Chapter 4 and will return to it in Chapter 9, so we might regard this type of sensitivity analysis for a constant-sum game as an exercise in applying sensitivity analysis in linear programming.

For variable-sum games, it is less clear how to proceed, since linear programming does not play a role. We could at least formulate some basic questions, such as if we have a unique Pareto-efficient Nash equilibrium (so the game is strictly determined) and change one of the payoffs, how much change will the game admit while remaining strictly determined? Even though linear programming is not used to solve variable-sum games, we would guess that answering such questions would probably still involve solving systems of linear equations or inequalities and perhaps some matrix algebra.

However, the latter type of sensitivity analysis, when we vary the assumptions of the model, is vastly more interesting and also more useful, because games in real life are often not played under the restrictive assumptions of Chapters 6 and 7. If we want to have the most flexible and realistic models possible, it is imperative to ask what happens when we relax those assumptions of play. So this is the type of sensitivity analysis we study over the next few sections.

8.1.1   What if Play Is Not Simultaneous?

Our first assumption to vary is the assumption that moves are made simultaneously and independently. What is affected by this change?

First consider constant-sum games. One place where we used the assumption of independent choice of strategies was when we derived the formula $E = pAq$ for expected payoffs. If the strategies are not chosen independently, then the probability of an outcome of the game (which is an intersection of two strategy choices) is not necessarily the product of the individual probabilities of the strategy choices, so the derivation of this formula $E = pAq$ breaks down. However, this will not turn out to be a be a problem, because on any one play of the game, we still assume both players are rational and skillful and have the same information about the payoffs. Thus, whoever moves second will know what strategy his or her opponent used in that play of the game and can respond using the expected value principle on each play. It won't matter if the player going first is using a mixed strategy in the long run. Furthermore, whoever is going first knows that the second player will respond this way, and so this will help him or her decide the best strategic choice even though he or she is moving first.

Here is another way to look at it. The concept of a mixed strategy was proposed to deal with the situation that if you used a pure strategy, that was tantamount to tipping your hand about what you were going to do in repeated play. Mixing the strategies on different plays was a way to keep your choice a secret so that your opponent could not take advantage of you. But if the choices are not made simultaneously, whoever goes second always knows what strategy the other player has used, so repeated play is no different than playing the game once. The second player can make his choice contingent on what the first player does, and the player going first knows that. Thus, the entire notion of a mixed strategy is more or less useless.

We have already observed in Chapter 6 that if a constant-sum game is strictly determined, secrecy is not necessary, and both players should just go for the saddle point no matter what. Thus, we suspect that non-simultaneous play will have no effect on how these games are played. However, in games without saddle points, we have already observed that secrecy is necessary (which is why mixed strategies were developed).

If you remove simultaneity, then you are removing any hope of secrecy for whoever moves first. Thus, constant-sum games that are not strictly determined should be played differently than with simultaneous play, and whoever goes second should expect to benefit from the loss of simultaneity. Let's illustrate this with some examples.

Consider a zero-sum game with payoff matrix:

$$A = \begin{bmatrix} 5 & -1 & 0 \\ -1 & -3 & 2 \\ -3 & -2 & -4 \end{bmatrix}.$$

In this game, there is a saddle point $v = -1$ in the first row and second column. Suppose the row player goes second. If the column player chooses the first column, then the row player will choose the first row to get the 5 payoff (which is what the expected value principle recommends). If the column player chooses the second column, the row player again chooses the first row according to the expected value principle. If the column player chooses the third column, the row player chooses the second row. But now the column player knows all this and is skillful and rational. What should he do? In the first case, he loses 5; in the second he gains 1; and in the third, he loses 2. Thus, he should choose the second column, precisely because it is the best of these worst-case scenarios. The rational outcome is indeed the saddle point.

The reader can no doubt make a similar analysis assuming the row player moves first, and again, the outcome is the saddle point. This illustrates that strictly determined constant-sum games are played the same whether the moves are simultaneous or not. The next example is not strictly determined:

$$A = \begin{bmatrix} 0 & -1 & 1 \\ 1 & 0 & -1 \\ -1 & 1 & 0 \end{bmatrix}.$$

The reader should recognize that this is the payoff matrix for rock paper scissors. We already know the result of simultaneous play. It's a fair game ($v = 0$) with optimal mixed strategies for both players to mix all three options with equal probability. But if the either player goes first and plays rock, the other player obviously should play paper to win. If the first player plays paper, the second responds with scissors to win, and if the first plays scissors, the second responds with rock to win. On any one play of the game, no matter what the first player does, he loses. Again, the second player responds according the expected value principle. This time, since the worst-case scenarios are all a tie with outcome -1, there is no reason for whoever goes first to favor any of the strategies. Thus, the first player could always play the same thing, or she could vary the plays in any mix. It doesn't matter, she always will lose a dollar.

The same argument works for any matrix without a saddle point. Whoever goes second will always use the expected value principle to select the best payoff, regardless of what his opponent does. Thus, whoever goes first knows he will always wind up at the worst payoff in his row or column. Thus, unless all these worst cases are a tie as in rock paper scissors, he should choose the strategy that has the best of the worst-case payoffs.

Our final example illustrates this:

$$A = \begin{bmatrix} 1 & -2 & 1 & 1 \\ 2 & 1 & 0 & -1 \\ 0 & 3 & -4 & 3 \end{bmatrix}.$$

Suppose the column player goes first. Then he knows the row player will force him to lose 2 if he chooses the first column, to lose 3 in the second column, to lose 1 in the third column, and to lose 3 in the fourth column. Rationality dictates that he chooses the third column, so the outcome is the 1 in the first row and third column. If the row player goes first, she knows that choosing the first row will result in a loss of 2, choosing the second will result in a loss of 1, and choosing the third will result in a loss of 4, so rationality dictates that she chooses the second row, and the outcome is the -1 in the second row and fourth column.

What about variable-sum games? As usual, things are less clear. It's still true that the player who moves second should use the expected value principle. However, it is not necessarily the case that this player benefits. Consider the example:

$$A = \begin{bmatrix} (3,2) & (4,6) \\ (5,3) & (0,1) \end{bmatrix}.$$

We analyzed this game for simultaneous play in Section 7.4.3. We observed that the $(4,6)$ outcome and the $(5,3)$ outcome were both Nash equilibriums but that they were not equivalent or interchangeable. So with simultaneous play, each player tries to get his or her preferred Nash equilibrium (both of which are Pareto efficient), but the unfortunate result is the outcome is $(0,1)$, the worst-possible outcome. However, if either player goes first, he or she can choose the strategy containing his or her preferred outcome. Then the player going second would choose according to the expected value principle, and the results are $(5,3)$ if the row player goes first and $(4,6)$ if the column player goes first. Thus, regardless of who goes first, both players benefit, the result is a Nash equilibrium and is Pareto efficient, and whoever moves first gets his or her preferred Nash equilibrium. Playing the game simultaneously was something of a dilemma and a disaster, but playing it sequentially removed all the ambiguity and difficulty – except in deciding who moves first. Both players obviously want to do so!

Consider this example:

$$A = \begin{bmatrix} (5,7) & (6,2) & (8,1) \\ (3,5) & (5,6) & (7,8) \end{bmatrix}.$$

In this game, the first row dominates the second, but there are no dominance relations for the columns. By higher-order dominance or by drawing the movement diagram, one sees the outcome in the first row and first column with payoff $(5,7)$ is a Nash equilibrium. The reader should draw the payoff polygon to see that this equilibrium is not Pareto efficient (Exercise 2). So the game is not strictly determined, but it still seems likely that under simultaneous play, the outcome is $(5,7)$, because the column player would expect the row player to choose her dominant strategy. This does not change if the column player goes first. No matter what the column player does, the row player would respond by choosing the first row by the expected value principle because that row is dominant. Then the column player, anticipating this, would choose the first column because the 7 payoff for him is the best of the three in the first row. So the outcome is still the Nash equilibrium with payoff $(5,7)$.

However, if the row player can go first, she can choose the second row (ignoring her own dominant strategy), knowing that the column player would respond by choosing the third column to get his payoff of 8, the best possible in the second row. So the outcome would be the $(7,8)$ payoff in the second row and third column. Notice that even though this point is not an equilibrium point, it is Pareto efficient (in fact, it has the highest sum of payoffs possible, so it satisfies the stronger first condition on benefiting the group). Furthermore, both players have improved from their $(5,7)$ outcome, so this is a win-win situation. Both players benefit from the row player going first, but neither benefits from the column player going first.

We'll look at one more example:

$$A = \begin{bmatrix} (9,5) & (5,7) \\ (4,4) & (4,6) \\ (0,1) & (2,3) \end{bmatrix}.$$

In this game, the reader should verify (Exercise 3) that the first row dominates both of the other rows, and the second column dominates the first. Thus, the outcome with payoff $(5, 7)$ is a Nash equilibrium at the intersection of the two dominant strategies. The reader should draw the payoff polygon and see that this is the only Pareto-efficient Nash equilibrium, so the game is a strictly determined variable-sum game. So the $(5, 7)$ outcome is certainly what would happen under simultaneous play. However, because each player has a dominant strategy, whoever goes first must still assume his or her opponent will stick with his or her dominant strategy by the expected value principle, just as in the previous example for the row player. Anticipating that, whoever goes first can do no better than to stick with his or her own dominant strategy, so the outcome is the same as it was using simultaneous play.

Notice that the outcome with payoff $(9, 5)$ is also Pareto efficient (and, in fact, is the highest-possible payoff sum of all the outcomes). Even though it is not an equilibrium point, the row player sure would like to have that outcome, in order to obtain the 9 payoff instead of the 5 payoff at the equilibrium point. We have just seen that this is not possible, regardless of who goes first, but don't forget we are still operating under all of our other assumptions, such as the players not communicating. If you combine sequential play with certain limited forms of communication, the row player *can* obtain the $(9, 5)$ payoff. See Exercise 4 for the details and a reference.

There seems to be little that we can say for variable-sum games played sequentially in general. Each game should be looked at carefully to see how moving sequentially affects the outcome.

8.1.2   Game Trees and Reverse Induction

In the previous subsection, we looked at single-move games and replaced simultaneous play with sequential play. But many games are sequential by virtue of the fact that they consist of several moves (often alternating between players) rather than one move. Examples include chess and checkers, board games such as Monopoly and Risk, many card games, and combinatorial games such as nim or tictacktoe. When the game consists of a sequence of moves for each player, it is more natural to display the game by using extensive form rather than normal form – that is, by using a game tree instead of a payoff matrix.

When the game is single-move, it is obvious that there is an equivalence between normal and extensive forms. We pointed this out in Chapter 6, when we analyzed the Battle of the Bismarck Sea. The same information was in the game tree as in the payoff matrix. Surprisingly, the same is true even for multi-move games if we use the proper definition of a "strategy" in such a game. Thus, in theory, there is no loss of generality by studying single-move games – every game can be suitably viewed as a single-move game.

We illustrate this idea with the following example.

Example 8.1 In the (fictitious) board game Without Frontiers, Hans and Enrico are in charge of rival civilizations who are colonizing unoccupied planets in space. Terran planets are good for establishing human colonies, while mineral-rich planets are good for setting up factory bases. The entire game is quite long and complicated, with Hans and Enrico alternating moves, during which the players have various options open to them, including waging war on each other. We will model only one small part of the game. Both Hans and Enrico have a fleet of ships in a particular star system that has both a Terran planet and a Mineral-rich planet. Hans has already established a factory base on the mineral-rich planet in a previous move. There are three steps before the players earn their payoffs for this star system. First, Hans must decide whether to use the factory base to build attack ships or to fortify the factory base with defensive weapons. Enrico does not know what Hans has decided. Enrico then has two options: He must either attack Hans's factory base or use his turn to establish a colony on the Terran planet. If he attacks Hans, Hans only has one option – to defend the factory base (both with attack ships of his own and/or any defensive weapons he has built to protect the base). On the other hand, if Enrico chooses to colonize the Terran planet, then Hans has a second choice to make – he can either build more factories and expand his industrial base on the mineral-rich planet or he, too, can colonize the Terran planet, which is large enough to support both civilizations. The payoffs are registered as points, and both having colonies and having factories contribute to the points. Having ships and weapons does not give points directly, but since these items defend colonies and factories, they can indirectly ensure that the players reap points. Finally, if there is a battle, both sides can lose points by having ships and/or factories destroyed and/or population killed.

The tree diagram in Figure 8.1 outlines the sequence of moves before the payoffs are awarded, and the payoffs are ordered pairs with Enrico's points in the first coordinate and Hans's points in the second coordinate. The payoffs listed for each outcome are based on the size of the fleets each player has and the number of factories Hans has built, in addition to the decisions made by each player. The reader need not worry about how the actual payoffs were determined.

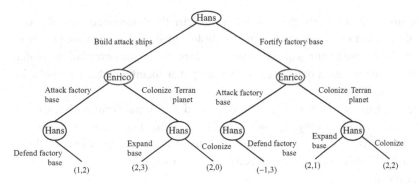

Figure 8.1 Partial payoffs in a round of the space colonization game Without Frontiers.

Now, we must define strategies in such a way that the game is essentially a one-move game. The key to this is for each player to specify, in advance, what he would do in any situation that arises. This includes what he would do in later stages of the game, which are contingent on what happens before then. Then he can be regarded as having chosen his "moves" all at once. In this example, consider Enrico first (we will make him the row player). The only thing that precedes his choice to attack is Hans's decision. However, since he is unaware of the choice Hans makes, he cannot make his choice contingent on it. Thus, Enrico really only has two possible strategies – attack Hans or colonize the Terran planet. So the payoff matrix will have two rows. Hans has more choices for his strategies. First, he must decide whether to build attack ships or to fortify his defenses. Then, should Enrico decide *not* to attack him, he has two other options: either to expand his factory base or to colonize the Terran planet. Of course, if Enrico attacks him, he has no other choice than to defend himself. Thus, since Hans has to make at most two sets of two decisions, he has $2 \times 2 = 4$

different strategies he can use in the game. His second choice can be contingent on his first, so we can list the strategies for Hans as the

AE – First build attack ships, and
      if Enrico colonizes, then expand the factory base.
AC – First build attack ships, and
      if Enrico colonizes, then colonize also.
FE – First fortify the factory base,
      and if Enrico colonizes, then expand the factory base.
FC – First fortify the factory base, and
      if Enrico colonizes, then colonize also.

Thus, we set up the 2×4 payoff matrix as follows.

		Hans			
		AE	AC	FE	FC
Enrico	Attack factory base	$(1,2)$	$(1,2)$	$(-1,3)$	$(-1,3)$
	Colonize Terran planet	$(2,3)$	$(2,0)$	$(2,1)$	$(2,2)$

In the exercises, you are asked to verify that this game is a strictly determined variable-sum game and to give the solution, but the main point of the example is that even though the play was sequential and some decisions were contingent on previous events, we could nevertheless model it as a single-move game by having each player select a strategy that specified what he would do in any situation ahead of time.

Therefore, any game in extensive form can also be converted to normal form with a payoff matrix. However, while this is true in theory, it is not very useful in practice, because the payoff matrices will be quite large. After all, in any game where there are more than two or three sequential steps, the number of different ways to specify what you would do in any eventuality quickly becomes astronomical.

For example, consider the rather simple game of tictacktoe. The first player has nine possible strategies on the first move. The player moving second would have to specify which of the remaining eight positions he would choose for each of the nine choices made by the first player, just to specify a single strategy for his first move. By numbering the squares from 1 to 9, we could write down such a strategy as an ordered 9-tuple. For example, the symbol $(2,3,6,2,9,5,5,1,6)$ would be the strategy "If the first player selects square 1, I would choose square 2, and if she selects square 2, I would select square 3, and if she selects square 3, I would select square 6, etc." The only stipulation is two players cannot both select the same square, so in such a 9-tuple (a_i), we must have $a_i \neq i$, so there are eight choices for each coordinate. That means there are $8 \times 8 \times \ldots \times 8 = 8^9 = 134{,}217{,}728$ different strategies for just the second player's first move! Then for the first player's second move, she would have to say which of the seven remaining squares she would take for each of the $9 \times 8 = 72$ outcomes of the first two moves – so $7^{72} \approx 7 \times 10^{60}$ possible strategies for her second move! Nobody could possibly analyze a payoff matrix of that size, and we are still less than halfway through the game!

So, although normal form games and extensive form games are equivalent, the size of the payoff matrix would make it impractical or even impossible to analyze a sequential game when it is in its normal form. So that raises the question if we can just analyze it in its extensive form. That is, can we predict the outcome and optimal strategies by working directly from the game tree diagram?

Going back to our example, Enrico has a bit of a problem. Since Enrico doesn't know what decision Hans made in the first step, he actually doesn't know at which of the two nodes labeled "Enrico" he is located when it is time to make his decision. Whenever this happens, the nodes in question are said to be in the same *information set*. In Straffin (1993), this is indicated by connecting the nodes by a dashed line, and in Luce and Raiffa (1957), it is indicated by circling all the nodes in the same information set. This makes it difficult to analyze the tree. Chapter 8 of Straffin (1993) contains a beautiful example of a game tree with various information sets, which is an application to a business marketing model. We highly recommend the reader look it up and read it. However, the games there are analyzed not by looking at the game tree but by converting them to normal form.

There is one case, though, where analyzing the game directly using the game tree is possible and easy. That is the case where all the information sets are single nodes, or, another way to say it, each player knows exactly where on the tree diagram he or she is located when it is time to make a move. Games that have that property are said to be *games of perfect information*.

For example, in the game Without Frontiers, suppose that Enrico does know what decision Hans makes in the first step. That has two effects on the game. First, it makes it possible for Enrico to make his choice contingent on what Hans does, so that Enrico now has four strategies instead of two, and the payoff matrix becomes 4×4. The second change is that now Enrico's two nodes are in distinct information sets so that the game becomes a perfect information game. Thus, it is possible to analyze it by looking at the tree diagram directly in lieu of the payoff matrix. Both approaches are explored in the exercises.

The procedure for analyzing a game tree when the game is of perfect information is known as *reverse induction* or *backwards induction*. You start at the end nodes of the game (called the "leaves" of the tree) and work your way back (toward the "root" of the tree – usually drawn on the top level as in Figure 8.1). At each stage, you decide which branch the player at that node would follow (based on the rational behavior assumption) and eliminate the other branches. We illustrate the process with an example.

Example 8.2 Mr. Brooker rents the Salty Dog Tavern from PH Royal Zonophone Industries (PHRZ) but wants to complain to the landlord that the plumbing is substandard. PHRZ has a choice of ignoring the complaints, fixing the plumbing, or selling the building and leaving the commercial rental business. If PHRZ sells the building, Mr. Brooker considers it a wash (0 points) since he has no idea if the new landlord will respond more favorably or not, while PHRZ regards it as a slight gain (2 points). If PHRZ fixes the plumbing, it is an expense for the landlord, but in the long run, it is better for the company, so both Mr. Brooker and the company each regard this as a 1 point gain. If PHRZ ignores the complaints, then Mr. Brooker thinks he has four options. He can back down from the complaint and just live with the substandard plumbing, a loss of 2 points for him but a gain of 6 points for PHRZ. He can move out and take his business to the Pale White Whale Corporation down the street, which is unexpected work for him and a loss of income for PHRZ, so each regards that option as 0 points payoff. He can file suit against PHRZ. If PHRZ settles, that's 3 points for Mr. Brooker and -1 points for PHRZ; however, if PHRZ fights the suit, then the time wasted in litigation helps the tenant more than the landlord, so the payoffs are 2 points for Mr. Brooker and 1 point for PHRZ. The last option is for Mr. Brooker to withhold rent payments. If this happens, PHRZ can either countersue Mr. Brooker for nonpayment, resulting in a payoff of -2 to each party, or PHRZ can evict Mr. Brooker, which is a payoff of -1 for Mr. Brooker and 0 for PRHZ. On the other hand, if Mr. Brooker never complains about the plumbing in the first place, he regards that as a minor inconvenience with a payoff of -1, while PHRZ gets away with a score of 5 points. What is the predicted outcome of this game?

The tree diagram in Figure 8.2 depicts the situation.

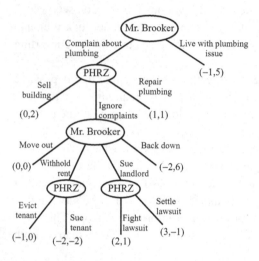

Figure 8.2 Payoffs to (Mr. Brooker, PHRZ).

For the purposes of this example, we assume each player knows the payoffs to the other player so both players have the same information. Furthermore, both parties know exactly where they are in the tree diagram – this is a game of perfect information. Now we illustrate backwards induction. Start at the last node(s), where a decision is made – in this case, the two PHRZ nodes near the bottom. Consider the one on the right. If you are PRHZ and are being sued by Mr. Brooker, you can fight or settle. The payoffs to you are the second coordinates in the ordered pairs at the end of the branches. You would clearly prefer the 1 payoff to the -1 payoff, so you would choose to fight the lawsuit. Similarly, if Mr. Brooker withheld the rent, your choices are to evict him or to sue him. You prefer the payoff of 0 for eviction to the payoff of -2 for a lawsuit, so you would evict in that circumstance.

To indicate that you would not take the branches labeled "settle lawsuit" or "sue tenant," you could cross those branches off or erase them. We will indicate this by a line crossing out the path not taken. So the bottom of the diagram now looks like what is shown in Figure 8.3.

We can now ignore the payoffs $(-2, -2)$ and $(3, -1)$ at the ends of the branches we crossed off. Both players know that PHRZ would never choose these branches. Now move up the tree to Mr. Brooker's node with four possible courses of action. Look at the payoffs to Mr. Brooker, which are the first coordinates in the ordered pairs. Choose the best one, knowing that if he sues PHRZ, he'll end up with a payoff of 2 because they will fight the lawsuit, and if he withholds the

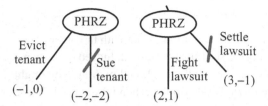

Figure 8.3 Pruning the game tree.

rent, he'll be evicted and end up with a payoff of −1. Also, if he moves out, he earns 0, and if he backs down, he earns −2. The best outcome for Mr. Brooker is the 2 that comes from suing the landlord. Now cross off the other three branches, and move up the tree to the next PHRZ node with three courses of action. If PHRZ sells the building, the company gets a payoff of 2, and if it repairs, the plumbing, it gets a payoff of 1. If it ignores the complaint, the company knows it will be sued and will have to fight it for a payoff of 1. Since the 2 is the highest payoff, PHRZ will sell the building. Finally, back up to the root node, which is Mr. Brooker's node. If he lives with the plumbing issue, the payoff is −1, while if he complains, he knows PHRZ will sell the building and he'll get a payoff of 0. Since the 0 is highest, he will opt to complain.

Thus, the prediction is Mr. Brooker complains about the plumbing, and PHRZ responds by selling the building and leaving the commercial rental business. We hope the new landlord is more sympathetic to Mr. Brooker's plumbing concerns. Figure 8.4 shows the entire tree with the relevant branches crossed off and the final outcome circled.

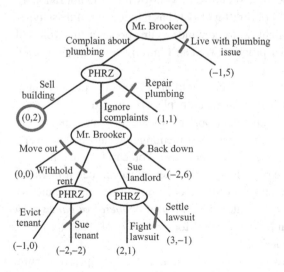

Figure 8.4 The outcome of reverse induction is the solution of the game.

The game tree for any game of perfect information can be analyzed in the same way and the optimal outcome determined without converting the game to normal form. Nevertheless, it is good practice for the reader (see Exercise 7) to convert the game to normal form and check that one obtains the same solution.

8.1.3   What if Communication Is Allowed?

In a constant-sum game, there is no advantage in being able to communicate with the other player before selecting your strategy, assuming all the players are skillful and rational. Indeed, if the game is strictly determined, then both players know to go for the saddle point, and nothing they might say should change their strategies. If the game is not strictly determined, then they should choose their strategy secretly in the optimal mix. It would only hurt you if you gave away your choice in any individual play of the game. You might lie in an effort to mislead your opponent, but if he or she is skillful and rational, this should have no effect on his or her choice.

However, if the game is variable-sum, then communication can affect the play of the game. First of all, since there may be outcomes that are "win-win" situations, it may be in the interest of the players to cooperate so they can coordinate their choices to obtain better payoffs than they might by playing noncooperatively. What if they are not interested in cooperation? We can divide variable-sum game theory into two types of play. On the one hand, we have noncooperative play. In this case, there still may be some limited ability to affect the outcomes of a game, especially if the choices are not simultaneous. The player going second might say things to the player going first to try to influence his or her decision. Even if the play is simultaneous, each player may try to say things to the other before the moves are selected to try to influence his or her choices.

These comments to the other players usually take the form of threats that if they select a certain strategy, you will retaliate by doing something harmful (which is also not a rational choice, otherwise you would have no reason to say anything as they would expect you to follow a rational course of action). Or, instead of a threat, it might be a promise that if they select a certain choice, you will reward them (again, by making an unexpected choice of your own that is not rational). See Exercise 4 for an example of this. Chapter 14 of Straffin (1993) has a treatment of these types of threats and promises, including what you might do to convince your opponent that you will indeed follow through on your threat and/or your promise. We can't improve on what is covered there, so we will say no more about this case.

The other case is when the players are open to partial cooperation. That is, the players might be interested in negotiating a solution that is beneficial to both players. In order to help the negotiations, it might be useful if you are able to give your opponent some of your winnings. In other words, we assume that the payoffs (whatever they are) are something that can be divided and shared among the players. Certainly not every game has that property. You can't share a military victory or a jail sentence, for example. But if the payoffs have a monetary value, then it might be possible for one player to make payments to the other in order to coax him or her into a mutually beneficial arrangement. This type of game theory is called the theory of *cooperative games with side payments*. This theory was pioneered in von Neumann and Morgenstern (1944) and forms the basis of the use of game theory in modern economics.

Of course, real-life economics has attributes of both cooperative and noncooperative game theory. For example, picking up the November 13, 2019, edition of the *New York Times* and turning to the Business section, we find the headline that the US president "Mixes Optimism with Threats Over China – He says a trade deal is 'close' but warns more tariffs are possible." The same is true for other types of game model applications, such as actions involving international diplomacy and politics. If one considers almost any international event, from the Cuban Missile Crisis to nuclear arms treaties to the Paris Climate Agreement, there is always a combination of threats, promises, and negotiations between the players before any action is taken (with a notable exception being some acts of terrorism).

One common way to approach negotiation is to begin with the point where both players play their (prudential) strategy where they just look at their own payoffs and play as if it were a zero-sum game. We have already pointed out that this mode of play does not violate the rational behavior assumption, and the payoffs the players obtain, called their security levels, are guaranteed minimum payoffs. Even though this outcome may not be an equilibrium point, one can argue that in a sense, this is the least each player can win by not playing cooperatively at all. Then, one seeks solutions that can improve one or both players' payoffs by negotiation, so the additional winnings can be regarded as a sort of bonus that comes from cooperation. But it is not always straightforward to decide on the best outcome, or even a fair outcome, even with this reasonable beginning. For example, consider the game with payoff matrix

$$A = \begin{bmatrix} (3,5) & (6,1) \\ (7,2) & (5,7) \end{bmatrix}.$$

We considered this game in Section 7.4.2, and in Exercise 7 of that section, the reader verified that the security levels for the players are $(v_R, v_C) = (5.4, 3.667)$. The outcomes that would improve the payoffs for at least one of the players are to the right and above this point in the payoff polygon, shaded in Figure 8.5.

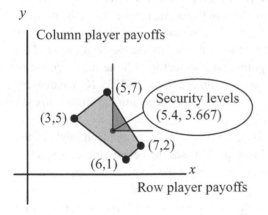

Figure 8.5 The are shaded above and to the right of the security levels indicate possible improvements in the payoffs that might come from cooperation.

So any of these points is an improvement over the security level payoffs, but which one is the "best" or "fairest" solution? Since the whole point of negotiation is to improve things for the group rather than for the individual player, it certainly makes sense to choose a solution that is Pareto efficient, so on the line segment between $(5,7)$ and $(7,2)$, but only those points in the dark region on this segment. Some authors call this set of points the *negotiation set*. The line connecting $(5,7)$ and $(7,2)$ has equation $y = -\frac{5}{2}x + \frac{39}{2}$, as is easily checked. So the extreme points on the negotiation set are $(5.4, 6)$ when $x = 5.4$, and $(6.333, 3.667)$ when $y = 3.667$.

Suggesting either one of these extreme points as a solution probably would not work, as the entire additional benefit over the security levels goes to just one of the players. A fair solution is somewhere in between these points on the segment that is the negotiation set, but which one? One suggestion is to just take the midpoint of this segment. That seems fair, right? In this case, the midpoint is $(5.866, 4.833)$. Would you feel this was fair if you were the row player? The additional payoff to you due to the negotiation is 0.466, while the additional amount to the column player is 1.1667. You might feel this proposed solution favors the column player unfairly. After all, with noncooperative play, your security level is 5.4, compared to only 3.667 for the column player. That seems to indicate a strategic advantage for the row player, and the midpoint solution appears to make the row player give up too much of their advantage.

Perhaps it would be more appropriate to select a Pareto-efficient point somewhere below the midpoint in order to preserve a little of this advantage. But again, which point? Should a point be chosen that is $\frac{3.667}{5.4}$ of the distance between the top point $(5.4, 6)$ and the bottom point $(6.333, 3.667)$? That sounds fair because the additional payoff coming from the negotiation is closer to being proportional to the actual payoffs. Another suggestion would be to take the point that is the intersection of the negotiation set with the radial line through the origin and the security level point. Then their relative payoffs would be exactly proportional to what they are without the negotiation. But wait – if you translate a game, shouldn't that produce a game with the same "optimal" solution? Translating a game might change those proportions, so maybe trying to make the additional payoff from negotiation proportional to the ratio of the security levels is not a good idea. Of course, one could argue that translating the game would also change the strategic advantage that one player has over the other by virtue of receiving a bonus or paying a penalty due to the amount the payoffs were translated, so maybe the proportional solution is fine after all?

Yet another suggestion would be to just split the additional payoff that comes from the negotiation evenly (rather than trying to split the total payoff evenly). In other words, start at the security level point, and move in the direction of the $(1, 1)$ vector until one hits the negotiation set. That is, take the line with slope 1 through the security point and intersect it with the line segment of the negotiation set. In our example, the intersection has coordinates $(6.067, 4.333)$. Notice that this still preserves something of the row player's strategic advantage, and the additional amount to each player over the security level is equal – namely 0.667.

One can think of some problems that might arise using this method. For example, it might be the case that the line through the security level point with slope 1 misses the negotiation set entirely!

There does not seem to be any compelling argument that one of these suggested approaches is the best way to proceed. Faced with a multiplicity of approaches, one possible way out would be to try to specify in advance some _conditions of fairness_ that you would want a negotiation procedure to satisfy. Mathematicians call these conditions _axioms_. Then you can search for a negotiation procedure that satisfies these axioms.

This is exactly what John Nash did to come up with his own method of solution by negotiation, also known as the _Nash arbitration scheme_. Those interested can read about it in chapter 16 of Straffin (1993), which contains a proof that Nash's scheme is the _unique_ method that satisfies his axioms. We will present Nash's method (without proof) in the exercises at the end of this section.

We close this subsection with one final observation. The stronger condition we mentioned that benefits the group is trying to obtain a payoff for which the total is as large as possible. In the preceding example, that point is $(5, 7)$, where the total is 12. No other point in the payoff polygon has that high a sum, including the potential solutions proposed earlier. If we are assuming that the payoffs can be combined and shared, why not propose that point as the solution? Then the problem becomes a problem in fair division – how should the players split their winnings of 12? Observe that no matter how the winnings are split, unless the point is $(5, 7)$ itself, the payoffs are larger than any point on the payoff polygon, since they are on the line $x + y = 12$, which meets the payoff polygon only at the point $(5, 7)$. So, if the players can come to an agreement of how to split the payoffs, the solution would be superior to any payoff in the negotiation set! Of course, fair division problems are another type of mathematical problem that are not addressed in this text, so it is still something of a mystery how to negotiate the best solution. Nevertheless, we hope the reader sees the importance of the negotiation problem and the potential ways to resolve it.

8.1.4 What if Your Opponent Is Indifferent? Games against Nature

In all of the variations of game theory we have considered, we have always assumed the players behave rationally, which means they act in a way to increase their own payoffs as much as possible. For constant-sum games, maximizing one's own payoff means minimizing your opponent's payoff so that your opponent striving to help himself or herself aligned with striving to hurt you. The minimax theorem shows that there is a best way to play under these assumptions. However, when we consider variable-sum games, even if we keep the rationality assumption, helping yourself no longer aligns with hurting your opponent. This nonalignment causes some problems; for certain games, it was not clear what the optimal solution was. Should you strive for a Nash equilibrium? A Pareto-efficient outcome? Should you ignore your opponent's payoffs and play prudentially, or not?

At least in a variable-sum game against a rational player, you had some idea of what is good for your opponent and perhaps some insight into how he or she might play the game. But what if we go further and assume your opponent is not even trying to maximize his or her own payoff? What if he or she is just plain indifferent to your actions and your payoffs and to their own payoffs? You might imagine that under such a relaxation of the rationality assumption, there is even more ambiguity about how to play optimally than there was in a variable-sum game.

One example of an indifferent opponent is nature, so these types of games are often called *games against nature*. The first thing to note is that since your opponent is indifferent to its own payoff, but you retain the rational goal of maximizing your own payoff, there is no reason for you to pay any attention to what the payoffs to nature are – after all, nature is not paying any attention either! Thus, when writing the payoff matrix down, you only care about your own payoffs, so there is no need to use ordered pairs to write the payoffs. Furthermore, since we are not bothering to even record payoffs to nature, the entire distinction between "constant-sum" and "variable-sum" is meaningless.

Here is an extremely simple example to illustrate the ideas. Mr. Holly is about to leave his house in the morning to walk to the bus stop to catch his ride to commute to work. The bus stop is about a half mile away. It may or may not rain that day, and Mr. Holly must decide whether or not to carry an umbrella. If it rains, he would be quite happy to have it with him, especially since it is large enough to shield both him and the (very smart and pretty) girl he likes, who also waits at the bus stop. He regards the payoff in that case as 5 points. If it does not rain, and he carries the umbrella, it is a minor inconvenience – he must use one of his hands to carry the thing and he might look a bit silly carrying it on a bright sunny morning. He regards that as -1 point. Now, if he does not carry it, and it does not rain, he is pleased with himself for not burdening himself with it unnecessarily, but not as pleased as having it when he needs it. So he regards that as 3 points. Finally, if he does not have it, but it rains, not only is he wet and miserable, but he may catch a cold as well, and he certainly does not make a favorable impression on the girl at the bus stop, so he regards this as a payoff of -6. Thus, making Mr. Holly the row player, the payoff matrix is:

$$A = \begin{array}{c} \\ \text{Umbrella} \\ \text{No umbrella} \end{array} \begin{array}{c} \text{Rain} \quad \text{Sun} \\ \left[\begin{array}{cc} 5 & -1 \\ -6 & 3 \end{array} \right] \end{array}$$

Mr. Holly wants to know if he should bring the umbrella or not, or how often he should. Notice his "opponent" in this game is nature, and nature is completely indifferent to Mr. Holly's travails. Nature is not "out to get him" by raining on him when he is without his umbrella, nor does nature's gain (whatever it might be) by raining on Mr. Holly play a role in his decision to use an umbrella or not, so we don't care about payoffs to nature.

There is one situation where we know exactly how to play. That situation is when Mr. Holly knows what strategy (pure or mixed) nature is using. Then, according to the expected value principle, he should respond using the pure strategy that gives him the highest expected payoff. Notice that the expected value principle is valid whether or not your opponent is rational. You don't need to know how and why they chose his or her strategy – you just need to know what it is. In fact, when Mr. Holly checks the weather forecast for the day and discovers there is a 30% chance of rain, what he is doing is, in fact, finding out the mixed strategy that nature likely is using. So, according to Corollary 6.3, with a 30% chance of rain, the expected payoffs for Mr. Holly's pure strategies are

$$Aq = \begin{bmatrix} 5 & -1 \\ -6 & 3 \end{bmatrix} \begin{bmatrix} 0.3 \\ 0.7 \end{bmatrix} = \begin{bmatrix} 0.8 \\ 0.3 \end{bmatrix}.$$

So, clearly, the best strategy on such a day is to carry the umbrella.

The difficulty arises if Mr. Holly has no idea what strategy nature is using. Notice that to say nature is indifferent is *not* the same thing as saying nature acts randomly. But one possible way to proceed if you do not know nature's strategy mix is to *assume* it is random, which is to say assume that each outcome is equally likely, and then use the expected value principle. In this case, since you are assuming that $q = \begin{bmatrix} \frac{1}{2} \\ \frac{1}{2} \end{bmatrix}$, it is clear that the entries in Aq are simply the averages of the numbers in the rows of A, so choosing the row with the highest expected payoff means choosing the row with the highest average. We mentioned this approach to playing a game back when we discussed the Battle of the Bismarck Sea. At that time, we pointed out that choosing the row with the best average payoff is not how to play against a rational opponent, but it is an accepted approach to playing against nature or an indifferent opponent. Notice also that choosing the row with the highest average is the same as choosing the row with the highest sum. In Chapter 10 of Straffin (1993), the author suggests that this approach goes all the way back to the mathematician Laplace. This is no surprise, since Laplace was one of the pioneers of probability theory, and this approach is just a special case of the expected value principle. In this example, Laplace would recommend that you carry the umbrella, as the row sum of that strategy (4) is better than the row sum of not taking the umbrella (-3).

But, Laplace's is not the only approach to a game against nature. Just as choosing the best of the worst-case scenarios led to a possible way of playing a variable-sum game (it led to the prudential strategey and its associated security level), doing the same might be useful here. After all, even though an indifferent opponent is not "out to get you," the oponent may accidentally choose a strategy that hurts you by putting you in the worst case of the row you choose. So it might make sense to select the row with the highest minimum, which is the best of the worst cases.

In this case, the first row has minimum -1 while the second row has minimum -6, so this approach (named after its originator Wald in Straffin (1993)) would suggest choosing the first row. That is, always carry your umbrella! This maximin approach does make sure you never get a payoff lower than -1, which sounds pretty good. On the other hand, it recommends that you always assume the worst could happen, and some days are just obviously a good bet not to carry an umbrella, so maybe there is a more optimistic approach? One such approach is presented in Straffin (1993) and will be explored in Exercise 12d at the end of this section.

There are other approaches possible. One is based on the idea of minimizing the regret you might feel if you've made the wrong choice. For another, you might say by decree that whenever you play a game against nature, you always simply pick the first row! Let's call that the *lazy* approach. Now, in this example, the lazy approach, Wald's approach and Laplace's approach all recommend the same thing – to bring the umbrella. But that is not what happens in general. It is easy to construct payoff matrices where each of these methods gives a different recommendation, as in Chapter 10 of Straffin (1993).

So, the conclusion seems to be that games against nature are as ambiguous as variable-sum games or as arbitration schemes. But some methods seem better than others. Wouldn't you trust Wald's or Laplace's method more than the lazy method? Can we characterize the "good" methods?

Just as Nash set up a set of axioms for what he considered to be "good" arbitration schemes, one can try to set up axioms for "good" methods to play games against nature. The mathematician John Milnor did exactly that in Milnor (1954). He gives 10 axioms there, but unfortunately, he was not as lucky as Nash was with his axioms for arbitration, because Milnor shows that *no method* could possibly satisfy all 10 axioms! Six of those axioms are reproduced in Straffin (1993), but even with just six axioms, each of the methods we suggested violates one or more of the axioms. Interested readers can seek the details there, as we do not intend to reproduce the axioms here. We just wanted to mention the axiomatic approach. Even though no method satisfies all the axioms, an individual can select a method that satisfies the axioms most important to him or her.

Exercises

1. The following payoff matrices are for zero-sum games. In each case, predict the outcome if the row player plays first, and repeat the exercise if the column player goes first. Give reasons for your answers.

   a.

   $$\begin{bmatrix} 6 & 1 & 0 & -2 \\ 3 & 5 & 2 & 7 \\ -2 & -3 & -1 & 6 \end{bmatrix}$$

   b.

   $$\begin{bmatrix} -4 & 3 & 1 \\ 2 & 0 & 4 \\ 1 & -1 & -3 \end{bmatrix}$$

c.

$$
\begin{bmatrix}
-2 & -1 & 0 & 1 & 2 \\
10 & 8 & 5 & -1 & -9 \\
5 & -9 & 3 & -8 & 7
\end{bmatrix}
$$

2.   In the text, we considered the variable-sum game with payoff matrix:

$$
A = \begin{bmatrix}
(5,7) & (6,2) & (8,1) \\
(3,5) & (5,6) & (7,8)
\end{bmatrix}.
$$

Draw the payoff polygon, and verify that the pure-strategy Nash equilibrium is not Pareto efficient so that the game is not strictly determined. Verify that if the column player goes first, the predicted outcome is the Nash equilibrium, but if the row player goes first, both players can improve on their outcomes.

3.   In the text, we considered the variable-sum game with payoff matrix:

$$
A = \begin{bmatrix}
(9,5) & (5,7) \\
(4,4) & (4,6) \\
(0,1) & (2,3)
\end{bmatrix}.
$$

Verify the claims that were made in the text. That is, both players have dominant strategies, the game is strictly determined, and if either player moves first, the natural outcome is the same as under simultaneous play.

4.   Consider the payoff matrix of Exercise 3. In the text, we suggested that if we allowed communication between the players, and the column player went first, that the row player could obtain the $(9,5)$ outcome, even though it is not an equilibrium point. We investigate this here.

a.   Suppose that the column player moves first, but before he moves, the row player says to him, "If you choose the second column, I will choose the third row." Assume that the column player believes her. What is the resulting outcome of the game? Explain.

b.   Suppose that the column player moves first, but before he moves, the row player says to him, "If you choose the second column, I will choose the second row." What is the resulting outcome, again, assuming the column payer believes her? Explain.

c.   These types of remarks that the row player makes to the column player can be called *threats*. A threat is a statement about action you will take, depending on what your opponent does; that is, it is contingent on your opponent. (If you do this, I'll do that in response.) Furthermore, your response should certainly hurt your opponent. Observe that in both Part a and Part b, if the column player ignores the row player's threat, then his payoffs go down from the predicted outcome without communication. Does a threat hurt or help the row player? However, a threat does not always work. What would be a necessary condition on the payoffs in the second column in order for there to be a threat that would have the desired effect?

REMARK 8.3   *In a game where a threat will not help the row player, it is possible that a* promise *might, or perhaps a combination of a threat and a promise. For more information on the difference between a threat and a promise, and to see more examples, refer to chapter 14 of Straffin (1993).*

5. Consider the round of the game Without Frontiers that we introduced in the text. Verify that the game in normal form is a strictly determined varible-sum game, and determine the solution.

6. a. If we vary the rules of Without Frontiers so that Enrico knows what decision Hans makes on the first step, then the number of strategies open to Enrico doubles from two to four. Write down Enrico's four strategies.

   b. Convert the game to normal form, assuming Enrico knows what decision Hans makes on the first step. Solve the game.

   c. When Enrico knows Hans's initial decision, we mentioned that the game becomes a game of perfect information. In that case, analyze the game by performing reverse induction on the game tree. Your solution should agree with the one you obtained in Part b.

7. We solved the Salty Dog Tavern problem (Example 8.2) by analyzing the game tree. Instead, convert the game to normal form, and analyze it via considering the payoff matrix. Your answer should agree with the one we got in the text.

8. Consider the variable-sum game with payoff matrix:

$$A = \begin{bmatrix} (4,0) & (1,6) \\ (-1,3) & (2,-1) \end{bmatrix}.$$

   a. Draw the payoff polygon. Verify that the game is not strictly determined.

   b. Find each player's prudential strategies and security levels. Mark the point with the security levels as $(v_R, v_C) = (x_0, y_0)$ on the payoff polygon. Determine the negotiation set, including the coordinates of the two endpoints.

   c. One suggestion in the text for the negotiated solution point was the midpoint of the negotiation set. Find the coordinates of that point.

   d. Another suggestion in the text for the negotiated solution point was to preserve the proportions of $\frac{v_C}{v_R}$, which means finding the point in the negotiation set that also lies on the radial line through (x_0, y_0) and the origin. Find the coordinates of that point.

   e. A third suggestion was to find the line through (x_0, y_0) with slope $m = 1$ and take the point where this line intersects the negotiation set. Find the coordinates of that point.

9. In the text, we mentioned that John Nash discovered a unique arbitration scheme that satisfied certain axioms. Here is his method, which we present without discussion of the axioms involved and without proof: The suggested negotiated solution point is the point (x, y) in the negotiation set, which maximizes the product $(x - x_0)(y - y_0)$. We'll call that point the *Nash point*. Find the coordinates of the Nash point for the payoff matrix in Exercise 8.

10. Consider the variable-sum game with payoff matrix

$$A = \begin{bmatrix} (8,4) & (3,9) \\ (6,7) & (4,0) \end{bmatrix}.$$

   a. Verify that this game is not strictly determined, and draw the payoff polygon.

   b. Determine the prudential strategies for each player and the corresponding security levels. Mark the point with the security levels as $(v_R, v_C) = (x_0, y_0)$ on the payoff

polygon. Determine the negotiation set, including the coordinates of the two endpoints.

c.   Find the coordinates of the Nash point for the negotiated game.

11.   The following payoff matrix represents a game against nature, where you are the row player.

$$A = \begin{bmatrix} 1 & 2 & 3 & 0 & 1 \\ 2 & 1 & 2 & 1 & 1 \\ 7 & 0 & 0 & 2 & 1 \\ 0 & 2 & 0 & 0 & 2 \end{bmatrix}.$$

a.   Suppose you know that nature is using the mixed strategy $q = \begin{bmatrix} 0.1 \\ 0.3 \\ 0.2 \\ 0.3 \\ 0.1 \end{bmatrix}$. What row

should you choose?

b.   Suppose you know that nature is using $q = \begin{bmatrix} 0 \\ 0.3 \\ 0 \\ 0.1 \\ 0.6 \end{bmatrix}$. What row should you choose?

12.   For the payoff matrix in Exercise 11, what row is recommended by the following methods?

a.   The Laplace method
b.   The Wald method
c.   The lazy method
d.   The Hurwicz method (see chapter A of Straffin (1993)). This method is similar to
Wald's method, but instead of just looking at the minimum in each row, we take a
weighted average of the minimum and the maximum in each row as follows. Choose
a weight α between 0 and 1. For each row, compute the weighted average:

$$\alpha(\text{row minimum}) + (1 - \alpha)(\text{row maximum}).$$

Then choose the row with the highest weighted average. Notice that if $\alpha = 1$, this
reduces to Wald's method, looking only at the worst that can happen in each row. At
the other extreme, when $\alpha = 0$, you are looking only at the best that can happen
in each row. You are free to choose α, but the closer to 0 α is, the more you are
weighting the best that can happen, so this is a more optimistic outlook. That's why
α is called a *coefficient of optimism*. Determine which row this method recommends
for the matrix in Exercise 11 if $\alpha = 0.9$. What about $\alpha = 0.7$?

13.   Don Condo, a NASA scientist, is studying a newly discovered planet, which has been named
D'Rhonda. Planet D'Rhonda is close enough to Earth to make it feasible to send probes and
rockets there if we want to. The probes could be robotic or a human exploratory mission
to set up a small, scientific base there, or even a larger mission to set up a sustainable
permanent human colony there, if the conditions are right. Of course, we could also be
content to study D'Rhonda from a distance using telescopes and other remote instruments.

The benefits to humankind depend on what we find there. D'Rhonda could be (*A*) a barren planet with no water, no useful minerals, and no life; (*B*) barren of water and life, but rich in minerals; (*C*) a Terran ("Earthlike") planet with water, minerals, and possibly life; (*D*) populated by an advanced civilization of benevolent aliens; or (*E*) populated by a race of hostile aliens. When NASA weighed the costs and possible benefits of each option, it came up with Table 8.1 of payoffs to humankind (don't worry about the units).

Table 8.1 Payoffs to NASA.

	A	*B*	*C*	*D*	*E*
Study remotely with telescopes only	2	1	0	0	3
Send robotic probe	1	2	3	4	1
Human exploratory scientific base	0	1	5	7	−3
Human colony	−2	0	6	10	−5

a. Suppose NASA estimates the probabilities of each of the five cases $A - E$ as
$$q = \begin{bmatrix} 0.4 \\ 0.55 \\ 0.03 \\ 0.01 \\ 0.01 \end{bmatrix}$$. If NASA is correct, what form of exploration should we undertake?

b. Suppose NASA has no idea what the probabilities of $A - E$ are. What form of exploration should we undertake according to Laplace?

c. What form of exploration should we undertake according to Wald's method?

d. What form of exploration should we undertake according to the Hurwicz method (see Exercise 12d) with $\alpha = 0.4$.

14. Consider the following hypothetical game of politics, which we model as a sequential game. The US president is a member of one party – say, the "Coffee" Party – which controls the Senate. The House of Representatives is controlled by the opposite party – say, the "Tea" Party. The president recently signed the Health Care Bill, which is now a federal law. Inexplicably, the Tea Party is angry about this law and would like to repeal it.

Currently, the federal government needs to pass a budget for next year, and this budget contains money for social assistance programs such as food stamps and other programs that the Tea Party is against. The Tea Party's main goal is to repeal the Health Care Law, and their secondary goal is to lower the budget allocations for food stamps and other programs.

The budget is now being considered by the House of Representatives. The Tea Party has three options. They can pass the budget as it is written, which means they get neither of their desires. This would be considered the end of the game (in a clear victory for the Coffee Party). The second choice for the Tea Party is to compromise and accept the Health Care Law but try to negotiate a reduced budget for food stamps and other social programs. If they do this, the Coffee Party will agree to a slight reduction in the budget in order to get it passed. The Coffee Party would regard this as a victory even with the reduced budget, so this choice would also end the game. The third choice for the Tea Party is to take a hard line and demand that the Health Care Law be repealed before they negotiate anything in the budget proposal. If they choose this option, the game does not end, and the Coffee Party must respond.

The Coffee Party has two choices. The first is to give in to the Tea Party demands and repeal the Health Care Law in order to pass the budget. The Tea Party would regard this as a big victory since they get their highest priority, even if they ended up passing the food stamp budget. The other choice for the Coffee Party is to resist the Tea Party demands. That is, they will simply ignore the demands. After all, the Health Care bill is already a law, so the time is past for negotiations about it. As federal law, everyone is required to follow it, and to demand to repeal it before negotiating a budget puts the Tea Party on shaky legal ground, plus it makes them look like impetuous bullies just trying to force their will in a battle they already lost, which is bad for their image and will probably lead to lost seats in the next election. Thus, if the Coffee Party resists, the ball is back in the court of the Tea Party.

Now the Tea Party has three options. First, they can give up their demand to repeal the Health Care Law. We call this option "give in," and it will end the game. A second option is to negotiate with the Coffee Party. If they do so, the Health Care Law will remain intact, but the Coffee Party will compromise and give the Tea Party a reduction in the budget for food stamps (probably a larger reduction than if the Tea Party hadn't made their demand in the first place). Their third option is to stick to their demand and refuse to do anything. In this case, the government shuts down because no budget is passed. Although the tree diagram might conceivably continue beyond this point, we'll assume that if the government shuts down, the game ends in order to keep the tree small.

In reality, this is not a zero-sum game. For example, if the government shuts down, everyone is hurt (including the Tea Party) due to the detrimental effect this would have on the economy. However, we would like to model it as a zero-sum game, and we do this by considering the payoff to be the number of congressional seats gained by either party in the next mid-term elections. If one party gains three seats, the other party loses three seats, so with this interpretation it is a zero-sum game. According to studies of which both parties are aware, the payoffs in the next elections are very likely to be as follows: If the Tea Party passes the budget as is, they lose six seats to the Coffee Party. If they compromise and negotiate a reduced food stamp budget, they lose three seats to the Coffee Party. If they demand a repeal to the Health Care Law, and the Coffee Party gives in to this demand, the Coffee Party will lose six seats to the Tea Party. If the Coffee Party resists these demands, and the Tea Party gives in, the Tea Party will lose five seats to the Coffee Party. If the Coffee Party resists and the Tea Party and then negotiates, neither party will lose seats to the other. Finally, if the government shuts down, the Tea Party will lose seven seats to the Coffee Party.

a. Draw the game tree for this game, and label each terminal node with the payoff to the Coffee Party.
b. Notice this game has perfect information. Analyze the game by reverse induction, and determine the outcome.
c. Convert the game to normal form with the Coffee Party as the row player, and solve the game by considering the payoff matrix. (Your answer should agree with Part b.)
d. The conclusions of Parts b and c are based on assuming that both players know all the payoffs ("perfect information" implies this). But the payoffs are in the future and

perhaps difficult to predict. Suppose that the Tea Party mistakenly thought that they would gain seats if the government shut down. For example, they thought they would gain five seats instead of losing seven seats. Then the two players have different information – another basic assumption about game theory we can vary for a sensitivity analysis. In this case, the Coffee Party has the correct information that they will gain seven seats if the government shuts down, while the Tea Party erroneously thinks they will gain five seats in this case. How would that change the outcome of the game?

8.2 Ordinal Games

One problem with the example about Mr. Holly's umbrella is that the payoff points didn't mean anything. Is a payoff of 6 really worth twice what a payoff of 3 is worth? There are many games where the payoffs are either subjective or difficult to measure numerically. In this section, we'll see that we can still retain something of the dynamics of the game if, instead of actual numerical payoffs, all we know is in what order we prefer the outcomes of the game.

So assume there are two players, and we retain all the other assumptions (rational behavior, both players have the same information about the payoffs, and the players move simultaneously and independently). Suppose the row player has m strategies, and the column player has n strategies, so that there are $r = mn$ outcomes in the game. Each player ranks the r outcomes in order of his or her preference. Mimicking what is done in Taylor and Pacelli (2008), we will assign the most preferred outcome with the highest number r, because we are already accustomed to thinking that a higher number is better for a player. Thus, the most preferred outcome is denoted r, the next most preferred is denoted $r - 1$, etc., until the least preferred outcome, which is labeled 1. We assume there are no two outcomes that the players prefer equally – no ties in the rankings. Then we form ordered pairs of these preference rankings and use them to fill in the payoff matrix.

For example, suppose the row player has two strategies and the column player has three. The payoff matrix might look like this:

$$\begin{bmatrix} (3,5) & (2,6) & (4,1) \\ (6,2) & (1,4) & (5,3) \end{bmatrix}.$$

So, for example, in the $(1, 1)$ position, the "payoff" $(3, 5)$ indicates that this outcome is the row players fourth choice (because the order of preference is $6, 5, 4, 3, 2, 1$), while it is the column player's second choice. The $(4, 1)$ "payoff" indicates it is the row player's third choice and the column player's last choice, etc. Because "first" (choice), "second" (choice), "third" (choice), etc., are ordinal numbers, we call these games *ordinal games*.

8.2.1 Dominance and Nash Equilibrium Points

First, we see what concepts from the usual theory carry over to ordinal games. Note that the notions of "zero-sum" or "variable-sum" are meaningless. The "payoffs" are preference rankings – so ordinal numbers – for which an addition is not even defined. However, some concepts carry over just fine. The notion of a movement diagram, for example, still makes sense and is drawn the same way as for variable-sum games. Vertical arrows point to the highest first coordinate in a column, and horizontal arrows point to the highest second coordinate in a row. The movement diagram for the matrix is shown in Figure 8.6.

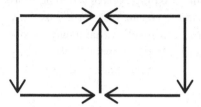

Figure 8.6 Movement diagram for an ordinal game.

The concept of one strategy dominating another also carries over.

DEFINITION 8.4   *A strategy S dominates a strategy T if the player prefers every outcome in S more than the corresponding outcome in T.*

For the row player, that means every first coordinate of S is higher than the corresponding first coordinate in T; for the column player, every second coordinate in S is higher than the corresponding coordinate in T. In the preceding example, neither row dominates the other. However, the middle column dominates both of the other two columns. As usual, you can read dominance in the movement diagram since all the arrows point to the entries in the dominant row or column. (It's not obvious directly from the movement diagram, however, that neither the first column dominates the third nor the reverse. One must look more closely at the payoff matrix.)

Similarly, the way to detect stability carries over as well.

DEFINITION 8.5   *An outcome of an ordinal game is called a* Nash equilibrium *if neither player can improve the preference ranking of their payoff by a unilateral change of strategy.*

As usual, Nash equilibria are evident from the movement diagram by virtue of being an outcome to which both vertical and horizontal arrows point. In the preceding example, the outcome in the $1, 2$ position, with payoff $(2, 6)$ is a Nash equilibrium.

The concept of a payoff polygon does not carry over because the payoffs are not real numbers. Nevertheless, the concept of Pareto efficiency still makes sense:

DEFINITION 8.6   *An outcome of an ordinal game is* Pareto efficient *if there is no other outcome that improves both players' preferences.*

Notice that in this definition it is not necessary to consider the case of just one player improving because we have assumed there are no ties in the preference rankings for each player.

In the game just described, the Nash equilibrium is necessarily Pareto efficient because one of the players obtains his or her most preferred outcome 6, so it is not possible to improve his or her ranking.

8.2.2   Prisoners' Dilemma and Other Dilemmas

Unfortunately, some of the difficulties we had with variable-sum games also carry over. There are ordinal games where Nash equilibria are not Pareto efficient and games where they are not equivalent and interchangeable even if they are Pareto efficient. These difficulties are already present in the 2×2 case, so we'll focus on that case for now. The famous prisoner's dilemma illustrates the former issue.

Example 8.7 *The Prisoners' Dilemma.* Chris and Tina were arrested for a crime. This is not a true story, and in most narrations, the crime is said to be a bank robbery. The exact crime doesn't matter. It could be robbing a bank; it could be burning down a house. To make the dynamics of the dilemma work, what is required is that they were arrested only with some puzzling circumstantial evidence and cannot be convicted without a witness account or a confession. They can be convicted for a lesser offense for which the evidence is solid, such as carrying a concealed weapon or driving a stolen vehicle. District Attorney Jerry Byrne has the offenders kept in separate cells when he tells each one that he or she will go to jail for a short time for the lesser offense (say, one year) unless he or she turns state's evidence against his or her partner and gives a witness account for the more serious crime. If he or she does that, the charges for the lesser crime will be dropped. However, if the partner gives a witness account for his or her role in the larger offense, he or she will have to serve time for that. How much time depends on what he or she does – if he or she refuses to cooperate and do not give a witness account against the partner, he or she will have the book thrown at them and serve a long sentence (say, eight years), while the partner goes free for having turned state's evidence against him or her. If they both give witness accounts against each other, they both will serve time, but a shorter sentence (say, three years), for cooperating with DA Byrne.

Each prisoner wants most to go free, which happens when they talk and their partner does not, so that is the most preferred outcome for them. Their second most preferred outcome is to spend the one year in jail for the lesser crime, which happens when they both remain silent. The third preferred outcome is the short sentence they get when they both talk, and their least preferred outcome is the long sentence they receive if they remain silent but their partner talks and goes free. The payoff matrix is as follows.

		Tina	
		Remain silent	Talk
Chris	Remain silent	$(3,3)$	$(1,4)$
	Talk	$(4,1)$	$(2,2)$

For each person, the strategy of providing a witness account against the partner ("talk") is dominant. If both partners do so, they will convict each other and end up with the $(2,2)$ outcome, which is a Nash equilibrium. Obviously, they would have done better to both remain silent and end up with the $(3,3)$ outcome! The $(2,2)$ outcome is not Pareto efficient. However, the $(3,3)$ outcome is highly unstable – not only does each person stand to gain by a unilateral change from this outcome, but each one stands to lose a lot if he or she does not change unilaterally but the opponent does, because when one person obtains his or her first preference, the other obtains his or her last preference, sometimes called the "sucker's payoff." It would take a great deal of trust in the other person to not switch away from the $(3,3)$ outcome.

So, just as with variable-sum games, the interests of the individual player may not align with the interests of the group, making it difficult to say what rational behavior is. This particular dilemma, although it is beautifully illustrated by this ordinal game, is exactly the dilemma that shows up in many games with real number payoffs. In fact, we could model this story as a variable-sum game by using the number of years of a jail sentence as a payoff (the payoff being negative). Indeed, here is the payoff matrix.

		Tina	
		Remain silent	Talk
Chris	Remain silent	$(-1, -1)$	$(-8, 0)$
	Talk	$(0, -8)$	$(-3, -3)$

In chapter 12 of Straffin (1993), a general form is given, with conditions the payoffs must satisfy, for a variable-sum game to be a prisoner's dilemma, which is attributed to Rapoport and Chammah. In a sense, the real number payoffs obscure things – the essence of the conflict between the individual and the group, the fear of getting stuck with the sucker's payoff, and the Pareto-inefficient Nash equilibrium that arises as the intersection of dominant strategies are all clearly visible in the ordinal game.

The very first variable-sum game we looked at in Section 7.4.1 had the same qualitative features exactly with this payoff matrix:

$$A = \begin{bmatrix} (5,5) & (2,11) \\ (10,2) & (3,4) \end{bmatrix}.$$

So we would be inclined to call this matrix a prisoner's dilemma also, even though it does not have the exact form specified by Rapoport and Chammah. As we'll see later, the dynamics of the prisoner's dilemma come up a lot in real-life situations of conflict, so if anyone were to be able to determine the best way to resolve the dilemma, a lot of problems would be solved.

Before looking at applications of the prisoner's dilemma, let's consider some other well-known dilemmas that can be modeled as 2×2 ordinal games. Unlike the prisoner's dilemma, the next examples do not have dominant strategies. However, they have Pareto-efficient Nash equilibria that, however, fail to be equivalent and interchangeable, thus causing problems in the execution of the games. Consider a payoff matrix of an ordinal game of the form:

$$A = \begin{bmatrix} (b,b) & (a,4) \\ (4,a) & (c,c) \end{bmatrix}.$$

where a, b, c must be one of $1, 2, 3$. We seek a model where neither player has a dominant strategy. Since $4 > b$ no matter what b is, only the second row and the second column could possibly be dominant. For that to fail to happen, we would need $a > c$. We would like the outcomes that contain the 4 payoffs to be Nash equilibria. In order for that to happen, we also need $a > c$, so the bottom arrow on the movement diagram points to the left and the vertical arrow on the right points up.

Let's consider this case by case. If $c = 3$, then $a > c$ implies $a = 4$, which is impossible since each player has already used his or her four preferences. Thus, c cannot be 3. If $c = 2$, then $4 > a > c$ implies $a = 3$, and therefore, $b = 1$. This leads to the following payoff matrix:

$$A = \begin{bmatrix} (1,1) & (3,4) \\ (4,3) & (2,2) \end{bmatrix}.$$

The reader can check that the outcomes with $(4,3)$ and $(3,4)$ are both Pareto-efficient Nash equilibria. However, the row player prefers $(4,3)$, and the column player prefers $(3,4)$. If they both try for their preferred Nash equilibrium, the outcome is $(2,2)$, which is not an equilibrium and not Pareto efficient. Thus, this payoff matrix is also a dilemma in the sense that it is not clear what the optimal choice is. For example, the row player might reason that her opponent wants the $(3,4)$ and decide to choose the first row because she thinks a payoff of 3 is not too bad and wants to avoid the $(2,2)$ outcome. However, if the column player thinks the same way and they both choose the first strategy, they end up at $(1,1)$! Is there a story that this ordinal game models like the story of Chris and Tina, which is modeled by the prisoner's dilemma? Yes, there is – it's called the *Battle of the Sexes*, and it is explored in Exercise 9 at the end of this section.

Now we examine the case when $c = 1$. Then there are two solutions, one has $b = 2$ and $a = 3$; the other has $b = 3$ and $a = 2$. The former of these leads to a payoff matrix similar to the Battle of the Sexes, except the $(1,1)$ and $(2,2)$ outcomes switch places:

$$A = \begin{bmatrix} (2,2) & (3,4) \\ (4,3) & (1,1) \end{bmatrix}.$$

It has a very similar dynamic to the Battle of the Sexes – each player has a Pareto-efficient Nash equilibrium, but they are not equivalent and interchangeable, and if each player tries for his or her preferred equilibrium, the outcome is not an equilibrium point, nor is it Pareto efficient. In fact, in this case, they end up at $(1,1)$, the worst possible outcome. This dilemma is called *follow the leader*, and a story that illustrates it is explored in Exercise 10.

The final dilemma of this type is $b = 3$ and $a = 2$:

$$A = \begin{bmatrix} (3,3) & (2,4) \\ (4,2) & (1,1) \end{bmatrix}.$$

In this case, there is a similar dynamic to both of the previous games except that if both players decide to not pursue their preferred equilibrium, this time they end up at a pretty good outcome of $(3,3)$. Also, this game is similar to the prisoner's dilemma as well, with the difference being where the 1 and 2 payoffs are located. In this game, each player gets his or her least preferred outcome simultaneously, while in the prisoner's dilemma, one player gets his or her worst payoff at the outcome where the opponent gets his or her best payoff.

This game is called the *chicken* dilemma. In the (ridiculous!) game of chicken, the two players are driving cars directly toward one another, each one waiting for the other to "chicken out" and swerve out of the way. In the preceding matrix, the first row/column is the 'swerve out of the way' strategy, and the second row/column is the "do not swerve" strategy. Each player most wants the other driver to swerve while he or she doesn't not swerve, winning the game by displaying the most bravery. If they both swerve, the $(3,3)$ outcome is second best – they both display an equal amount of bravery and avoid crashing as well. The $(1,1)$ payoff if neither swerves is a disaster – the cars crash and the drivers are injured and possibly die. Who cares how brave they were now? Sometimes this game is called *brinkmanship* because the players are pushing to the brink of disaster in an effort to get the other player to back down first.

Because there are only four possible preferences in a 2×2 ordinal game, there are only a finite number of different such games (unlike games with real number payoffs). In fact, it can be shown that there are essentially only 78 different such games, once you account for symmetries in the payoffs and switching the players, etc. We have just seen that four of them represent dilemmas where it is not clear how to play the game. But not all of them lead to dilemmas. For example, the following game has an easy solution:

$$A = \begin{bmatrix} (4,3) & (3,2) \\ (2,4) & (1,1) \end{bmatrix}.$$

In this game, the first row and column are both dominant, and the natural outcome has payoff $(4,3)$, which is the only Pareto-efficient Nash equilibrium. Here, it is clear what rational behavior is – each player chooses his or her dominant strategy, and the result is both stable and Pareto efficient. In this game, what is good for the individual aligns with what is good for the group. For this reason, just as in the case for variable-sum games, we say an ordinal game is *strictly determined* if there is a unique Pareto-efficient Nash equilibrium. Notice the other case, where several different outcomes that could be equivalent Nash equilibria cannot happen for ordinal games because we stipulated that each player ranks the outcomes in order of preference without ties.

What about larger ordinal games? Just as in the 2×2 case, some are strictly determined, and some are dilemmas. The dilemmas arise for the same reasons they do in the 2×2 case. Either there are Pareto-inefficient Nash equilibria or more than one Pareto-efficient Nash equilibria that are not equivalent or interchangeable. We'll mostly stick to the 2×2 case then, since all of the difficulties are already present there. With just two strategies, many authors like to use a common term to name the strategies. The terms are "cooperating" or "defecting," and what these terms mean depends on the actual game. For example, in the prisoner's dilemma, cooperating (with the other player – your partner, not with the police!) means remaining silent, while defecting means to talk. In chicken, cooperating means to swerve while defecting means to drive straight on, etc. Some authors (see Taylor and Pacelli (2008), for example) use the term "noncooperating" instead of "defecting."

Finally, we have made no mention in these games of mixed strategies. A moment's reflection will reveal why not. The concept of a mixed strategy is still valid – in repeated play, sometimes play one strategy, sometimes another. Or in a single-play game, play a strategy with a certain probability. However, since we are not measuring payoffs with real numbers, we have lost all ability to compare different mixed strategies. The formula $E = pAq$ that we used in games with real number payoffs no longer applies. For this reason, we can't say much of anything useful about mixed strategies, with one exception. In a repeated game, we might try to use our choices of strategies as a way of sending a "message" to the other player if actual communication is not possible. We'll come back to this point later.

8.2.3   Some Applications

During the Cold War, the USA and the USSR built up their nuclear arms stockpiles to the point that each side had enough weapons to destroy the world several times over. Why would each side continue to build arms after they had enough to destroy the world once? Does not that behavior seem wasteful at best and absolutely insane at worst? Each country has the option each year to continue to build arms or to maintain their current stockpiles (or even disarm to an extent). Since it seems safest for the world if we do not increase the nuclear stockpiles, we'll think of the maintain/disarm strategy as cooperating and the strategy to build more arms as defecting. Since it is difficult to measure what each side gains in either case, we'll simply try to give the preference rankings of the outcomes and model the arms race as an ordinal game.

If one side continues to build while the other maintains or disarms, the country that builds is seen to be gaining a military superiority over the other. Each side most wants this military superiority and least wants the other side to have it. If both countries build arms, or if neither does, then both sides maintain a military parity. Given a military parity, both countries prefer to not build arms over building, since building arms is expensive and weakens the economy, so mutual maintain/disarm is each country's second preferred outcome, while mutual buildup is each country's third preferred outcome. Thus, the payoff matrix is this.

		USSR	
		Cooperate (maintain/disarm)	Defect (build arms)
USA	Cooperate (maintain/disarm)	$(3,3)$	$(1,4)$
	Defect (build arms)	$(4,1)$	$(2,2)$

The reader should recognize at once that this is a prisoner's dilemma. Each side has a dominant strategy to defect, leading to a Nash equilibrium at the mutual arms buildup. This was an especially stable outcome during the Cold War, because (at least in the early stages of the Cold War) neither side trusted the other enough to try disarming and risk losing the military edge (the sucker's payoff in this case). Once you realize the arms race is governed by the dynamic of a prisoner's dilemma, it becomes easier to understand why each country kept building arms. This does not justify mutual buildup to the point of annihilating the world many times over, but it does explain why it happened.

Of course, we have simplified things a lot. This game is not played just once. Each year, both countries allocate money for arms buildup, so the game is repeated. Everyone might agree that mutual disarmament is better for both countries (not to mention safer for the entire world) than mutual buildup. But how does one initiate mutual cooperation without risking losing one's military edge? Otherwise put, how should one play the iterated version of the prisoner's dilemma?

In 1980, Professor Robert Axelrod of the University of Michigan ran two tournaments. Teams of computer scientists, psychologists, and game theorists were invited to submit algorithms for playing the iterated prisoner's dilemma (a version with real number payoffs rather than the ordinal game version, so scores could accumulate), and these algorithms were pitted against each other in a computer-run tournament. Some of the algorithms were complicated, but the winner of both tournaments was the following very simple algorithm submitted by Anatol Rapoport.

ALGORITHM 8.8 Tit for Tat. *In the first move, cooperate. Then, on each subsequent move, do whatever your opponent did the previous move.*

Some game theorists believe this is an optimal strategy. It is not an invincible algorithm – it can lose to some other algorithms, but overall, it does quite well. The attributes that seem to make it do so well include a willingness to cooperate and instant retaliation if your opponent does not cooperate. Furthermore, this retaliation is short-lived and quick to forgive – if your opponent cooperates once, you immediately go along and cooperate also. This is what we meant when we suggested using your play to "send a message" in a game if no communication is possible.

Of course, in the real arms race, actual communication did take place. After a few years of mutual buildup, both sides eventually negotiated arms treaties that limited further buildup. History shows that both sides broke these treaties after a time. The reasons perhaps include a deep mistrust of the other side and the fear of losing the military edge, but they may also include other factors as well, such as the development of new technology and/or using the breaking of a treaty to punish the other side for some other perceived political infraction that was not directly related to the arms race.

It is well known that the arms race can be modeled as a prisoner's dilemma, but these types of dilemmas arise in many other areas of real life. The following example shows that we can model grade inflation as a prisoner's dilemma.

· Graham is a math professor at Cooper College, and Alice is a math professor at Nash Academy. Both schools are highly selective, competitive undergraduate institutions. The faculty at both schools are concerned about grade inflation – they feel that if most students get an A, then the grade does not mean anything. Both schools would like to uphold the strictest of standards in an effort to fight grade inflation. However, both Graham and Alice teach students who intend to go on to graduate study, and the schools are competing to send students to the best graduate programs (which obviously helps their schools recruit new students). Each professor has the option of cooperating, and upholding the school's intended standards for stricter grading, or defecting, and being more forgiving in awarding grades. If a student earns mostly A grades, that does not guarantee a spot in graduate school, but earning many grades of C likely diminishes the student's chances for acceptance. Let's suppose that the most important objective of each teacher is to have students go on to graduate study, and their secondary goal is to fight grade inflation. From this perspective, the best outcome is when many of their students earn A's while the students graduating from the other institution have lower grades, which is the worst outcome from the point of view of the other school whose students have lower grades. If both professors grade the same way, then their students stand roughly equal chances of acceptance in graduate school, but in that case, each would prefer to uphold the strict grading standards than to not. Thus, we have the payoff matrix, another prisoner's dilemma.

		Graham	
		Cooperate (strict grading)	Defect (easy grading)
Alice	Cooperate (strict grading)	(3, 3)	(1, 4)
	Defect (easy grading)	(4, 1)	(2, 2)

In reality, of course, admission into graduate school depends on a lot more than grades, including letters of recommendation, scores on the GRE exam, and how the students impress the interviewers at the graduate programs. That's okay – remember a mathematical model is supposed to be simpler than the real-life situation that it models. But the simplified model reveals some of the basic dynamics of grade inflation. Professors may want to fight grade inflation by holding their students to strict standards, but at the same time, they do not want to disadvantage their students in applying to graduate school. They are caught in the same dilemma that makes the arms race spiral out of control.

An interesting aspect of this example is that the same conclusions can be drawn under some different assumptions. There are other motivations for professors that will make easy grading dominant besides a concern for their students' chances at admission to graduate school. For example, if a professor gets a reputation as a tough grader, that may drive some students away. The result could be enrollments small enough to cause the dean to cancel the professor's classes. Another possibility is the instructor may want to minimize the chance of getting poor student evaluations or their course, as such evaluations may play a role in tenure and promotion decisions for the faculty member. In each case, whatever the motivation for easy grading, the payoff matrix is the same prisoner's dilemma.

This example is also a good reminder that situations of conflict arise with many players and not all games are between two players. The grading decision is made by all of the professors at all schools, not just Alice and Graham. The same conflict between individual benefit and group benefit is evident. Regardless of how many teachers there are, each one has a dominant strategy to grade less strictly, but if everyone does, everyone is worse off than if nobody did. So the conflict arising in the prisoner's dilemma appears also with many players. We briefly consider n-player games later in this chapter.

In Exercise 14 at the end of the last section, a political conflict was analyzed in the United States Federal System trying to pass a budget. This was a fictional example, but it was inspired by the actual federal government shutdown of 2013. In the exercise, our fictional conflict was modeled sequentially, using a tree diagram, and the payoffs were real numbers (expected numbers of seats in congress gained or lost in the next election). But the essence of the conflict can be simplified by using an ordinal game model. The central issue of the real conflict in 2013 was funding for the already-passed Patient Protection and Affordable Care Act, commonly referred to as "Obamacare." Key features of Obamacare were scheduled to start on October 1, 2013, the same day that the new federal fiscal year started. During September, Congress was debating the 2014 Continuing Appropriations Resolution (the budget bill). The two chambers of Congress could not agree on funding levels, causing a "funding gap," which triggered a 16-day government shutdown.

Specifically, on September 20, the House of Representatives (controlled by the Republican Party), urged by its most conservative members, added resolutions to the funding bill, which delayed funding and/or defunded parts of the Obamacare program. These resolutions added by the House were unacceptable to the Democratic-controlled Senate. The Senate removed the extra resolutions and passed a revised version on September 27 with its own added resolutions to maintain funding for Obamacare at current levels with no additional conditions. The House refused to pass the Senate version, and no compromise was reached by September 30, whence there was no federal budget for the 2014 fiscal year beginning the next day. Therefore, the government "shut down," meaning that only essential services continued to operate. About 800,000 government workers were furloughed without pay, and many more were required to work without knowing when and if they would be paid, until the conflict was resolved 16 days later on October 16. It was the third-longest shutdown of the US government in history.

For our simplified model, the two players are the House and the Senate. Each side had an option to negotiate a compromise with the other side, as they eventually did on October 16. We'll call this the cooperative strategy. The other option was to refuse to compromise and support only the version of the continuing appropriations resolution sponsored by its own chamber, the noncooperation or defection strategy. Each side would most prefer to defect and hold on to its own version of the bill while the other side relented. The second preference would be mutual cooperation so that a compromise can be worked out. The third preference for each chamber is to give in to the other's demands and pass the bill sponsored by the other chamber. For both chambers, the least preferred outcome is mutual defection, since it results in a shutdown that hurts all Americans (especially the government employees who are furloughed or not paid). Therefore, the payoff matrix is as follows.

		Senate	
		Cooperate (negotiate/ compromise)	Defect (refuse to compromise)
House	Cooperate(negotiate/compromise)	(3, 3)	(2, 4)
	Defect (refuse to compromise)	(4, 2)	(1, 1)

As the reader can see, this is not a prisoner's dilemma but a chicken dilemma. Each side wants the other side to back down first and is willing to drive the conflict to the brink of disaster (and in this case, past the brink) before compromising themselves. Again, this model oversimplifies the conflict. In reality, the players did not act simultaneously, and communication took place between the players and with outside parties such as the president. This model predicts mutual defection, leading to a government shutdown, which is indeed what happened, so even though it is a simple model, it seems to capture the essence of the conflict.

On the other hand, the shutdown did not last forever. With each passing day, both parties lost support from their constituents for continuing the shutdown. (One survey suggested that 53% of Americans held the Republicans responsible in the end.) Eventually, a compromise was reached. There is no provision in our model for a player *changing their mind and switching strategies*, but if we could build that in, the model might be more realistic.

Such examples of brinkmanship appear in many conflicts. Every shutdown of the federal government due to a failure of Congress to pass a budget can be modeled the same way, including the 21-day 1996 shutdown during the Clinton administration over opposition to major spending cuts and the 35-day shutdown in 2018 during the Trump administration over funding for an expansion of the US-Mexico border wall, the longest shutdown in US history.

Much has been written about conflicts of international diplomacy that can be modeled as chicken dilemmas, such as the behavior of some European countries just prior to World War I. See Snyder (1971) for a general discussion and the references there in for specific examples.

Probably the most famous example of a diplomatic chicken dilemma was the Cuban Missile Crisis of 1962. Here is a very simplified outline of the crisis. The USSR installed medium-range missiles tipped with nuclear warheads in Cuba in the fall of 1962. The USA detected them via spy plane photography almost immediately, even before the missiles were operational. This discovery precipitated the crisis, which lasted about two weeks in October 1962. At first, the USA kept the discovery a secret while they formed an "Executive Committee" to weigh options and decide what to do. Since the missiles could reach targets in the USA (including Washington, DC) within five minutes of being launched, the presence of these missiles gave the USSR "first strike" capabilities, and leaders in the USA feared that was the intended purpose of the missiles. Once the news about the missiles was shared, tensions rose correspondingly throughout the world due to the increased likelihood of nuclear war between the superpowers, which was later estimated by

President Kennedy to have been between $\frac{1}{3}$ and $\frac{1}{2}$. It is believed to be the closest the world has come to nuclear war.

There are a number of ways to model this conflict as a game. In Straffin (1993), a simplified version is set up in extended form, but the analysis is incomplete, as the payoffs are not specified. Note that one could still analyze the game tree using reverse induction even if the payoffs are not given as real numbers and all one knows are the preference rankings for the outcomes. This suggests that ordinal games can be played in extensive form as well as normal form.

In Brams (1985a and 1985b), political scientist Steven Brams explicitly adopts the ordinal game format. There is some flexibility in deciding what the strategies are, but the resulting payoff matrix is a chicken dilemma. In each model proposed by Brams, the USSR has the same two options. The "cooperate" strategy is to withdraw the missiles from Cuba, while the "defect" strategy is to maintain the missiles there. The options for the USA are somewhat less clear. For any pair of options the USA adopts as strategies, the less aggressive option will be designated as "cooperate" and the more aggressive as "defect." In one of Brams's models, to cooperate meant to set up a naval blockade around Cuba to prevent shipments of weapons to that island from the USSR (something the United States ultimately did), while to defect meant to attack the missile bases with an air strike to remove them.

In another of his models, the defection strategy was even more aggressive – a full-scale invasion of the island with the goal of removing not only the missile bases but also the Communist dictator Fidel Castro (presumably with Castro gone, Cuba would no longer be a USSR satellite, thus ensuring no future arms buildup on the island). In this model, the cooperation strategy for the USA would be to give up the option to invade Cuba, presumably publicly. This is also something the USA did, and it would be an outcome highly valued by the USSR. Indeed, the previous year, the USA did try to invade the island but failed (the "Bay of Pigs Invasion" of 1961). There is some speculation that the reason the USSR put the missiles there in the first place was as a reaction to the failed invasion, in the hopes of protecting Cuba from future invasions.

In each case, the first preference of each superpower would be to defect while the other side cooperates. This would demonstrate to the world the military superiority and steadfast resolve of the "winning" country. If that can't happen, then the second choice would be mutual cooperation for each country. The third preference would be to be on the losing end – to cooperate while the other side defected. The fourth preferred option for each side would be mutual defection. In this case, with the missiles being maintained in Cuba and being attacked by air strike or by invasion or possibly both, the conflict would certainly escalate to nuclear war, which would be a disaster of unprecedented proportions. Thus, the payoff matrix is indeed a chicken dilemma, regardless of which pair of strategies the USA uses (blockade vs. air strike or invade vs. promise not to invade) as shown in the following table.

		USSR	
		Cooperate (withdraw missiles)	Defect (maintain missiles)
USA	Cooperate (blockade or give up option to invade Cuba)	(3, 3)	(2, 4)
	Defect (air strike or invade Cuba)	(4, 2)	(1, 1)

These two simplified models show very clearly the tension of the times. In Brams (1985a and 1985b), these models are expanded by considering that in the real crisis, the moves were not simultaneous, and communication between the players did take place. There is also the possibility that the executive committee that decided on the US moves considered other options besides a blockade or an invasion. For example, the committee very likely discussed doing both – an air strike followed later by an invasion. There was also undoubtedly some discussion about what to say to the USSR ahead of either option. Should they give them an ultimatum – "remove the missiles or else we invade," which would allow for the possibility that the USSR might cooperate before things got worse? Some probably felt that was a poor move, since if the USSR knows the USA plans to attack, the USSR could decide to use the missiles rather than lose them and attack the USA first. On the other hand, if the USA did not warn the USSR ahead of time and simply attacked the missile bases, the USA would be seen by much of the world as ethically not being in the right. A superpower using a "sneak attack" on a tiny country is exactly the kind of behavior for which the USA had long criticized the USSR.

Yet other options may have been considered. In the end, the USA made a promise (not made public until months after the crisis) to the USSR to remove its Jupiter missiles from Turkey. These missiles were already obsolete, so it was not that much of a concession for the USA to do so, yet a trade of missiles in Cuba for missiles in Turkey may have helped both sides decide to back down from an aggressive stance without either side losing face. Perhaps offering to make this trade was a possible move considered by the executive committee (although such a move would not have been favored by the "hawks" in the US government and on the committee, so it is no surprise that it was kept secret at first).

Before leaving this example, we note that a chicken model predicts mutual defection, since each side is trying for the Nash equilibrium that best helps it. Of course, in real life, that is not what happened. By the end of the crisis two weeks later, we had fortunately evolved to mutual cooperation and averted a nuclear confrontation. Thus, while the chicken model does show the dynamics and the tensions of the conflict, it is not accurate as a predictive model. As noted, the reasons for this were probably because the game is sequential in nature – moves were not made simultaneously and independently – and because communication was a part of the real-life game. When Brams considered these modifications to the chicken model in his articles and books, he was executing exactly the kind of sensitivity analysis that should be done with any model.

Are there other modifications that can be made in the basic ordinal game model aside from making the moves sequential and allowing communication? One possibility that will be explored in the following exercises is to consider that since the USA had more than two options, to make the payoff matrix larger than 2×2. It turns out that change does not really modify our conclusions. A better modification, also based on the work of Brams (1994), is described in the next section.

Exercises

For Exercises 1–8, for each ordinal payoff matrix, draw the movement diagram, and find all dominant strategies (if any) and all Nash equilibria (if any). If possible, predict the outcome if the game is played as usual with the players choosing simultaneously and independently. State if the game is strictly determined.

1.
$$\begin{bmatrix} (1,2) & (2,3) \\ (3,4) & (4,1) \end{bmatrix}$$

2.
$$\begin{bmatrix} (2,3) & (3,2) \\ (1,1) & (4,4) \end{bmatrix}$$

3.
$$\begin{bmatrix} (2,3) & (3,2) \\ (1,4) & (4,1) \end{bmatrix}$$

4.
$$\begin{bmatrix} (2,4) & (1,3) \\ (3,1) & (4,2) \end{bmatrix}$$

5.
$$\begin{bmatrix} (6,3) & (1,6) & (4,1) \\ (2,5) & (5,4) & (3,2) \end{bmatrix}$$

6.
$$\begin{bmatrix} (8,3) & (3,2) & (5,1) \\ (4,7) & (9,8) & (2,9) \\ (6,4) & (1,5) & (7,6) \end{bmatrix}$$

7.
$$\begin{bmatrix} (8,3) & (3,7) & (5,5) \\ (4,6) & (9,2) & (2,9) \\ (6,1) & (1,8) & (7,4) \end{bmatrix}$$

8.
$$\begin{bmatrix} (1,3) & (4,2) & (8,6) \\ (6,1) & (7,5) & (3,7) \\ (9,8) & (2,4) & (5,9) \end{bmatrix}$$

9.   The following story illustrates the *Battle of the Sexes dilemma*. James and Carly have plans to meet after work for a date. James wants to go to see his favorite football team, the Steamrollers, in their final game of the season, while Carly would prefer to go to the cinema to see the film *Anticipation*. Of course, they both have a strong preference to be together rather than alone, and they left for work this morning without finalizing their plans. Carly's phone battery has died, so they cannot communicate before going to either event, both of which start at 6 p.m. – leaving just enough time to travel to each location from work. Model this as a 2 × 2 ordinal game, and show that the payoff matrix is indeed the Battle of the Sexes dilemma. If each person tries for his or her preferred Nash equilibrium, what is the predicted outcome of the game?

10.   The following story illustrates the *follow the leader dilemma*. Bruce and Clarence simultaneously pull up to stop signs across from each other on a busy intersection of Tenth Avenue. When a gap finally appears in the traffic, each driver has a choice to either drive directly into the gap or else to concede the right of way to the other driver. If they both drive, they could crash, which would be the last preference of both drivers. If they both concede, then each driver is delayed, and they may both lose the opportunity to drive into the gap before it vanishes, so that is the second-to-last preference for both drivers (a "Tenth Avenue Freeze Up"). However, if one driver goes and the other concedes, the one that goes gets his first preference, and there may still be time for the one who concedes to follow the leader into the gap before it vanishes. Model this situation as a 2 × 2 ordinal game, and show that the payoff matrix is indeed the follow the leader dilemma. If each driver tries for his or her preferred Nash equilibrium, what is the predicted outcome?

11.   Consider the following situation. Mr. Browne's car is in need of service, and his local garage down on Main Street has only two openings – a Friday appointment and a Saturday appointment – after which it is several weeks before they have another opening. Thus, Mr. Browne needs to make an appointment for one of those two days. Meanwhile, Mrs. Browne has an eye infection, and her eye doctor has openings also only on the same two days, Friday and Saturday. We suppose that each person will make his or her own appointment, but Mr. and Mrs. Brown are not able to communicate with each other ahead of time about their choices, and we assume they make the calls to schedule the appointments simultaneously. They both need the car to travel to their appointments. The purpose of this exercise is to set up the situation as an ordinal game under various assumptions about the preferences of the players, and in each case, we assume that each person is aware of the preferences of the other person.

   a.   Assume that both players, first preference is for the appointments to be on different days because it would be difficult (but not impossible) to works things out if they are the same day. Given that, each person has a minor preference for Friday over Saturday – Mr. Browne because the garage charges more for work done over the weekend, and Mrs. Browne because she is anxious to treat her infection as soon as possible. Write the payoff matrix as an ordinal game. If the matrix happens to be one of the four dilemmas we have studied, state which one. If the game is strictly determined, say what the predicted outcome is.

   b.   Suppose that each person's highest preference is to have a Friday appointment (for the reasons given in Part a), and the secondary preference for both players is to schedule the appointments on different days. Follow the same directions as in Part a.

   c.   Suppose that Mr. Browne most prefers a Saturday appointment so he will not have to leave work early on Friday, while Mrs. Browne most prefers a Friday appointment. For each of them, a secondary preference is to schedule the appointments on different days. Follow the same directions as in Part a.

   d.   Suppose that Mr. Browne most prefers a Saturday appointment and Mrs. Brown most prefers a Friday appointment, but for each of them, a secondary preference is to schedule the appointments on the same day so they can do all their errands at once. Follow the same directions as in Part a.

   e.   Suppose that for Mr. Browne, his main preference is for a Saturday appointment, and his secondary preference is to have appointments on different days. Mrs. Browne most prefers to have the appointments on the same day (so Mr. Browne can drive her and she does not have do any driving with her compromised eye), and she has a secondary preference for a Friday appointment since it is less time to wait to treat her infection. Follow the same directions as in Part a.

f.   Both players' first preference is for the appointments to be on different days rather than the same day. Both players feel their second preference is for a Friday appointment if they go different days but or Saturday if they go on the same day. Follow the same directions as in Part a.

12.   Angie and her sister Kiki are selecting colleges. They each have been accepted at a prestigious school in Boston and at one in New York City. All they need to do is decide which school to attend, which – for the purposes of this exercise – we assume they do simultaneously and independently without talking with each other. After spending all that time together during their childhood operating their famous lemonade stand, they each feel a strong preference to spend some time apart and attend different schools. However, they each have a secondary preference to live in New York City over living in Boston. Model this situation as an ordinal game. Which of the dilemmas is the payoff matrix?

13.   Show that, in Exercise 12, if the strength of the girls' preferences is reversed so that their highest preference is to live in New York City and their secondary preference is to attend different schools, then the game becomes strictly determined. What is the predicted outcome?

14.   Suppose that Angie and Kiki have been accepted at three schools: Harvard, Yale, and Princeton. Angie ranks them in exactly that order, while Kiki ranks them in the reverse order – Princeton, Yale, then Harvard. Suppose each girl's first priority is to go to a school she has ranked as high as possible, but they each have a secondary priority to be located as close as they can be to each other (so attending the same school is ideal from this persepctive). For the purposes of deciding the preferences for the outcomes, note that Princeton and Harvard are quite far apart, Yale is closer to both, but somewhat closer to Harvard than it is to Princeton. Model this situation as a 3×3 ordinal game. Draw the movement diagram, find all Nash equilibria, and state if the game is strictly determined. What is the predicted outcome?

15.   Show that, in Exercise 14, if the strength of the girls, preferences is reversed (but not their rankings of the schools) so that each girl's highest priority is to be as close as possible to her sister, while each girl's second priority is to attend a school she has ranked as high as possible. Then the game is not strictly determined; in fact, there are three Nash equilibria that are all Pareto efficient, but no two of them are equivalent or interchangeable. How would you resolve their dilemma?

16.   Make up a payoff matrix for a 2×2 ordinal game where all four outcomes are Pareto efficient, and the game is strictly determined, and also make up one that is not strictly determined.

17.   Johnny Gunn and Peter Rivers are spies working for the CIA. They have infiltrated a hostile foreign spy organization under cover and are gathering inside information on the group. They have become aware of a plan to rob a bank in order to obtains funds for a secret arms deal. Agent Gunn and Agent Rivers each are considering exposing the plot, which would mean exposing their cover and leaving the organization to return home. Both most

want to be credited for exposing the plot themselves, since they will advance in their jobs and get a nice raise. However, if one exposes the plot, the organization will be alerted to the existence of double agents, which will put the other agent in danger. Thus, the worst outcome for each is if the other one exposes the plot. However, if they both expose the plot, they each will have a blown cover and will no longer be able to operate as secret agents within the organization, thereby cutting off the flow of information to the CIA, so each regards the outcome of neither one exposing the plot as better than both exposing it. Model the situation as an ordinal game. Is the game is strictly determined or not, and if not, which dilemma is it?

18. Consider the Cuban Missile Crisis discussed in the text. Suppose that we try to model this as an ordinal game that is larger than 2×2.

   a.   Suppose the USSR stays with just two strategies – either maintain the missiles, or withdraw them. For the USA, suppose there are three possible strategies – maintain the missiles, withdraw them, or an invasion, in order of increasing aggression. Set this up as a 3×2 ordinal game. As in the text, we assume each side most prefers to be aggressive while the other player is not aggressive, since it demonstrates power and resolve, and each side least prefers any outcome that will lead to nuclear war. From this perspective, it seems that if the USSR maintains the missiles, both an air strike and an invasion are equally bad, since it is probable that either outcome will lead to war. For the purposes of setting up the payoff matrix, just assume that an invasion is worse if the USSR maintains the missiles since it is more likely to lead to war than an air strike. In fact, it is worse in any case since the USSR might regard an attack against Cuba as an attack against it, and this could lead to war anyway, even if the missiles are withdrawn. Is the resulting game strictly determined or a dilemma?

   b.   Suppose the USSR considers three options – withdraw the missiles, maintain the missiles, or use the missiles (a first strike against the United States, which, of course, would lead to nuclear war for sure, as the USA would retaliate with nuclear weapons). In this model, assume the US options are just the two options, invade or promise to not invade. Follow the same directions as in Part a.

   c.   Do you think that having a larger payoff matrix brings any insights to the conflict that were not already present in the 2×2 models?

8.3   Sequential Games: The Theory of Moves

In this section, we engage in a little sensitivity analysis with ordinal games. Our starting point is that while the chicken dilemma modeled the essential dynamics of the Cuban Missile Crisis, it is not necessarily a good model to use for predictive purposes. Indeed, if each player tries for the Nash equilibrium that most benefits him or her, the prediction is the disastrous $(1, 1)$ outcome. As we know, that did not happen in 1962. As noted before, in Brams (1985a and 1985b), Brams considered the model obtained by allowing communication between the players, and he also considered the sequential nature of the game when the players do not move simultaneously.

Furthermore, we noted earlier that in the chicken dilemma model of the US federal government shutdown, while the $(1, 1)$ outcome was indeed obtained, it did not last because players were allowed to change their strategies. We suggested that a model that allows changes of strategies in a sequential game might be more realistic. Also, if players are allowed to switch strategies, that implies there is some starting point from which initial changes might be made.

When one modifies 2×2 ordinal game models to incorporate sequential moves, a starting position, and the ability to switch strategies, one arrives at the *theory of moves*, a technique first proposed in Brams (1994) and slightly modified in Taylor and Pacelli (2008). We use the version in Taylor and Pacelli (2008) here.

The idea is that each 2×2 ordinal payoff matrix can be used to create eight different sequential games, four when the row player has the first option to switch strategies (depending on which of the four outcomes is the initial position) and four when the column player has the first move option. Each such sequential game is modeled in extensive form, and, assuming the tree diagram is finite, one can analyze it via reverse induction to see what would actually happen in each of the eight cases. Note that the assumption that it can be analyzed by reverse induction implies we do not relax the intelligence assumption of game theory – we continue to assume both players know and agree on all the preference rankings in the payoff matrix.

Here are the rules for creating the tree diagram. We give the rules for when the row player moves first. If the column player moves first, the rules are similar:

ALGORITHM 8.9 *Suppose you are given a 2×2 ordinal payoff matrix, and a sequential game is to be played with the row player moving first. The steps taken to set up the tree diagram for the Theory of Moves are as follows: .*

1. *Each player makes a simultaneous and independent choice of a strategy. The resulting outcome on the payoff matrix is the* starting *position.*
2. *The row player makes a choice either to* switch *to the other strategy or to* stay *with her current strategy. In each case, draw a branch of the tree, and label it with the resulting outcome. For any branch that is not a* terminal *node, go to step 3. If every node is terminal, then stop – the tree is complete.*
3. *The column player then has a choice to either* switch *to the other strategy or to* stay *with his current strategy. For any branch that is not a* terminal *node, go to step 2. If every node is terminal, then stop – the tree is complete.*

The conditions for a *terminal node* are as follows:

1. If it is the row player's turn to move, and the current outcome is $(4, -)$, except on the row player's first move.
2. If it is the column player's turn to move, and the current outcome is $(-, 4)$.
3. If either player elects to *stay*, except a stay on the row player's first move.

The rules ensure that both players have at least one chance to switch strategies if they want, which is why an initial move by the row player (or whoever goes first) never ends the game. Furthermore, if one player places the other in an outcome, then presumably, he or she is happy with that outcome, but if it happens to be the first choice of the other player (the 4 preference), then – since both players are satisfied – there should be no further switches. The same holds if either player elects to stay with their current choice, as long as both players have had a chance to switch. Once the game tree is set up, the outcome of the game is determined using backwards induction.

When a game is played under simultaneous moves, the "stable points" are called Nash equilibria. For the theory of moves, we have the following definition:

DEFINITION 8.10   *If an outcome of the game has the property that whenever it is the starting position, it is also the final position when the row player moves first, it is said to be a non-myopic equilibrium when row goes first. Similarly, if it has the property that whenever it is the starting position, it is also the final position when the column player moves first, it is said to be a non-myopic equilibrium when column goes first. If an outcome has both properties, we drop the modifying phrases and just say it is a* non-myopic equilibrium.

The name stems from the idea that by analyzing the tree diagram for the sequential moves, one is looking ahead, or being "far-sighted" by anticipating what the other would do in each outcome.

We now analyze the sequential games generated by the chicken dilemma payoff matrix, where we denote by C the cooperation strategy (swerve out of the way) and by D the defection strategy (drive straight on).

		Column player	
		C	D
Player	C	$(3,3)$	$(2,4)$
	D	$(4,2)$	$(1,1)$

Suppose that $(3,3)$ is the starting position, and the row player goes first. In our trees, at each node, the left branch will indicate "stay" and the right branch will indicate "switch." Terminal nodes are boxed off. The reader should verify that according to the preceding rules, we obtain the game tree shown in Figure 8.7.

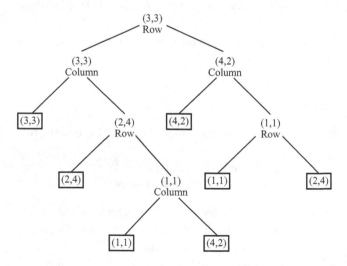

Figure 8.7 The game tree for the chicken dilemma, assuming (3,3) is the staring position.

At this point, neither player has actually made any moves, but the tree outlines the possibilities. We apply reverse induction to analyze the tree in order to determine what to do in the actual play of the game. This is done just as we did for games with real number payoffs. Crossing off the branches leading to undesired outcomes, we obtain the game tree shown in Figure 8.8.

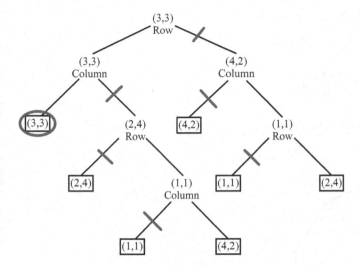

Figure 8.8 Determining the outcome via backwards induction.

The final outcome is the circled node. Thus, if the starting position is $(3,3)$ and the row player goes first, then the ending position is also $(3,3)$, which means that $(3,3)$ is a non-myopic equilibrium when row goes first. By symmetry, it is also a non-myopic equilibrium when column goes first (see Exercise 1), and so it is a non-myopic equilibrium.

We now consider the other possible starting positions. Suppose $(2,4)$ is the starting position (and row goes first.) This leads to the tree diagram shown in Figure 8.9.

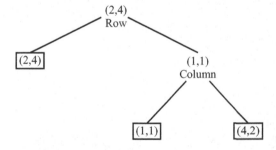

Figure 8.9 The game tree assuming $(2,4)$ is the starting position.

Reverse induction now gives the tree diagram shown in Figure 8.10.

With this starting position, there is no symmetry, so when column moves first, we obtain the tree diagram shown in Figure 8.11.

Reverse induction gives the tree diagram shown in Figure 8.12.

It is interesting to note that the final outcome of the game in this case depends on who moves first! Starting at $(2,4)$, if the row player moves first, the final outcome is $(4,2)$, while if the column player moves first, the final outcome is $(3,3)$. In particular, $(2,4)$ is not a non-myopic equilibrium for either player moving first. In Exercise 2, we ask the reader to check that the reverse is true if

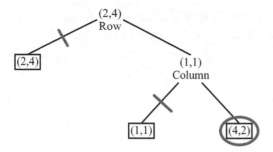

Figure 8.10 The outcome when (2,4) is the starting position.

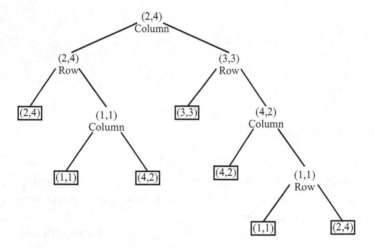

Figure 8.11 The game tree if (2,4) is the starting position, but the column player moves first.

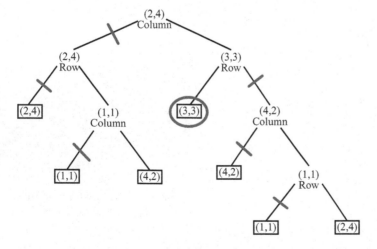

Figure 8.12 The outcome when (2,4) is the starting position and the column player moves first.

the starting position is $(4,2)$. That is, if the column player moves first, the final outcome is $(2,4)$, while if the row player moves first, the final outcome is $(3,3)$.

Finally, let's check when $(1,1)$ is the starting position, and the row player goes first. The game tree is shown in Figure 8.13.

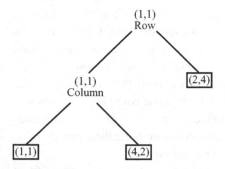

Figure 8.13 The game tree when $(1,1)$ is the starting position and the row player moves first.

Reverse induction gives the game tree shown in Figure 8.14.

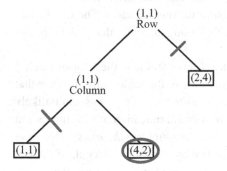

Figure 8.14 The outcome when $(1,1)$ is the starting position and the row player moves first.

Thus, the final position is $(4,2)$. In Exercise 3, we ask the reader to check that if the column player moves first, then the final position is $(2,4)$.

Summarizing, Table 8.2 gives the final positions for each starting position for each of the players moving first. Note that $(3,3)$ is the only non-myopic equilibrium.

Table 8.2 Theory of moves analysis of the chicken dilemma

	Final position	
Starting position	Row moves first	Column moves first
$(3,3)$	$(3,3)$	$(3,3)$
$(2,4)$	$(4,2)$	$(3,3)$
$(4,2)$	$(3,3)$	$(2,4)$
$(1,1)$	$(4,2)$	$(2,4)$

How can this be applied in real-life situations? Consider first the Cuban Missile Crisis. The initial choice of the USSR (the column player) is to put the missiles in, which is the defection strategy. Even though that move came before the US response, I would argue that until the US response, there was no game at all. When the USA finally responded, it was with the blockade option, which is the less aggressive (cooperation) strategy. If we think of that as the US initial choice, then the starting position of the game is $(2,4)$, and it is the USSR who moves first. According to Table 8.2, the natural outcome is $(3,3)$, which corresponds to the USSR removing the missiles. This was the actual result, so the theory of moves version of the chicken dilemma appears to be a more realistic model of the crisis than the simultaneous moves version.

However, we must be careful drawing conclusions. After all, it was more than the blockade that led to the missile withdrawal. We have already mentioned that a promise not to invade Cuba was made in return of the withdrawal, as well as a clandestine agreement to remove the US missiles from Turkey. Still, even if it is not perfect, the sequential version is a more realistic model, since the simultaneous moves version makes the wrong prediction of the outcome.

Partly, the realistic conclusion we made is because of the particular starting position of $(2,4)$ and the assumption that the USSR moves first (that is, it responds to the blockade). Suppose the USA moves first instead of the USSR. According to Table 8.2, the result should be $(4,2)$. A look at the tree diagram reveals why – if the USA switches to an agressive strategy in a first move, then the USSR has no choice but to acqueisce if they want to avoid nuclear war. But is that realistic? In this case, the agressive defection strategy is an air strike against the missile bases. The USSR may have felt locked into a response to that aggression (in fact, they said publically that if the missiles were attacked that they would retaliate).

So it seems that in this case that it is more likely that the $(1,1)$ outcome is the final one rather than the $(4,2)$ outcome. In fact, it is likely that of the eight cases in the table, only the four that lead to the $(3,3)$ outcome are reasonably realistic. The outcomes of $(4,2)$ or $(2,4)$ seem unlikely without building more into the model. Perhaps looking at how communication affects things can resolve this. If the USA could convince the USSR that it was *committed to the air strike option before actually embarking on it,* perhaps the $(4,2)$ outcome is more realistic – except, of course, that once the missiles are removed, there is no need to follow through with the air strike, so we end up at $(3,3)$ after all.

What about other applications? Notice that in none of the eight positions of the table is $(1,1)$ the final outcome. So when we apply the theory of moves version of the chicken dilemma to the federal government shutdown, no matter what the starting position is and regardless which chamber of Congress moves first, it never predicts an actual shutdown of the government, whereas the simultaneous moves version does. Since the real-life game is a sequential game, and it is hard to imagine that the simple simultaneous moves model is truly accurate, one has to look elsewhere to see why the sequential version of chicken did not give the "correct" prediction. We can see two possibilities: (1) The chicken dilemma was the wrong matrix, which means that the preferences are not correct. In particular, perhaps one or both of the chambers of Congress did not really regard a shutdown as the worst-case scenario. (2) The fear of nuclear war in the Cuban Missile Crisis is much stronger than the fear of a shutdown, so even though the $(1,1)$ preference ranking of a shutdown is correct, it is not nearly as compelling as the $(1,1)$ outcome of the Cuban Missile Crisis.

In the first case, one might try experimenting with other payoff matrices. For example, if each chamber thought the worst outcome was if it caved in to the other's demands, rather than a shutdown, which would then be the second-worst outcome, then the matrix is a prisoner's dilemma, not a chicken dilemma. In the exercises, we ask the reader to do a theory of moves analysis of the prisoner's dilemma.

In the second case, it seems that it is not enough to give the preference rankings, and we should resort back to some sort of real number payoffs to distinguish between the two situations. It certainly seems possible to model such games sequentially in a way similar to the theory of moves. We find this idea intriguing, but if we believe that the model would be more flexible if we could incorporate *probabilities of crossing off a branch* as opposed to a binary yes/no choice. The idea is that, in the case of the Cuban Missile Crisis, the $(1, 1)$ outcome is much less likely to be attained than it is in the government shutdown crisis because the payoff would be much worse in the Cuban Missile Crisis.

Another possibility is that one might want to make the preference rankings *functions of time* rather than constants. The idea is that government shutdown may not have been the least preferred outome of one or both of the chambers of Congress at first, but after the shutdown dragged on, more people were hurt by it, and more blame was assigned to the lawmakers for causing it, at some point, it became the least preferred outcome, and then the players changed their strategies.

Clearly, subtleties abound, and we can make our models quite sophisticted by invoking time-dependent preferences and/or probabilistic models in an effort to become more realistic. However, our purpose in this section is only to explore the theory of moves, so we leave these more sophisticated modifications for another time and place.

In conclusion, there are certainly advantages to using the theory of moves, and there are situations where it does bring additional insights, even if it does not completely capture every nuance of a situation of conflict. For example, in Taylor and Pacelli (2008), there is an application of the theory of moves version of the prisoner's dilemma that better explains the superpower decisions to not get involved in the Yom Kippur War of 1973 than does the simultaneous moves version. Also, in the exercises at the end of this section, the reader is asked to give a theory of moves analysis of the follow the leader dilemma, and this definitely gives a more realistic interpretation of traffic flow than does the simultaneous moves version.

We conclude this section with modification to the theory of moves that may help it be a useful tool for ordinal payoff matrices that are larger than 2×2. It is possible to extend the theory of moves for ordinal games to larger payoff matrices with no changes at all in the rules. If a player has four strategies, for example, and it is his or her turn to move, he or she would have the option to stay at his or her current strategy (which would end the game if it wasn't the first move since it is a "stay"), or he or she could switch to any of the other three options. Thus, there would be four branches drawn from the current node – one "stay" and three different "switch" branches. In principle, this is the only change needed to extend the model to larger matrices. However, based on some examples we have looked at, it seems very likely that the tree will be infinite with so many options for switches. In fact, it is possible for the tree to be infinite even with a 2×2 preference matrix. To avoid this, we need to somehow allow for the possibility of more terminal nodes.

For example, consider the 3×2 payoff matrix for the Cuban Missile Crisis.

	USSR	
	Withdraw missiles	Maintain missiles
USA Blockade	(4, 5)	(3, 6)
Air strike	(5, 4)	(2, 2)
Invasion	(6, 3)	(1, 1)

However, suppose we reason as follows: one or both of the players may be happy with his or her second-favorite outcome rather than hold out for his or her first preference. Therefore, a node with a $(5, \_)$ or $(\_, 5)$ payoff should end any further switches (except for the first move, as usual). In fact, with many outcomes in an $m \times n$ payoff matrix, there may be several acceptable outcomes to end the game.

Therefore, we make the following modifications to the definition of a terminal node. Each player specifies a *threshold value* (r for the row player and c for the column player), and then items 1 an 2 of the terminal node definitions are replaced by the following:

1. If it is the row player's turn to move, and the current outcome is $(x, -)$, where $x \geq r$, except on the row player's first move.
2. If it is the column player's turn to move, and the current outcome is $(-, y)$, where $y \geq c$.

The third rule for a terminal node (a "stay," except on the first move) is unchanged.

Now, consider the 3×2 payoff matrix for the Cuban Missile Crisis. Let's specify the thresholds as follows. The USA would find acceptable any outcome where the USSR removes the missiles. So the USA would end the game at any outcome in the first column, even though it may prefer some of these outcomes more than others. Thus, we set the threshold at $r = 4$. Suppose the USSR is happy with any outcome where the USA is no more aggressive than a blockade; that is, they would accept either outcome in the first row. Thus, we set the USSR threshold at $c = 5$.

Now suppose we are at the $(3, 6)$ initial position and the USSR moves first. The reader should verify that the tree diagram is now finite, as is shown in Figure 8.15.

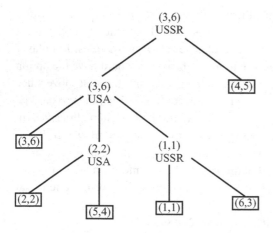

Figure 8.15 The game tree for the 3×2 version of the Cuban missile crisis with the specified thresholds for terminal nodes.

Now applying reverse induction, we obtain $(4, 5)$ as the final outcome (Exercise 8). This seems like a fairly realistic model for the crisis, and it does predict the "correct" outcome. Of course, one would have to assume that each player knows the other player's threshold in order for the analysis to be correct. Note also that the final position need not be an outcome that meets both players' thresholds, as it does in this example (see Exercise 9).

Exercise

1. Show that in the theory of moves version of the chicken dilemma, $(3, 3)$ is a non-myopic equilibrium when the column player moves first.
2. Draw the two tree diagrams in the theory of moves version of the chicken dilemma when the starting position is $(4, 2)$, and determine the final outcomes. Your answers should agree with the table in this section.
3. Draw the tree diagram for the theory of moves version of the chicken dilemma when $(1, 1)$ is the starting position and the column player moves first. Show that the final outcome is $(2, 4)$.
4. Do a complete theory of moves analysis of the prisoner's dilemma (this means drawing all eight tree diagrams and making a table similar to the one we made for the chicken dilemma). Show that there are two distinct non-myopic equilibria.
5. Do a complete theory of moves analysis of the follow the leader dilemma. Show that in this game there are no non-myopic equilibria. However, show that whoever moves first gets his or her most preferred outcome, and whoever moves second gets his or her second-most preferred outcome (in other words, whoever moves first gets his or her preferred Nash equilibrium point!). Note that the simultaneous play version of this dilemma predicts $(1, 1)$ as most likely. Thinking of the story we used to illustrate this dilemma (see Exercise 10 in Section 8.2) and considering your experience of drivers in similar situations, which model is more realistic?
6. It is possible that the tree diagram of a theory of moves version of a payoff matrix is not finite. For example, if after several branches you arrive at an outcome you have seen higher up in the tree, and with the same person's turn to move, the cycle will repeat forever. The following example, taken from Exercise 35 in chapter 4 of Taylor and Pacelli (2008) illustrates this. Assume $(2, 3)$ is the starting position and the row player moves first.

| | | Column | |
		C	D
Row	C	(2, 3)	(3, 1)
	D	(4, 2)	(1, 4)

7. Verify that the game tree is infinite for the 3×2 version of the Cuban Missile Crisis using the usual ending rules for the theory of moves. Assume $(3, 6)$ is the starting position, and either player moves first.
8. a. Apply reverse induction to the game tree obtained from the Cuban Missile Crisis ((3×2) version with the given thresholds of 4 for the USA and 5 for the USSR)

obtained in the text, and verify the final outcome is $(4, 5)$, the "mutual cooperation" outcome.

  b.   Complete the analysis for every possible starting position and for each player moving first. Determine if any outcomes are non-myopic equilibria.

9.   Give a theory of moves analysis of the 3×2 Cuban Missile Crisis, assuming the thresholds for both players are 5. Assume $(3, 6)$ is the starting position and the USSR moves first.

8.4   A Brief Introduction to *n*-Player Games

In this section, we discuss games with more than two players. We will see that the fundamental questions for such games are usually different than for two-player games, and therefore, the mathematical techniques required to analyze them are different than what we have learned for two-player games.

8.4.1   Dominance and Nash Equilibrium Points

We begin with the following example.

Example 8.11 Robert has a nickel, Colleen has a dime, and Paige has a quarter. They are sitting at a round table, with Robert first, then Colleen, and then Paige, in clockwise rotation. They agree to play a game with their coins. Each will place his or her coin secretly on the table with either heads up or tails up, and the coins will be revealed simultaneously. The rules are if there is one tail and two heads, the person who played tails wins both of the other players' coins. If there are two tails and one head, the two players who played tails must exchange their coins with each other, and the person who played heads keeps his or her own coin. If three tails are revealed, the players all pass their coins to the person next in rotation (clockwise) at the table. If all three are heads, no exchange is made, and each player keeps his or her own coin. Model this as a strategic game.

The first thing to do is decide how to write down the payoffs. Since there are three players, we will use ordered triples, with the first coordinate for Robert, the second for Colleen, and the third for Paige. There are eight possible outcomes, so we can list all the possible payoffs in a table. We order the notation for the outcomes Robert, Colleen, and then Paige; so HTT, for example, means Robert played heads, Colleen and Paige both played tails.

Outcome	Payoff to (Robert, Colleen, Paige)
HHH	$(0, 0, 0)$
HHT	$(-5, -10, 15)$
HTH	$(-5, 30, -25)$
THH	$(35, -10, -25)$
TTH	$(5, -5, 0)$
THT	$(20, 0, -20)$
HTT	$(0, 15, -15)$
TTT	$(20, -5, -15)$

The reader should check that the payoffs in the table are correct. Notice that this game is zero-sum – the sum of the payoffs of the three players is zero in each outcome.

Next, what about a payoff matrix? A matrix has a row player and a column player, but in this game, there are three players. That and the fact that the payoffs are ordered triples suggest that instead of a matrix, we need a three-dimensional table. Since we cannot easily draw a three-dimensional table, we can draw a set of matrices, one for each of Paige's strategies, as follows.

		Paige selects *H*	
		Colleen	
		H	*T*
Robert	*H*	$(0, 0, 0)$	$(-5, 30, -25)$
	T	$(35, -10, -25)$	$(5, -5, 0)$

		Paige selects *T*	
		Colleen	
		H	*T*
Robert	*H*	$(-5, -10, 15)$	$(0, 15, -15)$
	T	$(20, 0, -20)$	$(20, -5, -15)$

We can think of each matrix as being on a separate "page" or "layer" of a three-dimensional payoff table. Robert chooses the row, Colleen chooses the column, and Paige chooses the page. Remember that the players choose simultaneously and independently. Just because we have drawn the table with separate pages does not mean Paige chooses first!

The concepts of movement diagrams, one strategy dominating another, and Nash equilibria all carry over to this example and to games with even more players and/or more strategies. For example, we could draw the movement diagram of this game as a cube, with arrows as the edges. Instead, we "flatten" it out so it becomes a planar diagram.

The "back" page when Paige selects *T* is the inner square shown in Figure 8.16, and the "front page," when Paige selects *H*, is the large square on the perimeter. All the horizontal or vertical arrows are thus within one page and are drawn as usual, pointing to the best outcomes for Robert (the highest first coordinate) for the vertical arrows and the best outcomes for Colleen (the highest second coordinate) for the horizontal arrows. The diagonal arrows are the ones that go from one page to the other and point to the best outcomes for Paige (the highest third coordinate).

The fact that all four of the vertical arrows point downwards indicates that *T* is a dominant strategy for Robert. Neither of the other players have a dominant strategy. However, with arrowheads pointing to them from all three directions, the two circled outcomes are both Nash equilibria. If any of the three players moved unilaterally away from those outcomes, their payoffs would go down.

Of the two equilibria, Robert prefers the $(20, 0, -20)$ outcome, but it does not matter, since both of the equilibria are outcomes of his dominant strategy – he must choose *T* to obtain either one. Colleen prefers the $(20, 0, -20)$ also, so trying for that outcome, she should choose *H*. However, Paige prefers the $(5, -5, 0)$ and to obtain that, she must choose *H* also. Therefore, if each player tries for his or her preferred equilibrium, the most likely outcome is *THH*, which has payoff

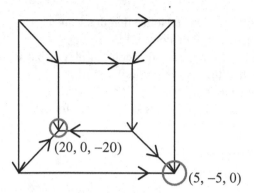

Figure 8.16 The movement diagram for the game in Example 8.4

$(35, -10, -25)$. This is at an outcome that is not an equilibrium so it is unstable. The fact that Robert does so well is basically an accident, and it would not hold up under repeated play since it is not a Nash equilibrium.

This example shows that, even if the game is zero-sum, a game with three or more players can have the same kind of problems that two-player variable-sum games have. That is, we can have Nash equilibria that are not equivalent and interchangeable (even though they may be Pareto efficient, as all outcomes are in a zero-sum game).

One could, just we did for two-player variable-sum games, define a concept of a *strictly determined* game, which is one where either the Pareto-efficient Nash equilibria are unique or else they are all equivalent and interchangeable. It would be clear how to play such games. However, if the game is not strictly determined, it is unclear exactly what rational behavior would entail, and not much could be said in general. We also point out that although we are using real number payoffs, the case of ordinal games for three or more players leads to similar conclusions.

There is a reason that we encounter the same difficulties in the three player case for zero-sum games that we do for two-player variable-sum games, as noted in Proposition 8.12.

PROPOSITION 8.12   *Any n-player (variable-sum) game is equivalent to an (n + 1)-player zero-sum game.*

Proof   Begin with an *n*-player game. Introduce a new player (called the *banker*). The banker has only one strategy (in other words, when the other players are making a choice of strategies, the banker doesn't make any choice at all). However, by definition, his payoff is the negative of the sum of the payoffs of all the other players. Including his payoffs now makes the game zero-sum.   ∎

Think of the board game Monopoly. Perhaps there are three or four active players rolling dice and moving around the board and buying properties, etc. There is nobody who "plays" the banker in Monopoly. The players may have portfolios worth vastly different amounts, but when you add all their values up, including the money left in the bank and the undeveloped property, you always get a constant sum, which is the total amount of money and property in the box that comes with the game! This example is very illustrative, and that is why we are calling the extra player we can adjoin to any game the *banker*.

For another example, consider the game of the Saturday chores at the Anderson household from Chapter 6. At the time we introduced the game, we specified that Mr. Anderson was not himself a player – just the three children were players in the first version, and just two of the children in the second version. But if we include Mr. Anderson as the banker for these games, then they become zero-sum (even though the first version was already constant-sum, when the banker is included, it is actually zero-sum). Note that when the children choose their chores, Mr. Anderson does not make any choice – the banker does not actively affect the play of the other players.

In conclusion, unless the game happens to be strictly determined, there is very little we can say in general for any game with three or more players in terms of optimal strategies, even if it is zero-sum.

8.4.2 Cooperative Games: Games in Characteristic Function Form

Part of the reason we cannot say too much about *n*-player games unless they are strictly determined is that we have been assuming that "rational play" means each person is playing for himself or herself. This is totally reasonable when there are two players, but for games with more plyers, real life shows many examples where (if communication is allowed among the players) some of the players form "teams" or "coalitions" and the members of the coalition coordinate their play so that they act as if the entire team was a single player.

Anybody who has played the board game Risk has seen this. It is a game of world domination, and players have armies that fight each other in order to take control of land until one player rules the world. Quite frequently, two (or more) of the players make a temporary truce so they can both attack a different player in an effort to eliminate this other player from the game. These truces are inevitably broken in the long run, but if they last long enough to eliminate some players, they have served a purpose.

Even in other board games, such as Monopoly, certain coalitions form naturally. Two brothers playing in a group of other players might give each other "breaks" such as taking a cheap property for payment instead of collecting the full amount of a rent on a property with a hotel, for example, or letting another player "owe" him or her money for a few moves to allow that player to obtain the needed funds to pay. (These sorts of breaks are presumably against the rules of Monopoly, but we've all seen people bend the rules to help a player.)

Even in the coin game example of the last section, Colleen and Paige do not do well at all if they play as individuals because they can't coordinate their moves to obtain either Nash equilibrium. But if they talked it over, they might decide how to move in every play of the game so that they alternate which Nash equilibrium they obtain. That way, Robert will not just win $0.35 by accident – he will alternately win $0.05 and $0.20 only. There is nothing he can do to counter the collaboration of Colleen and Penny – he must always play his dominant strategy of T or risk doing even worse. Of course, what is to prevent Robert from forming a coalition with one of Colleen or Paige against the other player? After all, even with alternating the Nash equilibrium outcomes, neither Colleen nor Paige is doing very well against Robert. It might occur to one or both of them that they might do better by teaming up with Robert instead of each other.

We can try to analyze what would happen if each possible coalition formed. First, suppose it's Robert vs. a team of Colleen and Paige. We'll make the team the column player of the resulting two-player game. Since the game is zero-sum, we'll only list Robert's payoffs in the payoff matrix.

		Colleen and Paige			
		HH	HT	TH	TT
Robert	H	0	-5	-5	0
	T	35	20	5	20

In this game, the only saddle point is at TTH, where Robert wins 5 and the team loses 5. Earlier, we suggested that Colleen and Paige coordinate to alternate which Nash equilibrium to obtain, but on closer analysis, the outcome THT where Robert wins 20 is only a Nash equilibrium in the $2 \times 2 \times 2$ game when they play as individuals, not in the 2×4 game when Colleen and Paige team up. Otherwise put, the Nash equilibrium where the team loses 20 is worse for the team than the one where they lose 5, so they would never play that one.

Of course, note that the actual payoffs are $(5, -5, 0)$ at this outcome, so in reality, Colleen is bearing the entire loss of the team, while Paige breaks even. Paige may think this is an acceptable outcome, but Colleen may object to it and may decide not to team up with Paige. Can Paige provide any incentive to Colleen to stay on her team? Well, she might offer to compensate Colleen for some of her losses. For example, offer to give Colleen 2 each time the game is played, so that, effectively, the payoffs are $(5, -3, -2)$ instead of $(5, -5, 0)$. Without trying to answer whether or not this is enough of an incentive for Colleen, just observe that what is being proposed here is that when a coalition forms, it may be possible to decide how to split the winnings (or loss) from the partnership other than the actual outcomes.

This presumes that (1) the value of an outcome is actually transferable from one player to the next, and (2) such "side payments" are allowed by the rules of the game. In an n-player game, if we allow communication, allow cooperation so that coalitions may form, assume transferability of assets between players, and allow such side payments between the players in a coalition, those are very different assumptions for game theory than the ones we posed for two-player games.

As mentioned in Section 8.1.3, game theory under these assumptions is called the theory of _cooperative games with side payments_ and was first developed in von Neumann and Morgenstern (1944). The economic arena is an area where these conditions are frequently met. Money is the ultimate transferable asset. Mergers of companies, which are the coalitions, often take place, or even looser coalitions are possible, such as "associations" that might consist of a set of companies that partially cooperate for mutual benefit, even if they do not merge and share all their assets. For example, companies might join professional associations that can negotiate cheaper insurance rates or medical benefits than they could get individually. It should be no surprise that this type of game theory plays a big role in economics and the its development led to the Nobel Prize in Economics in 1994 for some of the developers of game theory.

With such games, the focus is not so much on the actual strategies played but on new questions that arise in these types of games, such as the following:

- Which coalitions should form (or are likely to form)?
- How should a coalition split its winnings?

The second question is within the realm of "fair division" problems, which are not addressed in this book. We will spend just a little more time on the first question before moving on.

To get back to the previous example, what if Robert and Paige team up against Colleen? We make Colleen the row player and find the payoff matrix is the following.

		HH	HT	TH	TT
		Robert and Paige			
Colleen	H	0	−10	−10	0
	T	30	15	−5	−5

In this game, Colleen does not have a dominant strategy. However, the second column *HT* dominates the first, and the third column dominates the fourth. Thus, Robert and Paige should play different sides of their coin, and crossing off the dominated columns leads to the following.

		Robert and Paige	
		HT	TH
Colleen	H	−10	−10
	T	15	−5

Now the second row dominates the first, so by higher-order dominance, the solution is *TTH* (in the notation of the original game – Colleen plays *T*, and the Robert/Paige team plays *TH*) with a payoff of $(5, -5, 0)$ (which happens to be one of the Nash equilibria of the original $2 \times 2 \times 2$ game). This time, it is Robert who does better than Paige in the coalition, so it may be in Robert's interest to offer a side payment to Paige to keep the coalition. On closer inspection, Robert can already win 5 on his own, against a coalition of the other two players, so why would he offer such a side payment? There seems to be no advantage to Robert to participate in this coalition.

Finally, what if Robert and Colleen team up against Paige? Making Paige the row player, the matrix is the following.

		Robert and Colleen			
		HH	HT	TH	TT
Paige	H	0	−25	−25	0
	T	15	−15	−20	−15

Analyzing by higher-order dominance as in the last game, or just checking for saddle points directly, we see the solution is *THT* (Paige plays *T*, and the Robert/Colleen team plays *TH*) with an actual payoff of $(20, 0, -20)$, which happens to be the other Nash equilibrium of the original game. Again, Robert gains the most, but this time, in order to obtain the 20 payoff that he could not get in a coalition with Paige, he does have to offer Colleen some incentive as a side payment to keep her on the team.

Of the three coalitions, it seems the third (Robert and Colleen) is the most likely to form. Can we codify why?

We can set up a *value* for each coalition, which measures what the coalition can win if it coordinates play against all the other players. For example, consider Robert as an individual player. If we assume he is playing against a coalition of every other player (Colleen and Paige in this example) but can still obtain a guaranteed minimum payoff of 5 against this coalition, which is guaranteed by the minimax theorem, we set the value of the singleton coalition {*Robert*} at 5. We are assuming that if not every player joins the coalition against him, then Robert may actually do better. In other words, the worst off he will be is if everyone else coordinates her actions against him. We'll come back to this point later. Thus, the value of 5 truly is the best of his worst-case scenarios and a guaranteed minimum payoff. Following Straffin (1993), we call it Robert's *security level*, just as we did for variable-sum games.

Similarly, the values of the singleton coalitions for the other players are their security levels, so the value for {*Colleen*} is -5, and the value for {*Paige*} is -20. What about coalitions of two players? Its value should be the minimum it can guarantee itself when playing against a coalition of all remaining players. With three players, the "remaining players" are just the third player, so in our example, these security levels are just the values of the games stated earlier (which are negatives of the values of the individual players because the game is zero-sum). Furthermore, we assign a value of 0 to the empty set – a coalition with no members cannot guarantee itself any winnings (or incur any losses). Finally, what if all three players collaborate together (this is called the *grand coalition*) and coordinate their plays? Well, in this example, since it is a zero-sum game, the team of all three wins a total of 0 (they just have to decide how to divide it). Summarizing, we have the following table.

Coalition	Value
∅	0
{*Robert*}	5
{*Colleen*}	−5
{*Paige*}	−20
{*Colleen, Paige*}	−5
{*Robert, Paige*}	5
{*Robert, Colleen*}	20
{*Robert, Colleen, Paige*}	0

This table will do two things for us. First, it will clarify how to decide which coalitions might form. For each person, draw an arrow to the coalitions that represent the most improvement they can make by joining the coalition. For Robert, who can make 5 on his own, the only coalition that would improve this is {*Robert, Colleen*}, which will give the team winnings of 20 (again, they still need to decide how to split it). For Colleen, the same is true. She can lose 5 on her own or in a coalition with Paige (even if Paige offers her a small side payment), but she can win 20 in a coalition with Robert, so {*Robert, Colleen*} is her best coalition as well. For Paige, who loses 20 on her own, she can improve to losing 5 in a coalition with Colleen, improve to breaking even by joining the grand coalition, and improve to winning 5 in a coalition with Robert. For her, the most desired coalition is {*Robert, Paige*}. In this example, each "best" coalition is a two-person partnership, and Figure 8.17 shows each player's preferred partner.

Since Robert and Colleen prefer each other and nobody prefers Paige, it seems that the most likely solution is that Robert teams up with Colleen against Paige in this game. While we were able to answer the question of what coalitions should form, note that it will not always be this easy. It seems possible that with three players, it is possible that *A* prefers *B*, who prefers *C*, who prefers *A*. Then it is not clear what coalition actually forms (at least not without looking at specific side payment offers). If the game has more than three players, it seems possible that the best coalitions

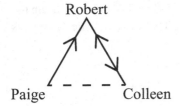

Robert

Paige   Colleen

Figure 8.17 Preferred coalition partners in the coin game.

may not just be two-person partnerships, which may complicate things further. In any case, at least in this example, the values of the coalitions let us decide what coalitions should form.

The second thing this table of values does for us is make us realize that to answer the question of which coalition should form, all we needed was the value of each coalition. The specific strategies of the game were not important. This leads to a generalization of what me mean by a "game."

DEFINITION 8.13   *A game in characteristic function form is a pair (N, v), where N is a set of players, and v is a characteristic function, which assigns to each coalition (that is, to each subset A of N), a real number value, v(A).*

That's it! Whenever we start with a strategic game coming from a payoff matrix, we can obtain a game in characteristic function form from it, by using the security level for each coalition as its value. However, there may be games in characteristic function form that do not come from a payoff matrix, showing that the new definition is more general than the old.

DEFINITION 8.14   *A game in characteristic function form is said to be* superadditive *if whenever there are two* disjoint *coalitions, A and B, the value function satisfies*

$$v(A \cup B) \geq v(A) + v(B). \tag{8.1}$$

Disjoint means that there are no players who are members of both coalitions. The terminology comes from set theory, which is reviewed briefly in Section A.1. The definition says that the amount that the coalitions could win by joining forces is at least as much as the sum of their individual winnings. In other words, larger coalitions are at least as good or better at coordinating their efforts and gaining value than by splitting into smaller coalitions acting independently. This is why, when we consider a game coming from a payoff matrix to find the value of any coalition A, we assume the rest of the players $(N - A)$ coordinate their efforts against A. The rest of the players (the *complement* of A – see Section A.1 for terminology) coordinating their actions against A is the most difficult configuration of adversaries that A could be pitted against. This is the point we said we'd come back to: any game coming from a payoff matrix is automatically superadditive.

The simplest game in charactersitic function form that does not come from a payoff matrix is called *divide the dollar*, and was first presented in von Neumann and Morgenstern (1944).

Example 8.15 Three people are told that they will be given a dollar if they can decide by majority vote how they will split it.

Let $N = \{A, B, C\}$ be the three players. Any coalition with two or more members is a winning coalition since it is a majority rule. The value of any such coalition is 1, representing the dollar won. Any coaliton of one or zero members is not majority rule, so it will not be sufficient to win the dollar, giving that coalition a value of 0. That's it – the entire game. The charactertistic function is

$$v(\varnothing) = v(\{A\}) = v(\{B\}) = v(\{C\}) = 0$$
$$v(\{A, B\}) = v(\{A, C\}) = v(\{B, C\}) = v(\{A, B, C\}) = 1.$$

So this is a game in characteristic function form, but it does not come from a payoff matrix. Clearly it is variable-sum, and the reader can check that it is superadditive (Exercise 4). In this game, each player wants to pair up with someone, but it does not matter who, by symmetry. So it is not clear if the grand coalition will form (meaning that the three players agree how to split the dollar) or if some partnership of two players will form and, by majority vote, decide to cut out the third person altogether. If so, which pair will form? One would have no clue without looking at specific offers for side payments. There is a very nice discussion of this game in Straffin (1993), as well as other games in characteristic function form not coming from a matrix game.

The next section consists of an introduction to legislative voting systems, which are specific examples of games in characteristic function form. This is all we want to say about *n*-player games in general. The reader who wants to learn more can consult Straffin (1993), von Neumann and Morgenstern (1944), and Taylor and Zwicker (1999).

Exercises

1.   Consider a coin game similar to the one in the text. Again, Robert, Colleen, and Paige are sitting clockwise around a round table. This time, Robert has a dime, Colleen has a half-dollar, and Paige has a dollar coin. Again, they simultaneously reveal whether their coins are heads up or tails up. The payoff rules are slightly different this time. If all heads are up, no exchange is made. If one tail is up, that person wins both of the other coins. If two tails are up, each player with tails up wins half of the remaining coin. If three tails are up, they pass their coins counterclockwise to the next person.

   a.   Construct a table of payoffs and use it to draw the movement diagram of the game. Show that all three players have T as a dominant strategy, which leads to a unique Nash equilibrium. Thus, the game is strictly determined. What is the outcome?
   b.   Find the security level for each player and each coalition, and thereby convert the game to characteristic function form.
   c.   Which coalition is likely to form? Show that the solution is the same outcome as predicted in Part a using individual play.
   d.   In this example, we can answer the question of how the coalition in Part c should split its winnings. Explain why there is only one reasonable way to split the winnings for this coalition. [**Hint**: Argue that if the proposed split was anything other than the obvious one, one or the other player in the coalition would refuse to join the coalition.]

2.   In the text, we did not consider what happens if the players use mixed strategies. Since the coin game in the text is not strictly determined, it seems reasonable to try mixed strategies. One difficulty is the lack of a quick formula to replace the formula $E = pAq$ we used for

two-player games. Another is the lack of a version of the minimax theorem. Nevertheless, let's see if we can remedy the former difficulty and try some examples.

a. As usual, let $p = \begin{bmatrix} p_1 & p_2 \end{bmatrix}$ be the the row player's (Robert's) mixed strategy, so p_1 is the probability of playing H, and p_2 the probability of playing T. Similarly, $q = \begin{bmatrix} q_1 \\ q_2 \end{bmatrix}$ is the column player's (Colleen's) mixed strategy, and let $r = \begin{bmatrix} r_1 & r_2 \end{bmatrix}$ be Paige's mixed strategy. Suppose that $p = \begin{bmatrix} \frac{1}{2} & \frac{1}{2} \end{bmatrix}$, $q = \begin{bmatrix} \frac{2}{3} \\ \frac{1}{3} \end{bmatrix}$, and $r = \begin{bmatrix} \frac{1}{5} & \frac{4}{5} \end{bmatrix}$. By using a table of the eight possible outcomes, compute the expected payoff E_R to Robert. For example, argue that the probability of the HTH outcome is the product $\left(\frac{1}{2}\right)\left(\frac{1}{3}\right)\left(\frac{1}{5}\right) = \frac{1}{30}$ because the players choose their strategies independently. Do the same for each outcome, multiply by the appropriate values, and add. Repeat for E_C and E_P, the expected payoffs for Colleen and Paige.

b. Let $A_1 = A_H$ be the payoff matrix, given that Paige chooses H (the "front square" of the three-dimensional payoff table), and let $A_2 = A_T$ be the payoff matrix, given that Paige chooses T (the "back square" of the payoff table). Show by direct comparison to the calculations in Part a that E_R can be obtained by considering just the payoffs to Robert in A_i ($i = 1, 2$) and performing the following matrix product:

$$ r \cdot \begin{bmatrix} pA_1q \\ pA_2q \end{bmatrix} = [r_1, r_2] \cdot \begin{bmatrix} pA_1q \\ pA_2q \end{bmatrix}. \tag{8.2}$$

Show that we can also recover E_C (resp., E_P) with the same formula, but using the payoffs for Colleen (and, respectively, Paige) for the entries in A_i.

c. Use the Eq. (8.2) to find the expected payoffs to the players if $p = \begin{bmatrix} 0.3 & 0.7 \end{bmatrix}$, $q = \begin{bmatrix} 0.4 \\ 0.6 \end{bmatrix}$ and $r = \begin{bmatrix} 0.9 & 0.1 \end{bmatrix}$.

d. Eq. (8.2) generalizes to three-player games of arbitrary size $m \times n \times t$. In this case, since Paige has t strategies, the three-dimensional payoff table has t pages, and each page A_i is an $m \times n$ matrix. Thus, if $p = \begin{bmatrix} p_1 & p_2 & \cdots & p_m \end{bmatrix}$, $q = \begin{bmatrix} q_1 \\ \vdots \\ q_n \end{bmatrix}$, and $r = \begin{bmatrix} r_1 & \cdots & r_t \end{bmatrix}$, then the expected payoff is $E = r \cdot \begin{bmatrix} pA_1q \\ \vdots \\ pA_tq \end{bmatrix}$. Suppose we have a $2 \times 4 \times 3$ game, where

$$ A_1 = \begin{bmatrix} (1,2,4) & (2,-3,5) & (-2,-3,7) & (5,0,-5) \\ (0,-2,6) & (3,3,4) & (-4,4,0) & (2,8,1) \end{bmatrix} $$

$$ A_2 = \begin{bmatrix} (-1,0,4) & (3,2,1) & (0,10,0) & (2,-5,8) \\ (2,1,1) & (1,0,3) & (4,5,-4) & (3,4,0) \end{bmatrix} $$

$$ A_3 = \begin{bmatrix} (0,3,-4) & (0,1,-4) & (2,1,2) & (-1,-2,-3) \\ (4,2,-1) & (2,6,-3) & (0,0,-2) & (-2,3,5) \end{bmatrix}. $$

Find the expected payoff for each player if the following mixed strategies are used:

$$p = \begin{bmatrix} \frac{1}{3} & \frac{2}{3} \end{bmatrix}$$

$$q = \begin{bmatrix} \frac{1}{10} \\ \frac{2}{10} \\ \frac{3}{10} \\ \frac{4}{10} \end{bmatrix}$$

$$r = \begin{bmatrix} \frac{2}{5} & \frac{1}{5} & \frac{2}{5} \end{bmatrix}.$$

(Observe that this game is variable-sum, but that does not affect the formula for E.)

3. Consider the following variation of the Anderson household chore game: The three players are the children Ian, Jon, and Laurie. There are only two chores to be done – the dishes D and the raking R. In this version, each child secretly writes down D or R on a piece of paper. When the papers are revealed, each child works on the chore he or she chose. If more than one child elects the same chore, that is fine; each will get a partial payment. If a chore is not chosen by any of them, it is assigned to the three of them to split (which only happens if all three choose the other chore, so in this case, all three children share both chores).

Mr. Anderson has some asymmetry built into the payoffs so that if one or both of the boys choose the dishes, they are rewarded with extra pay, and if Laurie chooses the raking, she is paid more than the boys. However, if nobody chooses a chore, when it is assigned, the pay is less for that chore than it would be if somebody chose it. The basic pay is $6 for the dishes and $12 for the raking. However, if Ian and Jon choose dishes, each earns an extra $1. If only one of them chooses dishes, he earns the entire extra $2. If Laurie chooses raking, she earns an extra $3 if she does the job alone, $2 if she shares the job with one other person, and $1 if all three are raking. If the dishes have to be assigned because nobody chose that chore, the pay is $1 per person (for a total of $3 instead of $6). If the raking has to be assigned because nobody chose it, the pay is $3 per person (for a total of $9 instead of $12). Note that, unlike previous versions of the game, choosing no chore or both is NOT a possible strategy – the players' paper must say D or R only.

a. Model this as a $2 \times 2 \times 2$ game. Write down the payoff table (that is, the payoff matrices A_i for each choice that Laurie makes). Also write down the movement diagram. Determine all instances of dominance, if any, and any Nash equilibria. Is the game strictly determined?

b. Convert the game to characteristic function form. Note that the game is not constant-sum, so for each coalition, to compute its security level, you must use the prudential strategy (use your own payoffs, ignore your opponent's payoffs, and play as if it were zero-sum).

c. Can you determine which coalition is likely to form?

4. Verify that the divide the dollar game is superadditive.

5. In the friendly town of Snowblind, there are three snow removal businesses operating. Latimer's Snow Goose Plowing Service plows driveways and small parking lots. Harrison's Blow Away, Inc., uses snow blowers to clear sidewalks and decks. McGuire's Eaves of Congestion, Ltd., removes snow and ice from roofs. Because some customers want more

than one service and because they might be able to save by sharing equipment and overhead costs, the companies are exploring the possibility that some or all three of them may merge. They've hired a consultant from Midnight Sun and Hot Springs Marketing Firm to study the situation. The consultant has provided the following data: In a typical winter month, with the companies working separately, the average profit for Snow Goose Plowing is $12,000, the average profit for Blow Away is $8,000, and the average profit for Eaves of Congestion is $5,500. They estimate that the average monthly profit for Snow Goose and Blow Away working together would be $22,000, the average profit for Blow Away and Eaves of Congestion working together would be $15,000, and the average profit of Snow Goose and Eaves of Congestion working together together would be $18,000. Finally, they estimate that if all three companies join forces, the average monthly profit would be $3,000.

Model this as a game in characteristic function form, and show that it is superadditive. Can you predict the most likely coalition to form?

6. Consider the following game in charactertistic function form: $N = \{a,b,c,d\}$, and characteristic function given by the following table.

Value function $v(S)$ for $S \subseteq N$

$v(\varnothing) = 0$

$v(a) = v(b) = 1$, $v(c) = 2$, $v(d) = 3$

$v(\{a,b\}) = 2$, $v(\{a,c\}) = 3$, $v(\{a,d\}) = 4$

$v(\{b,c\}) = 3$, $v(\{b,d\}) = 5$, $v(\{c,d\}) = 6$

$v\{(a,b,c)\} = 4, v(\{a,b,d\}) = 6$, $v(\{a,c,d\}) = 6$, $v(\{b,c,d\}) = 6$

$v(\{a,b,c,d\}) = 8$

Determine whether or not the game is superadditive.

8.5 Legislative Voting and Political Power

When it comes to mathematics applied to politics, there are two distinct types of voting that can be studied. When voting for candidates for an election, that is called *social choice voting*. There is much that can be said about the mathematics of social choice voting, and it is a fascinating area of study. We don't have the space here to cover it, so we refer the interested reader to Hodge and Klima (2005), Robinson and Ullman (2011), Taylor and Pacelli (2008), and chapter 20 of Straffin (1993).

In this final section of the chapter, we focus on *legislative voting*, where a legislative body (group of people) votes to pass a proposition, resolution, or law, based on its rules of passage. This type of voting is also called *yes–no voting*. We introduce this type of voting because such systems are examples of games in characteristic function form. Ours is a brief introduction. More details can be found in Taylor and Pacelli (2008), chapters 26–28 of Straffin (1993), and Taylor and Zwicker (1999), although the last text is for mathematically mature readers.

8.5.1   Legislative Voting Systems

We begin with a hypothetical example.

Example 8.16 The Holding Company Big Sister has an executive committee that decides the company's policies. The executive committee has three members: Janis, James, and Sam. Janis started the company, so her votes carries more weight than the other members; her weight is the combination of the other two members' weights. So James and Sam each have one vote, and Janis has two votes (or think of it as one vote with a weight of 2). In order for a proposal to pass, it must have the support of at least three votes.

This is an example of a *weighted voting system*. In such a system, each voter has a weight attached to his or her vote, and passage requires a coalition whose total weight equals or surpasses a fixed quota q. Such a voting system can always be regarded as a game in characteristic function form. The set N is the set of voters, and if the total weight of a coalition X is $\geq q$, so it is a winning coalition, we assign $v(X) = 1$; otherwise, it has value 0.

REMARK 8.17   *We make the assumption throughout Section 8.5 that in any voting system, the empty coalition is losing (so $v(\varnothing) = 0$ as usual) and the grand coalition is winning, so $v(N) = 1$.*

Here is the characteristic function for the Big Sister example, where $q = 3$.

Coalitions and their values

$v(\varnothing) = 0$
$v(Janis) = v(James) = v(Sam) = 0$
$v(\{James, Sam\}) = 0, \ v\{Janis, James\} = v\{Janis, Sam\}) = 1$
$v(Janis, James, Sam\}) = 1$

We define two voting systems to be *equivalent* if they have the same value function v. This is the same thing as saying they have exactly the same winning coalitions. As an example, suppose the weights in the preceding example are 5 for Janis, 2 for James, and 3 for Sam, and we reset the quota at $q = 6$. It should be clear that this weighted system has exactly the same winning coalitions as the preceding example – namely, $\{Janis, James\}$, $\{Janis, Sam\}$, and the grand coalition of all three voters. Thus, even though the weights and quota are different, this modified system is equivalent to the one in Example 8.16.

In fact, the system need not even be described as a weighted system at all. Suppose we stipulate that in the executive committee, passage requires the support of at least two voters, but Janis has *veto power*, which means Janis must be a part of every winning coalition. If she does not support a proposal, it doesn't pass. Again, we find the exact same three winning coalitions, so this system is equivalent to the example. So although this third case is not described as a weighted system (passage in a weighted system *only* requires enough weight to exceed the quota and does not have any extra stipulations such as veto power), it is actually equivalent to a weighted system. There are real-life examples of voting systems which are not described as weighted, but they are equivalent to a weighted system, such as the United Nations Security Council (see Exercise 1 at the end of this section).

So the first question that comes to mind is: is every voting system equivalent to a weighted system?

The answer to that question is "no"! Not every system is weighted. A simple condition for a voting system to be weighted is discussed in Taylor and Pacelli (2008) and proved in full in Taylor and Zwicker (1999). We explore this (and a related weaker condition) in Exercises 2–6.

Notice that the divide the dollar game is a voting system. We have three players, and since the winning coalitions are decided by majority rule, it is a weighted system – each voter gets one vote and the quota is $q = 2$.

Example 8.18 *The US Federal Legislature.* When a federal law is proposed, there are three possible ways it can pass and become law. It can have the support of a simple majority of the Senate, a simple majority of the House of Representatives, and the support of the president. Or it can have the support of a simple majority of the House of Representatives, the support of exactly 50 of the 100 senators, the vice president (considered as the "tie-breaking vote" in the Senate), and the president. Or, if the president does not support it ("Presidential veto"), it can still pass with a $\frac{2}{3}$ majority in both chambers (the "congressional override" of the presidential veto).

This is a complicated system, and is certainly not described as a weighted system. The voters N include the president, the vice president, 100 senators, and 435 representatives, for a total of 537 voters. There are far too many coalitions to write down the entire value function v, but we can describe it.

- If X contains 51 or more senators, 218 or more representatives, and the president, or
- If X contains 50 senators, 218 or more representatives, the president and the vice president, or
- If X contains 67 or more Senators, 290 or more representatives,

then $v(X) = 1$; otherwise, $v(X) = 0$.

We would probably guess, after some reflection, that this is not a weighted system. One clue is that if a coalition X only contains 217 or fewer representatives, then no matter what the weight of a senator or the president is, we can't make this a winning coalition, even if the entire Senate joined the coalition. Another clue is that the vice president's vote is not always counted, but in a weighted system, the weight of any voter in a coalition is always counted. But this is just a guess at this point. (See Exercise 4 at the end of the section.)

There is some standard notation for a weighted system. If the weights are $w_1, w_2, \ldots, w_n$, written in non-increasing order $w_1 \geq w_2 \geq w_3 \ldots$, for the n voters and the quota is q, we abbreviate this by the symbol

$$[q; w_1, w_2, \ldots, w_n]. \tag{8.3}$$

Thus, the divide the dollar game is represented as $[2; 1, 1, 1]$, and the Big Sister Example 8.16 is $[3; 2, 1, 1]$. The modified version of the Big Sister example can be represented as $[6; 5, 3, 2]$.

Consider the following example of a voting system. $N = \{A, B, C\}$ (three voters), with the following list of winning coalitions:

$$\{A\}, \{A, B\}, \{A, B, C\}.$$

This is a valid system (because specifying the list of winning coalitions is sufficient to determine the characteristic function v), but there is something about it that is strange. Namely, $\{A, C\}$ is not a winning coalition. This is odd because $\{A\}$ is a winning coalition, and usually, when more voters join a winning coalition, it remains a winning coalition.

DEFINITION 8.19   *A voting system is said to be* monotone *if whenever $X \subseteq Y$ and X is a winning coalition, then Y is also a winning coalition.*

So the system just described is not monotone. Here it is in characteristic function form: $N = \{A, B, C\}$, and the following.

$v(\varnothing) = 0$
$v(A) = 1, \ v(B) = v(C) = 0$
$v(\{A, B\}) = 1, \ v(\{A, C\}) = v(\{B, C\} = 0$
$v(A, B, C\} = 1$

Notice that in this example, we have $0 = v\{A, C\} = v(\{A\} \cup \{C\})$, but $v(A) = 1$ and $v(C) = 0$, so $v(A) + v(C) = 1 + 0 = 1$. This shows $v(\{A\} \cup \{C\}) < v(A) + v(C)$, and so v is not superadditive. Can we generalize this?

We have the following proposition.

PROPOSITION 8.20   *A voting system, regarded as a game in characteristic function form, is superadditive if and only if it is monotone and has the property that there do not exist disjoint winning coalitions.*

Proof   Suppose the system is not monotone. That means there are coalitions $X \subseteq Y$ where X is winning and Y is losing. In particular, X and Y cannot be the same coalition, so there are voters in Y that are not members of X. Let $Z = \{$voters in Y who do not belong to $X\}$. Then Z is nonempty (as we just pointed out), and Z and X are disjoint by definition. Finally, note that $X \cup Z = Y$; in other words, Y is formed by combining two disjoint coalitions X and Z. Now since X is winning, we have $v(X) = 1$, and Y is losing, so $v(Y) = 0$. But $v(Y) = v(X \cup Z)$. If this were a superadditive value function, then $v(X \cup Z) \geq v(X) + v(Z)$. In other words,

$$0 \geq 1 + v(Z).$$

This implies $v(Z) < 0$, but this is a contradiction because, for *any* coalition, its value is either 0 or 1. It follows that v is not superadditive. We have shown that "not monotone" implies "not superadditive," which is equivalent to the contrapositive statement that superadditive implies monotone.

Furthermore, if there were disjoint winning coalitions X and Y, then $v(X) = v(Y) = 1$, so $v(X) + v(Y) = 1 + 1 = 2$, but $v(X \cup Y)$ is at most 1. (In fact, it is equal to 1, since we showed the system is monotone, and so $X \cup Y$ is winning.) So we must have $v(X \cup Y) < v(X) + v(Y)$, a contradiction to the hypothesis that v is superadditive. It follows that there are no such disjoint winning coalitions.

Conversely, suppose the system is monotone and does not have disjoint winning coalitions. We claim that it must be superadditive. So let X and Y be disjoint coalitions. We must show that

$$v(X \cup Y) \geq v(X) + v(Y).$$

The assumption that there do not exist disjoint winning coalitions means that $v(X)$ and $v(Y)$ cannot both be 1, so $0 \leq v(X) + v(Y) \leq 1$. Since the value of any coalition is 0 or 1, if the right side is 0, there is nothing to show. If the right side is 1, the inequality can fail only if $v(X \cup Y) = 0$, which means $X \cup Y$ is a losing coalition. But since X and Y are both subsets of $X \cup Y$, it follows that both X and Y are losing also, since if one were winning, the assumption of that the system is monotone would imply that $X \cup Y$ is also winning. But then $v(X) = v(Y) = 0$, whence the right side $v(X) + v(Y) = 0 + 0 = 0$, so, again, the inequality holds.   ∎

This proves the proposition, which characterizes voting systems that are superadditive games. In the next subsection, and in the exercises that follow, we always assume a voting system is superadditive unless otherwise indicated.

8.5.2 Political Power

In a weighted voting system, the idea is that the voters with more weight should have more of a say over passage than a voter with less weight. That is, more weight means more "political power." However, the idea of political power is elusive and not directly measured by weight. We have already seen, for example, that different weights and quotas might be an equivalent voting system with exactly the same winning coalitions! Furthermore, we are not even addressing other more nebulous types of "political power" such as the ability of a politician to argue, persuade, and win people over to his point of view or a person's network of connections to others in positions of power. The only type of "political power" we can try to measure mathematically is a voter's ability to be part of a winning coalition to pass a proposal.

We reiterate that power is not directly measured by weight. For example, it should be intuitively clear that Sam and James have the exact same political power in the Big Sister executive committee, even in the modified version where they do not have the same weight. Furthermore, for systems that are not weighted, such as the US Federal Legislature, one might suppose that different voters might have different amounts of power. For example, even though the presidential veto can be overridden, one might speculate that the president has more power than a senator.

So if weights are misleading in measuring political power and may not even exist if the system is not weighted, then just how can we measure political power? There are various known attempts to answer this question, and since a voter's "power" has something to do with the winning coalitions to which she belongs, all the known approaches deal with looking at those winning coalitions.

One of the earliest and most used methods for computing political power is called the *Shapley–Shubik power index*, introduced in 1954 in Shapley and Shubik (1954). This measure of power is related to a general concept of a game in characteristic function form known as the *Shapley value* of the game. Here we will describe the power index only, and refer readers who want to know more about the Shapley value to chapters 26 and 27 of Straffin (1993).

We need the notion of a *pivotal voter* first. Imagine lining up the voters N in some fixed order. Now start with the empty set, and move from left to right, adding a voter to a growing coalition one voter at a time in the order given. Eventually, when we reach the end of the lineup, we have attained the grand coalition, which is a winning coalition by the remark in Section 8.5.1. Since we started with a losing coalition, there is one voter who, when he or she joins the coalition, converts it from losing to winning, and since we assume superadditivity, the coalition remains winning as the rest of the voters join it. That person who first converts the coalition to a winning one is said to be the *pivotal voter for that ordering of the voters in N*. Every ordering has some voter that is pivotal. For a voter a, we define the Shapley–Shubik power index of a to be

$$SSI(a) = \text{(number of orderings of } N \text{ for which } a \text{ is pivotal)/(total}$$

$$\text{number of orderings of } N).$$

This is a fraction between 0 and 1 and is the proportion of orderings of the voters for which a is pivotal. You can also interpret it this way: if each ordering of the voters is considered to be equally likely, $SSI(a)$ is the probability that a is a pivotal voter. If there are n voters altogether, the number in the denominator is $n!$ (factorial – see Appendix). Also, if we add up the SSI of each voter, we obtain 1 as a total (as is the case when we sum the probabilities of the outcomes in any probability experiment – see Appendix).

Let's compute the SSI for each voter in the Big Sister executive committee. We have three voters – $N = \{\text{Janis, James, Sam}\}$ – so there are $3! = 6$ different possible orderings. In Table 19 we state which voter is pivotal. Recall that the quota is $q = 3$ and Janis is the only voter with weight 2, as shown in Table 8.3.

Table 8.3 Pivotal voters in Example 8.6.

Ordering	Pivotal voter
Janis, James, Sam	James
Janis, Sam, James	Sam
James, Janis, Sam	Janis
James, Sam, Janis	Janis
Sam, Janis, James	Janis
Sam, James, Janis	Janis

According to the definition, we have

$$SSI(Janis) = \frac{4}{6}$$

$$SSI(James) = SSI(Sam) = \frac{1}{6}.$$

So, according to the Shapley–Shubik measure, Janis has about $\frac{2}{3}$ of the power of this voting system, while the others have $\frac{1}{6}$ each. So, although Janis has twice the weight of the other voters, she has four times their power, at least according to this particular way of measuring power.

Let's do one more example.

Example 8.21 King Heart, Ltd., is a company that manufactures generators and other electrical equipment. All business decisions are made by the board consisting of Peter (P) with eight votes, Hugh (H) with six votes, Guy (G) with four votes, and David (D) with one vote. The quota for passage is set at $q = 10$. Thus, the system is the weighted system $[10; 8, 6, 4, 1]$. Let's compute the SSI for each voter. Since there are $n = 4$ voters, there are $4! = 24$ possible orderings. We create Table 8.4 to indicate the pivotal voter for each ordering.

Counting up, we see Peter is pivotal in eight of the orderings, Hugh is pivotal in eight, and Guy is pivotal in eight. Thus, Peter, Hugh, and Guy each have $\frac{8}{24} = \frac{1}{3}$ of the power, and David has no power at all – he is never a pivotal voter. A voter who has no power is called a *dummy voter*.

Table 8.4 Pivotal voters in Example 8.8.

Ordering	Pivotal voter	Ordering	Pivotal voter
$PHGD$	H	$GPHD$	P
$PHDG$	H	$GPDH$	P
$PGHD$	G	$GHPD$	H
$PGDH$	G	$GHDP$	H
$PDHG$	H	$GDPH$	P
$PDGH$	G	$GDHP$	H
$HPGD$	P	$DPHG$	H
$HPDG$	P	$DPGH$	G
$HGPD$	G	$DHPG$	P
$HGDP$	G	$DHGP$	G
$HDPG$	P	$DGPH$	P
$HDGP$	G	$DGHP$	H

Again, we see the ratio of power is not the same as the ratio of weights, and although Peter has the most weight of 8 (close to the quota!), he has no more power than either Hugh or Guy.

Before discussing other approaches to power, we observe that $n!$ grows very fast, so unless the voting system is very small, listing all the orderings is not a feasible approach to finding the Shapley–Shubik index. One can develop careful ways of counting the possible orderings for which a voter is pivotal without listing them all. In this example, note that since no voter has weight more than 9, the first voter is never pivotal. Also, since any three voters' weights total at least 10, the last voter is never pivotal. So in every ordering, the pivotal voter is either the second or the third voter.

We can note at the outset that David is a dummy voter (see Exercise 9). Given that, there is no way the second voter can be pivotal if David is first or second. However, if David is third or fourth, then the first two voters have weights that sum to at least 10, so in these cases, the second voter must be pivotal. In all other cases, the third voter is pivotal. Of the 24 orderings, half of them, or 12, have David in third or fourth place. Of these 12, clearly each of the other voters appears second in the same number of orderings, or 4. But the same argument works if David is first or second and the pivotal voter is whoever is third. There are 12 of these orderings, and each of the other voters is third the same number of times, or four. Thus, David is never pivotal, and each of the other voters must be pivotal $4 + 4 = 8$ times, for $\frac{1}{3}$ of the total power. If the reader is familiar with the enumeration techniques (reviewed in the Appendix), he or she can handle more complicated voting systems.

One thing that seems odd about the Shapley–Shubik index is that often votes are cast in secret ballots and revealed all at once, so ordering of the voters doesn't really matter. Is there a way to try to measure power by simply looking at the winning coalitions after they have formed, without regard to what order the voters join them?

Let's take a very naive approach and just count, for each voter, the total number of winning coalitions for which they are a member. It seems reasonable that the more winning coalitions that you belong to, the more power you have. One problem with this is that the answer is an integer, not a fraction, and it seems to depend on the total number of winning coalitions. If a voting system has 100 winning coalitions and you are a member of 10 of them, that is less power than being a member of 10 winning coalitions in a system where there are only 15 winning coalitions. So we can create a fractional index by taking the number of winning coalitions to which a member belongs and dividing by the total number of winning coalitions.

For example, in the Big Sister example, there are just three winning coalitions. Janis belongs to all three (because she has veto power), and James and Sam each belong to two coalitions. Thus, our "Very Naive Power Index" is $VNPI(Janis) = \frac{3}{3}$, and $VNPI(James) = NPI(Sam) = \frac{2}{3}$. We hope the reader immediately sees the problem here: these fractions do *not* add up to 1. Can we fix this?

The problem is that each coalition is counted more than once in the numerators, but not in the denominators. For example, the coalition {Janis, James} counts once for Janis and once for James, but in the denominator, it is just one of the three coalitions listed there. So we can fix this if we do the following: for the denominator, instead of the total number of coalitions, use the total number of voters – that is, the sum of the number of voters in each winning coalition.

This leads to the following provisional definition: For each voter a, the "Less Naive Power Index" of a, denoted $LNPI(a)$, is the fraction where the numerator is the number of winning coalitions containing a and the denominator is the sum of the number of voters in all the winning coalitions.

So, in the Big Sister example, we have the denominator is $2 + 2 + 3 = 7$, so $LNPI(Janis) = \frac{3}{7}$, while $LNPI(James) = LNPI(Sam) = \frac{2}{7}$. Now we have fractions that add up to 1, like with the Shapley–Shubik power index. But notice this power index assigns quite a different answer than the SSI does. With the SSI, Janis had four times the power of the other voters. With the naive index, she has just 1.5 times the power of the other voters.

Which index is "right"? Well, the point of our construction of the LNPI is that, since we haven't precisely said what me mean by "political power," there may be more than one valid way to try to measure it.

What about the King Heart, Ltd., example? The first step is to compile a complete list of winning coalitions. In Exercise 9, you are asked to verify that the following list of eight winning coalitions is complete:

$$PH, PG, \ HG$$
$$PHD, PHG, PDG, HGD$$
$$PHGD.$$

Thus, the denominator is $3(2) + 4(3) + 4 = 22$. We see P, H, and G are members of six of these, while D is a member of four. Thus, our "less naive" index assigns powers as

$$LNPI(P) = LNPI(H) = LNPI(G) = \frac{6}{22} = \frac{3}{11}$$
$$LNPI(D) = \frac{4}{22} = \frac{2}{11}.$$

So, again, the answers differ from the SSI. In particular, notice that David has power with this index, so he is no longer a dummy voter. Does that seem correct to you? We can't answer that until we give a precise definintion of "dummy voter." So far, we have only mentioned it in the context of the SSPI, but there is a better definition (see Definition 8.9) that doesn't depend on the SSPI.

Again, our made-up index, the less naive index, was invented just to illustrate that more than one way to measure power is possible, so you do not think that the SSI is the only "correct" answer. Notice that our index, like the SSI, does detect that Peter, Hugh, and Guy all have equal power, despite having different weights. However, it doesn't seem to correctly detect dummy voters.

Can this be fixed or improved? We can think of a couple of ways to do so. We'll derive one improvement now, and then we'll give a better solution, which is well known.

Recall that we are assuming our system is monotone. So, looking at the list of eight winning coalitions, note that those in the first row are subsets of all the others, but no other winning coalitions are subsets of them. In other words, if any voter were to leave the coalitions in the top row, he or she would no longer be winning.

DEFINITION 8.22 *A coalition X is said to be a* minimal winning coalition *if X is winning, but if any voter left the coalition, it becomes losing.*

In other words, looking at the charactertistic function, $v(X) = 1$, but $v(Y) = 0$ for any proper subet Y of X. In some sense, the minimal coalitions are the most important ones. Once it is winning, who cares if more voters join?

We can now make precise what we mean by a dummy voter.

DEFINITION 8.23 *A voter in a voting system who is not a member of any minimal winning coalition is said to be a* dummy voter.

In the King Heart example, that is exactly why David is considered a dummy voter.

So what if, in our naive index, we decided to count only the minimal winning coalitions? Then our denominator is 6, not 22, and notice that D is not a member of any of them, while each of the other voters belong to two of the minimal winning coalitions. With this modification, we get modifed values of $\frac{2}{6} = \frac{1}{3}$ for Peter, Hugh, and Guy, and 0 power for David. This happens to agree with the SSI of the voters, but that is probably just a coincidence.

Thus, we modify our provisional definition above to the following.

DEFINITION 8.24 *For a voter a, the* Naive Power Index *of a, $NP(a)$, is a fraction where the numerator is the number of minimal winning coalitions of which a is a member, and the denominator is the sum of the number of voters in each minimal winning coalition.*

In the Big Sister example, there are only two minimal winning coaltions, and they each have two voters. Thus, the denominator is $2 + 2 = 4$. Janis belongs to both, while James and Sam each belong to one of them. Thus:

$$NP(\text{Janis}) = \frac{2}{4} = \frac{1}{2},$$

$$NP(\text{James}) = NP(\text{Sam}) = \frac{1}{4}.$$

In the late 1960s, there was a lawsuit involving the county board of Nassau County, New York. The attorney John F. Banzhaf III argued that certain districts were not represented correctly. To make his point, he introduced a new power index, today called the *Bahzhaf power index* of a voter.

Banzhaf's idea was almost what we did above when we corrected our naive index by considering only minimal winning coalitions. However, rather than consider just winning coalitions that become losing when *any* voter leaves (the minimal ones), he instead considers winning coalitions that become losing when *specific* voters leave. For instance, if we consider the winning coalition PHD in the preceding example, it is not minimal because if D leaves, what remains is still winning. However, if P or H leave, then what remains is losing, so somehow these coaltions should count in measuring P's power or H's power, but not in measuring D's power. This leads to the following definition:

DEFINITION 8.25   *If a voter b is a member of a winning coalition X and the coalition becomes losing if b leaves it, we say b has a* critical defection *from coalition X.*

Notice that a coalition X is minimal if and only if every member has a critical defection from it. We can also state this via the characteristic function. The voter a has a critical defection from coalition X if and only if $v(X) = 1$ and $v(X - \{a\}) = 0$, and it is minimal if and only if $v(X - \{a\}) = 0$ for all $a \in X$.

Banzhaf's idea is to count for each voter the total number of critical defections; following Taylor and Pacelli (2008), we call this the *total Banzhaf power* of the voter, which we define as follows.

DEFINITION 8.26   *Suppose a voting system $N = \{a_1, a_2, \ldots, a_n\}$. Denote the total Banzhaf power of a voter a_i by $TBP(a_i)$. Then the* Banzhaf index *of voter a_i is*

$$BI(a_i) = \frac{TBP(a_i)}{TBP(a_1) + TBP(a_2) + \cdots + TBP(a_n)}.$$

Consider the Big Sister example. The winning coalitions are the two minimal ones {Janis, James} and {Janis, Sam}, where each player has a critical defection, and the grand coalition {Janis, James, Sam}, where only Janis has a critical defection. So we have $TBP(\text{Janis}) = 3$, and $TBP(\text{James}) = TBP(\text{Sam}) = 1$.

Thus,

$$BI(\text{Janis}) = \frac{3}{3 + 1 + 1} = \frac{3}{5}$$
$$BI(\text{James}) = BI(\text{Sam}) = \frac{1}{5}.$$

Compare this to SSI of the voters and to the naive index of the voters where we only consider the minimal winning coalitions. We now have three different power indices. Observe that the only difference between the naive index and the Banzhaf index in this example is that the grand coalition is counted in the Banzhaf index for Janis only, because she has a critical defection from it. But in the naive index, we only considered the minimal coalitions for all three of the voters.

This is not the end of the story. There are other power indices that have been proposed. The reader can find two more: the *Johnston power index* and the *Deegan-Packel power index* (both proposed in 1978) in Taylor and Pacelli (2008). More can also be found by doing an Internet search, and websites exist that can quickly calculate the various power indices for a given voting system. We will be content with the examples already covered as our goal was just to introduce the ideas, not make a detailed study.

However, we make two more observations. First, the situation of power indices is reminiscent of the situation of playing games against nature. We mentioned that one approach to sorting out all the different methods for playing such games is to try to create a set of axioms that any good method should satisfy. We wonder of the same might be done here? Is there a set of axioms that any good power index should satisfy? For example, one obvious axiom would be that a dummy voter should have power 0. We leave this as an open question for the reader to ponder.

The axiom just suggested makes sense because we now have a definition of dummy voter based just on coalitions (instead of on the SSI). That is the correct way to define concepts in voting theory. For example, we can now clarify two other concepts.

DEFINITION 8.27   *a. A voter a has* veto power *if a is a member of every winning coalition.*

b. A voter a is a dictator *if a is a member of every winning coalition and is a member of no losing coalition.*

Thus, as we already mentioned, Janis from Big Sister has veto power. But she is not a dictator because the coalition {Janis} is a losing coalition with Janis as a member. The idea is that someone with veto power can force a proposal to fail to pass by themselves, but a dictator is someone with veto power who can also force a proposal to pass by themselves.

As a second observation, note that even though all the examples done in this subsection were weighted systems, the definitions work fine for any voting system. However, if the system is not weighted, the very notion of a numerical power index is misleading in a sense. For example, in the US federal system, what if there was a power index that assigned a power of 0.001 to a representative and a power of 0.003 to a senator. Doesn't that suggest that a senator is three times as "powerful" as a representative? And yet, if there are 217 representatives and 51 senators and the president in a coalition, no matter how many senators join your coalition, it is still losing, whereas all you need is one representative to make it winning.

The problem is that by using a numerical measure, it seems to imply a linear ordering of voters. But in the US federal system, sometimes you need a senator and sometimes you need a representative, so a numerical measure of power doesn't seem to work in all cases. What we should do, perhaps, is similar to what we did when we passed from real number payoffs to ordinal games. Instead of trying to assign a real number measuring a voter's power, we compare two voters and just say which one is more "helpful" when he or she joins your coalition. That is, we simply rank them in order of preference and allow for the possibility that maybe not all voters can be compared to each other, like the senator and the representative in the earlier example (so the ordinal version of political power would be what is known as a *partial order* rather than a linear order). This notion of *ordinal power* is a fascinating one, and the reader can learn more about it in Taylor and Pacelli (2008) or at a more advanced level in Taylor and Zwicker (1999).

This concludes our brief introduction to legislative voting systems, and also our discussion of game theory.

Exercises

1.   In the United Nations Security Council, there are 15 countries that are the voters in the system. Five of the countries are called *permanent members*. They are China, England, France, Russia, and the United States. The remaining 10 nonpermanent members rotate periodically among the other member countries of the UN. If the Security Council wants to pass a proposal, it requires the support of at least nine of the 15 countries on the Council. However, any one of the five permanent members has veto power – a vote of "no" from one of them will kill the proposal. (We ignore the possibility of abstentions.) With this

description, the UN Security Council is not described as a weighted system. Show that it is equivalent to a weighted system. [**Hint**: Give each nonpermanent member a weight of 1. The five permanent members seem to have equal power, so give them all the same weight, called x. Now just make x and q large enough so that no coalition is winning unless all five permanent members are in the coalition. You can do this by trial and error, or you can try to be more efficient and systematic about it by setting up some inequalities to solve.]

2. Consider the following weighted voting system: Velour Secret is a fashion clothing company, and all production decisions are made by a committee of department managers: Lou has eight votes, Christa has eight votes, John has seven votes, Maureen has six votes, Sterling has five votes, and Doug has 3 votes. To pass a production idea and start production, a quota of 20 votes is required.

   a. Consider the following two coalitions: $X = \{$Lou, John, Sterling$\}$ and $Y = \{$Lou, Christa, Maureen, Doug$\}$. Show that both coalitions are winning.
   b. Consider what happens when the two coalitions swap John for Christa. Do both coalitions, just one, or neither remain winning?
   c. Answer the same question as if the two coalitions swap Sterling for Doug.
   d. Is it possible for these coalitions to swap two voters that will render both coalitions losing? Explain.

3. The considerations of Exercise 2 lead to the following definition (explored in Taylor and Pacelli (2008) and Hodge and Klima (2005)): a voting system is said to be *swap robust* if whenever you have two winning coalitions X and Y, and a one for one trade (a "swap") is made between X and Y, then at least one of the resulting coalitions must be winning. (The voters that are traded must not belong to the intersection of the two coalitions.) Prove that every weighted system is swap robust. [**Hint**: Let q be the quota and let X and W be two winning coalitions, so that $w(X) \geq q$ and $w(Y) \geq q$, where $w(X)$ means the sum of the weights of the voters in X. Now select x in X and y in Y to trade, and consider what happens to the weights of the coalitions in two cases. First, what if the weights of x and y are the same, and second, what if the weights of x and y are different?]

4. In Exercise 3, you showed a weighted voting system must be swap-robust. It follows that if a voting system is *not* swap robust, then it cannot be weighted. Show that the US federal system is not swap robust, and, therefore, it is not weighted, as suggested in the text. This shows that voting systems exist that are not weighted.

5. The condition of swap-robustness does not completely characterize weighted systems. It's true that every weighted system is swap robust, but there are also non-weighted systems that are swap robust as well. For example, given that the following voting system is not equivalent to a weighted system, show that, nevertheless, it is swap robust.

   The Windy Association is an investment firm has a board of directors consisting of six people. Each potential investment is voted on. A particular investment is made if it receives at least three of the six possible votes, subject to the restriction that the people voting for the investment must have a total of at least 25 years experience with the firm between them. The number of years experience of each board member is given by Table 8.5.

   Show that this voting system is swap robust. [**Hint**: There are two conditions for a coalition to be winning. It needs a minimum of three voters and a minimum of 25 years

Table 8.5 A passing vote requires a minimum total number of years experience of those who vote yes.

Board member	Years experience
Mary	18
Terry	12
Jules	8
Brian	5
Jim	4
Ted	3

experience. If there are two winning coalitions and a swap is performed, consider separately how the swap affects each of the two conditions.]

6. A condition that a voting system may satisfy that is stronger than swap-robustness is called *trade-robustness*. Its definition is similar to swap-robustness, except that the trade of voters can be an arbitrary trade between any number of winning coalitions instead of a one-for-one swap between just two winning coalitions. To be precise, a voting system is *trade-robust* if, whenever you have two or more winning coalitions and these coalitions perform an arbitrary trade of voters, then at least one of the resulting coalitions must still be winning.

Show that the Windy Association board of directors system from Exercise 5 is not trade-robust. [**Hint**: Try to make up two winning coalitions X and Y, such that a two-for-one trade renders them both losing.]

REMARK 8.28 *In 1992, Alan Taylor and William Zwicker proved that the condition of trade-robustness completely characterizes a weighted system – that is, a voting system is weighted if and only if it is trade robust. The Windy Association voting system in Exercise 5 is not weighted because, in Exercise 6, you showed it is not trade-robust, even though it is swap-robust. This characterization theorem for weighted systems is discussed in Taylor and Pacelli (2008), and a full proof appears in Taylor and Zwicker (1999) and Taylor and Zwicker (1992).*

7. Suppose that the six New England states form a voting system for regional economic development. Passage on any initiative requires at least three of the six votes, subject to the requirement that the states supporting the initiative must constitute at least 50% of the New England population, and also subject to the condition that either Maine or Connecticut supports the measure. Assume the population percentages are given by Table 8.6.

Show that this system is not swap-robust and, therefore, it is not a weighted system.

8. Consider the weighted voting system $[8; 4, 3, 3, 1, 1]$. List all the winning coalitions (there are 11 of them). Which ones are minimal?

9. a. Explain why David is a dummy voter in the King Heart, Ltd., example from the text.
   b. Verify the list of winning coalitions for this example given in the text.

10. Consider the weighted voting system $[9; 4, 3, 3, 1]$. Explain why the voter with one vote is a dummy voter. Do any of the voters have veto power? Is any voter a dictator?

11. Suppose we lower the quota in the weighted system in Exercise 10 from 9 to 8.

Table 8.6 Percentage of total New England population in each state.

State	Population percentage
Connecticut	20%
Massachusetts	26%
Maine	19%
Rhode Island	15%
New Hampshire	10%
Vermont	10%

a.   Compute the Shapley–Shubik index for each voter.

b.   Answer the same question if we lower the quota to 7 instead of 8.

12.   Compute the Banzhaf index for the voters of the system in Exercise 8.

13.   Compute the Banzhaf index for the voters in the Windy Association board of directors in Exercise 5. (You will first need to list the winning coalitions.)

14.   Compute the Banzhaf index for the voters in the voting system of New England states in Exercise 7. (First list the winning coalitions.) This exercise shows that the Banzhaf index makes sense for non-weighted systems and is calculated the same way.

15.   A voting system can have more than one voter with veto power (see Exercise 1). Can it have more than one dictator? Explain.

For more exercises on the Shapley–Shubik power index, see the exercises after Section A.2.

9   More Linear Programming

To complete our study of the simplex algorithm in linear programming, we need to discuss what happens if the initial point (the origin) is not feasible. This calls for modified pivoting rules, called Phase I pivoting, and with it, we can solve maximization problems that are not in standard form due to mixed constraints. Then, we discuss how Phase I pivoting gives an alternate approach to minimizations problems. Finally, we revisit the notion of sensitivity analysis and show how to deal with minor changes in the setup of a problem via the simplex algorithm, compute stable ranges, and handle integer programming problems.

9.1   Phase I Pivoting

Consider the following linear programming problem:

Example 9.1

$$\text{Maximize } P = 2x + 3y$$

subject to:

$$5x + 6y \leq 120$$
$$5x + 3y \leq 90$$
$$x + y \geq 10$$
$$2x + 3y \geq 24$$
$$x \geq 0; \ y \geq 0$$

This problem has mixed constraints, so it is not a standard form maximization problem. If we multiply the last two structural inequalities through by a factor of -1, all the inequalities will become less-than-type inequalities. These inequalities are

$$-x - y \leq -10$$
$$-2x - 3y \leq -24.$$

With the structural inequalities all written as the less-than type, the problem becomes what we previously called *semi-standard form*. But it is still not standard form because the constants on the right side of these last two inequalities are negative. This means that the origin is not a feasible point. (In Exercise 1, you are asked to solve this problem by graphing in the decision space, and you can see the origin is not feasible in your graph.) If you were to construct the initial tableau, the way one normally does for Phase II pivoting, the capacity column would contain negative entries, meaning that two of the basic variables S_3 and S_4 are negative, a violation of the feasibility conditions.

Therefore, we need Phase I pivoting rules first to pivot the tableau until it becomes a tableau representing a feasible point, after which we can finish the problem by the usual Phase II rules. In this section, we learn how to perform Phase I pivoting. Actually, in our approach to Phase I, it is not so much that the rules for pivoting are different; it is that the initial tableau is set up differently. There are several different but equivalent ways to do this. We follow the approach in Calvert and Voxman (1989). Here is how it works:

Since we don't allow any slack variables S_i to be negative, it is not sufficient to convert the third and fourth inequalities to an equation by the addition of a slack variable. Instead, we add two variables to each such inequality – the first is called a slack variable S_i, as usual, and the second is called an *artificial variable* A_i. Actually, we also want $A_i \geq 0$, so we add the slack variable and subtract the artificial one. Thus, the third and fourth constraints are replaced by the following equations:

$$-x - y + S_3 - A_3 = -10$$
$$-2x - 3y + S_4 - A_4 = -24.$$

In this way, at the initial point, which is the origin, we can set $x = y = 0$ and set $S_3 = 0 = S_4$ as well, while $A_3 = 10$ and $A_4 = 24$. That way, all the variables are happily nonnegative. But while the variables A_3 and A_4 start out with nonzero values (making them basic variables for the origin), in the end, we want them to have a value of 0, so that by the time we reach the feasible region, all the A_i are nonbasic. That is why they are called *artificial* – we know in advance that we want them to be 0 in any basic, feasible solution. They only start out temporarily as nonzero. We have an immediate goal of making them disappear from all the equations (that is, making them nonbasic). This is referred to as "driving the artificial variables from the solution." Once an artificial variable is 0, we will never make it an incoming variable again, so we will never pivot on an column labeled with an artificial variable.

In summary, we are replacing the system of equations:

$$5x + 6y + S_1 = 120$$
$$5x + 3y + S_2 = 90$$
$$-x - y + S_3 = -10$$
$$-2x - 3y + S_4 = -24$$

with a different system – namely, this one:

$$5x + 6y + S_1 = 120$$
$$5x + 3y + S_2 = 90$$
$$-x - y + S_3 - A_3 = -10$$
$$-2x - 3y + S_4 - A_4 = -24.$$

Obviously, with more variables, the second system is *not* equivalent to the first. Regardless, if we take any solution of the second system in which $A_1 = A_2 = 0$, and the other variables are nonnegative, then that will be a basic, feasible solution of the first system. Once such a solution is reached, we finish by using Phase II.

So how do we go about forcing each A_i to become 0? Note that, since we are assuming $A_i \geq 0$ for all i, it follows that each $A_i = 0$ if and only if their sum is 0. Invent a new variable (called w) that is the sum of all the artificial variables. In this problem, $w = A_3 + A_4$. Since the $A_i \geq 0$, so is their sum w. So, if we try to minimize w and arrive at a tableau where $w = 0$, then we know each $A_i = 0$ at this point. However, we prefer to phrase the problem as a maximizing problem, since we are already familiar with how to solve those. Minimizing w is the same as maximizing its negative, so define $z = -w = -A_3 - A_4$. Now we seek to pivot the tableau first to maximize z (in other words, our preliminary objective becomes to maximize z, and we hope that the maximum value of z is 0, which is why we are calling it z – short for "zero"). If we succeed, then Phase I is complete, and we can move on to our original objective of maximizing P.

Thus, we are going to annex an additional row at the bottom of the initial tableau that represents z. However, although we know $z = -A_1 - A_2$, we need to express it in terms of the other variables. We can do this by solving the third and fourth equations for the negatives of the artificial variables and adding. We obtain

$$-A_3 = x + y - S_3 - 10$$
$$-A_4 = 2x + 3y - S_4 - 24$$

adding gives:

$$z = 3x + 4y - S_3 - S_4 - 34.$$

Thus, the initial tableau for Phase I looks like this.

	x	y	S_1	S_2	S_3	S_4	A_3	A_4	Capacity
S_1	5	6	1	0	0	0	0	0	120
S_2	5	3	0	1	0	0	0	0	90
$-A_3$	−1	−1	0	0	1	0	−1	0	−10
$-A_4$	−2	−3	0	0	0	1	0	−1	−24
P	−2	−3	0	0	0	0	0	0	0
z	−3	−4	0	0	1	1	0	0	−34

The reader should note the following:

- We have labeled rows 3 and 4 with the artificial variables, not the slack variables, because at the origin, the A_i are the basic variables. They are labeled with negative signs because the capacities are negative (as a result of putting the inequalities into semi-standard form). Equivalently, we could multiply through these last two rows by −1 and label them with A_3 and A_4 without the negatives in the labels. We converted to semi-standard form so that the negative capacities serve as a red flag that we need Phase I. However, having the rows labeled with artificial variables can serve as the red flag if you prefer the other approach.
- Although our first goal is to maximize z, not P, we still include a row for P so that, with each pivot, we can express P in terms of whatever the basic variables are for that particular basic solution. Just remember never to pivot on that row.

• When we filled out the row for z, we negated each coefficient just as we did for the coefficients of P. However, the bottom-right corner is filled in with -34 rather than 0, because at the origin, z has a value of -34 (x, y, S_3, and S_4 are all nonbasic variables with value 0 at the origin).

• As noted before, once an artificial variable is 0, we never want it to be an incoming variable, so we never pivot on the columns labeled A_i

• We now pivot as usual, seeking to maximize z, and hope that we obtain a basic solution where $z = 0$. If that happens, at this point, the artificial variables are all 0 as well, so the solution is a basic feasible solution to our original problem. We can then delete the columns labeled A_i (why keep them if we will never pivot on them?), delete the last row labeled z, and finish pivoting the reduced tableau using Phase II until P is maximized.

We illustrate the process with the preceding example. In the first pivot, we select y as incoming (since the most negative entry in the z row is -4). The corresponding ratios for the rows are $\frac{120}{6} = 20$, $\frac{90}{3} = 30$, $\frac{-10}{-1} = 10$, and $\frac{-24}{-3} = 8$. Since 8 is the smallest nonnegative ratio, the fourth row is the pivot row, and the outgoing variable is A_4. Pivoting on this row (check this!) leads to the next tableau.

	x	y	S_1	S_2	S_3	S_4	A_3	A_4	Capacity
S_1	1	0	1	0	0	2	0	-2	72
S_2	3	0	0	1	0	1	0	-1	66
$-A_3$	$-\frac{1}{3}$	0	0	0	1	$-\frac{1}{3}$	-1	$\frac{1}{3}$	-2
y	$\frac{2}{3}$	1	0	0	0	$-\frac{1}{3}$	0	$\frac{1}{3}$	8
P	0	0	0	0	0	-1	0	1	24
z	$-\frac{1}{3}$	0	0	0	1	$-\frac{1}{3}$	0	$\frac{4}{3}$	-2

This tableau represents a point that is still not feasible (since $z \neq 0$). We pivot on the first column (according to Bland's rule). The smallest nonnegative ratio is 6, in the third row. Thus, the third row is the pivot row, and the incoming variable x replaces A_3, the outgoing variable. Pivoting leads to the next tableau.

	x	y	S_1	S_2	S_3	S_4	A_3	A_4	Capacity
S_1	0	0	1	0	3	1	-3	-1	66
S_2	0	0	0	1	9	-2	-9	2	48
x	1	0	0	0	-3	1	3	-1	6
y	0	1	0	0	2	-1	-2	1	4
P	0	0	0	0	0	-1	0	1	24
z	0	0	0	0	0	0	1	1	0

Since the bottom row entries are all nonnegative, the maximum value of z has been reached. Since this maximum value is $z = 0$, we know this is a feasible basic solution with the artificial variables all nonbasic with value 0. Thus, we are finished with Phase I. Deleting the appropriate row and columns leads to this tableau.

	x	y	S_1	S_2	S_3	S_4	Capacity
S_1	0	0	1	0	3	1	66
S_2	0	0	0	1	9	-2	48
x	1	0	0	0	-3	1	6
y	0	1	0	0	2	-1	4
P	0	0	0	0	0	-1	24

We see that although this tableau represents a feasible point, it is not optimal. We pivot on the S_4 column. Ignoring the two negative ratios, the pivot row is the third row, so S_4 replaces x. There is, in fact, one more pivot needed after that one. The reader can check (Exercise 2) that we arrive at the following final tableau.

	x	y	S_1	S_2	S_3	S_4	Capacity
S_3	$-\frac{1}{6}$	0	$\frac{1}{6}$	0	1	0	10
S_2	$\frac{5}{2}$	0	$-\frac{1}{2}$	1	0	0	30
S_4	$\frac{1}{2}$	0	$\frac{1}{2}$	0	0	1	36
y	$\frac{5}{6}$	1	$\frac{1}{6}$	0	0	0	20
P	$\frac{1}{2}$	0	$\frac{1}{2}$	0	0	0	60

Thus, the solution is

$$x = 0$$
$$y = 20$$
$$P = 60 \text{ (maximized)}$$
$$S_1 = 0 \text{ and } M_1 = \tfrac{1}{2}$$
$$S_2 = 30 \text{ and } M_2 = 0$$
$$S_3 = 10 \text{ and } M_3 = 0$$
$$S_4 = 36 \text{ and } M_4 = 0,$$

and, of course, as expected, $A_3 = A_4 = 0$.

Two or three questions remain. The first is, what happens if an artificial variable is still listed as basic at the end of Phase I? The second is how do we know this method always works? And the third is how do we handle constraints that are exact equalities instead of inequalities? We will address the first question now and save the second for after the third.

If, at the end of Phase I, an artificial variable A is still listed as a basic variable in the row labels, note that because $z = 0$, the value of A is nevertheless 0. We have seen before that when a basic variable is 0, this means the problem is degenerate. Suppose that, in the row with the artificial variable label, there is some nonzero entry a in a column that is headed by a decision variable or a slack variable. In this case, pivot on the a position. This will replace the variable A with the designated incoming decision or slack variable. Furthermore, since the capacity column has a 0 in this row, after the pivot, none of the numbers in the capacity column will change (even if $a < 0$), so not only does the solution remain feasible; it actually represents the same corner of the feasible region X, just using a different set of basic variables to describe the solution (as is typical with degenerate problems).

On the other hand, suppose the only entries in the row that are under decision or slack variables are 0. In that case, when we delete the slack variable columns from the tableau, we obtain a row of all zeros, which means the corresponding constraint is redundant (it is a linear combination of other constraints), and we can ignore or delete the entire row from all future pivots. As these two cases are mutually exclusive and exhaustive, this handles the situation of rows labeled by artificial variables after z attains a value of 0. Strictly speaking, getting rid of these artificial variable labels on the rows are part of Phase I, so you are not really finished with it until all these row labels are gone. We illustrate with Example 9.2.

Next, what about constraints that are exact equalities? Here is one approach. If you are given $ax + by = c$ as a constraint, you can replace it with the pair of inequalities:

$$ax + by \leq c$$
$$ax + by \geq c.$$

Clearly, the equation is satisfied if and only if the two inequalities are satisfied. Then you could handle the second inequality as we did before, by putting in both a slack and an artificial variable. This method will always work. However, it increases the number of rows in the tableau by one. If you have three such equations as constraints, that would mean adding three rows to the tableau. Each additional row adds a lot of work to the pivoting process, so we would prefer to avoid adding rows if possible. The way out is to treat the equality constraint the same way we treat a greater-than-type constraint – with an artificial variable. We'll negate the equation first to be consistent with the way we treat inequalities (this is not essential), and then we'll subtract an artificial variable that starts out basic but must be driven out of the solution, just as we did for greater-than-type inequalities. We do not add or subtract a slack variable, however, because in the solution, there is no slack in an exact equation. Then proceed as before.

Example 9.2 Consider the following linear programming problem with mixed constraints:

$$\text{maximize } P = 2x + 5y$$
$$\text{subject to:}$$
$$x + y \leq 72$$
$$4x + 6y = 240$$
$$4x + 3y \geq 120$$
$$x \geq 0, \ y \geq 0.$$

Incidentally, this is the canonical way to write problems with mixed constraints – all the $\leq$ type inequalities first, followed by the exact equations, followed by the $\geq$ type constraints. We convert the exact equation to $-4x - 6y - A_2 = -240$, and the third inequality to $-4x - 3y + S_3 - A_3 = -120$. Then

$$z = -A_2 - A_3 = 8x + 9y - S_3 - 360,$$

and the initial tableau is the following:

	x	y	S_1	S_3	A_2	A_3	Capacity
S_1	1	1	1	0	0	0	72
$-A_2$	-4	-6	0	0	-1	0	-240
$-A_3$	-4	-3	0	1	0	-1	-120
P	-2	-5	0	0	0	0	0
z	-8	-9	0	1	0	0	-360

The first pivot is the second row and second column. Leaving the details for the reader, the next tableau is the following:

	x	y	S_1	S_3	A_2	A_3	Capacity
S_1	$\frac{2}{6}$	0	1	$-\frac{1}{6}$	0	0	32
y	$\frac{4}{6}$	1	0	$\frac{1}{6}$	0	0	40
$-A_3$	-2	0	0	$\frac{3}{6}$	1	-1	0
P	$\frac{8}{6}$	0	0	$\frac{5}{6}$	0	0	200
z	-2	0	0	$\frac{9}{6}$	1	0	0

The fact that $z = 0$ means it is maximized, so this tableau must represent a basic feasible solution to our original problem. Indeed, when $x = 0$ and $y = 40$, it is easy to verify that $P = 200$, $S_1 = 32$, $S_3 = 0$, and both of the artificial variables are indeed 0. This seems odd, because there is still a negative entry in the bottom row. However, that is because the problem is degenerate, as indicated by the presence of a basic artificial variable. The smallest nonnegative row ratio is the 0 in the third row. If you pivoted there (third row and first column), none of the numbers in the capacity column would change since we would be just reassigning which variables are basic at the current corner of the feasible region X. In Exercise 3, the reader is asked to finish the problem via this approach.

However, here we take a different approach. Since $z = 0$, we know the artificial variables have been driven out of solution. Thus, we can delete the columns headed by artificial variables and the bottom row. We are left with the following tableau.

	x	y	S_1	S_3	Capacity
S_1	$\frac{2}{6}$	0	1	$-\frac{1}{6}$	32
y	$\frac{4}{6}$	1	0	$\frac{1}{6}$	40
$-A_3$	-2	0	0	$\frac{3}{6}$	0
P	$\frac{8}{6}$	0	0	$\frac{5}{6}$	200

Now we are in the position mentioned before that we have an artificial variable still labeled as basic. According to our rules to drive out that variable, we can pivot on either the -2 in that row (first column) or on the $\frac{3}{6}$ (the fourth column). We illustrate the former here and ask the reader to try the latter in Exercise 3. So pivoting on the -2 yields the following tableau.

	x	y	S_1	S_3	Capacity
S_1	0	0	1	$-\frac{1}{12}$	32
y	0	1	0	$\frac{1}{3}$	40
x	1	0	0	$-\frac{1}{4}$	0
P	0	0	0	$\frac{7}{6}$	200

Having driven out the artificial variable, we are now done with Phase I. Note that the corner did not change on the last pivot – we stayed at $(0, 40)$ – but we merely reassigned x as the basic variable with value 0 instead of the artificial variable. Now, on to Phase II – but everything in the P row is nonnegative, so this point must be also optimal, and there is no further pivoting to do. The solution is

$$x = 0$$
$$y = 40$$
$$P = 200 \text{ (maximized)}$$
$$S_1 = 32$$
$$S_2 = 0$$
$$S_3 = 0,$$

and, of course, $A_2 = 0 = A_3$.

So we have done examples with exact equalities and an example where an artificial variable remains basic at the "end" of Phase I. We did not display an example of the other case, where after deleting the artificial variable columns, there is still an artificial variable row that however, has all 0 entries. As noted before, we just delete the entire row in that case, because that particular constraint is redundant. An example is shown in Calvert and Voxman (1989) for any reader who is interested.

We can now outline the entire simplex method.

ALGORITHM 9.3   The simplex algorithm *(maximization problems).*

1.   *Add a nonnegative slack variable S_i to each less-than-type inequality to convert it to an equation. If the problem is a standard form maximization problem (all constants on the right side of the inequalities are nonnegative and no other constraints exist besides less-than-type inequalities, then the origin is a feasible point. In this case, construct the initial tableau and proceed directly to* Phase II. *Otherwise go to* Phase I:

2.   Phase I

     a.   *For each exact equality constraint, add a nonnegative artificial variable A_i. (If desired, you may negate the entire equation, as done earlier.)*

     b.   *For each greater-than-type constraint, negate the inequality to put it in semi-standard form. Then add a nonnegative slack variable S_i and subtract a nonnegative artificial variable A_i to convert it to an equation.*

     c.   *Let z be the sum of the negatives of all the artificial variables, so that $z \leq 0$. Express z in terms of the basic and decision variables, by solving the equations you obtained in steps 2a and 2b for the artificial variables and adding their negatives.*

    d.   *Find a basic, feasible solution.*

        i.   *Construct the initial tableau, including rows for both z and the original objective function. Label the rows in the initial tableau as follows: for any row corresponding to a less-than-type inequality, label the row with the corresponding slack variable; for any row corresponding to a greater-than-type inequality, label the row with the negative of the corresponding artificial variable; and for any row corresponding to an exact equality constraint, label the row with the corresponding artificial variable, or its negative if you negated the entire equation in step 2a. (The purpose of the negative signs on the labels is so that we do not violate the nonnegativity constraints. Remember that each decision, slack, and artificial variable is nonnegative. So the negative row labels will correspond to any row with a negative number in the capacity column.)*

        ii.   *Apply* Phase II *pivoting to maximize z. If the maximum value of z is negative, then* stop. *The problem has no feasible solution. If the maximum value of z is 0, then we have found a basic, feasible solution.*

        iii.   *Remove any remaining artificial variables from the basis. If a row is labeled with an artificial variable, then either the problem is degenerate or that constraint is redundant. If there is a nonzero entry in that row in a column labeled with a decision variable or a slack variable, then the problem is degenerate. The artificial variable can be driven out from the solution by pivoting on any such entry. If every entry in that row in a column labeled with a decision or slack variable is 0, then that constraint is redundant. In this case, the entire row can be deleted from the tableau.*

        iv.   *Delete from the tableau the row corresponding to z and any columns labeled with artificial variables. The tableau that remains is a basic feasible solution, and* Phase I *is complete. Proceed to* Phase II.

3.   Phase II: *See Algorithm 5.2 in Section 5.2.*

Finally, what about the question of how we know this procedure always works? The answer is given by the following theorem.

THEOREM 9.4   *A maximization linear programming problem has a feasible solution if and only if all of the artificial variables have value 0 at the end of Phase I.*

The proof of this theorem, for interested readers, is not difficult at all, and it appears in Calvert and Voxman (1989).

We remarked earlier that there were several different but equivalent approaches to Phase I pivoting. We went through the approach using artificial variables because it is conceptually clear what is happening with this approach. However, because of the addition of a row and possibly several columns to the initial tableau, the result makes the pivoting a bit more tedious. Is it possible to approach Phase I with a technique that avoids adding any rows or columns to the tableau? We conclude this section by presenting one such approach now, but without proof. This approach is sometimes called *dual pivoting*:

ALGORITHM 9.5   Alternate Phase I (dual pivoting, taken from section 6.5 of Calvert and Voxman (1989)):

1. *Determine the pivot* row *first. This can be any row with a negative capacity value. It is usual to take the row with the most negative entry in the capacity column.*

2. *Determine the pivot* column *next. If no row with a negative capacity column entry has any other negative entries, then stop. The problem has no feasible solution. Otherwise, for each column with a negative entry in the pivot row, form the ratio of the objective row entry to the pivot row entry. Choose the column with the* maximum *of these ratios, breaking ties, as usual, by Bland's rule.*

3. *Pivot on the entry that is at the intersection of this row and column (the pivot number.)*

4. *If all entries in the capacity column are nonnegative, then you have found a feasible solution. Phase I is over, so proceed to Phase II. Otherwise, return to step 1 and iterate.*

There are other variations on Phase I as well. For example, the procedure in section 4.4 of Tomastik (1994) is a variation of dual pivoting. Furthermore, it is not discussed by Calvert and Voxman (1989) how to handle an exact equality constraint in Algorithm 9.5. Presumably, one is supposed to replace the single equation with two opposite inequalities as mentioned prior to Example 9.2. That will work, but it does add a row to the tableau. Since we would like to avoid that, we suggest a hybrid of the two methods. Use Algorithm 9.5, but for exact equality constraints, use an artificial variable as we did in the first approach. Negate the equation first so that the capacity column will be negative. Then, in step 1 of Algorithm 9.5, put the highest priority on pivoting rows with artificial variables (that is, we take care of exact equalities first). Once an artificial variable is outgoing, one never wants to make it incoming again, so it is not even necessary to add a column for it – we would never pivot on that column.

We illustrate by solving Example 9.2 again. using this approach. Here is the initial tableau.

	x	y	S_1	A	S_3	Capacity
S_1	1	1	1	0	0	72
$-A$	-4	-6	0	1	0	-240
S_3	-4	-3	0	0	1	-120
P	-2	-5	0	0	0	0

Although we said it is not necessary to include the A column, we include it here for completeness. According to the hybrid version of Algorithm 9.5, we first pivot on the second row. The ratios of the objective row element divided by the pivot row element for those columns with negative entries are $\frac{-2}{-4} = \frac{1}{2}$ for the x column and $\frac{-5}{-6} = \frac{5}{6}$ for the y column. The maximum ratio occurs on the y column, so the pivot number is -6 in the a_{22} position. The result of the pivot is the following tableau.

	x	y	S_1	A	S_3	Capacity
S_1	$\frac{1}{3}$	0	1	$\frac{1}{6}$	0	32
y	$\frac{2}{3}$	1	0	$-\frac{1}{6}$	0	40
S_3	-2	0	0	$-\frac{1}{2}$	1	0
P	$\frac{4}{3}$	0	0	$-\frac{5}{6}$	0	200

Since we never pivot on a column with an artificial variable, we see this tableau must be optimal! The solution agrees with the one obtained before, but with fewer pivots – indeed, it only took one pivot in this case.

Exercises

1. Solve the problem in Example 9.1 by graphing in the decision space. Note that the origin is not feasible.

2. Finish doing the pivots in Example 9.1, verify that the final tableau is as claimed in the text, and verify that the answer is the same as the one you obtained in Exercise 1. On your graph in Exercise 1, follow and mark the corners we covered in the pivots for this example so you can see which corners were visited on the way to the feasible set and, once there, to the optimal point.

3. a. Solve the problem in Example 9.2 by graphing in the decision space, and verify the answer is the same as the one obtained in the text.

   b. As suggested in the text, continue pivoting after the second tableau to drive out the artificial variable.

   c. As suggested in the text, in the third tableau, after we delete the artificial variable columns, finish the problem by pivoting in the third row, fourth column, instead of the choice we made in the text. Verify that the solution point is still the same, but with a different basic variable with value 0.

4. Solve the following problem:

$$\text{maximize } P = 5x + 4y$$
$$\text{subject to:}$$
$$6x + 4y \leq 240$$
$$x + 2y \leq 96$$
$$x + y = 50$$
$$x \geq 0, \ y \geq 0.$$

5. Solve the problem:

$$\text{maximize } P = x + 4y$$
$$\text{subject to:}$$
$$4x + 7y \leq 140$$
$$5x + 22y = 228$$
$$3x + 2y \geq 36$$
$$x \geq 0, \ y \geq 0.$$

6. In Section 3.4, you solved the following diet mix problem via a graphical method. Now solve it by the simplex method. Kerry's Kennel is mixing two commercial brands of dog food for its canine guests. A bag of Dog's Life Canine Cuisine contains 3 pounds of fat, 2 pounds of carbohydrates, 5 pounds of protein, and 3 ounces of vitamin C. A giant-size bag

of Way of Life Healthy Mix contains 1 pound of fat, 5 pounds of carbohydrates, 10 pounds of protein, and 7 ounces of vitamin C. The requirements for a week's supply of food for the kennel are that there should be at most 21 pounds of fat, at most 40 pounds of carbohydrates, and at least 21 ounces of vitamin C. How many bags of each type of food should be mixed in order to design a diet that maximizes protein?

7. Solve the following linear programming problem from Section 3.4 via the simplex algorithm. The Jefferson Plastic Fantastic Assembly Corporation manufactures gadgets and widgets for airplanes and starships. Each case of gadgets uses 2 kilograms of steel and 5 kilograms of plastic. Each case of widgets uses 2 kilograms of steel and 3 kilograms of plastic. The profit for a case of gadgets is $360, and the profit for a case of widgets is $200. Suppose Jefferson Plastic Fantastic Assembly Corporation has 80 kilograms of steel available and 150 kilograms of plastic available on a daily basis and can sell everything it manufactures. How many cases of each should the corporation manufacture if it is obligated to produce at least 10 cases of widgets per day? The objective is to maximize daily profit.

8. Solve the following linear programming problem from Section 3.4 via the simplex algorithm. Mr. Cooder, a farmer in the Purple Valley, has at most 400 acres to devote to two crops: rye and barley. Each acre of rye yields $100 profit per week, while each acre of barley yields $80 profit per week. Due to local demand, Mr. Cooder must plant at least 100 acres of barley. The federal government provides a subsidy to grow these crops in the form of tax credits. The government credits Mr. Cooder four units for each acre of rye and two units for each acre of barley. The exact value of a "unit" of tax credit is not important. Mr. Cooder has decided that he needs at least 600 units of tax credits in order to be able to afford to his loan payments on a new harvester. How many acres of each crop should he plant in order to maximize his profit?

9. Solve the following linear programming problem from Section 3.1. Arnold and Penny Layne have up to $60,000 to invest for a one-year period. There are three investment options they are considering: certificates of deposit, which have an expected annual return of 3%; municipal bonds (the proceeds of which go to the cleaning and upkeep of local buses, police cars, and fire engines), which have an expected annual return of 5%; and stocks on women's apparel, with an expected annual return of 11%. They would like to maximize their expected return for the year but are adhering to the following guidelines suggested by their financial adviser:

- The amount invested in municipal bonds should be at least $10,000.
- Because of stock volatility, the amount invested in stocks should be at most $10,000 more than the amount invested in bonds.
- Because of the reliability of certificates of deposit, at least $\frac{1}{3}$ of the total investment should be in certificates of deposit.

10. a. Solve the following problem:

$$\text{maximize } P = 2x + 3y + 5z$$
$$\text{subject to:}$$
$$x + y + 2z \le 60$$
$$2x + y + z \ge 12$$

$$x + 2y + z \geq 10$$
$$7x + 8y + 5z \geq 54$$
$$x \geq 0, \ y \geq 0, \ z \geq 0.$$

b. Show that the fourth constraint is a linear combination of the second and third and is, therefore, redundant. (An alternate approach to solving the problem is to delete the fourth constraint altogether, before starting any pivoting.)

11. Solve the brainteaser in Exercise 21 of Section 3.1 using the simplex algorithm.
12. Solve Problem 4 using Algorithm 9.5 (dual pivoting).
13. Solve Problem 6 using Algorithm 9.5 (dual pivoting).
14. Solve Problem 7 using Algorithm 9.5 (dual pivoting).

9.2 Alternate Approach to Minimization Problems

In Section 5.3, we decided that the way to solve a minimization problem via the simplex method was to convert it to a maximization problem so that we would not have to learn separate pivoting rules for minimization. There were two ways to convert a maximization to a minimization. One was to consider the dual problem. However, because in that chapter we had not yet learned Phase I, that technique was helpful only if we started with a standard form minimization problem.

The other method for converting was to change minimizing C to maximizing $-C$ (i.e., multiply through by -1). In fact, we have already exploited this idea in the last section, when we decided that minimizing the sum w of the artificial variables was the same as maximizing $z = -w$. In Section 5.1, this method was no help because, again, we had not yet learned Phase I. Now that we have learned Phase I, we can use this technique to solve any minimization problem, whether it is standard form or not.

In summary, the method is this:

- Multiply the objective function C through by -1, and maximize the negative of C.
- Any inequality of the $\geq$ type must also be multiplied through by -1 to covert the problem to semi-standard form.

The result is you have converted your minimization problem to a maximization problem (with constraints that almost always require artificial variables.)

Then solve this problem using Phase I.

That's it! Here's an example.

Example 9.6 Solve the following minimization problem:

$$\text{minimize } C = 12x + 8y$$

subject to:

$$12x + 5y \geq 120$$
$$3x + 5y \geq 75$$
$$x \geq 0, \ y \geq 0.$$

This is the same as solving this problem:

$$\text{maximize} -C = -12x - 8y$$
$$\text{subject to:}$$
$$-12x - 5y \le -120$$
$$-3x - 5y \le -75$$
$$x \ge 0, \ y \ge 0.$$

According to Phase I techniques, we replace the two inequalities with the following equations:

$$-12x - 5y + S_1 - A_1 = -120$$
$$-3x - 5y + S_2 - A_2 = -75.$$

Then we have a preliminary objective of maximizing

$$z = -A_1 - A_2 = 15x + 10y - S_1 - S_2 - 195. \tag{9.1}$$

Thus, the initial tableau is as follows.

	x	y	S_1	S_2	A_1	A_2	Capacity
$-A_1$	-12	-5	1	0	-1	0	-120
$-A_2$	-3	-5	0	1	0	-1	-75
$-C$	12	8	0	0	0	0	0
z	-15	-10	1	1	0	0	-195

The first pivot row and column are shaded. Performing the pivot yields the following tableau, where we have also shaded in the next pivot row and column.

	x	y	S_1	S_2	A_1	A_2	Capacity
x	1	$\frac{5}{12}$	$-\frac{1}{12}$	0	$\frac{1}{12}$	0	10
$-A_2$	0	$-\frac{45}{12}$	$-\frac{3}{12}$	1	$\frac{3}{12}$	-1	-45
$-C$	0	3	1	0	-1	0	-120
z	0	$-\frac{45}{12}$	$-\frac{3}{12}$	1	$\frac{15}{12}$	0	-45

Performing the next pivot yields the following tableau.

	x	y	S_1	S_2	A_1	A_2	Capacity
x	1	0	$-\frac{5}{45}$	$\frac{5}{45}$	$\frac{5}{45}$	$-\frac{5}{45}$	5
y	0	1	$\frac{3}{45}$	$-\frac{12}{45}$	$-\frac{3}{45}$	$\frac{12}{45}$	12
$-C$	0	0	$\frac{36}{45}$	$\frac{36}{45}$	$-\frac{36}{45}$	$-\frac{36}{45}$	-156
z	0	0	0	0	1	1	0

Since $z = 0$, we have found a basic feasible solution, and Phase I is done. Deleting the appropriate row and columns leads to the following tableau to start Phase II.

	x	y	S_1	S_2	Capacity
x	1	0	$-\frac{5}{45}$	$\frac{5}{45}$	5
y	0	1	$\frac{3}{45}$	$-\frac{12}{45}$	12
$-C$	0	0	$\frac{36}{45}$	$\frac{36}{45}$	-156

However, since there is nothing negative in the bottom row, this tableau is already optimal! No further pivoting is required. Since $-C$ is maximized, then C is minimized. The solution is

$$C = 156 \text{ (minimized)}$$
$$x = 5$$
$$y = 12$$
$$S_1 = 0, \qquad M_1 = \tfrac{36}{45} = \tfrac{4}{5}$$
$$S_2 = 0, \qquad M_2 = \tfrac{4}{5},$$

and, of course, $A_1 = A_2 = 0$. In Exercise 1, we ask the reader to verify the solution using other techniques.

Our example was a standard form minimization. We should illustrate one that is not standard form:

Example 9.7 The World Health Organization has asked the Sunday Noir Medical Labs to develop a vaccine for the deadly iron-paranoid virus. They have found two natural rain forest sources for the vaccine. Each gram extracted from Sweet Leaf Vine contains 1 milligram of chemical A, an antiviral agent that must be taken in moderation because of unpleasant side effects; 5 milligrams of a weakened strain of the iron-paranoid virus; and 1 milligram of chemical B, a buffer and stabilizer that prevents deterioration of the other ingredients. Each gram extracted from Black Moon Flower contains 1 milligrams of chemical A, 6 milligrams of the weakened virus, and 3 milligrams of chemical B. It costs \$3 to extract a gram from the Sweet Leaf Vine and \$5 to extract a gram from the Black Moon Flower. The labs have determined that an ideal dose of the vaccine must contain exactly 90 milligrams of the weakened virus. It also should contain at most 18 milligrams of the antiviral chemical A to keep the side effects in check, and at least 36 milligrams of the stabilizer, chemical B, to ensure longevity of the mixture. How many grams of each plant extract should be mixed to make the least expensive vaccine?

To solve this, let x be the number of grams of vine extract, and let y the number of grams of flower extract. Then we must

minimize $C = 3x + 5y$

subject to:

$x + y \leq 18$ (mg. chemical A)

$5x + 6y = 90$ (mg. weakened virus)

$x + 3y \geq 36$ (mg. chemical B)

$x \geq 0, \ y \geq 0$.

So we maximize $-C = -3x - 5y$, and after adding appropriate artificial variables to the second and third constraint, we obtain

$$-5x - 6y - A_2 = -90$$
$$-x - 3y + S_3 - A_3 = -36.$$

We can now define $z = -A_2 - A_3$, and from the preceding equations, we obtain z in terms of the other variables as

$$z = -A_2 - A_3 = 6x + 9y - S_3 - 126.$$

The initial tableau is then as shown in the following tableau, and we have indicated the first pivot row and column by shading.

	x	y	S_1		A_2	A_3	Capacity
S_1	1	1	1	0	0	0	18
$-A_2$	-5	-6	0	0	-1	0	-90
$-A_3$	-1	-3	0	1	0	1	-36
$-C$	3	5	0	0	0	0	0
z	-6	-9	0	1	0	0	-126

In Exercise 2, we ask the reader to check that after pivoting twice, we arrive at the following tableau.

	x	y	S_1		A_2	A_3	Capacity
S_1	0	0	1	$-\frac{1}{9}$	$-\frac{2}{9}$	$\frac{1}{9}$	2
x	1	0	0	$\frac{6}{9}$	$\frac{3}{9}$	$-\frac{6}{9}$	6
y	0	1	0	$-\frac{5}{9}$	$-\frac{1}{9}$	$\frac{5}{9}$	10
$-C$	0	0	0	$\frac{7}{9}$	$-\frac{4}{9}$	$\frac{11}{9}$	-68
z	0	0	0	0	1	1	0

Because $z = 0$, this represents a basic, feasible solution, so Phase I is finished. Deleting the z row and the two columns labeled with artificial variables yields the following.

	x	y	S_1		Capacity
S_1	0	0	1	$-\frac{1}{9}$	2
x	1	0	0	$\frac{6}{9}$	6
y	0	1	0	$-\frac{5}{9}$	10
$-C$	0	0	0	$\frac{7}{9}$	-68

But this table is optimal, since there are no negative entries in the last row, so there is no further pivoting needed for Phase II. The solution is

$$x = 6 \text{ mg sweet leaf vine extract}$$
$$y = 10 \text{ mg black moon flower extract}$$
$$C = \$68 \text{ (minimized)}$$
$$S_1 = 2 \text{ (mg short of the 18 mg limit for chemical } A)$$
$$S_3 = 0 \text{ (exactly 36 mg chemical } B),$$

and, of course, $A_2 = 0 = A_3$, and the mix has exactly 90 mg weakened virus.

This concludes this section.

Exercises

1. a. Solve Example 9.1 by graphing in the decision space, and verify your answer agrees with what we got before.
   b. Since Example 9.1 is standard form, you can also solve it by using duality. Do so, and verify that you obtain the same answer as we did before.
2. a. Solve Example 9.2 by graphing in the decision space, and verify the answer we obtained above.
   b. Complete the pivoting in Example 9.2.
3. Solve Exercise 4 from Section 3.4, reproduced here, using the method of this section:

$$\text{minimize: } C = 20x + 12y$$
$$\text{subject to:}$$
$$5x + 2y \geq 30$$
$$5x + 7y \geq 70$$
$$x \geq 0, \ y \geq 0.$$

4. Solve Exercise 8 from Section 3.4, reproduced here, using the method of this section.

   Toys in the Attic, Inc., operates two workshops to build toys for needy children. Mr. Tyler's shop can produce 36 Angel dolls, 16 Kings and Queens board games, and 16 Back in the Saddle rocking horses each day it operates. Mr. Perry's shop can produce 10 Angel dolls, 10 Kings and Queens board games, and 20 Back in the Saddle rocking horses each day it operates. It costs $144 per day to operate Mr. Tyler's shop and $166 per day to operate Mr. Perry's shop. Suppose the company receives an order from the Kids Dream On charity foundation for at least 720 Angel dolls, at least 520 Kings and Queens board games, and at least 640 Back in the Saddle rocking horses. How many days should the company operate each shop in order to fill the order at the lowest possible total cost?

5. Solve the following Exercise from Section 3.4 using the method of this section. The Poseidon's Wake Petroleum Company operates two refineries. The Cadence Refinery can produce 40 units of low-grade oil, 10 units of medium-grade oil, and 10 units of high-grade oil in a single day. (Each unit is 1,000 barrels.) The Cascade Refinery can produce 10 units of low-grade oil, 10 units of medium-grade oil, and 30 units of high-grade oil in a single

day. The company receives an order from the Mars Triangle Oil retailers for at least 80 units of low-grade oil, at least 50 units of medium-grade oil, and at least 90 units of high-grade oil. If it costs Poseidon's Wake $1,800 to operate the Cadence Refinery for a day and $2,000 to operate the Cascade Refinery for a day, how many days should the company operate each refinery to fill the order at lowest cost?

6. Solve the following problem from Section 3.4 using the method of this section:

$$\text{minimize: } z = 15x + 51y$$
$$\text{subject to:}$$
$$x + y \leq 100$$
$$3x + 10y \geq 300$$
$$18x + 10y \geq 900$$
$$-6x + 50y \geq 240$$
$$x \geq 0,\ y \geq 0.$$

7. Solve the following problem from Section 3.4 using the method of this section.

   As a result of a federal discrimination lawsuit, the town of Yankee, NY, is required to build a low-income housing project. The outcome of the lawsuit specifies that Yankee should build enough units to be able to house at least 44 adults and at least 72 children. The town also must meet a separate requirement to be able to house at least 120 people altogether. They have available up to 54,000 square feet on which to build. Each townhouse requires 1,800 square feet and can house 6 people (two adults and four children). Each apartment requires 1,500 square feet and can house four people (two adults and two children). Each townhouse costs $100,000 to build and each apartment costs $80,000 to build. How many of each type of housing unit should Yankee build in order to minimize the total cost?

8. Consider the diet problem in Exercise 16 of Section 3.1. Suppose that Derek decides to ignore his constraints on carbohydrates and salt. Solve the modified problem, using the method of this section.

9. Consider the diet problem in Exercise 18 of Section 3.1. Suppose that Phil decides to ignore the sugar constraint. Solve the modified problem, using the method of this section.

10. Solve the transportation problem in Exercise 20 of Section 3.1, using the method of this section. [**Remark**: Be very careful with your arithmetic! This problem is quite tedious. The tableaux are large – 7 × 15 matrices, and it takes nine pivots (invoking Bland's rule several times to break ties) to arrive at the final tableau. But the good news is the matrices have entries that are mostly 0 or 1 or −1. This makes it much easier to pivot than would be the case otherwise. All transportation problems share this unusual feature in the setup of the problem. We were able to do all the pivoting by hand without hurrying in about 45 minutes. On the other hand, the problem has only two supply points and three destinations. Most real-life transportation problems are much larger, making the pivoting by hand nearly impossible and even making the solution by computer slow and expensive. Because of this, special techniques have been developed for solving transportation problems that do not rely on the simplex algorithm. However, a down side is many of these other methods find a solution that feasible and near optimal but often not actually optimal. We do not cover these methods in this book. The interested reader can explore some of the more well-known of these methods – such as the *northwest corner method, the least cost method,*

and *Vogel's approximation method* – in more specialized books on linear programming, including Calvert and Voxman (1989) or Loomba (1976). A search online will lead you to other methods for solving transportation problems and even video tutorials on how to implement these methods.]

9.3 Sensitivity Analysis and the Simplex Algorithm

In this section, we consider some issues in sensitivity analysis. When the simplex algorithm is the method of solution, certain questions can be easily answered. We explore this in Sections 9.3.1– 9.3.3. Section 9.3.1 considers how to handle problems when we impose the constraint that the solution must be integers. Although this is a broad topic that merits its own chapter, we include it here to emphasize that since we are modifying the basic assumptions in our model by requiring integer solutions, it really is a part of sensitivity analysis.

9.3.1 Changes in Capacity

In this subsection, we address changes in the capacity column of the initial tableau. Let's revisit a known example to illustrate the process.

Example 9.8 In Angie and Kiki's lemonade stand problem, suppose the supply of lemons increased from 60 to 65. If there are no other changes to the problem, how does the solution change?

Of course, as pointed out in a previous chapter, one way to answer this would be to solve the problem over again, using 65 instead of 60. However, we seek a shortcut that will allow us to obtain the modified solution quickly. How can we arrive at the solution to the modified problem without pivoting from the beginning? Let's take a look at the initial and final tableaux for this problem, which we solved in Chapter 5. The initial tableau is the following.

	x	y	S_1	S_2	S_3	C
S_1	3	4	1	0	0	60
S_2	1	2	0	1	0	28
S_3	2	1	0	0	1	30
R	-1.25	-1.5	0	0	0	0

The final tableau is the following.

	x	y	S_1	S_2	S_3	C
x	1	0	$-\frac{1}{5}$	0	$\frac{4}{5}$	12
y	0	1	$\frac{2}{5}$	0	$-\frac{3}{5}$	6
S_2	0	0	$-\frac{3}{5}$	1	$\frac{2}{5}$	4
R	0	0	0.35	0	0.10	24

Now, the only change we are making is that in the initial tableau, the 60 is replaced by 65. We would like to know how this change affects the final tableau.

There are three places earlier in this book where clues for how to solve this problem can be found. The first clue is when we discussed sensitivity analysis in Chapter 4 via the graphical method. There, we discovered that as long as the change is within what we called the stable range for lemons, then at the new optimal point, the basic variables remain the same (even though their values might change). In fact, this is how we defined the stable range. Now, we don't know what the stable range is for lemons (well, we did find it graphically in Chapter 4, but let's ignore that). At this point, we do not know if 65 lemons is within this range, but let's proceed as if it is in the range and solve the problem, and then we can modify our approach later in cases where the change takes us outside the stable range.

So, assuming 65 is within the stable range for lemons, we seek a modified solution with the same basic variables $\{x, y, S_2\}$. It will be significant that the basic variables are listed in this order – the order they appear as row labels in the final tableau. For the next few paragraphs, we ignore the bottom row of the tableau.

The second clue for solving this problem was in Chapter 2, when we solved the bongo drum problem. There, we showed that if we change the amount of one of the supplies (that is, the capacity on the right side of the equation), then to find the modified solution, we merely have to perform a matrix multiplication,

$$X = A^{-1}C,$$

where C is the (modified) capacity column and A^{-1} is the inverse matrix of the coefficient matrix for the system of equations. That sounds like exactly what we need to do here, since we are changing the 60 to 65 and we want the new values of the basic variables. However, this does not apply to our situation directly because the coefficient matrix A in our case is the 3×2 matrix (listed under the columns for the decision variables in the initial tableau):

$$A = \begin{bmatrix} 3 & 4 \\ 1 & 2 \\ 2 & 1 \end{bmatrix}.$$

As we know from Chapter 2, A does not have an inverse because it is not a square matrix! So we need to get around this somehow. One approach would be to delete a row from A to make it 2×2, but which row? We could make an argument that it makes sense to delete the second row and just work with the remaining two rows (because S_1 and S_3 are the nonbasic variables, so we want to make sure they have value 0 at the new solution, so we keep those rows in our coefficient matrix). However, if we ignore the second row, our matrix multiplication $X = A^{-1}C$ will not include information about the new value of leftover limes. We'd rather not lose information, so we reject the approach of deleting a row from A to make it 2×2.

Now what? How about the reverse approach? Instead of deleting a row, how about if we add a column to change A to a 3×3 matrix? That way, we can keep the information we need about the leftover limes. What column should we add, and where? Well, we already stated the answer! We want to keep information about leftover limes – that's the variable S_2. It's as if we want to pretend that S_2 is a decision variable instead of a slack variable. In other words, since our set of basic variables is $\{x, y, S_2\}$, let's form a matrix called the *basic matrix B*, whose columns are the columns from the initial tableau with labels x, y, and S_2 – in that order:

$$B = \begin{bmatrix} 3 & 4 & 0 \\ 1 & 2 & 1 \\ 2 & 1 & 0 \end{bmatrix}.$$

This is a neat trick! We modify the coefficient matrix A to make it square, so now it may have an inverse B^{-1}. We wonder if we now use the matrix product

$$Y = B^{-1}C$$

whether or not the entries in Y will be the modified solution we seek. Consider the corresponding system of equations:

$$3x + 4y = 60$$
$$x + 2y + S_2 = 28$$
$$2x + y = 30.$$

We already know that $\{12, 6, 4\}$ is a solution to this system from the pivoting we already did to solve the original problem. Furthermore, when we pivoted, this solution appeared without free variables; that is, the solution is unique. It follows that the matrix B is invertible. Now we are exactly in the situation of the bongo drum problem example. We know that if we change 60 to 65 and perform the matrix product $Y = B^{-1}C$, then Y will indeed be the unique solution to the modified problem. This reduces our problem to the problem of finding B^{-1}, the inverse of the basic matrix B.

The third clue to which we alluded earlier, which is exactly what we need, appears in the series of Exercises 10–13 from Section 2.4. It was the alternate approach to finding the inverse of a matrix by pivoting. It might be a good idea to do those exercises now if you haven't already. In any case, here is a summary of the method:

Start with your square matrix B; annex an identity matrix of the same size $[B|I]$. Now pivot (by whatever steps you like) until the result is in the form $[I|M]$. Then $M = B^{-1}$. We can capture the essence of the method by the schematic:

$$[B|I] \longrightarrow \text{pivoting} \longrightarrow \left[I|B^{-1}\right].$$

Therefore, to find B^{-1}, we should pivot the following augmented matrix:

$$[B|I] = \begin{bmatrix} 3 & 4 & 0 & 1 & 0 & 0 \\ 1 & 2 & 1 & 0 & 1 & 0 \\ 2 & 1 & 0 & 0 & 0 & 1 \end{bmatrix}$$

until we see a 3×3 identity matrix in the first three columns. Then B^{-1} appears in the last three columns. Then we simply find the matrix product $B^{-1} \begin{bmatrix} 65 \\ 28 \\ 30 \end{bmatrix}$, and we have our solution to the modified problem. The alert reader will object at this point because we said we wanted to avoid pivoting, and this solution does not sound any easier that solving the entire problem over from the beginning because it still involves pivoting. But here's the beauty of this approach: *we have already done the necessary pivoting!*

To see this, compare $[B|I]$ to the initial tableau (continuing to ignore the last row). They are almost the same – the only difference is there is an extra column in $[B|I]$, but this column is a duplicate of the S_2 column. Before, we suggested we were pretending that S_2 was a decision variable. So if we duplicate that column and insert it in the appropriate spot, we have exactly $[B|I]$ (with the bottom row and the final column C included).

	x	y	S_2	S_1	S_2	S_3	C
S_1	3	4	0	1	0	0	60
S_2	1	2	1	0	1	0	28
S_3	2	1	0	0	0	1	30
R	-1.25	-1.5	0	0	0	0	0

Here, we have shaded in the added column. Notice that $[B|I]$ now appears in this modified initial tableau. Suppose we were to pivot this tableau, using the steps we already followed to find the optimal tableau. Then we know the result of every column except the new one. But a moment's reflection reveals that since the added column is a duplicate of another column, the exact same row operations performed on both columns will ensure that they remain duplicates in all future tableaux, including the final one. Thus, the result of pivoting this tableau must be the following.

	x	y	S_2	S_1	S_2	S_3	C
x	1	0	0	$-\frac{1}{5}$	0	$\frac{4}{5}$	12
y	0	1	0	$\frac{2}{5}$	0	$-\frac{3}{5}$	6
S_2	0	0	1	$-\frac{3}{5}$	1	$\frac{2}{5}$	4
R	0	0	0	$.35$	0	$.10$	24

But now we have a 3×3 identity matrix right where we want it! It follows that

$$B^{-1} = \begin{bmatrix} -\frac{1}{5} & 0 & \frac{4}{5} \\ \frac{2}{5} & 0 & -\frac{3}{5} \\ -\frac{3}{5} & 1 & \frac{2}{5} \end{bmatrix}.$$

The reader can easily check by matrix multiplication that $BB^{-1} = I_3$. But the crucial observation is that B^{-1} is just the 3×3 block that already appeared in the final tableau under the columns labeled with slack variables. So we do not need to do any additional pivoting. What we did here always works. So we have the following theorem.

THEOREM 9.9　*In any maximization linear programming problem solved by the simplex algorithm, consider a matrix B, the basic matrix, formed by the columns of the initial tableau under the basic variables of the solution, written in the order of how they are listed as basic variables in the final tableau (ignoring the bottom row.) Then:*

a. In the final tableau, B^{-1} appears in the columns labeled with slack variables (above the bottom row).

b. B^{-1} has the property that for any column X of the initial tableau, $B^{-1}X$ is the corresponding column of the final tableau (ignoring the entry in the bottom row).

The property in Part b says simply that all the pivoting we did can be expressed as a single matrix multiplication. This is a consequence of the different points of view for a system of equations. (Students who know linear algebra can prove this for themselves because every elementary row operation can be realized by multiplication by an "elementary matrix," and B^{-1} is simply the product of these elementary matrices. It would be too long a diversion for us to explore that here, but interested readers can find a discussion of elementary matrices in Lay (2012).)

Part b, applied to the capacity column, is how we solve our modified problem:

$$B^{-1}C = \begin{bmatrix} -\frac{1}{5} & 0 & \frac{4}{5} \\ \frac{2}{5} & 0 & -\frac{3}{5} \\ -\frac{3}{5} & 1 & \frac{2}{5} \end{bmatrix} \begin{bmatrix} 65 \\ 28 \\ 30 \end{bmatrix} = \begin{bmatrix} 11 \\ 8 \\ 1 \end{bmatrix}.$$

Thus, the solution is $x = 11$, $y = 8$, $S_2 = 1$. What about the bottom row R? Our theorem doesn't cover the change in R, but this is easy to compute:

$$R = 1.25x + 1.5y$$
$$= 1.25(11) + 1.5(8) = \$25.75.$$

Alternately, observe that there is no change to the marginal values (because 65 is within the stable range for lemons), and we added five lemons, so

$$R = 24 + 5(0.35) = \$25.75$$

Let's do another example. Suppose, instead of increasing lemons by five, we decrease sugar. One day, the girls' mother runs out of sugar, and they only have 25 tablespoons available instead of 30. What is the new optimal data? According to Theorem 9.9, we compute

$$B^{-1}C = \begin{bmatrix} -\frac{1}{5} & 0 & \frac{4}{5} \\ \frac{2}{5} & 0 & -\frac{3}{5} \\ -\frac{3}{5} & 1 & \frac{2}{5} \end{bmatrix} \begin{bmatrix} 60 \\ 28 \\ 25 \end{bmatrix} = \begin{bmatrix} 8 \\ 9 \\ 2 \end{bmatrix}.$$

The modifed revenue is $R = 1.25(8) + 1.5(9) = \$23.50$ (which makes sense since we lose 10 cents for each tablespoon of sugar we don't have). Thus:

$x = 8$ glasses sweet lemonade

$y = 9$ glasses tart lemonade

$R = \$23.50$

$S_1 = 0$ leftover lemons; $M_1 \doteq \$0.35$ per lemon

$S_2 = 2$ leftover limes; $M_2 = 0$

$S_3 = 0$ leftover sugar; $M_3 = \$0.10$ per tablespoon

is the revised optimal data. Once a problem has been solved via the simplex algorithm, it is very easy the handle changes in capacity, provided the change stays within the stable range.

Let's consider an example where the change takes us out of the stable range. Suppose, in the lemonade stand problem, the number of lemons increases from 60 to 75, and, as usual, that is the only change. If we blindly follow the steps outlined in Theorem 9.9, we obtain

$$B^{-1}C = \begin{bmatrix} -\frac{1}{5} & 0 & \frac{4}{5} \\ \frac{2}{5} & 0 & -\frac{3}{5} \\ -\frac{3}{5} & 1 & \frac{2}{5} \end{bmatrix} \begin{bmatrix} 75 \\ 28 \\ 30 \end{bmatrix} = \begin{bmatrix} 9 \\ 12 \\ -5 \end{bmatrix}.$$

So the problem is the -5. That indicates the value of the basic variable S_2. But a negative basic variable is an infeasible point (indeed, we are five limes short of being able to make nine glasses of sweet and 12 glasses of tart lemonade.) However, we can still salvage the situation. The full tableau we would obtain at this point is the following:

	x	y	S_1	S_2	S_3	C
x	1	0	$-\frac{1}{5}$	0	$\frac{4}{5}$	9
y	0	1	$\frac{2}{5}$	0	$-\frac{3}{5}$	12
S_2	0	0	$-\frac{3}{5}$	1	$\frac{2}{5}$	-5
R	0	0	0.35	0	0.10	29.25

We have manually computed the value of the lower-right corner $R = 1.25(9) + 1.5(12) = \$29.25$. The negative in the capacity column indicates that the corner is not feasible. However, rather than solving the entire problem over from the start, why not start with this infeasible tableau and pivot from there? Presumably, in most cases, this infeasible point is fairly close to the new optimal solution, so we should reach it with a minimum of effort, maybe only one or two further pivots.

Of course, because this point is an infeasible point, further pivoting would involve Phase I. We have a choice to use either Algorithm 9.3 or 9.5 (dual pivoting) for Phase I. If we use Algorithm 9.3, we would begin by writing out the equation for the third constraint:

$$-\frac{3}{5}S_1 + S_2 + \frac{2}{5}S_3 = -5.$$

As usual with exact equality constraints, we subtract an artificial variable, but we do not negate the equation first (since there already is a negative in the capacity column in this row):

$$-\frac{3}{5}S_1 + S_2 + \frac{2}{5}S_3 - A = -5.$$

Also, we determine the auxiliary objective function $z = -A = \frac{3}{5}S_1 - S_2 - \frac{2}{5}S_3 - 5$. Adding the appropriate rows and columns (and relabeling the third row with the negative of the artificial variable), we obtain the current tableau.

	x	y	S_1	S_2	S_3	A	C
x	1	0	$-\frac{1}{5}$	0	$\frac{4}{5}$	0	9
y	0	1	$\frac{2}{5}$	0	$-\frac{3}{5}$	0	12
$-A$	0	0	$-\frac{3}{5}$	1	$\frac{2}{5}$	-1	-5
R	0	0	0.35	0	.10	0	29.25
z	0	0	$-\frac{3}{5}$	1	$\frac{2}{3}$	0	-5

We pivot on the S_1 column. In forming ratios, ignore the negative ratio in the first row. The ratio in the second row is $12 \cdot \frac{5}{2} = 30$, while the ratio in the third row is $(-5)\left(-\frac{5}{3}\right) = \frac{25}{3} \approx 8.3333$. The smallest ratio indicates we pivot on the third row. Pivoting yields the following.

	x	y	S_1	S_2	S_3	A	C
x	1	0	0	$-\frac{1}{3}$	$\frac{2}{3}$	$\frac{1}{3}$	$\frac{32}{3}$
y	0	1	0	$\frac{2}{3}$	$-\frac{1}{3}$	$-\frac{2}{3}$	$\frac{26}{3}$
S_1	0	0	1	$-\frac{5}{3}$	$-\frac{2}{3}$	$\frac{5}{3}$	$\frac{25}{3}$
R	0	0	0	$\frac{7}{12}$	$\frac{1}{3}$	$-\frac{7}{12}$	$\frac{79}{3}$
z	0	0	0	0	0	1	0

We have arrived at a feasible tableau since $z = 0$, so we can delete the z row and the A column to obtain the tableau.

	x	y	S_1	S_2	S_3	C
x	1	0	0	$-\frac{1}{3}$	$\frac{2}{3}$	$\frac{32}{3}$
y	0	1	0	$\frac{2}{3}$	$-\frac{1}{3}$	$\frac{26}{3}$
S_1	0	0	1	$-\frac{5}{3}$	$-\frac{2}{3}$	$\frac{25}{3}$
R	0	0	0	$\frac{7}{12}$	$\frac{1}{3}$	$\frac{79}{3}$

But this represents an optimal tableau so no further pivoting is required. The solution is

$$x = \frac{32}{3} \approx 10.667 \text{ glasses sweet lemonade}$$

$$y = \frac{26}{3} \approx 8.6667 \text{ glasses tart lemonade}$$

$$R = \frac{79}{3} \approx \$26.33 \text{ (maximized)}$$

$$S_1 = \frac{25}{3} \approx 8.3333 \text{ leftover lemons; } M_1 = 0$$

$$S_2 = 0 \text{ leftover limes; } M_2 = \frac{7}{12} \approx \$0.58/\text{lime}$$

$$S_3 = 0 \text{ leftover sugar; } M_3 = \frac{1}{3} \approx \$0.33/\text{tablespoon.}$$

Using Algorithm 9.5 is slightly more efficient, since we do not have to introduce A or z into the problem. Looking at the starting tableau,

Algorithm 9.5 says to choose the third row as the pivot row (the only row with a negative capacity) and then the third column as the pivot column (the only column with a negative entry in the pivot row).

	x	y	S_1	S_2	S_3	C
x	1	0	$-\frac{1}{5}$	0	$\frac{4}{5}$	9
y	0	1	$\frac{2}{5}$	0	$-\frac{3}{5}$	12
S_2	0	0	$-\frac{3}{5}$	1	$\frac{2}{5}$	-5
R	0	0	0.35	0	0.10	29.25

The ratio for this column is $\frac{7}{20}\left(-\frac{5}{3}\right) = -\frac{7}{12}$, but there is no need to even compute it since it is the only ratio, so it is certainly the maximum ratio. Thus, we pivot on the $-\frac{3}{5}$ in the a_{33} position, which agrees with the pivot position using the other algorithm. The pivot yields the final tableau in one step.

	x	y	S_1	S_2	S_3	C
x	1	0	0	$-\frac{1}{3}$	$\frac{2}{3}$	$\frac{32}{3}$
y	0	1	0	$\frac{2}{3}$	$-\frac{1}{3}$	$\frac{26}{3}$
S_1	0	0	1	$-\frac{5}{3}$	$-\frac{2}{3}$	$\frac{25}{3}$
R	0	0	0	$\frac{7}{12}$	$\frac{1}{3}$	$\frac{79}{3}$

We, of course, obtain the exact same answer either way.

In our final example, we illustrate how to handle changes in capacity in mixed constraint problems.

Example 9.10 Consider the problem:

$$\text{maximize } P = 5x + 16y$$
$$\text{subject to:}$$
$$x + 3y \le 48$$
$$x + 6y = 66$$
$$5x + 3y \ge 60$$
$$x \ge 0, \ y \ge 0.$$

According to Algorithm 9.3, we negate the greater-than-type constraint, add a slack variable, and subtract an artificial variable to obtain $-5x - 3y + S_3 - A_3 = -60$. We also add an artificial variable to the exact equality constraint obtaining $x + 6y + A_2 = 66$. In Algorithm 9.3, we also said an option was to negate these exact equations if desired. However, if you plan to study changes in capacity via the matrix B^{-1}, do *not* negate these constraints. We'll see why in a moment. Then we write $z = -A_2 - A_3$ in terms of the other variables; in this problem, we obtain $z = 6x + 9y - S_3 - 126$. We now set up the initial tableau. In all of the examples we covered in Section 9.1, when setting up the initial tableau, we wrote columns for all the slack variables before all the columns for artificial variables. However, if we plan to study changes in capacity via B^{-1}, we must order the columns so that an identity matrix appears below the non-decision variables. This means listing all the slack variables for less-than-type inequalities first, followed by the artificial variables for the exact equations, followed by the slack variables for the greater-than-type constraints, followed by the artificial variables for the greater-than-type constraints. (That is, we keep the order the same as the way we write the constraints, $\le$ first, followed by $=$, followed by $\ge$.) In this problem, we obtain the following initial tableau.

	x	y	S_1	A_2	S_3	A_3	Capacity
S_1	1	3	1	0	0	0	48
A_2	1	6	0	1	0	0	66
$-A_3$	-5	-3	0	0	1	-1	-60
P	-5	-16	0	0	0	0	0
z	-6	-9	0	0	1	0	-126

You can see the 3×3 identity matrix in the shaded cells. We needed to order the columns this way and not to negate the A_2 row in order to obtain this identity matrix so we can apply Theorem 9.9. Note that, since the identity matrix involves the A_2 column, we need to keep this column through all pivoting – we do not delete it at the end of Phase I as we would normally do. In the exercises, you are asked to verify that after two pivots, we arrive at a feasible point as indicated in the tableau.

	x	y	S_1	A_2	S_3	A_3	Capacity
S_1	0	0	1	$-\frac{4}{9}$	$\frac{1}{9}$	$-\frac{1}{9}$	12
y	0	1	0	$\frac{5}{27}$	$\frac{1}{27}$	$-\frac{1}{27}$	10
x	1	0	0	$-\frac{1}{9}$	$-\frac{2}{9}$	$\frac{2}{9}$	6
P	0	0	0	$\frac{65}{27}$	$-\frac{14}{27}$	$\frac{14}{27}$	190
z	0	0	0	1	0	1	0

We have retained the shading because ultimately that is the block where B^{-1} appears. Now that Phase I is over, we can delete the z row. We could also delete the A_3 column, if desired, since it does not involve B^{-1}, nor will we ever pivot in that column. We will keep it for completeness. Remember we *must* keep the A_2 column. So, deleting the last row, we obtain a feasible, but not yet optimal, solution. One more pivot in Phase II leads to the optimal tableau.

	x	y	S_1	A_2	S_3	A_3	Capacity
S_3	0	0	9	-4	1	-1	108
y	0	1	$-\frac{1}{3}$	$\frac{1}{3}$	0	0	6
x	1	0	2	-1	0	0	30
P	0	0	$\frac{14}{3}$	$\frac{1}{3}$	0	0	246

The reader should note that the 3×3 shaded matrix is B^{-1}, and the marginal values appear in the last row below these columns as usual. The fact that the second (shaded) column is labeled A_2 rather than S_2 does not matter as far as interpreting the entries in the bottom row. Thus the final solution is

$$x = 30$$
$$y = 6$$
$$S_1 = 0; \quad M_1 = \frac{14}{3}$$
$$A_2 = 0; \quad M_2 = \frac{1}{3}$$
$$S_3 = 108; \quad M_3 = 0$$
$$P = 246 \text{ (maximized)}.$$

Now we can study changes in capacity. Suppose first that in the first constraint, we increase from 48 to 54. Find the revised optimal data. Second, what if, in the second constraint, we instead decrease from 66 to 63. Find the revised optimal data. We can solve both of these questions at once. We construct a two-column matrix C where in the first column we put the revised capacities for the former question, and in the second column we put the revised capacities for the latter question. Then the single matrix product $B^{-1}C$ answers both questions. Indeed:

$$B^{-1}C = \begin{bmatrix} 9 & -4 & 1 \\ -\frac{1}{3} & \frac{1}{3} & 0 \\ 2 & -1 & 0 \end{bmatrix} \begin{bmatrix} 54 & 48 \\ 66 & 63 \\ -60 & -60 \end{bmatrix} = \begin{bmatrix} 162 & 120 \\ 4 & 5 \\ 42 & 33 \end{bmatrix}.$$

Since all the entries in $B^{-1}C$ are nonnegative, both of the changes are within the stable ranges for their respective constraints. As for the values of P, in the first case, we increased capacity by six units, and the marginal value M_1 is $\frac{14}{3}$, so the objective value should increase by 28 to 274. In the second case, we lowered by three units, and the marginal value is $\frac{1}{3}$, so P should decrease by 1 to 245. Thus, the revised optimal data is as shown in Table 9.1.

Table 9.1 Optimal data for the revised problems.

Raise 48 to 54	Lower 66 to 63
$x = 42$	$x = 33$
$y = 4$	$y = 5$
$S_1 = 0; \quad M_1 = \frac{14}{3}$	$S_1 = 0; \quad M_1 = \frac{14}{3}$
$A_2 = 0; \quad M_2 = \frac{1}{3}$	$A_2 = 0; \quad M_2 = \frac{1}{3}$
$S_3 = 162; \quad M_3 = 0$	$S_3 = 120; \quad M_3 = 0$
$P = 274 \text{ (maximized)}$	$P = 245 \text{ (maximized)}$

In summary, all the hard work goes into finding the optimal tableau the first time. After that, small changes in capacity are quite easy to study with the simplex method – much easier than what we did graphically in Chapter 4 (and, unlike Chapter 4, the same techniques work no matter how many variables the problem has).

Exercises

1. a.   In the lemonade stand problem, compute the new optimal data if the number of lemons available decreases from 60 to 55.

b.   What would you expect to happen if you kept the amount of lemons and sugar at their levels of 60 and 30, respectively, but increased the supply of limes from 28 to 30? Verify your guess by using B^{-1}.

2.   a.   In Example 9.6, verify that pivoting leads to the final tableau indicated.

b.   Find the revised optimal data if the capacity in the first constraint increases from 48 to 60.

c.   Explicitly write down the matrix B, and verify that $BB^{-1} = I_3$.

3.   Recall the following exercise. Joni is putting her designing skills to work at the Court and Sparkle Jewelry Emporium. She makes two signature design bracelet models. Each Dawntreader bracelet uses 1 ruby, 6 pearls, and 10 opals. Each Hejira bracelet uses 3 rubies, 3 pearls, and 15 opals. She has 54 rubies, 120 pearls, and 300 opals to work with. If either model results in a profit of $1,800 for Joni, how many of each type should she make?

In Section 4.4, you answered the following questions using a graphical approach. Now answer them using the techniques of this section:

a.   What is the revised optimal data if the number of pearls available is increased to 150?

b.   What is the revised optimal data if the number of opals available increased to 330?

4.   Recall the following exercise. Green's Heavy Metal Foundry mixes three different alloys composed of copper, zinc, and iron. Each 100-pound unit of Alloy I consists of 50 pounds of copper, 50 pounds of zinc, and no iron. Each 100-pound unit of Alloy II consists of 30 pounds of copper, 30 pounds of zinc, and 40 pounds of iron. Each 100-pound unit of Alloy III consists of 50 pounds of copper, 20 pounds of zinc, and 30 pounds of iron. Each unit of Alloy I generates $100 profit, each unit of Alloy II generates $80 profit, and each unit of Alloy III generates $40 profit. There are 12,000 pounds of copper, 10,000 pounds of zinc, and 12,000 pounds of iron available. The foundry has hired Gary Giante, an outside consultant, to help it maximize its profit. Fortunately, Mr. Giante knows the simplex algorithm.

In Section 5.2, you solved this problem via the simplex algorithm. Determine the revised optimal data if the amount of copper available increased to 13,000 pounds.

5.   Recall the following exercise. The Spooky Boogie Costume Salon makes and sells four different Halloween costumes: the witch, the ghost, the goblin, and the werewolf. Each witch costume uses 3 yards of material and takes 3 hours to sew. Each ghost costume uses 2 yards of material and takes 1 hour to sew. Each goblin costume uses 2 yards of material and takes 3 hours to sew. Each werewolf costume uses 2 yards of material and takes 4 hours to sew. The profits for each costume are as follows: $10 for the witch, $8 for the ghost, $12 for the goblin, and $16 for the werewolf. If the Spooky Boogie Costume Salon has 600 yards of material and 510 sewing hours available before the holiday, how many of each costume should the shop make in order to maximize profit, assuming it sells everything it makes?

In Section 5.2, you solved this problem via the simplex algorithm. Determine the revised optimal data under the following changes:

a.   The amount of material available decreases from 600 yards to 420 yards due to a labor strike in the garment industry.

b.   The amount of sewing hours increases from 510 hours to 540 hours due to the purchase of a new sewing machine.

6. Recall the following exercise with mixed constraints. The Jefferson Plastic Fantastic Assembly Corporation manufactures gadgets and widgets for airplanes and starships. Each case of gadgets uses 2 kilograms of steel and 5 kilograms of plastic. Each case of widgets uses 2 kilograms of steel and 3 kilograms of plastic. The profit for a case of gadgets is $360, and the profit for a case of widgets is $200. Suppose Jefferson Plastic Fantastic Assembly Corporation has 80 kilograms of steel and 150 kilograms of plastic available on a daily basis and can sell everything it manufactures. How many cases of each should the corporation manufacture if it is obligated to produce at least 10 cases of widgets per day? The objective is to maximize daily profit.

   In Exercise 7 of Section 9.1, you solved this via the simplex algorithm. Now compute the revised optimal data if the amount of plastic increases to 160 kilograms per day.

7. In Exercise 5 of Section 9.1, you solved the following problem with mixed constraints:

   $$\text{maximize } P = x + 4y$$
   $$\text{subject to:}$$
   $$4x + 7y \leq 140$$
   $$5x + 22y = 228$$
   $$3x + 2y \geq 36$$
   $$x \geq 0, \ y \geq 0.$$

   Find the revised optimal data under the following changes:

   a. The first constraint decreases to 120.
   b. The second constraint increases to 236.

8. a. Solve the following problem:

   $$\text{maximize } P = 48x + 40y$$
   $$\text{subject to:}$$
   $$3x + y \leq 120$$
   $$x + 3y \leq 120$$
   $$x + y \leq 66$$
   $$7x + 5y \leq 350$$
   $$x \geq 0; \ y \geq 0.$$

   b. Find the revised optimal data if the capacity in the fourth constraint increases from 350 to 500.

9.3.2 Changes in Objective Coefficients

Now let's consider the last row of a simplex tableau, which in the initial tableau contains the objective coefficients. We want to address the question of how small changes in these coefficients affect the solution. Before we do that, however, we repair a small deficiency in what we did in the last section when working with the inverse B^{-1} of the basic matrix. When we constructed B, B^{-1}, and in the calculation $B^{-1}C$ of the changes in capacity, we always ignored the last row and had to compute the change in the objective function separately. If you want to include the last row, you will need B^{-1} to also have a row corresponding to the last row. But the identity matrix in the initial tableau that appears beneath the slack variables does not extend to the bottom row. We can fix that easily. What we do is annex a unit column labeled with the objective variable in the initial tableau. We will never pivot on that column, and, because it is a unit column, it never changes in any of the other pivots we perform. Thus, it also appears in the final tableau as part of the larger B^{-1}.

We illustrate with the lemonade stand problem. The (usual) initial tableau and final tableau are as follows:

	x	y	S_1	S_2	S_3	C
S_1	3	4	1	0	0	60
S_2	1	2	0	1	0	28
S_3	2	1	0	0	1	30
R	-1.25	-1.5	0	0	0	0

and

	x	y	S_1	S_2	S_3	C
x	1	0	$-\frac{1}{5}$	0	$\frac{4}{5}$	12
y	0	1	$\frac{2}{5}$	0	$-\frac{3}{5}$	6
S_2	0	0	$-\frac{3}{5}$	1	$\frac{2}{5}$	4
R	0	0	$\frac{7}{20}$	0	$\frac{1}{10}$	24

respectively, where we have written the marginal values in the last row as fractions rather than decimals. What we are proposing is the addition of a unit column labeled R immediately following the S_3 column in the initial tableau, resulting in a modified initial tableau.

	x	y	S_1	S_2	S_3	R	C
S_1	3	4	1	0	0	0	60
S_2	1	2	0	1	0	0	28
S_3	2	1	0	0	1	0	30
R	-1.25	-1.5	0	0	0	1	0

The shaded cells exhibit a 4×4 identity matrix now. Now when we perform the exact same pivots that result in the usual final tableau, the R column does not change (try it!), so the final tableau becomes the following.

	x	y	S_1	S_2	S_3	R	C
x	1	0	$-\frac{1}{5}$	0	$\frac{4}{5}$	0	12
y	0	1	$\frac{2}{5}$	0	$-\frac{3}{5}$	0	6
S_2	0	0	$-\frac{3}{5}$	1	$\frac{2}{5}$	0	4
R	0	0	$\frac{7}{20}$	0	$\frac{1}{10}$	1	24

The upshot is that the 4×4 matrix in the shaded cells now becomes the inverse of a modified basic matrix $\widehat{B}$, with columns taken from the initial tableau with labels $\{x, y, S_2, R\}$ in that order (all the labels of the rows, including R). That is, if we define

$$\widehat{B} = \begin{bmatrix} 3 & 4 & 0 & 0 \\ 1 & 2 & 1 & 0 \\ 2 & 1 & 0 & 0 \\ -1.25 & -1.5 & 0 & 1 \end{bmatrix},$$

then we have, by the same argument that led to Theorem 9.9, that

$$\widehat{B}^{-1} = \begin{bmatrix} -\frac{1}{5} & 0 & \frac{4}{5} & 0 \\ \frac{2}{5} & 0 & -\frac{3}{5} & 0 \\ -\frac{3}{5} & 1 & \frac{2}{5} & 0 \\ \frac{7}{20} & 0 & \frac{1}{10} & 1 \end{bmatrix}$$

and that $\widehat{B}^{-1}$ has the property that when it is multiplied by any column of the initial tableau, the result is the corresponding column of the final tableau, this time including the bottom row. Now we can automatically compute the change in R if we change a capacity along with all the other entries in the final column. For example, if the number of lemons increases to 65, we have, letting $\widehat{C}$ stand for the full capacity column, including the last row entry:

$$\widehat{B}^{-1}\widehat{C} = \begin{bmatrix} -\frac{1}{5} & 0 & \frac{4}{5} & 0 \\ \frac{2}{5} & 0 & -\frac{3}{5} & 0 \\ -\frac{3}{5} & 1 & \frac{2}{5} & 0 \\ \frac{7}{20} & 0 & \frac{1}{10} & 1 \end{bmatrix} \begin{bmatrix} 65 \\ 28 \\ 30 \\ 0 \end{bmatrix} = \begin{bmatrix} 11 \\ 8 \\ 1 \\ \frac{103}{4} \end{bmatrix}.$$

The reader will recognize the same answer we obtained in the previous section, but this time including $R = \frac{103}{4} = \$25.75$. We can always include the objective row in this manner, working with $\widehat{B}^{-1}$ instead of B^{-1} in Theorem 9.9. Furthermore, we do not have to actually insert the spurious column R at the beginning and go through the pivots with the column there, because we know that the final tableau also just has a unit column inserted at the appropriate spot. Thus, $\widehat{B}^{-1}$ always has the following "block" form:

$$\widehat{B}^{-1} = \begin{bmatrix} B^{-1} & Z \\ M & 1 \end{bmatrix},$$

where $M = \begin{bmatrix} M_1 & \cdots & M_m \end{bmatrix}$ is a row vector containing all the marginal values, and Z is a column vector of all zeros. This explains in another way why it works. When we multiply by $\widehat{C}$, only the entries in the B^{-1} block contribute only to the entries in the last column that are above the last row (because Z has 0 entries). Also, to obtain the revised objective function, because the last entry of the last column of the initial tableau is 0, the matrix product just yields $M \cdot C$, the dot product of the marginal values with the capacity numbers. If the reader recalls, in Section 4.5.2, when we discussed duality, we showed that we could obtain the optimal value of R in exactly this way (see the calculation immediately following Conclusion 4.1 in that section).

Therefore, in practice, once B^{-1} is found, one could just construct $\widehat{B}^{-1}$ as before, by annexing the marginal values M at the bottom and a unit column on the right. It is not necessary to adjust the pivoting at all. Arguably, working with $\widehat{B}^{-1}$ is a minor advantage for what we did in the previous section, since it is easy to compute the new objective function anyway.

However, knowing $\widehat{B}^{-1}$ is a major advantage in discussing changes in the objective coefficients, the main topic of this section. We remind the reader that in Chapter 4, we showed that for small changes in any objective coefficient (a change within the stable range for that coefficient), the optimal point does not change at all. Not only do the basic variables remain the same, but their values remain the same as well. Of course, what does change is the value of the objective function itself, and quite possibly the marginal values could change. If we knew what the marginal values changed to, we could find the revised objective function by just taking the dot product $M \cdot C$ as before. So the real goal becomes how to find the revised marginal values without any further pivoting.

This is alway possible, and we illustrate now how to do so by looking at the lemonade stand example. Suppose the girls are considering raising the price of sweet lemonade so that it also is $1.50 per glass instead of $1.25. How does this affect the optimal solution? Like in the previous section, we may not be sure if this change is within the stable range or not for that price, but let's presume it is and see what happens.

So this means we have changed the $\widehat{B}$ to the following matrix:

$$\widehat{B} = \begin{bmatrix} 3 & 4 & 0 & 0 \\ 1 & 2 & 1 & 0 \\ 2 & 1 & 0 & 0 \\ -1.5 & -1.5 & 0 & 1 \end{bmatrix}.$$

How does its inverse change? The first three rows do not change at all (remember we assume the change is within the stable range, so we should obtain the same values of x, y, and S_2.) But we have unknowns in the M block in the bottom row. That is,

$$\widehat{B}^{-1} = \begin{bmatrix} -\frac{1}{5} & 0 & \frac{4}{5} & 0 \\ \frac{2}{5} & 0 & -\frac{3}{5} & 0 \\ -\frac{3}{5} & 1 & \frac{2}{5} & 0 \\ M_1 & M_2 & M_3 & 1 \end{bmatrix}.$$

Actually, we know that $M_2 = 0$ because of complementary slackness (since S_2 remains at a value of $4 \neq 0$). But even without noticing that in advance, we can now proceed to find all of the M_i by just using the equation $\widehat{B}\widehat{B}^{-1} = I_4$. In particular, the values of M_i only affect the bottom row of the product matrix because the last column of B is a unit vector (all the entries of Z are 0). So, looking at the entry in the a_{41} position of the product, we have

$$(1.5)\frac{1}{5} - (1.5)\frac{2}{5} + (0)\frac{3}{5} + (1)\,M_1$$

$$\frac{3}{10} - \frac{6}{10} + M_1 = 0$$

$$M_1 = \frac{3}{10} = 0.3.$$

Thus, the marginal value of a lemon has decresed from \$0.35 to \$0.30 per lemon with this change in the price. Similarly, for the a_{42} position of the product:

$$-(1.5)0 - (1.5)0 + (0)1 + (1)M_2 = 0$$

$$M_2 = 0.$$

Thus, M_2 has remained at 0, as predicted before. Finally, for the a_{43} position:

$$-(1.5)\frac{4}{5} + (1.5)\frac{3}{5} + (0)\frac{2}{5} + (1)M_3 = 0$$

$$-\frac{12}{10} + \frac{9}{10} + M_3 = 0$$

$$M_3 = \frac{3}{10} = 0.3.$$

Thus, the marginal value of a tablespoon of sugar has increased from \$0.10 to \$0.30.

We now have the revised marginal values, and because they all came out to be nonnegative, that means the change in price was indeed within the stable range. We can now compute the change in the value of revenue by $M \cdot C$ as before:

$$R = \begin{bmatrix} 0.3 & 0 & 0.3 \end{bmatrix} \begin{bmatrix} 60 \\ 28 \\ 30 \end{bmatrix} = \begin{bmatrix} 27 \end{bmatrix}.$$

In summary, raising the price of sweet lemonade to \$1.50 led the the same optimal point but an increase in revenue to \$27. This is because the calculations we just did allow us to fill in what the entire final tableau would be had we done the pivoting with this change in price. In this example, we obtain the following.

	x	y	S_1	S_2	S_3	C
x	1	0	$-\frac{1}{5}$	0	$\frac{4}{5}$	12
y	0	1	$\frac{2}{5}$	0	$-\frac{3}{5}$	6
S_2	0	0	$-\frac{3}{5}$	1	$\frac{2}{5}$	4
R	0	0	$\frac{3}{10}$	0	$\frac{3}{10}$	27

Visibly, this tableau is still optimal. Furthermore, had the change not been within the stable range, one of the M_i would come out to be negative. We could still construct this tableau, but with the negative entry in the bottom row, it would no longer be optimal. However, instead of pivoting the entire problem over from the beginning, just continue pivoting from the (modified) current tableau.

Our final example illustrates this. Suppose the girls want to have a sale on the sweet lemonade and they lower the price to $1.00. We compute the revised marginal values as before:

$$(1)\frac{1}{5} - (1.5)\frac{2}{5} + (0)\frac{3}{5} + (1) M_1 = 0$$

$$M_1 = \frac{2}{5}$$

For M_2, we obtain 0 as expected for this corner point:

$$(1)0 - (1.5)0 + (0)1 + (1)M_2 = 0$$

$$M_2 = 0.$$

For M_3 we obtain

$$-(1)\frac{4}{5} + (1.5)\frac{3}{5} + (0)\frac{2}{5} + (1)M_3 = 0$$

$$M_3 = -\frac{1}{10}.$$

The negative value indicates the change is outside the stable range for the price of sweet lemonade. Nevertheless, we compute the revenue:

$$R = \begin{bmatrix} 0.4 & 0 & -0.1 \end{bmatrix} \begin{bmatrix} 60 \\ 28 \\ 30 \end{bmatrix} = \begin{bmatrix} 21 \end{bmatrix},$$

which in turn allows us to construct the current tableau, which is no longer optimal.

	x	y	S_1	S_2	S_3	Capacity
x	1	0	$-\frac{1}{5}$	0	$\frac{4}{5}$	12
y	0	1	$\frac{2}{5}$	0	$-\frac{3}{5}$	6
S_2	0	0	$-\frac{3}{5}$	1	$\frac{2}{5}$	4
R	0	0	$\frac{4}{10}$	0	$-\frac{1}{10}$	21

We continue pivoting. The pivot column is S_3, and the pivot row is S_2. We obtain this tableau.

	x	y	S_1	S_2	S_3	Capacity
x	1	0	1	-2	0	4
y	0	1	$-\frac{1}{2}$	$\frac{3}{2}$	0	12
S_3	0	0	$-\frac{3}{2}$	$\frac{5}{2}$	1	10
R	0	0	$\frac{3}{20}$	$\frac{1}{4}$	0	22

We have arrived at our revised optimal tableau. It only took one pivot. The solution is

$x = 4$ glasses sweet lemonade

$y = 12$ glasses tart lemonade

$R = \$22$ (maximized)

$S_1 = 0$ leftover lemons; $M_1 = \frac{3}{20} = \$0.15$ per lemon

$S_2 = 0$ leftover limes; $M_2 = \frac{1}{4} = \$0.25$ per lime

$S_3 = 10$ tablespoons leftover sugar; $M_3 = \$0$ per tablespoon.

So far, we have only illustrated the case of changing an objective coefficient for a basic decision variable. What if we want to change the objective coefficient for a nonbasic decision variable? There is not much difference in the procedure – in fact, since we are not changing the objective coefficients for the basic variables, there is no work to do to compute the marginal values M_i, as they do not change. However, one of the entries in the last row of the tableau will change, so we need to use $\widehat{B}^{-1}$ to find that changed entry. If the changed entry becomes negative, then further pivoting is necessary, as before.

To illustrate this, consider Exercise 5 from the last section, the Spooky Boogie Halloween Costume problem. The initial tableau for that problem is as follows:

	w	x	y	z	S_1	S_2	Capacity
S_1	3	2	2	2	1	0	600
S_2	3	1	3	4	0	1	510
P	-10	-8	-12	-16	0	0	0

Here w is the number of witch costumes, x is the number of ghost costumes, y is the number of goblin costumes, and z is the number of werewolf costumes. S_1 represents yards of material and S_2 represents hours on the sewing machine, and P is the profit from each costume. The reader has probably already solved this problem (or can check directly now) that the final tableau is the following.

	w	x	y	z	S_1	S_2	Capacity
x	1	1	$\frac{1}{3}$	0	$\frac{2}{3}$	$-\frac{1}{3}$	230
z	$\frac{1}{2}$	0	$\frac{2}{3}$	1	$-\frac{1}{6}$	$\frac{1}{3}$	70
P	6	0	$\frac{4}{3}$	0	$\frac{8}{3}$	$\frac{8}{3}$	2,960

Thus, the optimal solution is to make 230 ghost costumes and 70 werewolf costumes and nothing else, to the dismay of goblin and witch fans. This uses up all the resources and makes a profit of \$2,960 for the shop. In this case, we have

$$\widehat{B}^{-1} = \begin{bmatrix} \frac{2}{3} & -\frac{1}{3} & 0 \\ -\frac{1}{6} & \frac{1}{3} & 0 \\ \frac{8}{3} & \frac{8}{3} & 1 \end{bmatrix},$$

with the two entries $\frac{8}{3}$ being the marginal values for this corner point. Now, suppose we change the profit for a witch costume, improving it to $15, and want to determine the new optimal data. We ask the same question for increasing it to $18. Because the matrix $\widehat{B}$ does not involve the w column (w is nonbasic), the equation $\widehat{B}\widehat{B}^{-1} = I$, which determines the marginal values, shows that changing the profit coefficient for a witch costume has no effect on the marginal values. Therefore, if we pivot from the beginning with the adjusted profit for a witch costume to the current tableau, we have the exact same $\widehat{B}^{-1}$. If we want to write down the tableau corresponding to this corner, it will have the exact same rows above the objective function as our previous solution does. As for the objective row itself, remember that property b from Theorem 9.9 holds for the extended matrix $\widehat{B}^{-1}$. We multiply it by every column of the initial tableau to obtain the corresponding column of the current tableau (we hesitate to continue to call this tableau the "final" tableau since it may no longer represent an optimal point). But every column in the initial tableau is unchanged, save the w column. We conclude that the only change in the current tableau is the 6 in the bottom row of the w column. We compute the new value:

$$\begin{bmatrix} \frac{2}{3} & -\frac{1}{3} & 0 \\ -\frac{1}{6} & \frac{1}{3} & 0 \\ \frac{8}{3} & \frac{8}{3} & 1 \end{bmatrix} \begin{bmatrix} 3 \\ 3 \\ -15 \end{bmatrix} = \begin{bmatrix} 1 \\ \frac{1}{2} \\ 1 \end{bmatrix}.$$

The new value is 1, and the new "current" tableau is still the "final" one:

	w	x	y	z	S_1	S_2	Capacity
x	1	1	$\frac{1}{3}$	0	$\frac{2}{3}$	$-\frac{1}{3}$	230
z	$\frac{1}{2}$	0	$\frac{2}{3}$	1	$-\frac{1}{6}$	$\frac{1}{3}$	70
P	1	0	$\frac{4}{3}$	0	$\frac{8}{3}$	$\frac{8}{3}$	2,960

Thus, there is no change to the optimal solution. A profit of $15 is not sufficient to warrant manufacturing a witch costume. However, a profit of $18 is sufficient. Following the same steps with the change to $18 leads to

$$\begin{bmatrix} \frac{2}{3} & -\frac{1}{3} & 0 \\ -\frac{1}{6} & \frac{1}{3} & 0 \\ \frac{8}{3} & \frac{8}{3} & 1 \end{bmatrix} \begin{bmatrix} 3 \\ 3 \\ -18 \end{bmatrix} = \begin{bmatrix} 1 \\ \frac{1}{2} \\ -2 \end{bmatrix}.$$

The current tableau is no longer optimal.

	w	x	y	z	S_1	S_2	Capacity
x	1	1	$\frac{1}{3}$	0	$\frac{2}{3}$	$-\frac{1}{3}$	230
z	$\frac{1}{2}$	0	$\frac{2}{3}$	1	$-\frac{1}{6}$	$\frac{1}{3}$	70
P	-2	0	$\frac{4}{3}$	0	$\frac{8}{3}$	$\frac{8}{3}$	2,960

A further pivot is required, leading to the next tableau.

	w	x	y	z	S_1	S_2	Capacity
x	0	1	-1	-2	1	-1	90
w	1	0	$\frac{4}{3}$	2	$-\frac{1}{3}$	$\frac{2}{3}$	140
P	0	0	$\frac{8}{3}$	2	$\frac{7}{3}$	$\frac{10}{3}$	3,100

Thus, the new plan is to make 140 witch costumes and 90 ghost, a triumph for witch lovers everywhere, but still a disappointment to goblin fanatics.

We hasten to point out that if we did make a change to an objective coefficient for a basic decision variable so that the M_i change, care must be taken to check all the entries in the last row to make sure that none of them have become negative.

We close this section by noting that there is a connection between changes in the capacity and changes in objective coefficients that comes from the dual problem. Indeed, the capacity constants in the primal become the objective coefficients in the dual and vice versa. So it should be possible to exploit that connection. For example, to study changes in objective coefficients, convert to the dual, and study changes in the capacity in the dual. We will pursue the details only in Exercise 10 of the next subsection. We will not give a formal treatment of this topic, but we point out that in section 7.4 of Calvert and Voxman (1989), duality is explored in the context of sensitivity analysis more formally.

Exercise

1. In the lemonade stand problem, determine the revised optimal data if the girls decide to lower the price of tart lemonade to $1.30 per glass.
2. In the lemonade stand problem, determine the revised optimal data if the girls decide to raise the price of tart lemonade to $1.60 per glass.
3. In the lemonade stand problem, determine the revised optimal data if the girls decide to raise the price of tart lemonade to $2.00 per glass.
4. In the Spooky Boogie Halloween Costume problem, determine the revised optimal data if the profit of a goblin costume increases to $14.
5. In the Spooky Boogie Halloween Costume, determine the revised optimal data if the profit of a werewolf costume decreases to $12.
6. Recall the following problem: Joni is putting her designing skills to work at the Court and Sparkle Jewelry Emporium. She makes two signature design bracelet models. Each Dawntreader bracelet uses 1 ruby, 6 pearls, and 10 opals. Each Hejira bracelet uses 3 rubies, 3 pearls, and 15 opals. She has 54 rubies, 120 pearls, and 300 opals to work with. If either model results in a profit of $1,800 for Joni, how many of each type should she make? In earlier sections of the text, you solved this problem via the simplex algorithm. Suppose Joni's profit for a Hejira bracelet increases to $2,000. Find the revised optimal data.
7. The Brain One Lighting and Heavenly Music Corporation designs sophisticated ambient lighting and music for airports and other public spaces. Each airport contract requires 6 weeks with the design team, 4 weeks with the technical team, and a 2-week review with the legal team to check local building codes, etc. The profit for an airport contract is $8,000,000.

Each museum contract requires 5 weeks with the design team, 4 weeks with the technical team, and 1 week with the legal team and generates $6,000,000 profit. Each theater contract requires 3 weeks with the design team, 3 weeks with the technical team, and 1 week with the legal team and generates $5,000,000 profit. The design team works 48 weeks per year. The technical team works 44 weeks per year (and spends 4 weeks per year in research and development). The legal team works 30 weeks per year (and spends the rest of the year doing pro bono work lobbying to keep arts funding in public government and education). How many of each type of contract should they take on each year in order to maximize profits?

a.   Solve this problem via the simplex algorithm.
b.   Suppose the profit for a theater contract decreased to $3,000,000. Determine the revised optimal solution.

8.   Recall the following problem with mixed constraints: The Jefferson Plastic Fantastic Assembly Corporation manufactures gadgets and widgets for airplanes and starships. Each case of gadgets uses 2 kilograms of steel and 5 kilograms of plastic. Each case of widgets uses 2 kilograms of steel and 3 kilograms of plastic. The profit for a case of gadgets is $360, and the profit for a case of widgets is $200. Suppose the Jefferson Plastic Fantastic Assembly Corporation has 80 kilograms of steel available and 150 kilograms of plastic available on a daily basis, and can sell everything it manufactures. How many cases of each should the corporation manufacture if it is obligated to produce at least 10 cases of widgets per day? The objective is to maximize daily profit.

In Exercise 7 of Section 9.1, you solved this via the simplex algorithm. Now find the revised optimal data if the profit for a case of widgets drops to $120.

9.3.3   Stable Ranges

As the reader has worked through the examples and exercises in the past two sections, we hope that it has become evident that if a change in either a capacity or an objective coefficient is small enough to be within the stable range for that capacity number or coefficient, then we could find the revised optimal data with no additional pivoting, using nothing more than matrix multiplication with B^{-1} or $\widehat{B}^{-1}$. On the other hand, if the change was large enough to be outside of the relevant stable range, then in addition to matrix multiplication, additional pivoting was necessary (but a minimal amount of such pivoting – often just one additional pivot suffices). Furthermore, in each case where additional pivoting was necessary, this fact was detected by the occurrence of a negative entry in either the capacity column (in Section 9.3.1) or the bottom row (in Section 9.3.2), after using the inverse of the basic matrix to find the revised "current" tableau.

This is because a negative entry in the bottom row indicates the the current set of basic variables is no longer optimal with the changed objective coefficients, while a negative entry in the capacity column indicates that the current set of basic variables is no longer a feasible point with the changed capacity. Conversely, if the revised current tableau has all nonnegative entries in the right-hand column, it represents a solution that is still feasible, while if all the entries in the bottom row of the revised tableau are also nonnegative, it represents a solution that remains optimal with the current set of basic variables. It follows that whatever change is made to either the capacity or objective coefficients must have been within the stable range for that number.

Therefore, we have before us a method to explicitly find the stable range of any capacity number or objective coefficient. We simply use B^{-1} or $\widehat{B}^{-1}$ to find the amount of change we can make in a given objective coefficient or capacity number that will retain the nonnegativity of each number in the capacity column and each number in the objective row.

We illustrate first with the stable range of a capacity number. Consider the lemonade stand problem. The initial tableau has the following familiar capacity column: $C = \begin{bmatrix} 60 \\ 28 \\ 30 \end{bmatrix}$ and, as

calculated in Section 9.3.1, the inverse of the basic matrix is $B^{-1} = \begin{bmatrix} -\frac{1}{5} & 0 & \frac{4}{5} \\ \frac{2}{5} & 0 & -\frac{3}{5} \\ -\frac{3}{5} & 1 & \frac{2}{5} \end{bmatrix}$. If we are

interested in finding the stable range of lemons, say, we replace the 60 in the capacity column with a variable, say u, and use Theorem 9.9, Part b. The revised capacity column in the current tableau is determined by $B^{-1}C$, and we simply set each entry of this product to be nonnegative (so that the current tableau remains feasible and, hence, optimal). This gives a system of inequalities that the variable u must satisfy, the solution of which is the stable range in question.

Thus,

$$B^{-1}C = \begin{bmatrix} -\frac{1}{5} & 0 & \frac{4}{5} \\ \frac{2}{5} & 0 & -\frac{3}{5} \\ -\frac{3}{5} & 1 & \frac{2}{5} \end{bmatrix} \begin{bmatrix} u \\ 28 \\ 30 \end{bmatrix} = \begin{bmatrix} 24 - \left(\frac{1}{5}u\right) \\ \frac{2}{5}u - 18 \\ 40 - \left(\frac{3}{5}u\right) \end{bmatrix}.$$

The value u is within the stable range for lemons if and only if each entry is nonnegative:

$$24 - \frac{u}{5} \geq 0$$

$$\frac{2u}{5} - 18 \geq 0$$

$$40 - \frac{3u}{5} \geq 0.$$

The first inequality says $u \leq 120$, the second says $u \geq 45$, and the third says $u \leq \frac{200}{3} \approx 66.667$. The common solution of all three inequalities is the interval $45 \leq u \leq \frac{200}{3}$. This is the stable range for lemons, or $[45, \frac{200}{3}]$ in interval notation. As long as the supply of lemons remains between 45 lemons and $66\frac{2}{3}$ lemons, the optimal solution will be the one where the basic variables are $x, y,$ and S_2, which is precisely the definition of stable range we gave back in Chapter 4. In fact, we already determined the stable ranges for the resources in this problem using Excel in Chapter 5 as well, and this agrees with the answer we found there.

Similarly, for the stable range of limes, we replace the 28 with a variable u in the capacity C and repeat the calculation:

$$B^{-1}C = \begin{bmatrix} -\frac{1}{5} & 0 & \frac{4}{5} \\ \frac{2}{5} & 0 & -\frac{3}{5} \\ -\frac{3}{5} & 1 & \frac{2}{5} \end{bmatrix} \begin{bmatrix} 60 \\ u \\ 30 \end{bmatrix} = \begin{bmatrix} 12 \\ 6 \\ u - 24 \end{bmatrix}.$$

In this case, two of the entries are independent of u and are already positive. This is because the basic solution includes leftover limes, and we are merely asking the question: how many limes must I start with to ensure there are leftover limes? We already know that in any such solution, we must have $x = 12$ and $y = 6$. The answer is $u \geq 24$, from solving the third inequality $u - 24 \geq 0$, but we already knew that as well, since we used exactly 24 limes to make the solution of $(12, 6)$, and anything over that will be leftover. Thus, the stable range for limes is $[24, \infty)$.

Finally, for the stable range for sugar:

$$B^{-1}C = \begin{bmatrix} -\frac{1}{5} & 0 & \frac{4}{5} \\ \frac{2}{5} & 0 & -\frac{3}{5} \\ -\frac{3}{5} & 1 & \frac{2}{5} \end{bmatrix} \begin{bmatrix} 60 \\ 28 \\ u \end{bmatrix} = \begin{bmatrix} \frac{4}{5}u - 12 \\ 24 - \left(\frac{3}{5}u\right) \\ \frac{2}{5}u - 8 \end{bmatrix}.$$

Setting these to be nonnegative yields $u \geq 15$, $u \leq 40$, and $u \geq 20$. The common solution is the interval $20 \leq u \leq 40$, the stable range for tablespoons of sugar, and agrees with what we found via Excel in Chapter 5.

In short, finding the stable range of a capacity number is extremely easy when the problem has been solved via the simplex algorithm, because we have B^{-1} available. We now move on to illustrate stable ranges for objective coefficients.

Continuing with the lemonade stand problem, suppose we want to compute the stable range for the price of a glass of sweet lemonade. We recall from the previous subsection that

$$\widehat{B}^{-1} = \begin{bmatrix} -\frac{1}{5} & 0 & \frac{4}{5} & 0 \\ \frac{2}{5} & 0 & -\frac{3}{5} & 0 \\ -\frac{3}{5} & 1 & \frac{2}{5} & 0 \\ M_1 & M_2 & M_3 & 1 \end{bmatrix}.$$

Where the M_i can be determined (for the current corner point) using the equation $\widehat{B}\widehat{B}^{-1} = I$, where the last row in $\widehat{B}$ contains the modified objective coefficients. So, if we want to compute the stable range of an objective coefficient, we replace that coefficient in the bottom row of $\widehat{B}$ with a variable u. Thus:

$$\widehat{B}\widehat{B}^{-1} = \begin{bmatrix} 3 & 4 & 0 & 0 \\ 1 & 2 & 1 & 0 \\ 2 & 1 & 0 & 0 \\ -u & -1.5 & 0 & 1 \end{bmatrix} \begin{bmatrix} -\frac{1}{5} & 0 & \frac{4}{5} & 0 \\ \frac{2}{5} & 0 & -\frac{3}{5} & 0 \\ -\frac{3}{5} & 1 & \frac{2}{5} & 0 \\ M_1 & M_2 & M_3 & 1 \end{bmatrix}$$

$$= \begin{bmatrix} 1 & 0 & 0 & 0 \\ 0 & 1 & 0 & 0 \\ 0 & 0 & 1 & 0 \\ \frac{1}{5}u + M_1 - \frac{3}{5} & M_2 & M_3 - \left(\frac{4}{5}u\right) + \frac{9}{10} & 1 \end{bmatrix}.$$

In order for this to be the identity matrix, we first solve for the M_i in terms of u:

$$M_1 = \frac{3}{5} - \frac{u}{5} = \frac{3 - u}{5}$$

$$M_2 = 0$$

$$M_3 = \frac{4}{5}u - \frac{9}{10} = \frac{8u - 9}{10}.$$

Now, since these M_i end up in the bottom row for the current tableau, for the value of u to be within the stable range, we must have each $M_i \geq 0$ (so that the current tableau remains the optimal one). Thus, $3 - u \geq 0$ yields $u \leq 3$, and $8u - 9 \geq 0$ yields $u \geq \frac{9}{8}$. It looks as if the stable range for the price of a glass of sweet lemonade is $\frac{9}{8} \leq u \leq 3$. This is correct because there are no other nonzero entries in the bottom row at this corner other than the entries under the slack variables (that is, other than the M_i). The reason for this is because in this example, every decision variable is a basic variable, which means the entry in the last row under the decision variables are all 0 – the rules for pivoting were designed, remember, to place a 0 in every row other than the pivot row, including the bottom row. Thus, as long as u is in the range $\left[\frac{9}{8}, 3\right]$, every entry in the bottom row of the current tableau remains nonnegative, so the current solution remains optimal. The stable range for a glass of sweet lemonade is the range \$1.125 to \$3.00 per glass.

Similarly, we can compute the stable range for a glass of tart lemonade:

$$\widehat{B}\widehat{B}^{-1} = \begin{bmatrix} 3 & 4 & 0 & 0 \\ 1 & 2 & 1 & 0 \\ 2 & 1 & 0 & 0 \\ -1.25 & -u & 0 & 1 \end{bmatrix} \begin{bmatrix} -\frac{1}{5} & 0 & \frac{4}{5} & 0 \\ \frac{2}{5} & 0 & -\frac{3}{5} & 0 \\ -\frac{3}{5} & 1 & \frac{2}{5} & 0 \\ M_1 & M_2 & M_3 & 1 \end{bmatrix}$$

$$= \begin{bmatrix} 1 & 0 & 0 & 0 \\ 0 & 1 & 0 & 0 \\ 0 & 0 & 1 & 0 \\ -\frac{2}{5}u + M_1 + \frac{1}{4} & M_2 & M_3 + \left(\frac{3}{5}u\right) - 1 & 1 \end{bmatrix}.$$

This leads to

$$M_1 = \frac{2}{5}u - \frac{1}{4} = \frac{8u - 5}{20}$$

$$M_2 = 0$$

$$M_3 = 1 - \frac{3}{5}u = \frac{5 - 3u}{5}.$$

Setting the $M_i \geq 0$ and solving for u yields $\frac{5}{8} \leq u \leq \frac{5}{3}$; that is, the price of a glass of tart lemonade must lie between \$0.625 and \$1.66$\frac{2}{3}$ per glass in order for the current tableau to remain the optimal one. These stable ranges agree exactly with the ranges we found geometrically in Section 4.2.

The lemonade stand problem is a bit deceptive because all the decision variables are basic at the current tableau. Recall that, in general, if there are nonbasic decision variables, every entry in the last row, not just the marginal values, must be checked to see if it remains nonnegative. So, for our final example, we consider the Spooky Boogie Halloween Costume problem.

For that problem, we have the relevant matrices are, as computed in the previous section, given by

$$\widehat{B} = \begin{bmatrix} 2 & 2 & 0 \\ 1 & 4 & 0 \\ -8 & -16 & 1 \end{bmatrix}$$

$$\widehat{B}^{-1} = \begin{bmatrix} \frac{2}{3} & -\frac{1}{3} & 0 \\ -\frac{1}{6} & \frac{1}{3} & 0 \\ M_1 & M_2 & 1 \end{bmatrix}.$$

As we know, the values of M_i are both equal to $\frac{8}{3}$ at the current optimal corner, but we label them as variables like in the last example above. So, let's compute the stable range of the objective coefficients for the basic decision variables – that is, for the profit of a ghost costume and the profit of a werewolf costume. For the ghost costume, we replace the 8 in the bottom row of $\widehat{B}$ with a variable, u, and find the revised M_i:

$$\widehat{B}\widehat{B}^{-1} = \begin{bmatrix} 2 & 2 & 0 \\ 1 & 4 & 0 \\ -u & -16 & 1 \end{bmatrix} \begin{bmatrix} \frac{2}{3} & -\frac{1}{3} & 0 \\ -\frac{1}{6} & \frac{1}{3} & 0 \\ M_1 & M_2 & 1 \end{bmatrix}$$

$$= \begin{bmatrix} 1 & 0 & 0 \\ 0 & 1 & 0 \\ M_1 - \left(\frac{2}{3}u\right) + \frac{8}{3} & \frac{1}{3}u + M_2 - \frac{16}{3} & 1 \end{bmatrix}.$$

This leads to

$$M_1 = \frac{2}{3}u - \frac{8}{3} = \frac{2u - 8}{3}$$

$$M_2 = \frac{16}{3} - \frac{u}{3} = \frac{16 - u}{3}.$$

It certainly must be true that these M_i must be nonnegative, so we can solve for u as we did before. We obtain $4 \leq u \leq 16$. So we know that in this range, the M_i will remain nonnegative, as desired. Before we can conclude that we have found the stable range, we must show that all the entries in the bottom row, not just the M_i, remain nonnegative. We again use the fact that we can find any column in the current tableau by multiplying the original column in the initial tableau by $\widehat{B}^{-1}$. So, leaving the form of $\widehat{B}^{-1}$ with the marginal values expressed in terms of u, we simply multiply $\widehat{B}^{-1}$ times the entire initial tableau:

$$\begin{bmatrix} \frac{2}{3} & -\frac{1}{3} & 0 \\ -\frac{1}{6} & \frac{1}{3} & 0 \\ \frac{2u-8}{3} & \frac{16-u}{3} & 1 \end{bmatrix} \begin{bmatrix} 3 & 2 & 2 & 2 & 1 & 0 & 600 \\ 3 & 1 & 3 & 4 & 0 & 1 & 510 \\ -10 & -u & -12 & -16 & 0 & 0 & 0 \end{bmatrix}$$

$$= \begin{bmatrix} 1 & 1 & \frac{1}{3} & 0 & \frac{2}{3} & -\frac{1}{3} & 230 \\ \frac{1}{2} & 0 & \frac{2}{3} & 1 & -\frac{1}{6} & \frac{1}{3} & 70 \\ u-2 & 0 & \frac{u-4}{3} & 0 & \frac{2u-8}{3} & \frac{16-u}{3} & 230u + 1{,}120 \end{bmatrix}.$$

Now we see the full set of inequalities that u must satisfy. Not only must we have $4 \leq u \leq 16$, which comes from setting the marginal values nonnegative, but we also see that $u \geq 2$ and $u \geq 4$, which come from setting the entries in the first and third columns (the nonbasic decision variables) to also be nonegative. Combining all the inequalities, we see the common solution is still $4 \leq u \leq 16$, so we can now safely claim we have found the stable range for the profit of a ghost costume – it must lie in the interval $[4, 16]$. In the exercises, we ask the reader to find the stable range of the proifit for a werewolf costume, the other basic decision variable.

Finally, let's consider the stable range of the objective coefficient of one of the nonbasic decision variables, such as the profit for a witch costume. This case is actually easier. That's because, if we want to be within the stable range, then that means we want that decision variable to remain nonbasic. In turn, this means that the values of the marginal values do not change. Indeed, the coefficient in the last row that is varied (and replaced by u) doesn't even appear in the matrix $\widehat{B}$

(recall we constructed B and $\widehat{B}$ by using only the columns corresponding to basic variables!). So if $\widehat{B}$ is unchanged, then so is its inverse. Thus, we know

$$\widehat{B}^{-1} = \begin{bmatrix} \frac{2}{3} & -\frac{1}{3} & 0 \\ -\frac{1}{6} & \frac{1}{3} & 0 \\ \frac{8}{3} & \frac{8}{3} & 1 \end{bmatrix}.$$

We now repeat the calculation to check that every entry in the last row remains nonnegtive: $\widehat{B}^{-1}$ times any column of the initial tableau yields the corresponding column of the current tableau. But as $\widehat{B}$ has not changed at all, then, as we suggested in the last subsection, the only possible change is in the only column of the initial tableau where a change is made – in this case, the w column, because we are finding the stable range of the profit for a witch costume:

$$\begin{bmatrix} \frac{2}{3} & -\frac{1}{3} & 0 \\ -\frac{1}{6} & \frac{1}{3} & 0 \\ \frac{8}{3} & \frac{8}{3} & 1 \end{bmatrix} \begin{bmatrix} 3 \\ 3 \\ -u \end{bmatrix} = \begin{bmatrix} 1 \\ \frac{1}{2} \\ 16-u \end{bmatrix}.$$

Since we want the bottom entry to remain nonnegative, the only condition that u must satisfy is $u \leq 16$. Thus, the stable range for a witch costume is $(-\infty, 16]$. In general, when changing the objective coefficient of a nonbasic decision variable, just as before, there will only be one condition on u. So such stable ranges will always look like $[-\infty, r]$ or $[r, \infty]$ for some positive number r. In the exercises, we ask the reader to determine the stable range of the objective coefficient for the other nonbasic decision variable in this example – namely, the profit from a goblin costume.

In summary, finding the stable range of any objective coefficient is fairly easy (just involving matrix multiplication) using $\widehat{B}^{-1}$, just as finding the stable range of a capacity number is obtained by working with B^{-1}.

Exercises

1. In the Spooky Boogie Halloween Costume problem, find the stable range of the capacity for material.
2. In the Spooky Boogie Halloween Costume problem, find the stable range of the capacity for sewing time.
3. In the Spooky Boogie Halloween Costume problem, find the stable range for the profit generated by a goblin costume.
4. In the Spooky Boogie Halloween Costume problem, find the stable range for the profit generated by a werewolf costume.
5. In Exercise 1 in Section 4.2, you found the stable ranges for the profit of each type of bracelet in the Court and Sparkle Jewelry Emporium. In this problem, recall that Joni makes two signature design bracelet models. Each Dawntreader bracelet uses 1 ruby, 6 pearls, and 10 opals. Each Hejira bracelet uses 3 rubies, 3 pearls, and 15 opals. She has 54 rubies, 120 pearls, and 300 opals to work with. If either model results in a profit of $1,800 for Joni, how many of each type should she make? Now find the stable range for the profit of each type of bracelet, using the techniques of this section.

6. In the Court and Sparkle Jewelry Emporium problem, find the stable range for the capacity of each of the three resources.
7. In Section 4.2, you solved the following problem based on graphical arguments. Now solve the problem using the technques of this chapter.

    Kerry's Kennel is mixing two commercial brands of dog food for its canine guests. A bag of Dog's Life Canine Cuisine contains 3 pounds of fat, 2 pounds of carbohydrates, 5 pounds of protein, and 3 ounces of vitamin C. A giant-size bag of Way of Life Healthy Mix contains 1 pound of fat, 5 pounds of carbohydrates, 10 pounds of protein, and 7 ounces of vitamin C. The requirements for a week's supply of food for the kennel are that there should be at most 21 pounds of fat, at most 40 pounds of carbohydrates, and at least 21 ounces of vitamin C. How many bags of each type of food should be mixed in order to design a diet that maximizes protein?

    a. Suppose the recipe for Dog's Life Canine Cuisine changes so that the protein content of each bag increases to 12 pounds. Find the new optimal data.
    b. Find the stable range for the protein in a bag of Dog's Life Canine Cuisine.
    c. Find the stable range for the protein in a bag of Way of Life Healthy Mix.

8. a. Consider the Brain One Lighting and Heavenly Music Corporation problem (Exercise 7 in Section 9.3.2). The company is considering hiring more designers. Compute the stable range for the weeks available in a year with the design team (the capacity number that is currently 48). Repeat for the weeks available with the technical team (the capacity which is currently 44).
    b. Find these stable ranges by solving the problem with Excel, as we did in Chapter 5, and verify that your solution agrees with what you got in Part a.
9. a. In the Brain One Lighting and Heavenly Music Corporation problem, determine the stable range for the profit generated by a museum contract and for the profit generated by a theater contract.
    b. Find this stable range by solving the problem with Excel, as we did in Chapter 5, and verify that your solution agrees with what you got in Part a.
10. Consider the following minimization problem, which you have solved earlier in the text:

    The Poseidon's Wake Petroleum Company operates two refineries. The Cadence Refinery can produce 40 units of low-grade oil, 10 units of medium-grade oil, and 10 units of high-grade oil in a single day. (Each unit is 1,000 barrels.) The Cascade Refinery can produce 10 units of low-grade oil, 10 units of medium-grade oil, and 30 units of high-grade oil in a single day. Poseidon's Wake Petroleum Company receives an order from the Mars Triangle Oil retailers for at least 80 units of low-grade oil, at least 50 units of medium-grade oil, and at least 90 units of high-grade oil. If it costs Poseidon's Wake $1,800 to operate the Cadence Refinery for a day and $2,000 to operate the Cascade Refinery for a day, how many days should the company operate each refinery to fill the order at least cost?

    In an exercise in Chapter 4, some sensitivity analysis was done on this problem based on graphical techniques. In this exercise, we use the techniques of this chapter to do the same, while exploring the connections between stable ranges and duality.

    a. Solve the problem using Phase I pivoting as we did in Section 9.2, or look up your previous solution to Exercise 5 in Section 9.2.

b.   Instead, solve the problem using duality without Phase I, or look up your previous solution to Exercise 5 in Section 5.3.

c.   We are interested in finding the stable range in the number of units of high-grade oil that comprises part of the order from the Mars Triangle Oil Retailers (this number, currently 90, is a capacity number in the primal problem and also an objective coefficient in the dual problem). Find the stable range for this capacity number in the primal problem.

d.   Instead, use the dual problem and compute its stable range as an objective coefficient. Compare your answer to that of Part c.

e.   Suppose the Mars Triangle Oil lowers that part of its order from 90 units to 60 units. Determine the revised optimal data (you can use either the primal or the dual to answer this).

9.3.4   Integer Programming: Branch and Bound Method

Our final topic in sensitivity analysis is to consider what happens if we modify the constraints of a problem to require that some or all of the decision variables are integers. The method we use to handle these additional constraints is similar to what we have done earlier in the chapter. That is, ignore the integer constraints at first, and solve the problem without them! If we are lucky, the optimal solution will come out to be integers, and there is no further pivoting necessary. However, if the solution does not come out to be integers, we go back and modify the problem, which will necessitate additional pivoting, analogous to the way additional pivoting was required if a change in a capacity number or an objective coefficient went outside the stable range.

We illustrate the process, called the *branch and bound method*, with the following example.

Example 9.11 Dyer States Appliance Store is ordering refrigerators and color TVs for its floor display for its annual *Money for Nothing Rebate Sale*. This is a big event, which they schedule to coincide with the annual mattress sale at the Sultans of Springs store next door, so it can capitalize on customer traffic. The store is brightly decorated, and the store plays festive Hawaiian music all weekend. They take cash and credit cards, and if you need financing, the staff will do financial background checks for free. Each refrigerator takes up 8 feet of wall space for its display, and each color TV takes up 5 feet of wall space for its display. The store has 400 feet of wall space available for its displays. Also, the store limited to ordering a total of no more than 60 items altogether due to its contract with the supplier. If each refrigerator will generate $500 profit for the store (after the rebate) and each color TV will generate $400 profit for the store (after the rebate), how many of each item should Dyer States Appliance Store order to maximize its total profit?

We let x stand for the number of refrigerators and y stand for the number of color TVs. The reader can easily verify that the setup of the problem is

$$\text{maximize } P = 500x + 400y$$
$$\text{subject to:}$$
$$x + y \le 60 \text{ (contract requirement)}$$
$$8x + 5y \le 400 \text{ (space requirement)}$$
$$x \ge 0,\ y \ge 0.$$

Additionally, since fractions of a refrigerator or a TV do not make sense in this problem, there is the additional constraint that x and y be integers. Ignoring the integer constraints for the moment, we set this up as a standard form maximization problem with initial tableau.

	x	y	S_1	S_2	Capacity
S_1	1	1	1	0	60
S_2	8	5	0	1	400
P	-500	-400	0	0	0

The reader should also verify (see Exercise 1) that after two pivots, we arrive at the optimal tableau.

	x	y	S_1	S_2	Capacity
y	0	1	$\frac{8}{3}$	$-\frac{1}{3}$	$\frac{80}{3}$
x	1	0	$-\frac{5}{3}$	$\frac{1}{3}$	$\frac{100}{3}$
P	0	0	$\frac{700}{3}$	$\frac{100}{3}$	$\frac{8,2000}{3}$

Thus, ignoring the integer constraint leads to the solution:

$$x = \frac{100}{3} = 33\frac{1}{3} \text{ refrigerators}$$

$$y = \frac{80}{3} = 26\frac{2}{3} \text{ color TVs}$$

$$P = \frac{82,000}{3} = \$27,333.33 \text{ (maximized)}$$

$$S_1 = 0, \ M_1 = \frac{700}{3} = \$233.33 \text{ per item}$$

$$S_2 = 0, \ M_2 = \frac{100}{3} = \$33.33 \text{ per foot wall space.}$$

Now in the interpretation stage of our mathematical model, we find we must reject the solution as is because of the fractional answers. How should we proceed from here? One naive suggestion is to simply round off the answers, since the best integer solution is probably near this solution. So, you might guess $x = 33$ refrigerators and $y = 27$ color TVs is the best answer. The problem is when you round the answers, you are not guaranteed to remain in the feasible region. On the other hand, even if you do remain in the feasible region, what assurance do you have that the rounded solution is truly the optimal one satisfying the integer condition in addition to the constraints? In this case, the solution $(33, 27)$ exactly satisfies the contact constraint $x + y = 60$. What about the space constraint? Well,

$$8(33) + 5(27) = 399.$$

So, in this case, the rounded solution does remain feasible. But might not there be a different feasible point with integer coordinates with a higher profit? Since this is a 2×2 problem, you could easily solve it by a graphical method, and from there, it would probably be fairly easy to check all the nearby feasible points with integer coordinates. It might be that $(33, 27)$ is indeed the optimal integer solution. However, don't forget we seek a technique that will work with any $m \times n$

problem. Suppose, after rounding off the n decision variables, we obtain an integer solution that is not feasible?

Presumably, one might guess that some of the variables should be rounded down, even if the nearest integer is the next highest, in order to keep the point inside the feasible region. But which ones? With n variables, there are 2^n different ways to select to round up or round down. If $n = 10$, that is 1,024 different integer points that are "close" to the given fractional solution. Surely, we don't want to check them all. It would require $m2^n$ calculations (m for each of the 2^n points) to determine which of these integer points are feasible! Again, once we find all these feasible integer solutions, there is still the problem of making sure one of them is optimal.

Clearly, rounding the answer is not a good approach. Instead, we look at the solution and add new constraints that force the current "solution" to be infeasible. For example, to make sure that $x = 33\frac{1}{3}$ refrigerators is not a feasible point, we could add either the constraint $x \leq 33$ or the constraint $x \geq 34$. This leads to two separate problems that must be solved. This process is called *branching* – with each variable that is not integer in solution, we branch into two separate problems. The first of these is the following:

$$\text{maximize } P = 500x + 400y$$
$$\text{subject to:}$$
$$x + y \leq 60 \text{ (contract requirement)}$$
$$8x + 5y \leq 400 \text{ (space requirement)}$$
$$x \leq 33 \text{ (branch requirement)}$$
$$x \geq 0, \ y \geq 0.$$

The second of these is the following:

$$\text{maximize } P = 500x + 400y$$
$$\text{subject to:}$$
$$x + y \leq 60 \text{ (contract requirement)}$$
$$8x + 5y \leq 400 \text{ (space requirement)}$$
$$x \geq 34 \text{ (branch requirement)}$$
$$x \geq 0, \ y \geq 0.$$

We now solve these two problems. The initial tableau for the first is the following.

	x	y	S_1	S_2	S_3	Capacity
S_1	1	1	1	0	0	60
S_2	8	5	0	1	0	400
S_3	1	0	0	0	1	33
P	-500	-400	0	0	0	0

After two pivots (Exercise 1, again), we arrive at this tableau.

	x	y	S_1	S_2	S_3	Capacity
y	0	1	1	0	−1	27
S_2	0	0	−5	1	−3	1
x	1	0	0	0	1	33
P	0	0	400	0	100	27,300

Notice that this solution happens to be the one we guessed when we naively just rounded off the variables:

$$x = 33 \text{ refrigerators}$$
$$y = 27 \text{ color TVs}$$
$$S_2 = 1 \text{ foot leftover wall space}$$
$$P = \$27{,}300$$

We won't bother writing the values of the marginal values. While we were lucky in that the solution to this modified problem contains only integer values for both decision variables, we do not yet know it is the optimal point with that property. We know that no solution could have a profit of more than \$27,333.33, but it is possible an integer solution might have a profit that is more than \$27,300 but still less than \$27,333.33.

We need to check the second branched problem. This problem has mixed constraints, so it requires Phase I techniques to solve. We write the branch constraint as $-x \le -34$, and replace it with the equation $-x + S_3 - A_3 = -34$. Then, as usual, $z = -A = x - S_3 - 34$. Thus, the initial tableau is as follows.

	x	y	S_1	S_2	S_3	A_3	Capacity
S_1	1	1	1	0	0	0	60
S_2	8	5	0	1	0	0	400
$-A_3$	−1	0	0	0	1	−1	−34
P	−500	−400	0	0	0	0	0
z	−1	0	0	0	1	0	−34

After one pivot (Exercise 1), we arrive at a feasible tableau, since $z = 0$.

	x	y	S_1	S_2	S_3	A_3	Capacity
S_1	0	1	1	0	1	−1	26
S_2	0	5	0	1	8	−8	128
x	1	0	0	0	−1	1	34
P	0	−400	0	0	−500	500	1,7000
z	0	0	0	0	0	1	0

We now delete the z row and the A_3 column and continue with Phase II pivoting. Two further pivots (Exercise 1) lead to the following tableau.

	x	y	S_1	S_2	S_3	Capacity
S_1	0	0	1	$-\frac{1}{5}$	$-\frac{3}{5}$	$\frac{2}{5}$
y	0	1	0	$\frac{1}{5}$	$\frac{8}{5}$	$\frac{128}{5}$
x	1	0	0	0	-1	34
P	0	0	0	80	140	27,240

The solution at this tableau is

$$x = 34 \text{ refrigerators}$$

$$y = \frac{128}{5} = 25.6 \text{ color TVs}$$

$$S_1 = \frac{2}{5} \text{ (because the total order is 59.6 items)}$$

$$P = \$27{,}240$$

In the first problem we solved, both variables came out to be integers. We are not so lucky here, so from this problem, further branching would seem to be necessary. Thus, we consider two new problems, one where we add the constraint $y \leq 25$ and one where we add the constraint $y \geq 26$. Both problems have mixed constraints. They are

maximize $P = 500x + 400y$

subject to:

$x + y \leq 60$ (contract requirement)

$8x + 5y \leq 400$ (space requirement)

$x \geq 34$ (branch requirement)

$y \leq 25$ (branch requirement)

$x \geq 0, \ y \geq 0$

and

maximize $P = 500x + 400y$

subject to:

$x + y \leq 60$ (contract requirement)

$8x + 5y \leq 400$ (space requirement)

$x \geq 34$ (branch requirement)

$y \geq 26$ (branch requirement)

$x \geq 0, \ y \geq 0.$

If you were to try to solve these two problems, this is what you would find. The first has the following solution:

$$x = 34\tfrac{3}{8} \text{ refrigerators}$$

$$y = 25 \text{ color TVs}$$

$$z = \$27{,}187.50.$$

The second problem has no feasible solution. Again, since x is not an integer in the solution to the first of these, it would seem even further branching is necessary. This process appears to be long and complicated. It would help if we kept track of all the branching with a *binary tree*. At each node, we will put the solution, and if a decision variable x_i that should be an integer is not, we branch from that node into exactly two other problems (which is why it is a binary tree). On one branch, we add the constraint $x_i \leq k$, and on the other branch, we add the constraint $x_i \geq k + 1$, if k is the integer such that $k < x_i < k + 1$. A convenient way to label the nodes is node r will branch into nodes $2r$ and $2r + 1$. Node 1 (the root of the tree) is the solution obtained when we solve the problem ignoring the integer constraints.

For the preceding example, the diagram in Figure 9.1 is the binary tree for the branching we have done so far.

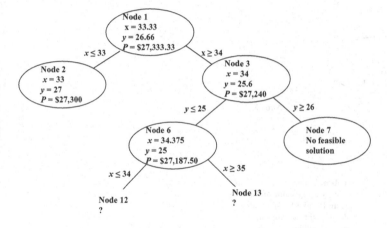

Figure 9.1 Branching a problem with integer constraints.

You should note that by our numbering convention, each node has a different number, but not all numbers are used, because some nodes are terminal and do not require further branching. For example, any node with a solution where all the required integer constraints are satisfied is terminal. So no branching is required from node 2, and thus, there is no node 4 or node 5 in this problem. Also, any node that has no feasible solution is terminal, such as node 7, so there is no node 14 and node 15.

The first question that comes to mind is how do we know this process will terminate? As long as the feasible set is bounded, there will only be a finite number of points inside it with integer coordinates, so this branching process certainly cannot go on forever. Second, is there any way to streamline the process? Well, observe that every time a new constraint is added to a problem to make a branch, the feasible set of the new problem is a (proper) subset of the original feasible set from its "parent" node. That means, for a standard maximization problem where the objective coefficients are all positive, the value of the objective function can only go down (or possibly stay the same). The reverse is true for a standard minimization problem with positive objective coefficients – the value can only go up (see Exercise 5). Observe that in the previous tree the value of P at node 6 is strictly less than the value of P at node 3, which in turn is strictly less than the value of P at node 1.

But notice that the value of P at the integer solution in node 2 is higher than the value in node 3: $27,240 < 27,300. By our previous observation, the value of P at any descendant of node 3 must then also be less than \$27,300. That includes node 6 ($P = \$27,187.50$), as well as its descendants, nodes 12 and 13 and any further descendants. If we had observed this right at node 3, we could have saved ourselves the work of any further branching from node 3!

This is the *bounding* part of the "branch and bound" method. If at any time, we arrive at a node B where the objective function is less than the objective function for some node A that satisfies all the integer constraints, it is not necessary to branch further from node B, because at any descendant of B, the value of the objective function is bounded by its value at B, which is less than the value at node A, so it cannot be an optimal solution even if it satisfied all the integer constraints. The solution we found at A is better.

Thus, in the example, we only needed to find the solutions at nodes 1, 2, and 3, as shown in Figure 9.2.

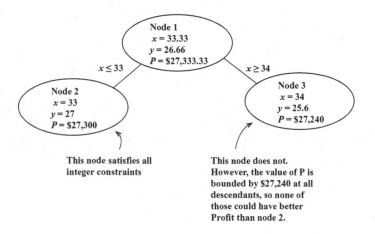

Figure 9.2 Using bounding to reduce the number of required nodes to check.

Thus, the optimal solution satisfying the integer constraints must be at node 2. No further branching is needed. We make the following observations:

1. In this example, the optimal solution happened to be the same solution as we obtained by naively just rounding off the original solution in node 1. But that does not happen all the time, as our next example illustrates.
2. We could have started branching from node 1 by adding constraints for y instead of x. In Exercise 2, you are asked to do this. You will arrive at the same solution. It can be shown that this always happens – so it does not matter which variable you start with.
3. In general, there might be more than one terminal node that satisfies the integer constraints. The optimal solution will be the one with the highest value of the objective function (or the lowest value if it is a minimization problem). Our next example illustrates this as well.

Example 9.12 Mr. Philip's studio shop, Glasswork Monolith Arts, makes a hanging ornament called a *Glass Tear*, which uses 1 ounce of glass and 1 ounce of quartz crystals to make and yields a $1 profit. The shop also makes two different paperweights. The Heart of Glass paperweight uses 1 ounce of glass and 2 ounces of quartz crystals and generates $2 profit. The Glass Onion paperweight uses 3 ounces of glass and 1 ounce of quartz crystals and generates $3 profit. Today Mr. Philip has 32 ounces of glass and 18 ounces of quartz crystals. How many of each item should he make in order to maximize profit?

Let t be the number of glass tear ornaments, h the number of heart of glass paperweights, and g the number of glass onion paperweights. These variables must all be integers, so this is an integer programming problem. The setup of the problem without the integer constraints is

$$\text{maximize } P = t + 2h + 3g$$

subject to:

$$t + h + 3g \leq 32 \text{ (ounces glass)}$$
$$t + 2h + g \leq 18 \text{ (ounces quartz crystals)}$$
$$x \geq 0, \ y \geq 0.$$

The initial tableau is as follows.

	t	h	g	S_1	S_2	Capacity
S_1	1	1	3	1	0	32
S_2	1	2	1	0	1	18
P	−1	−2	−3	0	0	0

In Exercise 3, the reader will check that after two pivots, we arrive at the optimal tableau.

	t	h	g	S_1	S_2	Capacity
g	$\frac{1}{5}$	0	1	$\frac{2}{5}$	$-\frac{1}{5}$	$\frac{46}{5}$
h	$\frac{2}{5}$	1	0	$-\frac{1}{5}$	$\frac{3}{5}$	$\frac{22}{5}$
P	$\frac{2}{5}$	0	0	$\frac{4}{5}$	$\frac{3}{5}$	$\frac{182}{5}$

The optimal data (ignoring the marginal values and slack values) for node 1 is

$$t = 0 \text{ Glass Tear ornaments}$$

$$h = \frac{22}{5} = 4.4 \text{ Heart of Glass paperweights}$$

$$g = \frac{46}{5} = 9.2 \text{ Glass Onion paperweights}$$

$$P = \frac{182}{5} = \$36.40 \text{ (maximized)}.$$

To create node 2, we'll add the constraint $h \leq 4$, and for node 3, we'll add $h \geq 5$. In Exercise 3, we ask the reader to verify the solutions for these two nodes. They are as follows. For node 2, we have

$$t = 0 \text{ Glass Tear ornaments}$$
$$h = 4 \text{ Heart of Glass paperweights}$$
$$g = \frac{28}{3} = 9\frac{1}{3} \text{ Glass Onion paperweights}$$
$$P = \$36.00 \text{ (maximized)}.$$

Because g is not an integer, we will require further branching from node 2. For node 3, the solution is

$$t = 0 \text{ Glass Tear ornaments}$$
$$h = 5 \text{ Heart of Glass paperweights}$$
$$g = 8 \text{ Glass Onion paperweights}$$
$$P = \$34.00 \text{ (maximized)}.$$

Since all the integer constraints are satisfied at node 3, this node is terminal. Next, we go back and branch from node 2. To create node 4, we will add the constraint $g \leq 9$ to the constraints in node 2. To create node 5, we add $g \geq 10$ to the constraints in node 2. In Exercise 3, the reader is asked to verify the solutions for these nodes are as follows. For node 4, we obtain

$$t = 1 \text{ Glass Tear ornament}$$
$$h = 4 \text{ Heart of Glass paperweights}$$
$$g = 9 \text{ Glass Onion paperweights}$$
$$P = \$36.00 \text{ (maximized)}.$$

Since this is an integer solution, it is a terminal node. For node 5, we obtain:

$$t = 0 \text{ Glass Tear ornaments}$$
$$h = 2 \text{ Heart of Glass paperweights}$$
$$g = 10 \text{ Glass Onion paperweights}$$
$$P = \$34.00 \text{ (maximized)}.$$

This is also a terminal node, so no further branching is required. Notice that in this example, we never used the bounding part of the process. We end up with the following binary tree (Figure 9.3):

In this example, all three terminal nodes have integer solutions. The optimal point must be node 4. Thus, the solution is

$$t = 1 \text{ Glass Tear ornament}$$
$$h = 4 \text{ Heart of Glass paperweights}$$
$$g = 9 \text{ Glass Onion paperweights}$$
$$P = \$36.00 \text{ (maximized)}.$$

In this problem, notice that the naive roundoff answer $t = 0$, $h = 4$, $g = 9$ was not one of the options.

There are other techniques that can be used to solve integer programming problems. For example, there is a method using "Gomoroy cutting planes," which uses additional constraints to find

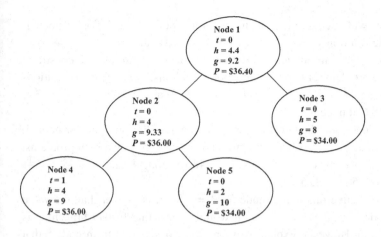

Figure 9.3 In this example, we never used the bounding process.

the smallest convex polytope Y inside the feasible region X that covers all the integer solutions inside the original region X. The region Y will have integer coordinates at all its corners, so it can be used as the feasible set for the problem including the integer constraints. A discussion of this method can be found in Calvert and Voxman (1989). The branch and bound method is probably the most commonly used technique.

Since our purpose was just to introduce integer programming and cast it as an episode of sensitivity analysis, this concludes our discussion of integer programming.

Exercises

1. Check all the pivoting in Example 9.1, the Money for Nothing Rebate Sale problem, and verify the solutions at each node.
2. Solve the Money for Nothing problem again, where at node 1, the branching is done by adding constraints that involve the variable y instead of x. Verify that you obtain the same answer as in Example 9.1.
3. Check all the pivoting in Example 9.2, the Glasswork Monolith Art Studio Shop problem, and verify solutions at each node.
4. This exercise illustrates that sometimes we can apply the bounding part of the process slightly more generally than explained in the preceding examples. The goal is to solve the following problem:

$$\text{maximize } P = 3x + 6y$$

subject to:

$$20x + 45y \leq 350$$

$$8x + 5y \leq 90$$

$$x \geq 0, \ y \geq 0 \text{ and both variables integer.}$$

a. Solve the problem without the integer constraints to obtain the solution at node 1. You should get the solution $x = \frac{115}{13} = 8\frac{11}{13}$, $y = \frac{50}{13} = 3\frac{11}{13}$, and $P = \frac{645}{13} = 49\frac{8}{13}$.

b. Add the constraint $x \leq 8$ to obtain the problem for node 2, and add the constraint $x \geq 9$ to obtain the problem for node 3. Solve both problems. Verify that the solutions are $x = 8$, $y = \frac{38}{9} = 4\frac{2}{9}$, and $P = \frac{148}{3} = 49\frac{1}{3}$ at node 2, and $x = 9$, $y = \frac{18}{5} = 3\frac{3}{5}$, and $P = \frac{243}{5} = 48.6$ at node 3.

c. Now branch at node 2. Add the constraint $y \leq 4$ to obtain the problem for node 4, and add the constraint $y \geq 5$ to obtain the problem for node 5. Solve both problems. Verify that the solutions are $x = 8$, $y = 4$, $P = 48$ for node 4, and $x = \frac{25}{4} = 6\frac{1}{4}$, $y = 5$, and $P = 48.75$ at node 5.

d. The value of the objective function at node 5, $P = 48.75$, is greater than the value at the terminal node 4, where $P = 48$. So it would appear that further branching is required from node 5. However, explain why any node descended from node 5 that satisfies all the integer constraints cannot have a value of P higher than 48. Therefore, no further branching is necessary at node 5. [**Hint**: The objective coefficients are integers!]

e. Similarly, no node satisfying the integer constraints that are descendants of node 3 can have a value higher than 48, even though the value of P at node 3 is higher than 48. Therefore, no further branching is required. Draw the binary tree for this problem, and state the final answer that satisfies the integer constraints.

f. Verify that $x = 6$, $y = 5$, and $p = 48$ is also a solution to this problem that satisfies the integer constraints. Since $P = 48$, which we already know is maximized by Part e, this is a different optimal solution. This illustrates that integer programming problems can have non-unique solutions. (Had we actually carried out the branching from node 5, which we decided was unnecessary, we would have found that solution as well.)

5. The branch and bound method also works with minimization problems. The only difference is that when a branch constraint is added to a problem and the new feasible set is a proper subset of of the one for the problem at the parent node, the value of the objective function must go up (or stay the same) instead of going down, so when you decide to ignore descendants of a node by the bounding part of the process, you bound from below rather than from above. In other words, if at node A, all the integer constraints are satisfied and if, at node B, they are not, but the value of the objective function at node A is less than or equal to the value at node B, you need not perform further branches from node B, since none of the descendant nodes will have an objective value lower than that at A. The following problem illustrates this.

Buttercup Builders is contracted by the White Stripes Foundation to build an apartment complex. Each studio apartment will cost eight units to build, each one-bedroom apartment will cost 11 units to build, and each two-bedroom apartment will cost 12 units to build, where a unit of cost is $10,000. The contract stipulates that at least 30 apartments must be included in the project, and at least $\frac{1}{4}$ of the total number of apartments must be two-bedroom apartments. How many apartments of each type should be built in order to minimize the total cost?

a. Set this up as an integer programming problem, and solve it without the integer constraints in order to find node 1.

b. Your solution to Part a should be $x = 22.5$ (studio apartments) $y = 0$ (one-bedroom apartments) and $z = 7.5$ (two-bedroom apartments) with a total cost of $C = 270$ units (which is $2,700,000). Branch on the variable x, so add $x \leq 22$ to create node 2 and $x \geq 23$ to create node 3. Solve both nodes.

c. Your solution to node 2 should be $x = 22$, $y = 0.5$, $z = 7.5$, and $C = 271.5$, and your solution to node 3 should be $x = 23$, $y = 0$, $z = 7.6666$, and $C = 276$. Neither node satisfies the integer constraints, but since this is a minimization problem, we choose to branch on node 2 and save node 3 for later. We will branch on the variable y. So create node 4 by adding the constraint $y \leq 0$, and create node 5 by adding the constaint $y \geq 1$. Solve both nodes.

d. Your solution to node 4 should be $x = 22$, $y = 0$, and $z = 8$ with $C = 272$. This is a terminal node. Your solution to node 5 should be $x = 21.5$, $y = 1$, and $z = 7.5$ with $C = 273$. Note that all nodes descending from node 5 will have a cost of $C \geq 273$ so no such descendant can have a better cost than node 4. Similarly, every descendant of node 3 will have a cost of $C \geq 276$, so no further branching is necessary. Draw the binary tree for this problem, and state the optimal integer solution. Can you explain why it was not even necessary to solve node 5?

6. To solve an integer programming problem using Excel, proceed as follows. Enter the data on the spreadsheet as usual. In the step where you are adding the constraints, you can add a constraint that a decision variable must be an integer by (after opening the "add constraint" pop-up) clicking on the cell that contains the value of that decision variable, then in the drop-down menu in the middle of the "add constraint" pop-up, select "INT." Then solve the problem with the usual steps. Use this method to solve the integer programming problems:

a. The Money for Nothing problem (Exercise 1).
b. The Glasswork Monolith Art Studio Shop problem (Exercise 2).
c. The Buttercup Builders problem (Exercise 5).

Appendix: A Rapid Review of Sets and Probability

In this appendix, we review set theory, enumeration, and probability theory. The purpose is to give quick coverage of these topics as a review. There is enough detail to introduce these topics to someone who has never seen them before, but the presentation is terse, with fewer examples and exercises than the rest of the book. Technically speaking, the reader should know most of the material in this chapter before studying the earlier chapters. Experience suggests that many students have seen this material and only need a reminder, so rather than putting this material first in the book and taking the time to cover it carefully, we include it as an appendix. By including sections on conditional probability and Bayes's theorem, we actually cover somewhat more probability than we need in the earlier chapters, but it makes the presentation on probability more complete.

A.1 A Review of Sets

A *set* S is any well-defined collection of objects. We take "object" as an undefined term. The objects are called *elements* of the set S. These elements can be physical objects such as the set of books in the Harper Vally Library, the set of elephants in the Happy Family Indoor Circus, or the set of people in a particular winning coalition of a voting system. Or they can be abstract constructions or ideas, such as the set of letters in the English alphabet, the set of natural numbers, the set of points in a circle in a plane, or the set of possible strategies in a game. The elements of a set could even be sets themselves, such as the set of circles in the plane or the set of all winning coalitions in a legislative voting system.

That a set S is *well-defined* is important. This means that given any potential object a, one can determine (in a finite number of steps) without ambiguity whether or not a belongs to S as an element. Thus, if we say "the set of all vowels in the English alphabet," that would be incorrect usage of the word "set," because the letter y is ambiguous – sometimes it acts like a vowel and sometimes like a consonant, so we cannot say whether or not the letter y belongs to this alleged "set." Some mathematicians also would object to a statement such as "the set of all real numbers with the property that every other decimal is an odd number." The reason for the objection is that a real number has an infinite number of decimal places to check, so it would take an infinite number of steps to verify whether or not a given real number belongs to this set, and we stipulated "a finite number of steps" in the definition.

If you are fascinated by questions such as this, you can learn more by taking a course or reading book in the foundations of mathematics or in logic. We'll assume the reader is comfortable with the notion of a set and its elements. Now for some notation and definitions.

Elements of a set are indicated by the symbol $\in$, so $17 \in A$ means 17 is an element of the set A. If a is not an element of the set A, we write $a \notin A$. We typically denote a set by a capital letter and elements of it by lower case letters. In mathematics, sets often arise as solutions to equations, and it is helpful to specify a domain in advance where the solutions might be found. This is termed a *universal set*, and often denoted U. For example, if you are interested in solutions to the equation $3x + 7 = 23$, even before solving the equation, it might be clear that any solution would have to be a rational number, so in your search for a solution, it would not help to look in a set that contains irrational or imaginary numbers, or letters of the alphabet, elephants in a circus, or books in a library (unless maybe the book is about set theory or solving equations?) So for this problem, the domain that is important, the universal set, is the set $\mathbb{Q}$ of rational numbers. If you are solving a problem about spelling an English word or listing letters that are always vowels, the universal set is the set of all letters – that is, the alphabet. If you are solving a problem concerning poker hands, the universal set would be the set of all possible poker hands, etc.

There are three standard ways to describe a set. They are the verbal method, the roster method, and via set-builder notation.

The *verbal* method is easy to understand, but due to the nuances and imprecisions of meanings in language, it may not be a good way to specify something that is supposed to be well defined. That is why we ran into trouble above when we suggested looking at the "set of vowels in the English alphabet." Of course, you can qualify your words to try to remove ambiguity. So one could say, for example, "the set of all letters in the English alphabet that could be vowels," and this is perfectly well defined, since the way it is phrased now clearly includes the letter y, or "the set of all letters in the English alphabet that are always vowels" if we clearly want to exclude the letter y. Still, except for the simplest sets, the verbal method is clumsy and imprecise.

To avoid ambiguities and to be precise as possible, we will usually opt for the roster method or the set-builder notation. The *roster method* is to simply list the elements of the set, enclosed by set braces { } and separated by commas (the order of the list does not matter). This avoids any ambiguity since the reader can simply see the elements. Thus, the following sets are clearly defined and unambiguous:

$$V = \{a, e, i, o, u\}$$
$$X = \{1, 2, 3, 4, 5, 6, 7\}.$$

The problem with the roster method is that some sets are very large, or even infinite, so it would be impractical or impossible to list all the members. For example, the set of all possible five-card poker hands has more than two million members, as we will see later.

The third method, *set-builder notation*, is the best method for most situations. The notation uses a letter, such as x, as a variable or a place-holder to stand for an arbitrary element of your set, and then you specify what rules or conditions that x must satisfy in order to be considered a member of the set. We use a vertical line as a separator to go between the name of the variable and the conditions. For example, the same two sets we defined using the roster method are

$$V = \{x \mid x \text{ is a letter of the English alphabet}$$
$$\text{and } x \text{ is always considered to be a vowel}\}$$
$$X = \{x \mid x \text{ is an integer and } 1 \leq x \leq 7\}.$$

The vertical bar is read as "such that" so this reads as "the set of all x such that ..." with whatever your conditions are filling in the ellipsis. Notice that in both examples, there is a universal set U lurking in the background. For the set V, the universal set is the English alphabet. You only apply the "always a vowel" criterion if you are starting with letters in the alphabet, not to numbers or animals or people. Similarly, for X, the universal set is the set $\mathbb{Z}$ of integers. You only check the inequality if x is an integer. So a variation on the set-builder notation is to specify the universal set to the left of the separating vertical bar. Thus,

$$X = \{x \in \mathbb{Z} | 1 \leq x \leq 7\}$$

has the same meaning as before. Similarly, if U is the English alphabet, then

$$V = \{x \in U | x \text{ is always considered a vowel}\}.$$

In fact, including the universal domain on the left is the proper way to do this, since it specifies where your variable x may "live." The shorter version we introduced first without specifying the universal set is a sloppy shortcut that borders on abuse of notation, for those concerned with the logical foundations of mathematics.

DEFINITION A.1   *a. A set with no elements is called the* empty set *or the* null set, *and is abbreviated by the symbol* {} *(empty set braces) or by the symbol* ∅._

b. If A and B are sets, we say B is a subset *of A if every element of B is also an element of A. We write this as $B \subseteq A$. In this case, we also say A is a* superset *of B.*

c. Two sets are equal, *written $A = B$, if $A \subseteq B$ and $B \subseteq A$. (This is tantamount to saying that A and B have exactly the same elements.)*

Notice that by the definition of a subset, if A is any set, then $A \subseteq A$ and $\varnothing \subseteq A$. These are called *trivial* subsets of A. Nontrivial ones are also called *proper* subsets of A.

Example A.2 Suppose $A = \{1,2,4,8\}$ and $B = \{1,2,3,4,5,6,7,8,9\}$. Then $A \subseteq B$. However, if $C = \{7,8,9,10\}$, then $C \not\subseteq B$ and $B \not\subseteq C$, because $10 \notin B$ and $1 \notin C$, respectively.

Example A.3 List all the subsets of $A = \{x, y, z\}$. We organize the subsets according to the number of elements they contain:

$$\varnothing$$
$$\{x\}, \{y\}, \{z\}$$
$$\{x,y\}, \{x,z\}, \{y,z\}$$
$$\{x,y,z\}.$$

The first and the last are the trivial subsets, and the ones in between are the proper subsets. Sets with one element, like those in the second row, are called *singleton* sets. Since any list of objects is a set, the collection of these eight subsets of A is again a set. It has a name (which will be explained in the next section):

DEFINITION A.4   *Given a set A, the set of all subsets of A is called the* power set *of A, written* $\mathcal{P}(A) = \{B | B \subseteq A\}$.

REMARK A.5   *Listing an element more than once does not change the set. Thus,* $\{1,2,3\} = \{1,2,3,3\}$, *as they are subsets of each other. However, sometimes, it is convenient to be able to list an element more than once. You can artificially do this, for example, by thinking of the elements that are repeated as being distinguishable in some way, such as having different colors or subscripts, etc. When you allow elements to be repeated, this is called a* multiset, *and multisets have their own theory. In this chapter, we mostly stick to ordinary sets (but see Exercise 5b in Section A.3)*

We close this subsection by recalling that there is a pictorial way of visualizing sets called a *Venn diagram.* In a Venn diagram, the universal set U is drawn as a rectangle, and all sets in U are drawn as closed regions inside the rectangle – most of the time they are circles but they don't have to be. Thus, typical Venn diagrams are depicted in Figure A.1.

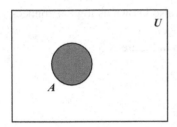
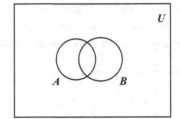

Figure A.1 Venn diagrams.

While sets are a fundamental concept underlying all of mathematics, in this book, they are specifically used later in this chapter in probability theory. They are also are especially useful in the last two sections of Chapter 8, because coalitions in an n-player game or a legislative voting system are subsets of the set N of players, the universal set in problems about such games.

A.1.1   The Algebra of Sets

In this subsection, we assume that all sets considered are subsets of some universal set U. We can create new sets from old using unary or binary operations.

DEFINITION A.6   *Given a set* $A \subseteq U$, *the* complement *of* A *is the set of all elements of* U *that are not in* A. *It is written* $A^c = \{x \in U \,|\, x \notin A\}$.

The Venn diagram illustraing the complement of A is the diagram on the right in Figure A.2.

We mention that other texts may use a different notation for the complement of A, such as $\overline{A}$ or A'.

DEFINITION A.7   *Given* $A, B \subseteq U$, *we can combine them in two distinct ways:*

a.   *The* union *of* A *and* B *is the set of all elements of* U *that belong to* A *or that belong to* B *(or both). It is written as* $A \cup B = \{x \in U \,|\, x \in A \text{ or } x \in B\}$.

b.   *The* intersection *of* A *and* B *is the set of all elements of* U *that belong to both* A *and* B. *It is written as* $A \cap B = \{x \in U \,|\, x \in A \text{ and } x \in B\}$.

c.   *A and B are said to be* disjoint *if* $A \cap B = \varnothing$.

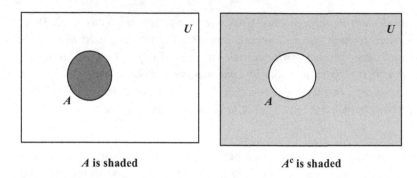

A is shaded A^c is shaded

Figure A.2 The complement of a set A.

Notice that the English words "not," "or," and "and," which are also known as logical connectives, are associated with complement, union, and intersection, respectively.

The Venn diagrams illustrating union and intersection are shown in Figure A.3.

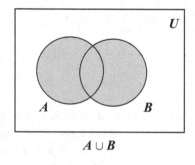

$A \cup B$

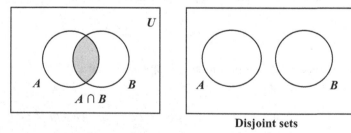

Disjoint sets

Figure A.3 Unions and intersections.

Example A.8 Let $U = \{1,2,3,4,5,6,7,8,9,10,11,12\}$, $A = \{1,2,4,8\}$, $B = \{2,3,4,5,6\}$, $C = \{3,5,9,11,12\}$ Find $A \cup B$, $A \cup C$, $B \cup C$, $A \cup (B \cup C)$, $(A \cup B) \cup C$, $A \cap B$, $A \cap C$, $(A \cup C) \cap B$, A^c, $(A^c)^c$, $A^c \cup B^c$, and $A^c \cap B^c$. Using the definitions, we find

$$A \cup B = \{1,2,3,4,5,6,8\}$$
$$A \cup C = \{1,2,3,4,5,8,9,11,12\}$$
$$B \cup C = \{2,3,4,5,6,9,11,12\}$$

$$A \cup (B \cup C) = \{1,2,3,4,5,6,8,9,11,12\} = (A \cup B) \cup C$$
$$A \cap B = \{2,4\}$$
$$A \cap C = \varnothing \text{ so they are disjoint sets}$$
$$(A \cup C) \cap B = \{2,3,4,5\}$$
$$A^c = \{3,5,6,7,9,10,11,12\}$$
$$B^c = \{1,7,8,9,10,11,12\}$$
$$(A^c)^c = \{1,2,4,8\} = A$$
$$A^c \cup B^c = \{1,3,5,6,7,8,9,10,11,12\}$$
$$A^c \cap B^c = \{7,8,9,10,11,12\}.$$

Notice that $(A^c)^c = A$, and $A \cup (B \cup C) = (A \cup B) \cup C$. These facts, and others like them, are no accident. We collect together several such rules for combining these set operations in the following theorem.

THEOREM A.9　*Let A, B, and C be subsets of U. Then:*

a.　$A \cup A = A$ *and* $A \cap A = A$
b.　$A \cup \varnothing = A$ *and* $A \cap \varnothing = \varnothing$
c.　$A \cup B = B \cup A$
d.　$A \cap B = B \cap A$
e.　$A \cup (B \cup C) = (A \cup B) \cup C$
f.　$A \cap (B \cap C) = (A \cap B) \cap C$
g.　$A \cup (B \cap C) = (A \cup B) \cap (A \cup C)$
h.　$A \cap (B \cup C) = (A \cap B) \cup (A \cap C).$

Formal proofs are covered in books and courses in mathematical logic. We will not give formal proofs here. The reader can, however, convince himself or herself that these are true in a couple of ways that do not constitute formal proofs. One way is to use Venn diagrams. For each of the equalities, draw two separate Venn diagrams – one shading in the set on the left side and one shading in the set on the right. You should end up with the same picture. We illustrate with a pictorial representation of Part g in Figure A.4.

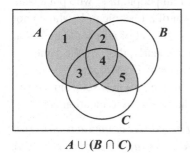

$$A \cup (B \cap C)$$

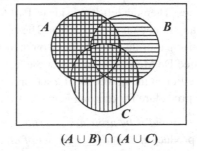

$$(A \cup B) \cap (A \cup C)$$

Figure A.4 Illustrating a distributive law.

On the left, A consists of the regions labeled 1, 2, 3, and 4, and $B \cap C$ consists of the regions labeled 4 and 5. Their union is all five regions (shaded). On the right side, $A \cup B$ is indicated with the horizontal thatched lines, and $A \cup C$ is indicated with vertical thatched lines. Their intersection is all the regions that contain both types of lines, which is exactly the regions 1–5 shaded on the left diagram. We emphasize this is not a rigorous proof, but hopefully it makes the claim clear. The reader is invited to similarly "prove" some of the other parts of the theorem this way in the exercises.

The second way to help see the equality of the sets in the claims is to look at an example with verbal descriptions. For example, if A is the set of all people who have visited Blueberry Hill, B is the set of all people who have visited Cherry Hill Park, and C is the set of all People who have visited Strawberry Fields, then, in words, $A \cup (B \cap C)$ is the set of all people who have either visited Blueberry Hill or who have visited both Cherry Hill Park and Strawberry Fields. On the other hand, $(A \cup B) \cap (A \cup C)$ is the set of all people who have visited either Blueberry Hill or Cherry Hill Park and have visited either Blueberry Hill or Strawberry Fields. (Think about it – if someone in the latter set has not visited Blueberry Hill, then he or she must have visited Cherry Hill Park and Strawberry Fields! So the two sets are the same.) The reader is invited to consider similar examples in the exercises.

Incidentally, parts c and d are termed *commutative laws*, parts e and f are termed *associative laws*, and parts g and h are termed *distributive laws*. They are similar to the laws of the same name in ordinary algebra, but notice the exception that there are two different distributive laws, unlike ordinary algebra.

The next result deals with complements.

THEOREM A.10   *Let A and B be subsets of U. Then*

a. $A \cap A^c = \varnothing$
b. $A \cup A^c = U$
c. $(A^c)^c = A$
d. $\varnothing^c = U$ *and* $U^c = \varnothing$
e. $(A \cup B)^c = A^c \cap B^c$
f. $(A \cap B)^c = A^c \cup B^c$.

We again omit the proofs. Just as with the previous theorem, the reader can look at Venn diagrams or verbal descriptions to get a better sense for why these are true. We illustrate Part f verbally here. Let U be the set of all people who like to eat a peach. Let A be the set of all people in U who put brown sugar on their peaches, and let B be the set of all people in U who put a taste of honey on their peaches. Then the left-hand side of Part f is the set of all people in U who do not put both brown sugar and honey on their peaches. The right side is the set of all people in U that either do not put brown sugar or do not put honey on their peaches. Clearly, they are the same set.

Parts e and f are called De Morgan's laws, after the logician Augustus De Morgan.

We have already mentioned the power set of a set, another way to create a new set from a given set. There is one more procedure for creating new sets from old that we wish to mention here.

DEFINITION A.11   *Let A and B be arbitrary sets (they need not be subsets of the same universal set U). The* Cartesian product *of A and B is the set of all ordered pairs (a, b) with elements of A in the first coordinate and elements of B in the second coordinate. This is usually written*

$$A \times B = \{(a, b) | a \in A \text{ and } b \in B\}.$$

Example A.12 Let $A = \{1,2,3\}$ and $B = \{a,b\}$. Then

$$A \times B = \{(1,a),(2,a),(3,a),(1,b),(2,b),(3,b)\}.$$

Similarly,

$$B \times A = \{(a,1),(b,1),(a,2),(b,2),(a,3),(b,3)\}.$$

Notice that $A \times B \neq B \times A$, even though they have the same number of elements. The ordered pairs in the latter set have letters in the first coordinate, while the ordered pairs of the former set have numbers in the first coordinate.

Of course, you could investigate other questions about the Cartesian product such as does the associative law hold? (It does, suitably interpreted.) Also, how does Cartesian product interact with union, intersection, and complements? We leave these questions for a more advanced treatment of sets and logic.

Cartesian products can be extended to more than two sets. For example, if we have three sets A, B, C, then $A \times B \times C$ is the set of all ordered triples (a,b,c), where $a \in A$, $b \in B$ and $c \in C$. If we have n sets, $A_1, A_2, \ldots, A_n$, then their Cartesian product is the set of all ordered n-tuples:

$$A_1 \times A_2 \times \cdots \times A_n = \{(a_1,a_2,\ldots,a_n)|a_i \in A_i, 1 \leq i \leq n\}.$$

Finally, we mention that if all A_i are the same set A, then we abbreviate the notation for the Cartesian product of n copies of A to A^n. The reader may already be familiar with this abbreviation, as $\mathbb{R}^2$ is a common notation for the Euclidean plane $\mathbb{R} \times \mathbb{R}$, $\mathbb{R}^3$ is Euclidean 3-space, etc. In Chapter 1, we freely used $\mathbb{R}^n$ for the set of n-tuples of real numbers.

Exercises

1. Let $A = \{a,b,c,d,e\}$, $B = \{a,c,e,f,g\}$, and $C = \{b,d,g,h\}$. Suppose $U = \{a,b,c,d,e,f,g,h\}$. Find the following sets:

    a. $A \cup B$ and $A \cap B$
    b. A^c, B^c, and C^c
    c. $(A \cap B) \cup C$
    d. $A^c \cap B^c$ and $A^c \cup B^c$
    e. $A \cap C \cap C$
    f. $A \cap B^c \cap C$

2. Suppose $U = \{1,2,3,4,5,6,7,8,9\}$, $E = \{x \in U | x \text{ is even}\}$, $F = \{x \in U | x \leq 7\}$, and $G = (x \in U | x \leq 2 \text{ or } x \geq 8\}$.

    a. Find $E \cup G$ and $E \cap G$.
    b. Find $E \cap F$ and $F \cap G$.
    c. Find $E^c \cup F^c$.
    d. Find $(E \cup F) \cap G$.
    e. Express G as a union of two sets with simpler definitions.

3.   Suppose $U = \{$all living musicians$\}$, $S = \{$singers in rock and roll bands$\}$, $G = \{$guitar players$\}$, $K = \{$keyboard and piano players$\}$, and $D = \{$drummers$\}$. Give verbal descriptions of the following sets:

    a. $S \cap (K \cup D)$
    b. $S \cap (G \cup K \cup D)$
    c. $(S \cap D) \cup (G \cap K)$
    d. $G \cap K^c \cap D$
    e. $S \cap G^c \cap K^c \cap D^c$

4.   Let $A = \{a, b, c\}$ and $B = \{x, y, z\}$. List the elements of $A \times B$ and $B \times A$.

5.   The King of Hollywood Auto Showroom is selling a new 2021 model four-wheel drive SUV called the California Eagle. The Eagle comes in the following colors: sunrise orange, witchy white, gone green, or peaceful purple. The following options for sound systems are available: Internet radio, CD player, or both.

    a.   Write down all the different possible choices as ordered pairs, with a color in the first coordinate and a sound system choice in the second. Thus, a Cartesian product represents all possible choices.

    b.   Suppose that in addition to a color and a sound system, the buyer may also choose between a six-cylinder engine, an eight-cylinder engine, or a hybrid gas/electric engine. Describe how the different models can now be described as a Cartesian product.

    c.   If the hybrid model does not come in the sunrise orange color, could you still use a Cartesian product to describe all the models?

6.   Let $A = (1, 2, 3, 4)$. Find $\mathcal{P}(A)$, the power set of A.

7.   There are other set operations, but the most important ones are easily expressed in terms of unions, intersections, and complements. For example, the *set difference* is defined as $A - B = \{x \in A | x \notin B\}$.

    a.   Use a Venn diagram to show that $A - B = A \cap B^c$.

    b.   Rewrite $A - (B - C)$ and $(A - B) - C$ in terms of unions, intersections, and complements. Are they the same?

    c.   Let $U = \{1, 2, 3, 4, 5, 6, 7, 8\}$, $A = \{2, 4, 6, 8\}$ and $B = \{1, 2, 3\}$. Find $A - B$ and $B - A$.

8.   The *symmetric difference* of sets is defined as $A \triangle B = (A - B) \cup (B - A)$.

    a.   Use a Venn diagram to show that $A \triangle B = (A \cup B) - (A \cap B)$. (Thus, the symmetric difference is the set theoretic connective that goes with the English "exclusive or.")

    b.   Find $A \triangle B$ for the sets in Exercise 6.

    c.   For any set, explain why $A \triangle A = \emptyset$.

A.2   Enumeration

Let S be a set and let $n(S)$ denote the number of elements in S. We assume S is a *finite* set, which means $n(S)$ is a natural number $n \in \mathbb{N}$. The number of elements of S is also called the *cardinality* of S. Cardinalities of infinite sets can be defined, and they are studied in more advanced books on sets and logic. *Enumeration* means finding $n(S)$ for some set S. However, since most sets are too large to describe by the roster method, one usually cannot just count the elements in S. So we seek techniques for finding $n(S)$ without listing all the elements. These techniques are part of a branch of mathematics called *combinatorics*.

Observe that the empty set is the only set with cardinality 0. We summarize some useful formulas for enumeration.

PROPOSITION A.13   *The cardinality of a union, also known as the addition principle.*

a.   *For arbitrary sets A and B, we have*

$$n(A \cup B) = n(A) + n(B) - n(A \cap B).$$

b.   *Special case: If A and B are disjoint sets, then*

$$n(A \cup B) = n(A) + n(B).$$

We will not give formal proofs. We illustrate Part a with a Venn diagrama as shown in Figure A.3a.

We are trying to count the elements in the shaded region. Adding $n(A) + n(B)$ accounts for the entire region; however, the elements of $A \cap B$ have been counted twice – once in $n(A)$ and once in $n(B)$ – so this sum overcounts the value we seek. Subtracting $n(A \cap B)$ corrects this. Notice Part b follows from Part a since if A and B are disjoint, then $n(A \cap B) = n(\varnothing) = 0$, so we obtain $n(A \cup B) = n(A) + n(B) - 0 = n(A) + n(B)$.

What if there are three or more sets? There is a general formula that extends the preceding proposition, which we hint at in the exercises. However, it may be easier to just work with the Venn diagrams. Here's an example.

Example A.14 McClean's Art School is conducting a survey of its current senior class to determine which art museums in New York City they have visited the past semester. They received 130 responses to the survey, and they learned that 87 of the students visited the Museum of Modern Art (MoMA), 58 students visited the Whitney Museum, and 46 students visited the Guggenheim Museum. Also, 45 students visited both MoMA and the Whitney, 30 students visited both MoMA and the Guggenheim, and 24 students visited both the Whitney and the Guggenheim. Finally, 20 students visited all three museums. Answer the following questions:

a.   How many visited the Whitney and the Guggenheim, but not MoMA?
b.   How many visited only MoMA?
c.   How many visited just one of the three museums?
d.   How many did not visit any of the three museums?

To solve this, let $U = \{$students who replied to the survey$\}$ be the universal set, and let $M = \{$those students who visited MoMA$\}$, $W = \{$ those students who visited the Whitney$\}$, and $G = \{$those students who visited the Guggenheim$\}$. The given information can all be expressed in terms of these sets. Specifically, $n(U) = 130$, $n(M) = 87$, $n(W) = 58$, $n(G) = 46$, $n(M \cap W) = 45$, $n(M \cap G) = 30$, $n(W \cap G) = 24$, and $n(M \cap W \cap G) = 20$.

One might try to answer the questions by expressing the questions in terms of the sets and using formulas. For example, the students in Part a are in the set $W \cap G \cap M^c$, and one can apply Part b of the proposition to the union: $(W \cap G \cap M^c) \cup (W \cap G \cap M) = W \cap G$. Since we know two of these and the sets on the left are disjoint, we obtain $n(W \cap G \cap M^c) + 20 = 24$, whence the answer to Part a is four students. However, this is a tedious approach. With only three sets involved, we can draw the Venn diagram, filling out the information we know by labeling each region with how many students are in it. Start with the the center $M \cap W \cap G$ and work your way outwards. The result is shown in Figure A.5.

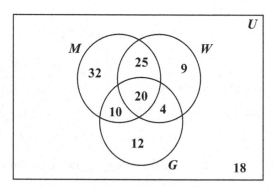

Figure A.5 Enumerating with the aid of a Venn diagram.

For example, we obtain the 10 and the 25 the same way we obtained the 4 – there were 30 students in $M \cap G$, but that included the 20 in the intersection of all three, leaving 10 for the region that represents $M \cap G$ but outside of W. Once those are filled in, we can determine the numbers for the regions that are just in one of the sets and, finally, the number outside of the union of the sets (18). After checking we have determined these correctly, the reader can easily read off the answers to the questions. They are 4 for Part a, 32 for Part b, $32 + 9 + 12 = 53$ for Part c, and 18 or Part d.

Next, we have the following proposition.

PROPOSITION A.15   *The cardinality of a Cartesian product, also known as the multiplication principle.*

a.   *Let A and B be sets. Then*

$$n(A \times B) = n(A) \cdot n(B).$$

b.   *General case: If $A_1, A_2, \ldots, A_k$ are k sets, then*

$$n(A_1 \times A_2 \times \cdots \times A_k) = n(A_1) \cdot n(A_2) \cdots \cdots n(A_k).$$

Again, we won't prove this formally, but we can illustrate it in a way that makes it seem clear if the reader is not already familiar with it. First of all, notice that we have already seen an example – Example A.12 from the last section. We have $n(A) = 3$ and $n(B) = 2$, and we observe that both $A \times B$ and $B \times A$ had six ordered pairs. Or think of it this way: if you have four pockets and six marbles in each pocket, how many marbles do you have? You can obtain the answer 24 by adding the marbles in each pocket, but *because the number of marbles is the same in each pocket*, you can multiply instead $4 \cdot 6 = 24$. This is really a Cartesian product – the first coordinate labels which pocket, and the second labels which marble in that pocket. So counting the marbles is really counting the ordered pairs.

Notice you could not multiply if one of the pockets had a different number of marbles; then you would be forced to add. But that is not a Cartesian product. When counting ordered pairs, there is always the same number of choices for the second coordinate for each choice of the first coordinate. The same idea extends to the general case. If you have four pockets and three bags in each pocket and five marbles in each bag, then you have $4 \cdot 3 \cdot 5 = 60$ marbles, this time using ordered triples instead of ordered pairs. Again, to use the multiplication principle, you need to know the that each pocket has the same number of bags and each bag has the same number of marbles. This idea extends to k-tuples as well.

This is as good a time as any to mention the power set principle.

PROPOSITION A.16 *The Power Set Principle. If A is a set then* $n(\mathcal{P}(A)) = 2^{n(A)}$.

This explains why we call it the power set. Its cardinality is a power of two – with the power being the cardinality of A itself. We also have seen an example of this in the last section. We listed the power set of $A = \{a, b, c\}$, and there were exactly eight subsets, because $2^3 = 8$.

Example A.17 Suppose there is a voting system N (in the sense of Section 8.5) with nine voters. How many different voting coalitions are there?

Since a coalition is just a subset of the set of voters, the question of how many coalitions is the question of how many subsets does N have? But this is exactly the cardinality of the power set of N. So by Proposition A.16, the answer is $2^{n(N)} = 2^9 = 512$ coalitions.

There is a second possible solution. Form an ordered 9-tuple with a coordinate assigned to each voter in N. Given a coalition, we can associate to it one of these 9-tuples as follows. If a voter belongs to the coalition, put a "1" in the corresponding coordinate. If that voter is not a member of the coalition, put a "0" in the corresponding coordinate. Then each coalition corresponds to a different 9-tuple with entries from the set $A = \{0, 1\}$. Thus, the number of coalitions is the same as the cardinality of A^9, which is 2^9 by Part b of Proposition A.15. (This argument is essentially an outline of the proof of Proposition A.16.)

Example A.18 Neil is betting on a the Saratoga Race track, in a race with eight horses. He wants to bet on a *trifecta* ticket, which means you have to correctly name the horse that comes in first place, second place, and third place. Notice that the order matters. For example, if Wildfire is first, Young and Crazy is second, and No-name is third, that is a different trifecta than No-name first, Wildfire second, and Young and Crazy third. How many different trifecta tickets are there?

Since order matters, we can describe each trifecta with an ordered triple. However, the same horse cannot be listed in more than one spot. That means while there are eight choices for the first coordinate, there are only seven for the second coordinate and only six choices for the third coordinate. Thus, by Proposition 10.3, Part b, there are $8 \cdot 7 \cdot 6 = 336$ different trifecta tickets.

While we could solve most problems with just the preceding propositions, we can streamline our work by noticing that problems like the trifecta ticket are very common. In problems like this, you are selecting k things from a set of n, but they must be distinct, and the order of selection matters. These are called permutations of k things from a set of n. There is a shortcut formula for how many such permutations there are, and if you can recognize that a counting problem is really about permutations, the formula will give you the answer quickly.

PROPOSITION A.19   *The number of permutations of k things from a set of n things is given by*

$$P(n,k) = n \cdot (n-1) \cdot (n-2) \cdots (n-k+1) = \frac{n!}{(n-k)!}.$$

The first equality follows immediately from Proposition A.15, Part b. Since we are counting ordered k tuples, we need k factors, with the number of choices decreasing by 1 in each factor, just like the trifecta problem. The second equality is because $n! = n \cdot (n-1) \cdot (n-2) \cdots 3 \cdot 2 \cdot 1$, and $(n-k)! = (n-k) \cdot (n-k-1) \cdots 3 \cdot 2 \cdot 1$, so when dividing the former by the latter, all the factors in the denominator cancel out, leaving just the ones in the proposition.

We remind the reader that by convention, $0! = 1$. With this convention, we have the following immediate corollary.

COROLLARY A.20   *The number of ways of ordering n things is n!*

This follows because the number of ways of ordering n things is $P(n,n) = \frac{n!}{0!} = n!$.

Examples: 1. How many different seven-digit phone numbers are there with no repeated digits?

2. How many different license plates can be made using three letters followed by four digits if none of the letters or digits are repeated?

3. How many ways are there to assign a president, a vice president, and a secretary to a club with 18 members, assuming a person can only do one job?

For the first one, we are simply choosing seven distinct digits from 10, and order matters, so the answer is $P(10,7) = \frac{10!}{3!} = 604,800$.

For the second one, we count the numbers and letters separately. There are $P(26,3)$ ways to choose the letters and $P(10,4)$ ways to choose the digits. What do we do with these numbers? We multiply them, by Part a of Proposition A.15. (Think if a license plate as an ordered pair with all the letters in the first coordinate and all the digits in the second coordinate.) Thus, the answer is $P(26,3) \cdot P(10,4) = \frac{26!}{23!} \cdot \frac{10!}{6!} = 78,624,000$ different license plates.

For the third example, we are selecting three people from a set of 18, and they must be distinct because each person can only do one job. Also, order matters, just like in the trifecta example. Thus the solution is $P(18,3) = \frac{18!}{15!} = 4,896$.

In each case, we could have obtained the answer just using Proposition A.15, Part b. For example, the second one is $26 \cdot 25 \cdot 24 \cdot 10 \cdot 9 \cdot 8 \cdot 7 = 78,624,000$. However, using permutations streamlines the work. We emphasize that we can only use the permutation formula if the objects selected are distinct; in other words, they are selected *without repetition*. The next example illustrates this.

4. How many different seven-digit phone numbers are possible if we allow digits to be repeated? Here, there are 10 choices for each of the seven coordinates, so Proposition A.15 Part b gives the answer as $10^7 = 10,000,000$, but it is not a permutation.

Now consider this following example.

Example A.21 *Pleasant Valley Monastery* has 18 monks. How many different ways are there to select a committee of four?

Notice again that the four people selected must be distinct (otherwise, it would not really be a committee of four). However, unlike the third example, this time, order of selection does not matter. The committee of Mike, Micky, Peter, and Davy is the same committee as Peter, Davy, Mike, and Micky. So this is not a permutation either. A committee is just a subset, so you might think this is related to the the power set instead. Well, it is related, but they are not the same question. The power set lists *all* the subsets of the club. We are only interested in listing subsets of a specific size – four members.

When selecting k things from a set of n, and they must be distinct but the order of selection does not matter, that is called a combination of k things from a set of n instead of a permutation. Combinations, too, have a shortcut formula because they come up so often.

PROPOSITION A.22   *The number of combinations of k things from a set of n things is given by*

$$C(n,k) = \frac{n!}{k!\,(n-k)!}.$$

Accepting the formula for the moment, the answer to Example 10.21 is $C(18,4) = \frac{18!}{4!(14)!} = 3{,}060$. The formula is easy to remember because the two numbers on the bottom sum up to n, and one of them is k. We remark that there are other accepted notations for $C(n,k)$ and $P(n,k)$. You should check whatever book you are using to know what notation is used. Some scientific calculators have built in buttons to compute permutations and combinations.

Example A.23 How many different five-card poker hands are there?

Although some games of poker involve revealing cards in sequence, we'll answer this for five-card stud, which means all five cards are dealt face down and no bets are made until all five cards are dealt. Thus, in making your hand, the order the cards were dealt does not matter. So we are selecting five distinct cards from a deck of 52 cards, and order does not matter, so the answer is

$$C(52,5) = \frac{52!}{5!\,47!} = 2{,}598{,}960.$$

Now, what about a proof of Proposition A.22?

Think of forming a permutation of k things taken from n as a two-step process. First, we must choose the k objects from n, without worrying about the order. Then, we order those k objects. Clearly, every permutation can be obtained in this manner, and the total number of such permutations we know is $P(n,k)$. But according to Part a of the multiplication principle, we can obtain this number as the product of the number of ways of doing each step. The number of ways of doing the first step is $C(n,k)$, and for each selection of k objects, we do have the same number of ways of doing the second step, and that number is $k!$ according to the corollary to the previous proposition. Thus,

$$P(n,k) = C(n,k) \cdot k!$$
$$C(n,k) = \frac{P(n,k)}{k!} = \frac{n!}{k!\,(n-k)!},$$

which proves the formula for combinations.

Our final example of the section is as follows.

Example A.24 The science department at Eight Miles High School has five chemistry teachers and seven biology teachers.

a.   How many ways can a committee of five can be formed?

b.   How many ways can a committee of five can be formed with exactly two chemistry teachers and three biology teachers?

c.   How many ways can a committee of five be formed with at most two chemistry teachers?

The solution of Part a is easy: $C(12,5) = \frac{12!}{5!7!} = 792$ different possible committees of five.

For Part b, we count the chemistry and biology choices separately, much like we did for the license plate example when we counted the letters and digits separately. (Since order does not matter, we use combinations instead of permutations, like we did in the license plate problem.) Since no matter how many ways there are of choosing the two chemistry teachers, for each such choice, we get the same number of choices of biology teachers, Part a of the multiplication principle tells us to multiply the answers. We obtain $C(5,2) \cdot C(7,3) = \frac{5!}{2!3!} \cdot \frac{7!}{3!4!} = 350$ different committees with two chemistry teachers and three biology teachers.

For Part c, the key is to break down the question into disjoint pieces. Note that "at most two" means "exactly two" or "exactly one" or "exactly none." Each one of these three cases is done like Part b. We know there are $C(5,2)C(7,3)$ ways for exactly two chemistry teachers, and similarly, there are $C(5,1)C(7,4)$ ways to select a committee with exactly one chemistry teacher, and $C(5,0)C(7,5)$ ways to select a committee of no chemistry teachers. Since these are connected by the word "or," that is a hint that we are taking a union of the separate cases. Furthermore, it is a disjoint union, because a committee cannot have both two chemistry teachers and one chemistry teacher simultaneously, etc. So according to Part b of the addition principle, we add these answers, obtaining

$$C(5,2)C(7,3) + C(5,1)C(7,4) + C(5,0)C(7,5)$$
$$= \frac{5!}{2!\,3!} \cdot \frac{7!}{3!\,4!} + \frac{5!}{1!\,4!} \cdot \frac{7!}{4!\,3!} + \frac{5!}{0!\,5!} \cdot \frac{7!}{5!\,2!}$$
$$= 350 + 175 + 21 = 546$$

committees with at most two chemists. This concludes our review of enumeration.

Exercises

1.   If $n(A) = 11$, $n(A \cup B) = 15$, and $n(A \cap B) = 4$, find $n(B)$.

2.   In the text, we showed that Part b of Proposition A.15 follows from Part a. Show that Part a follows from Part b, so that the two parts are actually equivalent. [**Hint**: Consider a Venn diagram.]

3.   Caravan Country Club is conducting a survey to determine the magazine subscription habits of their members. They receive a response from 90 golfers. Forty-two of them subscribe to *Paper Sun*, 21 of them subscribe to the *Grey and Pink Golf Digest*, and 30 of them subscribe to *Bemsha Swing Chronicle*. Also 12 subscribe to both *Paper Sun* and the *Golf Digest*, five subscribe to both the *Golf Digest* and *Bemsha Swing Chronicle*, and 15 subscribe to both the *Paper Sun* and *Bemsha Swing Chronicle*. Finally, five subscribe to all three magazines. Answer the following questions:

a.   How many subscribe to just *Paper Sun*?

b.   How many subscribe to exactly one of the three magazines?

    c.   How many subscribe to exactly two of the three magazines?

    d.   How many do not subscribe to any of the three?

4.   a.   Derive a formula for $n(A \cup B \cup C)$ in terms of the cardinalities of the sets A, B, C and the various intersections of these sets. [**Hint:** $A \cup B \cup C = D \cup C$, where $D = A \cup B$.]

    b.   Can you guess how this formula generalizes to four or more sets?

5.   The Night Hawks Diner is running a dinner special. For a single price, Tom can choose one of three appetizers (clam chowder, shrimp cocktail, or oysters), one of five entrees (shrimp scampi, flaming scallops, lobster a la hopper, strip steak, or surf and turf), one of four side dishes (rice, red potatoes, mashed potatoes, or stir-fried veggie mix), and one of three desserts (cheesecake, chocolate cream pie, or apple crisp). If he eats there once every day and orders a different dinner every night, how many days will Tom wait before he must order a duplicate?

6.   The Holistic House of Four Doors Healing Center has a staff of 11 therapists. In how many different ways can the center select four of the staff members to line up for a photograph for an article in the *Sardonicus Twelve Dreams Newsletter*?

7.   A license plate has three letters followed by five digits. The digit 0 is not allowed because it can be confused with the letter O. If the letters cannot be repeated, but the digits can, how many different license plates can be formed?

8.   In Exercise 5, how does the answer change if Tom is allowed to choose two different side dishes out of the four instead of just one?

9.   The Four Doors Healing Center from Exercise 6 must form some staff committees. The therapists include four medical doctors, five psychotherapists, and two acupuncturists.

    a.   How many different committees of five can be formed?

    b.   How many different committees of five can be formed with two doctors, two psychotherapists, and one acupuncturist?

    c.   How many different committees of five can be formed with exactly two doctors?

    d.   How many different committees of five can be formed with one psychotherapist and at least three doctors?

10.   A voting system has 20 members. How many different coalitions are there?

11.   a.   In Exercise 10, how many coalitions consisting of exactly k voters are there if $k = 2, 3, 4, 19, 20$?

    b.   Based on your answer to Part a, can you prove the following identity about combinations?

$$C(n, 0) + C(n, 1) + C(n, 2) + C(n, 3) + \cdots + C(n, n) = 2^n$$

The next exercises are for readers who have already studied Section 8.5. Recall the definition of the Shapley–Shubik power index. In the next few exercises, we will compute the SSI for the voters in the following voting system: Red Lizard Enterprises manufactures mellotrons, pianos, and other keyboard instruments. The company is controlled by the Steering Board, which is a weighted voting system. The weights of the voters are given by the following table, and the quota is set at $q = 12$.

Voter	Weight
Robert	4
Ian	4
Greg	3
Michael	3
Peter	1

12. a.   How many orderings of the voters are possible?

  b.   Consider the voter Robert. Explain why Robert is not pivotal unless he is fourth or fifth in an ordering.

  c.   How many different orderings are there with Robert listed fifth? Why is Robert pivotal in every such ordering?

  d.   If Robert is fourth, consider who is listed fifth. For each of the other voters A, find how many orderings there are with Robert pivotal and A fifth.

  e.   Add up all the orderings where Robert is pivotal, and determine $SSI(Robert)$.

13. a.   Consider the voter Greg. Explain why Greg is not pivotal unless he is fourth or fifth in an ordering.

  b.   Is Greg always pivotal when he is listed fifth? Count how many orderings there are in which Greg is fifth and pivotal.

  c.   Count how many orderings there are when Greg is fourth and pivotal. (The count depends on who is fifth.)

  d.   What is $SSI(Greg)$?

14. a.   Consider the voter Peter. Explain why Peter can only be pivotal if he is listed fourth.

  b.   Count all the orderings where Peter is pivotal.

  c.   Find $SSI(Peter)$, and show that, despite having only one vote, he has equal power with Greg and Michael.

15. The following example is taken from Taylor and Pacelli (2008). In 1958, the Treaty of Rome established the European Economic Community, a body that promoted common economic interests to its member countries. Later, this body was modified several times, until the European Union was established. However, back in 1958, there were only six member countries. Each country had a weight given by Table A.1, and the quota was set at $q = 12$.

Table A.1 The member countries of the 1958 Treaty of Rome and their weighted votes.

Country	Weight
France	4
Germany	4
Italy	4
Belgium	2
The Netherlands	2
Luxembourg	1

a. Explain why Luxembourg is never pivotal, and conclude it is a dummy voter.

b. Using arguments similar to the previous three exercises, determine the Shapley–Shubik index of each voter. [**Remark**: As an alternative to determining which place a voter is located in an ordering, you could argue by considering how many votes there are preceding the voter before that voter joins the coalition. That is how it is done in Taylor and Pacelli (2008). For example, for Belgium to be pivotal, there must be exactly 10 or 11 votes preceding Belgium. Count each case separately by breaking it into subcases depending on which countries precede it.]

A.3 Experiments, Sample Spaces, and Probability Models

The remainder of this chapter is concerned with probability theory.

A.3.1 Experiments and Sample Spaces

DEFINITION A.25 *An* experiment *is any activity with an observable outcome.*

DEFINITION A.26 *The* sample space *S of an experiment is the set of all possible outcomes.*

So a sample space plays the role of the universal set for a probability experiment

DEFINITION A.27 *An* event *is any subset of S. Singleton events (with just one outcome) are called* simple events.

For example, an experiment could be tossing a coin and recording whether it lands heads up H or tails T. Assuming these are the only possibilities, the sample space is $S = \{H, T\}$. Or the experiment could be rolling a six-sided die and recording the number on top. Then $S = \{1, 2, 3, 4, 5, 6\}$. The experiment could be a census worker knocking at the door of home to determine how many people live at an address. Then the sample space is some subset of the natural numbers.

Suppose the experiment is to toss a coin and roll a die and record both outcomes. Then the outcomes contain two bits of information, so they can be written as ordered pairs. The sample space is then the Cartesian product of the two individual sample spaces

$$S = \{(H, 1), (H, 2), \ldots, (H, 6), (T, 1), (T, 2), \ldots, (T, 6)\}.$$

Suppose the experiment is to roll a pair of dice and record the sum of the two numbers that come up. One student suggests this as the sample space: $S = \{2, 3, 4, 5, 6, 7, 8, 9, 10, 11, 12\}$, the set of all possible sums. Another student suggests the following set of ordered pairs should be used. (Think of the dice being distinguishable, such as being different colors. Then the first coordinate is the result of the red die, say, and the second the result of the white one.) The complete sample space is again a Cartesian product:

$$\{(1,1),(2,1),(3,1),(4,1),(5,1),(6,1),$$
$$(1,2),(2,2),(3,2),(4,2),(5,2),(6,2),$$
$$(1,3),(2,3),(3,3),(4,3),(5,3),(6,3),$$
$$(1,4),(2,4),(3,4),(4,4),(5,4),(6,4),$$
$$(1,5),(2,5),(3,5),(4,5),(5,5),(6,5),$$
$$(1,6),(2,6),(3,6),(4,6),(5,6),(6,6)\}$$

Which student is right? Well, they both are. Since our end goal is to record the sums of the two numbers, we don't really need to know what the individual numbers are. However, there are a couple of reasons to prefer the second way of writing the sample space. One reason is that it contains more information than the first way, and more information is usually a good thing. However, a second reason for preferring the second approach is more compelling. Namely, if the dice are fair, then each of the 36 outcomes in the second approach are equally likely to appear, whereas the sums in the first set are not equally likely.

Sample spaces where each outcome is equally likely are particularly easy to work with, as we'll see in the paragraphs that follow. They are called *uniform* sample spaces.

Consider the last experiment where two dice are rolled. Examples of events include $A = \{(1,1),(1,2),(2,1)\}$. In words, A is the event that the sum is at most 3. Remember, any subset of S is an event. Suppose B is the event that the sum is 9. Then, listing the outcomes in B, we have $B = \{(3,6),(4,5),(5,4),(6,3)\}$. If C is the event that the red die shows at least a 5 and the white die shows at most a 4, then $C = \{(5,1),(6,1),(5,2),(6,2),(5,3),(6,3),(5,4),(6,4)\}$. If D is the event of obtaining "doubles," then $D = \{(1,1),(2,2),(3,3),(4,4),(5,5),(6,6)\}$.

Since events are sets, they can be combined in the ways that sets are combined. Thus, we can speak of the complement of an event: the complement A^c in words is the event that the sum is at least 4 and consists of 33 different outcomes. We can also form the union or intersection of two events. Thus, $A \cup D$ means the sum is at most 3 or doubles are the outcome. $A \cap D$ means the sum is at most 3, and the outcome is doubles and is a simple event $A \cap D = \{(1,1)\}$. Observe that $A \cap B = \varnothing$. In words, the intersection says the sum is at most 3 and the sum is 9. Sounds impossible, right? That is why we make Part a of the following:

DEFINITION A.28   *a. The event $\varnothing$ is called the* impossible event.
   b. The event S (the entire sample space) is called the inevitable event.
   c. If $A \cap B = \varnothing$ (disjoint sets), we say that A and B are mutually exclusive events.

A.3.2   Probability Models

In a probability model, one tries to assign a numerical measure of how likely it is for an event to happen. We take this numerical measure to be a nonnegative fraction that is less than or equal to 1. The idea is that if E is an event, the closer $P(E)$ (the probability of E occurring) is to 1, the more likely it is to happen, and the closer it is to 0, the less likely it is to happen. We can formalize this with axioms as follows.

DEFINITION A.29   *Let S be a sample space. Assign to each event $E \subseteq S$ a real number $p(E)$ called the* probability *of E so that the following properties are satisfied:*

1. $0 \leq p(E) \leq 1$ *for any event E*
2. $p(\varnothing) = 0$
3. $p(S) = 1$
4. $p(E \cup F) = p(E) + p(F) - p(E \cap F)$.

Any such assignment is called a *probability model*. Other common terms are *probability function* and *probability distribution*. Before we give examples, we draw some consequences from the axioms. Although the definition applies to arbitrary sets S, in this book, we assume S is a finite sample space.

PROPOSITION A.30   *In a finite sample space S, property 4 of the definition is equivalent to the following alternative:*

Alternate 4. If E is any event and $E = \{x_1, x_2, \ldots, x_n\}$ so that the x_i's are the outcomes in E, then $p(E) = p(x_1) + p(x_2) + \cdots + p(x_n)$.

Proof   Note that $p(x_i)$ is an abbreviation of $p(\{x_i\})$ (recall that p is defined on subsets of S.) Assume property 4. We prove the alternate version by induction on n. Note that a special case of property 4 is that in case E and F are mutually exclusive events, then $p(E \cup F) = p(E) + p(F)$, because $p(E \cap F) = p(\varnothing) = 0$ by axiom 2. Now if $n = 2$, we have $E = \{x_1, x_2\} = \{x_1\} \cup \{x_2\}$, and the two sets are disjoint, so by the special case of property 4, we have $p(E) = p(x_1) + p(x_2)$, proving the alternate property 4 for $n = 2$. Assume the alternate property 4 holds for n, and we want to prove it for $n + 1$. Here, we have $E = \{x_1, x_2, \ldots, x_n, x_{n+1}\} = \{x_1, x_2, \ldots, x_n\} \cup \{x_{n+1}\}$, and again, these sets are disjoint. Therefore, by the special case of property 4, $p(E) = p(\{x_1, x_2, \ldots, x_n\}) + p(x_{n+1})$. But now, by the inductive hypothesis, the first part of this can be written as a sum of the individual probabilities, so we obtain $p(E) = p(x_1) + p(x_2) + \cdots + p(x_n) + p(x_{n+1})$, proving the alternate property 4 for all n.

Conversely, suppose the alternate property 4 holds. Then $P(E)$ is the sum of the $p(x)$ for $x \in E$, and $p(F)$ is the sum of the $p(x)$ for $x \in F$. Therefore, in the sum $p(E) + p(F)$, the $p(x)$ for $x \in E \cap F$ have all been added twice. Subtracting those which are listed twice will leave every $x \in E \cup F$ counted once, which is $p(E \cup F)$, and what has been subtracted is $p(E \cap F)$ by the alternate property 4. Thus, we end up with $p(E \cup F) = p(E) + p(F) - p(E \cap F)$, which is property 4.   ∎

PROPOSITION A.31   *In any probability model, we have*

1. $E \subseteq F$ *implies* $p(E) \leq p(F)$.
2. $p(E^c) = 1 - p(E)$.

Proof   Any event can be written as $F = F \cap S$, and since $S = E \cup E^c$ for any event, we have $F = F \cap (E \cup E^c)$. By the distributive law (Theorem A.9), we have $F = (F \cap E) \cup (F \cap E^c)$, and note that this is a disjoint union. Now assume $E \subseteq F$ so that $F \cap E = E$. Thus, $F = E \cup (F \cap E^c)$. By property 4, and because this is a disjoint union, we have $p(F) = p(E) + p(F \cap E^c)$. But axiom 1 implies $p(F \cap E^c) \geq 0$, so it follows that $p(F) \geq p(E)$. This proves part 1 of the preceding proposition.

For part 2 of the preceding proposition, note that $S = E \cup E^c$, and this is a disjoint union. So property 4 yields $p(S) = p(E) + p(E^c)$. By property 3, $p(S) = 1$, so solving for $p(E^c)$, we obtain $p(E^c) = p(S) - p(E) = 1 - p(E)$, as desired.   ∎

In light of Proposition A.3, it is very easy to give examples of probability models. All we need to do is specify a fraction for the probability of each simple event (outcome) and make sure these fractions sum to 1.

Example A.32 A fair die. The following table of probabilities sums to 1. Notice that each outcome is equally likely (because the die is fair).

Outcome	Probability
1	$\frac{1}{6}$
2	$\frac{1}{6}$
3	$\frac{1}{6}$
4	$\frac{1}{6}$
5	$\frac{1}{6}$
6	$\frac{1}{6}$

Each outcome is assigned a probability which is a fraction between 0 and 1 so property 1 holds. Property 2 holds by assumption. Property 3 holds because the probabilities all sum to 1. And property 4 holds because we can use the alternate property 4 to find $p(E)$ for any E, and it will be a sum of some or all the probabilities, which is ≤ 1 for all E. Thus, the table gives a probability model (which happens to be uniform).

Example A.33 A loaded die. This example is not a uniform space.

Outcome	Probability
1	$\frac{1}{10}$
2	$\frac{1}{10}$
3	$\frac{1}{10}$
4	$\frac{2}{10}$
5	$\frac{2}{10}$
6	$\frac{3}{10}$

Again, all the axioms are satisfied (with property 3 holding because the sum of the entries in the table is 1). So, again, this is a valid probability model. It is not, however, a uniform model because not every outcome is equally likely.

In the example of throwing a pair of dice, if the dice are fair, then each of the 36 ordered pair outcomes listed in the sample space in the last subsection is equally likely, but their probabilities sum to 1, so each outcome has a probability of $\frac{1}{36}$, and the model is uniform.

We conclude this section with examples of using probability models to compute probabilities of events.

In the Example A.33 of the unfair die, find the probabilities of the following events: A = the outcome is even, B = the outcome is at most 2, C = the outcome is between 2 and 5, inclusive. Also find $p(A \cap B)$, $p(A \cup B)$, and $p(B^c)$.

Solution: $A = \{2, 4, 6\}$, $B = \{1, 2\}$, $C = \{2, 3, 4, 5\}$. Thus, $p(A) = p(2) + p(4) + p(6) = \frac{1}{10} + \frac{2}{10} + \frac{3}{10} = \frac{3}{5}$ by the alternate property 4. Similarly, $p(B) = p(1) + p(2) = \frac{1}{5}$, and $p(C) = p(2) + p(3) + p(4) + p(5) = \frac{3}{5}$. We could calculate the other three in the same way, by listing the elements of the event and adding their probabilities. For example, $A \cap B = \{2\}$, so $p(A \cap B) = \frac{1}{10}$. But we can use property 4 instead to obtain $p(A \cup B) = p(A) + p(B) - p(A \cap B) = \frac{3}{5} + \frac{1}{5} - \frac{1}{10} = \frac{7}{10}$. Finally, Proposition A.15 gives $p(B^c) = 1 - p(B) = 1 - \frac{1}{5} = \frac{4}{5}$.

In the example of throwing a pair of fair dice, find the probabilities of the events A, B, C, D from the last subsection.

Solution: A is the event that the sum is at most 3, with simple outcomes $(1, 1)$, $(1, 2)$, $(2, 1)$. Since the dice are fair, each ordered pair has probability $\frac{1}{36}$. Therefore, $p(A) = \frac{1}{36} + \frac{1}{36} + \frac{1}{36} = \frac{3}{36} = \frac{1}{12}$. Similarly, $p(B) = \frac{1}{36} + \frac{1}{36} + \frac{1}{36} + \frac{1}{36} = \frac{4}{36} = \frac{1}{9}$, corresponding to the four outcomes for which the sum is 9. There seems to be a pattern here. Since each outcome is equally likely, it appears that for any event E, we have $p(E) = \frac{n(E)}{36}$. We'll discuss this formula and its consequences in the next section, but it is correct in this example. Thus, $p(C) = \frac{8}{36} = \frac{2}{9}$ because $n(C) = 8$, and $p(D) = \frac{6}{36} = \frac{1}{6}$.

Exercises

1. Suppose a sample space $S = \{w, x, y, z\}$. How many different events are there? List them all.

2. A penny, a nickel, and a dime are flipped, and it is noted whether the coins show heads H or tails T. Show that $\{H, T\}^3$ is the sample space (Cartesian product). List all the elements.

3. In the example of Exercise 2, list the outcomes in the event "The penny shows H." Also list the outcomes in the event "There are at least two coins showing T."

4. A jar contains lemon candies, mint candies, and chocolate candies. A candy is drawn at random, and its flavor is noted. Write down the sample space, and also write down the outcomes in the event "the candy drawn is not lemon."

5. a. In the candy jar of Exercise 4, suppose the experiment is to draw two candies in order. Now list all the outcomes in the sample space.

   b. This part is an example where multisets are useful instead of sets. Same question as Part a, except we do not draw the candies in order – we just pull out two candies at once.

6. Suppose $S = \{a, b, c, d, e\}$, and it is known that $p(a) = p(b) = p(c)$, but d is twice as likely as c, and e is as likely as $\{c, d\}$. Determine the probability model. [**Hint**: You are being asked to find the probabilities of each of the five possible outcomes. Let $p(a) = x$. Write all the probabilities in terms of x, them solve for x by using axiom 3.]

7. Suppose S is a sample space and E and F are events. Suppose we know $p(E \cup F) = 0.8$, $p(E) = 0.5$, and $p(F^c) = 0.6$. Find $p(E \cap F)$.

8. Suppose S is a sample space, and E and F are evens. Suppose we know $P(E) = 0.15$, $p(F) = 0.65$, and $p((E \cup F)^c) = 0.4$. Find $p(E \cap F)$.

9.   In the example of the loaded die given in the text, find p(outcome is at most 5) and p(outcome is 1 or 6).

10. a.   In the example of rolling two fair dice, find the probabilities of the following events: A = the sum is at most 4, B = {the numbers on the two dice differ by 1}, C = {the number showing on the white die is at least a 3}.

     b.   Also find $p(A \cap B)$, $p(A \cup B)$, $p(B \cup C)$, and $p(A \cap B \cap C)$.

A.4   Uniform Sample Spaces

In this section, we assume that S is a finite, uniform sample space. Thus, $S = \{x_1, x_2, \ldots, x_n\}$. Now let $p(x_1) = p$, and since the space is uniform, $p(x_i) = p$ for each i, $1 \le i \le n$. Applying the alternate property 4 together with property 3, we obtain $p + p + \cdots + p = 1$, or $np = 1$. Thus, $p = \frac{1}{n}$. Now, if $E = \{y_1, y_2, \ldots, y_r\}$ is any event, where the y_i are some of the x_j, then, again, the alternate property 4, gives $p(E) = p(y_1) + p(y_2) + \cdots + p(y_r)$, and since each $p(y_i) = \frac{1}{n}$, we obtain $p(E) = \frac{1}{n} + \frac{1}{n} + \cdots + \frac{1}{n}$ (r times). In other words, $p(E) = r\left(\frac{1}{n}\right) = \frac{r}{n}$. But notice that $n = n(S)$ and $r = n(E)$. We have proved the following.

THEOREM A.34   *In a finite, uniform sample space, if E is any event, then*

$$p(E) = \frac{n(E)}{n(S)}. \tag{A.1}$$

This formula makes computing probabilities easy for uniform spaces, since it reduces the calculation to an enumeration problem. Many students are familiar with this formula and know it by "the probability of E is the number of successes divided by the number of possibilities." Just remember that it works only for uniform spaces.

Example A.35 The battery for my wireless mouse just died. I reached in a box that contains 11 good batteries and 4 dead batteries, which look just like the good ones, and pick out a battery. What is the probability that I chose a dead battery?

The point of saying that the dead batteries look just like the good ones is to suggest that it is equally likely that any of the batteries will be chosen, so it is a uniform space. Since $n(S) = 15$ (the total number of batteries in the box) and $n(E) = 4$ (the four dead batteries), the probability I chose a dead battery is $\frac{4}{15}$ by Theorem A.9.

Example A.36 Suppose instead of one battery, I chose three, which I need for my TV remote. What is the probability I chose three good ones? What is the probability I chose exactly two good ones? The probability I chose at least one good one?

Even though it is the same box of 11 good batteries and four dead ones, the sample space for this problem is different. Since I'm choosing three batteries out of the box of 15, and order apparently does not matter, then $n(S) = C(15, 3)$, the number of different ways I can choose three from a set of 15.

In the first question, I want all three of them to come from the 11 good ones, so there are $n(E) = C(11, 3)$ ways to do that. Thus, the probability of getting three good batteries is, according to Theorem A.9, $\frac{n(E)}{n(S)} = \frac{C(11,3)}{C(15,3)}$. Numerically, this reduces to $\frac{11!}{8!3!} \cdot \frac{3!12!}{15!} = \frac{33}{91} \approx 0.36264$.

In the second question, the sample space is the same, but in the numerator, I'm choosing two from 11 good ones and one from four dead ones. By our counting techniques, we know the number of ways to do both is to multiply the two combinations. Thus, the probability of exactly two good batteries is $\frac{C(11,2)C(4,1)}{C(15,3)}$. This reduces to $\frac{55 \cdot 4}{455} = \frac{44}{91} \approx 0.48352$.

Finally, the probability of at least one good battery can be handled in two ways. One way is to break up the event "at least one good battery" into a disjoint union of cases, just like we did for the enumeration problems. Thus, "at least one" is the same as "exactly one" or "exactly two" or "exactly three." Each one of these is computed as in the second question, then we add the answers since this is a union of disjoint events. We obtain

$$\frac{C(11, 1)C(4, 2)}{C(15, 3)} + \frac{C(11, 2)C(4, 1)}{C(15, 3)} + \frac{C(11, 3)C(4, 0)}{C(15, 3)}$$
$$= \frac{66 + 220 + 165}{455} = \frac{451}{455} \approx 0.9912.$$

The second approach is easier. Note that the event "at least one good battery" is the complement of the event "no good batteries." The probability of no good batteries (so all three are dead) is $\frac{C(11,0)C(4,3)}{C(15,3)} = \frac{4}{55}$. Thus, the probability of at least one good battery is $1 - p$(three dead batteries) $= 1 - \frac{4}{455} = \frac{451}{455}$.

Example A.37 A seven-digit phone number is selected at random. What is the probability that there is at least one repeated digit?

This one again is simplified by considering the complementary event, which is no repeated digits, or seven distinct digits. Since the number is selected "at random," each phone number is equally likely, so we can use Eq. (A.1). However, with a phone number, order matters, so we will not be using combinations to count the sets. For the denominator, there are 10 choices for each of the seven digits, so there is a total of 10^7 different such phone numbers. For the numerator, we are looking to count the ways to select seven distinct digits out of 10, when order matters, so the answer is a permutation $P(10, 7)$. Thus, the probability of no repeated digits is $\frac{P(10,7)}{10^7}$, and the probability of at least one repeated digit is $1 - \frac{P(10,7)}{10^7} = 1 - 0.06048 \approx 0.93952$, almost a 94% chance!

Example A.38 A fair coin is flipped five times. What is the probability of obtaining exactly three heads H?

The sample space consists of all ordered 5-tuples of H or T; that is, it is the Cartesian product $\{H, T\}^5$. Thus, $n(S) = 2^5 = 32$. Since the coin is fair, all 32 outcomes are equally likely, so it is a uniform space. If E is the event of exactly three heads, it consists of all ordered 5-tuples with H in exactly three of the coordinates. But which coordinates? Well, there are five possible coordinates, and we select three of them for H. That completely determines the outcome because the remaining two coordinates must be T, so there are no more selections to make. Note that, although the coins are flipped in order (that's how we got ordered tuples in S), the order in which we select the coordinates does not matter (think about that!). Thus, $n(E) = C(5,3) = \frac{5!}{3!2!} = 10$. So the probability is $\frac{C(5,3)}{2^5} = \frac{10}{32} = \frac{5}{16}$.

Example A.39 In five-card stud poker, we know there are C(52 ,5) = 2,598,960 different possible poker hands. If the deck is well shuffled, find the probabilities of the following hands: a. four of a kind (four cards the same rank), and b. a flush (five cards the same suit but not all five in sequence).

Since the deck is "well shuffled," each hand is equally likely, so this is a uniform space, with $n(S) = C(52,5)$. So by Eq. (A.1), all we have to do is count the number of different hands for the numerator. In Part a, we have four cards of the same rank (there are 13 ranks – ace through king). So we must first choose a rank, and there are $C(13,1) = 13$ ways to do that. Once the rank is chosen, we must select all four of those cards, and there is only $C(4,4) = 1$ way to do that. But then we must also choose a fifth card from among the 48 remaining cards, and there is $C(48,1) = 48$ ways to do that. Thus, the probability of four of a kind is

$$\frac{13 \cdot 1 \cdot 48}{C(52,5)} = \frac{624}{2,598,960} \approx 0.00024.$$

For a flush, we must first choose a suit, and there are $C(4,1) = 4$ ways to do that. Once a suit is chosen, we must select five of the 13 cards in that suit, and there are $C(13,5)$ ways to do that. However, that includes all the straight flushes and royal flush, which are considered a different hands in poker. So we must subtract all these flushes where the cards are in sequence. Notice that one the bottom card is specified, the entire straight flush is completely determined. For example, if you tell me 8, I know the straight flush is 8, 9, 10, jack, queen. So to count all the straight flushes, we just have to count all the different possible bottom cards. For a royal flush, the bottom card is a 10, and for all the other straight flushes, the bottom card is ace–9, so altogether, there are ace–10 or 10 straight (or royal) flushes in each suit. Thus, the total number of flushes that are not in sequence is $4\{C(13,5) - 10) = 4(1,287 - 10) = 4(1,277) = 5,108$. Thus, the probability of a flush is

$$\frac{5,108}{2,598,960} \approx 0.0019654.$$

Thus, about one in every 4,000 hands is four of a kind, while about 1 in 500 is a flush. That's why four of a kind is considered a stronger hand - it's rarer. This concludes the section.

Exercises

1.   A fair coin is flipped 10 times. a. What is the probability of obtaining exactly six heads?

   b.   What is the probability of obtaining two or three heads?
   c.   What is the probability of obtaining at least two heads?

2.   Steven has a model of the solar system with a lightbulb sun. The bulb has burned out, so he reaches into a box of eight bulbs for a replacement. He remembers that three of the bulbs in the box are defective, so he draws two bulbs out at random, just in case. What is the probability that light is restored to his model solar system?

3. A jar contains 20 marbles, of equal size and weight, of which 10 are white, seven are green, and three are red. Without looking at the jar, four marbles are drawn at random.

    a. What is the probability they are all white?
    b. What is the probability that one is white, one is green, and two are red?
    c. What is the probability that at least one green but no red marbles are drawn?
    d. What is the probability that all four marbles are one color?
    e. What is the probability that all three colors are obtained?

4.   a. A full house in a game of poker is three of a kind plus a pair (of different rank), such as three queens and a pair of 7s, or three 4s and a pair of aces. In five-card stud, where just five cards are dealt face down at random from a well-shuffled deck, what is the probability of obtaining a full house?
    b. What is the probability of getting a hand that has two pairs? This means two pairs of different rank and a fifth card that is a different rank from both of the pairs, like a pair of 3s, a pair of jacks, and a 9.
    c. What is the probability of getting one pair?

5. A party has 30 attendees. What is the probability that at least two people at the party have the same birthday? Assume 365 days in a year, and each day is equally likely as a birthday. [**Hint**: See Example A.12. This is the well-known "birthday problem."]

A.5 Conditional Probability and Independent Events

Sometimes knowing that one event occurs affects the probability of another event. For example, suppose a survey of the households Putnam County revealed that the probability that the household annual income is more than $60,000 per year is 0.5. But what if you were told that in the household selected, the head of the household has a college degree? Would that suggest that the probability should be higher than 0.5? What if you were told that the head of that household was a doctor?

For another example, suppose the Major League Baseball World Series this year is between the Saratoga Sluggers and the Fargo Fielders, and the sports analytics experts suggest that the Fielders should win the series with a probability of 0.7. But then you were told that the Fielders lost the first three games in the series. Given that extra information, would you think the probability of .7 is too high and should be adjusted?

Conditional probability deals with questions like these. You are trying to find the probability of one event E, given that you know another event F has occurred. The symbol for conditional probability is $p(E|F)$, and is read as "the probability of E, given F has occurred," or more simply, "the probability of E, given F." We will give a precise definition later in this subsection.

In order to motivate the definition, we consider an example with a uniform space S.

Example A.40 Suppose Phil keeps a box of music CDs by his door so he can take one to play in his car when he leaves home. There are 100 CDs in his musical box, of which 30 are on the record label Charisma, 6 are by the band Genesis, including 5 on the Charisma label. One day Phil grabs a CD at random as he leaves to go shop around. What is the probability that it is a CD on the

Charisma label? What if he noticed that it was a Genesis CD – given that, what is the probability that it is on the Charisma label?

Let E be the event the CD is a Charisma CD, and F the event that it is a Genesis CD. The first question is to compute $P(E)$. Since it is a uniform space, we have $p(E) = \frac{n(E)}{n(S)} = \frac{30}{100} = 0.30$. In the second question, we are asked to find $p(E|F)$. The key to understanding conditional probability is to observe that once you are told that the event F has occurred, then the sample space has changed from all of S to just F. Since we already know F has occurred, to ask that E also occurs means that you are seeking for the probability of $E \cap F$. Thus, while the sample space has changed from S to F, at the same time, the event you consider a "success" has changed also – from E to $E \cap F$. In this example, that means once you know it's a Genesis CD, there are only six of those, and you know that five of those are on the Charisma label, so $p(E|F) = \frac{n(E \cap F)}{n(F)} = \frac{5}{6} \approx 0.83333$. So in this example, knowing that F occurred definitely affects the probability of E. The Venn diagram in Figure A.6 illustrates the general situation.

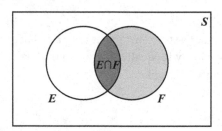

Figure A.6　Finding $p(E|F)$ – F becomes the sample space.

Thus, in the case of a uniform sample space, $p(E|F)$ becomes the ratio of $n(E \cap F)$ to the ratio of $n(F)$. But note that it is also the ratio of $p(E \cap F)$ to $p(F)$:

$$p(E|F) = \frac{n(E \cap F)}{n(F)} = \frac{\frac{n(E \cap F)}{n(S)}}{\frac{n(F)}{n(S)}} = \frac{p(E \cap F)}{p(F)}$$

because the $n(S)$ terms cancel. Now, if the space is not uniform, then the formulas involving $n(E \cap F)$ and $n(F)$ are useless, because Eq. (A.1) is not valid in that case. However, the expression on the right - the ratios of the probabilities, still makes sense, so we take that as our formal definition of conditional probability:

DEFINITION A.41　*In any sample space S, if E and F are events (with F nonempty), then the conditional probability of E, given that F has occurred is*

$$p(E|F) = \frac{p(E \cap F)}{p(F)}. \tag{A.2}$$

The requirement that F be nonempty is simply to avoid 0 in the denominator. It would not make sense to say the probability of E, given that F occurred if F is empty anyway, since in that case, F is the impossible event and cannot occur. Notice that $p(E|F)$ and $p(F|E)$ are two different probabilities. In one case, you know F has occurred, and you want the probability of E;

and in the other, you know that E has occurred, and you want the probability of F. In our example, $p(F|E)$ represents the probability that the randomly chosen CD is a Genesis CD, given that it is on the Charisma label. In this case, $p(F|E) = \frac{p(F \cap E)}{p(E)} = \frac{n(F \cap E)}{n(E)} = \frac{5}{30} \approx 0.16667$. In this case, the sample space is E, not F, so you should not expect the answer to be the same as $p(E|F)$. However, there is a relationship between the two conditional probabilities, which we explore in the next section.

We observe now that this definition also gives us a formula for the probability of an intersection of two events, which holds in general (if the reader has already studied Chapter 6, he or she may recall a formula there that only worked in a certain special case, which we clarify in a moment.) The general formula is as follows.

COROLLARY A.42 *Multiplication Rule. Let S be a sample space, and let E and F be two arbitrary events (with F nonempty). Then*

$$p(E \cap F) = p(F) \cdot p(E|F).$$

This follows immediately from the definition. Note if E is also nonempty, we could also write $p(E \cap F) = p(E) \cdot p(F|E)$. In words, the probability that both events occur is the probability of one of them (either one) times the probability that the second occurs, given that the first one occurred. This corollary makes it very easy to analyze certain probability problems that are illustrated by tree diagrams.

Example A.43 An urn has 10 balls in it – 5 red, 4 blue, and 1 green. Ruth selects two balls at random (without replacement) in sequence and the colors are noted. Find the probabilities of the following events:

a. p(first ball is red), p(first ball is blue), p(first ball is green).
b. p(second ball is red | first ball is red), p(second ball is green | first ball is blue), p(second ball is green | first ball is green)
c. p(first ball is red and second ball is blue), p(first ball is green and second ball is blue)
d. p(second ball is blue).

The tree diagram in Figure A.7 shows the color possibilities, and each branch is labeled with the appropriate probability.

Indeed, the initial branches corresponding to the first draw are labeled with the three answers needed for question a. p(first ball is red) $= \frac{5}{10}$ because there are 10 balls and five of them are red, using Eq. (A.1). Similarly, p(first ball is blue) $= \frac{4}{10}$ and p(first ball is green) $= \frac{1}{10}$. The probabilities on all of the second branches are conditional probabilities, because you are given what color was drawn on the first draw! Thus, p(the second draw is red | first draw is red) is $\frac{4}{9}$ because with one red ball missing from the first draw, there are nine balls left, and four of them are red. This is exactly the probability labeled on the first branch of the second draw, where R is on the first draw and where R is on the second. Similarly, the answers to question b can all be read off from the labels on the second-draw branches. We have p(second ball is green | first is blue) is the label on the branch corresponding to drawing B first then G second, which is $\frac{1}{9}$, and p(second ball is green | first ball is green) is the branch labeled 0. Indeed, given that the first draw is green,

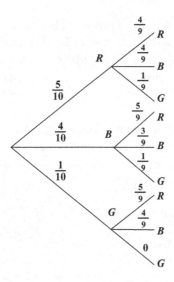

Figure A.7 A probability tree for drawing two balls from an urn in sequence.

there are no more green balls left for the second ball, so it is impossible for the second ball to be green, given that the first one is.

Thus, the probability labels on the tree diagram all correspond to the answers of a and b. What about c? Notice that the events in question are intersections, so they can be found by the multiplication rule in the corollary. Indeed, p(first ball is red $\cap$ second ball is blue) = p(first ball is red)$\cdot p$(second ball is blue | first ball is red). But these probabilities on the right are labeled on the tree. They appear on the path that follows the R for the first draw and B for the second draw, giving us $\frac{5}{10} \cdot \frac{4}{9} = \frac{2}{9}$. We use the entire path because we are considering both events. Similarly, to find p(first is green $\cap$ second is blue), we multiply the probabilities on the path that follows G first and then B: $\frac{1}{10} \cdot \frac{4}{9} = \frac{2}{45}$.

Finally, what about d? We suspect that, since there is no real difference between the first draw and the second draw, perhaps the answer is the same as the probability that the first ball is blue, which is $\frac{4}{10}$, but how can we prove that?

We'll give a rigorous explanation of the answer, and then we'll give a shortcut using the tree diagram. Let $B2$ be the event that the second ball is blue, let $R1$ be the event that the first ball is red, let $B1$ be the event that the first ball is blue, and let $G1$ be the event that the first ball is green. Notice that $R1$, $B1$, and $G1$ are pairwise mutually exclusive events whose union S is all possible colors on the first draw, because the first draw must be one of those three colors and no two colors, can be drawn simultaneously. Thus,

$$B2 = B2 \cap S = B2 \cap (R1 \cup B1 \cup G1)$$
$$= (B2 \cap R1) \cup (B2 \cap B1) \cup (B2 \cap G1)$$

by the distributive law. But these three events are disjoint (mutually exclusive), so we can find the probability of the union by adding the probabilities as usual:

$$p(B2) = p(B2 \cap R1) + p(B2 \cap B1) + p(B2 \cap G1).$$

But each of these is an intesection just like we illustrated in Part c, so each of these is a product (by Corollary A.42):

$$p(B2) = p(R1) \cdot p(B2|R1) + p(B1) \cdot p(B2|B1) + p(G1) \cdot p(B2|G1)$$
$$= \frac{5}{10} \cdot \frac{4}{9} + \frac{4}{10} \cdot \frac{3}{9} + \frac{1}{10} \cdot \frac{4}{9} = \frac{20 + 12 + 4}{90} = \frac{36}{90} = \frac{4}{10}.$$

We got the numerical values from the tree diagram. This is a completely rigorous answer using facts about disjoint events and the multiplication and addition rules for probabilities. But it may seem hard to memorize, so what is the shortcut? Simply this: since we are asking for the probability of the second ball being blue, we consider all three paths that end at B, going through the different possible first-draw vertices, and we add the results because they are mutually exclusive events. Visualizing these paths on the tree diagram seems easier than trying to reproduce the formula in terms of unions and intersections (and can actually help you reproduce the formula if you wanted to).

Thus, the answer is indeed the same as for the first draw, as we suspected. For another example, what is the probability that the second ball is green? We won't bother writing our the events as unions and intersections; we just refer to the tree diagram and add the products of the probabilites on all the paths that end with the second ball green:

$$\frac{5}{10} \cdot \frac{1}{9} + \frac{4}{10} \cdot \frac{1}{9} + \frac{1}{10} \cdot 0 = \frac{1}{10}.$$

We end this example by noting that we did not give this direction: find p(first ball is green | second ball is red). Notice that this particular conditional probability does not appear anywhere on the tree diagram (because you are given information about the second draw – not the first.) Nevertheless, we'll see in the next section that we can still use the tree diagram to answer this question.

The next definition is very important.

DEFINITION A.44 *In any sample space S, events E and F are said to be* independent *if* $p(E) = p(E|F)$.

In Chapter 6, we gave an intuitive definition of independent events when we said knowing that F occurs has no effect on $p(E)$. Observe that our formal definition here codifies exactly that notion. The probability of E, given that you know that F happened, is just the same as $p(E)$, without knowing anything about whether or not F occurred.

COROLLARY A.45 *If E and F are independent events, then*

a.　$p(E \cap F) = p(F) \cdot p(E)$
b.　*If* $E \neq \varnothing$, *then* $p(F|E) = p(F)$, *so that there is no difference between E and F being independent and F and E being independent, as the phrase suggests.*

Part a follows immediately from Corollary A.42, since $p(E|F) = p(E)$. We used this formula (when the events are independent) repeatedly in Chapters 6–8.

As for Part b, note that by the multiplication rule we have $p(E \cap F) = p(E) \cdot p(F|E)$, but by Part a, this becomes $p(E) \cdot p(F|E) = p(E) \cdot p(F)$, so unless $p(E) = 0$, we can cancel it to obtain $p(F|E) = p(F)$. (Note that if $E = \varnothing$, then $p(F|E)$ is not defined!).

In fact, observe that Part a of this corollary is equivalent to $p(F|E) = p(F)$, so that you can take the formula in Part a of this corollary as the definition of independent events if you want to. In fact, if you want to extend the definition of independent events to more than two events, the easiest way to do so is to use the product definition. We say that a (finite) set of events $\{E, F, G, H \ldots\}$ is *independent* if the probability of any intersection among them is equal to the product of the probabilities of the sets which are intersected. For example, $p(E \cap G \cap H) = p(E) \cdot p(G) \cdot p(H)$, etc.

We close this section with examples of how to determine if events are independent.

Example A.46 Which of the following events are independent?

a.   In Example A.2, E is the event the CD is a Charisma CD, and F the event that it is a Genesis CD.

b.   In Example A.12, E is the event the first ball is red, and F is the event the second ball is blue.

c.   Two cards are dealt in sequence from a well-shuffled deck. E is the event the first card is an ace; F is the event the second card is an ace.

d.   In Part c, let G be the event the first card is red. Are E and G independent?

e.   A fair coin is tossed twice. E is the event the first toss is H, and F is the event the second toss is H.

f.   In Part e, what if the coins was not fair, and $p(H) = 0.7$. Now are E and F independent?

In Part a, $p(E) = \frac{30}{100}$, and $p(F) = \frac{6}{100}$. Note that $p(E \cap F) = \frac{5}{100}$. Since $\frac{30}{100} \cdot \frac{6}{100} = \frac{9}{500} = 0.018 \neq 0.05$, the events are not independent by the corollary. In Part b, We know $p(F) = \frac{4}{10}$, as calculated before. But $p(F|E) = p(\text{second ball is blue } | \text{ first ball is red}) = \frac{4}{9}$, (it's one of the labeled branches of the tree diagram). Since these are different, the events are not independent by Definition A.11 (or Part b of the corollary).

For Part c, we encourage the reader to draw a tree diagram with branches "ace" and "not an ace." Then $p(\text{first card is an ace}) = \frac{4}{52} = \frac{1}{13}$. Also, $p(\text{second card is an ace } | \text{ the first card is an ace}) = \frac{3}{51} = \frac{1}{17}$. It seems that knowing E occurred has lowered the chances of F occurring so they do not seem to be independent. We can verify this by noting that $p(\text{second card is an ace}) = \frac{1}{13}$, just as in the urn problem, when the probability the second ball was blue was equal to the probability that the first ball was blue. Thus, $p(E) \cdot p(F) = \frac{1}{13} \cdot \frac{1}{13} = \frac{1}{169}$. However, $p(E \cap F) = \frac{1}{13} \cdot \frac{1}{17} = \frac{1}{221}$, by the multiplication rule or referring to your tree diagram (or, you can calculate this by the methods of the previous section. There are $C(52, 2)$ ways of dealing two cards, and $C(4, 2)$ ways of dealing two aces. Thus, $p(E \cap F) = \frac{C(4,2)}{C(52,2)} = \frac{6}{1326} = \frac{1}{221}$). However, you obtain the result, note that $\frac{1}{221} \neq \frac{1}{169}$, so again, the events are not independent.

For d, note that there are 26 red cards and two red aces. Thus, $p(E \cap G) = \frac{2}{52} = \frac{1}{26} = \frac{1}{13} \cdot \frac{1}{2} = p(E) \cdot p(G)$. So these two events are independent. For (e) $p(E \cap F) = p(\{HH\}) = \frac{1}{4}$ if the coins are fair. But $\frac{1}{4} = \frac{1}{2} \cdot \frac{1}{2} = p(E) \cdot p(F)$, so these events are also independent. (We leave it for the reader to verify that $p(F) = \frac{1}{2} = p(E)$.) For Part f, the statement that "$P(H) = 0.7$" seems to mean that the probability of obtaining heads is 0.7 on *any one toss*, regardless of what the previous outcome was. So with this understanding, $p(F) = 0.7 = p(F|E)$ and the events are still independent by definition, regardless of the coin being weighted. (This makes sense – regardless of how weighted the coin is, the result of one toss should not affect the outcome of the next toss.)

This concludes this section.

Exercises

1. Suppose A and B are events in a sample space, and suppose $p(A) = 0.3$, $p(B) = 0.6$, and $p(A \cap B) = 0.1$.

    a. Compute $p(A|B)$ and $p(B|A)$.
    b. Compute $p(A^c|B)$ and $p(A^c|B^c)$.
    c. Are A and B independent? Explain.

2. Suppose $p(E) = \frac{1}{3}$ and $p(F) = \frac{1}{4}$.

    a. Find the probability $p(E \cup F)$ if E and F are mutually exclusive.
    b. Find the probability $p(E \cup F)$ if E and F are independent.

3. Can two nonempty events C and D be simultaneously independent and mutually exclusive? Explain.

4. A jar contains 20 peppermint candies, 10 spearmint candies, and six tart apple candies, which all look alike. Two candies are drawn randomly in succession (without replacement). Draw a tree diagram and label the branches with appropriate probabilities. Use the diagram and/or the results of this section to answer the following:

    a. Find the probability that the first candy drawn is not apple.
    b. Find the probability that the second candy drawn is apple, given that the first one is spearmint.
    c. Find the probability that both candies are peppermint.
    d. Find the probability that the second candy is spearmint.

5. In Example A.46, Part c in the text, verify that the probability that the second card is an ace is indeed $\frac{1}{13}$.

6. Sheila is taking classes at Leibniz Academy. She estimates the probability that she passes her English class is 0.95, the probability that she passes her history class is 0.98, the probability that she passes her music class is 0.94. If these three events are independent, what is the probability that she will not pass at least one class?

7. The gender studies department at Leibniz Academy is conducting a study of the distribution of women students majoring in STEM disciplines (science, technology, engineering, and mathematics). Assume that at Leibniz Academy, there are no double majors, and four STEM disciplines have majors: mathematics, physics, chemistry, and biology. In the study, it is found that if a STEM major at Leibniz Academy is chosen at random, the probability that the student is a mathematics major is 0.2, the probability the student is a physics major is 0.1, the probability that the student is a chemistry major is 0.2, and the probability that the student is a biology major is 0.5. Among the math majors at the Academy, 50% are females; among the physics majors, 30% are female; among the chemistry majors, 40% are female; and among the biology majors, 60% are female.

    a. If a STEM major is randomly selected, what is the probability that the student is male, given that the student is a biology major?
    b. If a STEM major is randomly selected, what is the probability that the student is a female physics major?

    c.   If a STEM major is randomly selected, what is the probability that they are female? Overall, do you think that Leibniz Academy is having trouble attracting female STEM majors?

8.   A card is selected randomly from a well-shuffled deck. Determine whether or not the following events are independent:

    a. E is the event the card is red, and F is the event the card is a face card (jack, queen, or king).

    b. E is the event the card is a spade suit, and F is the event the card is a face card.

    c. E is the event the card is red, and F is the event the card is a heart suit.

    d. E is the event the card is a king, and F is the event the card is a face card.

9.   At the Nilsson Medical Clinic, patients are tested for Lyme disease by using two different diagnostic tests. The probability that test A correctly detects Lyme disease in a patient who has the disease is 0.6, the probability that test B correctly detects Lyme disease is in a patient with the disease is 0.7. Suppose that the probability that at least one of the tests will correctly detect the disease is 0.85.

    a.   Find the probability that both tests correctly detect the disease.

    b.   Find the probability that test A correctly detects Lyme disease, given that test B has correctly detected it.

    c.   Find the probability that test B correctly detects Lyme disease, given that test A has correctly detected it.

    d.   Are the events "test A correctly detects Lyme disease" and "test B correctly detects Lyme disease" independent?

A.6 Bayes's Theorem

Our final question in probability theory is what is the relationship between $p(E|F)$ and $p(F|E)$? The answer to this is known as Bayes's theorem. The theorem is named after Thomas Bayes (1701–1761), an English statistician, philosopher, and Presbyterian minister who discovered a special case of the theorem we are about to discuss.

    A form of the theorem is easy to derive from the multiplication rule. Since $E \cap F = F \cap E$, the multiplication rule implies

$$p(F) \cdot p(E|F) = p(E \cap F) = p(E) \cdot p(F|E).$$

Therefore, dividing both sides of this by $p(E)$ we obtain the following.

PROPOSITION A.47   *If E and F are nonempty events in a sample space S, then*

$$p(F|E) = \frac{p(F) \cdot p(E|F)}{p(E)}.$$

This is the essential idea in Bayes's theorem. However, what makes the theorem particularly useful is to rewrite the denominator in a particular way, and this revised version is what is known as Bayes's theorem. The reason for rewriting the denominator is that we very often want to apply

Bayes's theorem in examples where you are not given $p(E)$ directly but have to compute it, and the revised version of the theorem explicitly tells you how to compute it.

The situation that gives rise to the generalized version is this. Imagine breaking up the sample space S into a disjoint union of several subsets. That is, suppose there are mutually exclusive events F_1, F_2, ..., F_n such that $S = F_1 \cup F_2 \cup \cdots \cup F_n$. The terminology for this is that the family of sets $\{F_i\}$ form a *partition* of S. Now suppose E is any event. The Venn diagram in Figure A.8 illustrates (for $n = 4$) how we also obtain a partition of E from the family $\{F_i\}$.

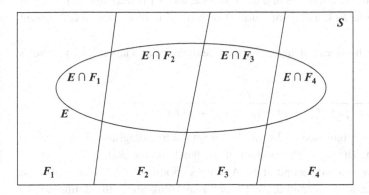

Figure A.8 A partition of S induces a partition of E.

In the diagram, E is the oval, and the F_i are separated by the diagonal lines. So $S = F_1 \cup F_2 \cup F_3 \cup F_4$ is a partition on S, and there is a corresponding partition on E formed by taking intersections with E:

$$E = (E \cap F_1) \cup (E \cap F_2) \cup (E \cap F_3) \cup (E \cap F_4).$$

Now, because this is a disjoint union, we compute the probabilities by adding:

$$p(E) = p(E \cap F_1) + p(E \cap F_2) + p(E \cap F3) + p(E \cap F_4).$$

However, each intersection on the right is now rewritten using the multiplication rule. For each i, we have $p(E \cap F_i) = p(F_i) \cdot p(E|F_i)$. Of course, there is nothing special about $n = 4$; this works for any n. Putting this all together, we have proved the following.

THEOREM A.48   Bayes's Theorem. *Let* $\{F_i | i = 1, 2, \ldots, n\}$ *be a family of mutually exclusive events which partitions* S, *and let* E *be any event in* S. *Then for each* i:

$$p(F_i|E) = \frac{p(F_i) \cdot p(E|F_i)}{p(F_1) \cdot p(E|F_1) + p(F_2) \cdot p(E|F_2) + \cdots + p(F_n) \cdot p(E|F_n)}.$$

There are a couple of things to note. First, observe that the terms on the bottom all look alike – just the index changes – and that one of those summands is also in the numerator. Second, given any F, we can always form a partition of S with $n = 2$ as: $S = F \cup F^c$, so we can always apply this theorem, even if we are given just two events E and F. Third, the expression for $p(E)$ that is in the denominator is computed exactly as we did in Part d of the urn problem in the last section when we computed the probability that the second ball was blue. We summed over all the paths on the tree diagram that ended in the event of the second ball being blue, and each summand was a product coming from the multiplication rule. The reader should review that example now. This means that Bayes's theorem will be especially useful in probability problems, like the urn problem, which are described by a tree diagram.

In fact, in the last section, we mentioned that we could use the tree diagram to find p(the first ball is green | the second ball is red), although this conditional probability did not appear directly on the tree diagram. Indeed, the partition here are the colors of the first draw. Thus, F_1 is the event that the first draw is red, F_2 is the event that the first draw is blue, and F_3 is the event that the first draw is green. Thus, $S = F_1 \cup F_2 \cup F_3$ is a disjoint union. The event E in this problem is the event that the second ball is red.

So we want to find $p(F_3|E)$. This does not appear on the tree diagram as a label, but by Bayes's theorem, we have

$$p(F_3|E) = \frac{p(F_3) \cdot p(E|F_3)}{p(F_1) \cdot p(E|F_1) + p(F_2)p(E|F_2) + p(F_3) \cdot p(E|F_3)}.$$

Now all the probabilities on the right side of this are labels on the tree diagram. Going back to the diagram, we can fill them in. The top is the product of the labels on the path from the initial vertex through a green first draw to a red second draw. And the denominator is the sum of these products for all paths that end at a red second draw. So one may think about the definitions of the probabilities to fill in the numerical values or just refer to the tree diagram to do it with less thinking about it:

$$p(F_3|E) = \frac{\frac{1}{10} \cdot \frac{5}{9}}{\frac{5}{10} \cdot \frac{4}{9} + \frac{4}{10} \cdot \frac{5}{9} + \frac{1}{10} \cdot \frac{5}{9}}$$

$$= \frac{5}{20 + 20 + 5} = \frac{5}{45} = \frac{1}{9}.$$

Let's look at two more examples.

Example A.49 Costello's Private Eyes is a multi-city private investigation firm. The Los Angeles office employs 40% of the detectives, the Chicago office employs 30%, the St. Louis office employs 20%, and the Boston office employs 10%. All the detectives are divided into specialties: "missing persons" cases, "divorce-related" cases, and "other" cases. In the Los Angeles office, 30% are missing persons specialists, 60% are divorce specialists, and 10% are other specialists. In Chicago, 20% are missing persons specialists, 30% are divorce specialists, and 50% are other specialists. In St. Louis, 25% are missing persons specialists, 50% are divorce specialists, and 25% are other specialists. In Boston, 10% are missing persons specialists, 50% are divorce specialists, and 40% are other specialists.

A detective selected at random is found to be a divorce specialist. What is the probability that he is based in Chicago?

We draw and label tree diagram. The percentages in each city are the ordinary probabilities on the first branches. The percentages of each specialty within each city are the conditional probabilities on the second set of branches. For example, p(detective works in St. Louis) = 0.2, and p (Missing Person | Los Angeles) = 0.3. The rest of them are as labeled in Figure A.9.

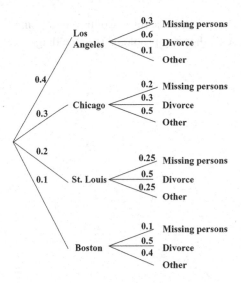

Figure A.9 Probability tree for Example A.21.

This is a Bayes's theorem problem. In the exercises, we ask the reader to write out the events and probabilities used, but here, we just use the tree directly. The numerator is the product of the probabilities on the path from the initial vertex through Chicago ending at divorce specialist, so $0.3 \cdot 0.3$. The denominator is the sum of these products over all paths that end in divorce specialist. Thus, p(based in Chicago | Divorce specialist) is

$$
\frac{0.3 \cdot 0.3}{0.4 \cdot 0.6 + 0.3 \cdot 0.3 + 0.2 \cdot 0.5 + 0.1 \cdot 0.5}
$$
$$
= \frac{0.09}{0.24 + 0.09 + 0.1 + 0.05} = \frac{0.09}{0.48} = \frac{3}{16}
$$

Our final example, although hypothetical, is an example of an important real life application of Bayes's theorem. Medical tests are not infallible.

Example A.50 *James Marshall Labs* has developed a diagnostic test to detect a certain disease. There are two kinds of mistakes to which the test can succumb – it can give a *false negative* (that is, it can fail to detect the disease if you have it), or it can give a *false positive* (that is, the test indicates you have the disease when you really don't.) Suppose that the probability of a false negative is 0.07, and suppose the probability of a false positive is 0.03. Suppose also that the probability that a random person has the disease is 0.006. Now, suppose you take the test and it gives a positive return indicating that you have the disease. What is the probability that you really have the disease?

The first thing to realize is that the false positive and false negative probabilities that are given are conditional probabilities. Indeed, if we denote a positive test result by P and a negative one by N, and if we also denote the event that a person has the disease by D, then $p(N|D) = 0.07$ (false negative), $p(P|D^c) = 0.03$ (false positive). Also $p(D) = 0.006$, so $p(D^c) = 0.994$, and $D \cup D^c = S$ is a partition of the sample space (all living people). Finally, the probability you are asked to find is also a conditional probability – it is $p(D|P)$, which is the reverse of the conditional probabilities which are given. Thus, this is a classic Bayes's theorem problem. The tree diagram is shown in Figure A.10.

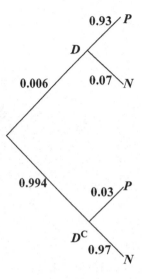

Figure A.10 Probability tree for Example A.22.

Notice that if there is a 0.07 chance of a false negative, then there must be a 0.93 chance the test correctly diagnoses the disease if it is present. Indeed, in any tree diagram like this, all the probabilities emanating from any vertex must sum up to 1. Now an easy application of Bayes's theorem gives

$$p(D|P) = \frac{p(D) \cdot p(P|D)}{p(D) \cdot p(P|D) + p(D^c) \cdot p(P|D^c)}$$
$$= \frac{0.006\,(0.93)}{0.006(0.93) + 0.994(0.03)}$$
$$= \frac{0.00558}{0.00558 + 0.02982} = 0.15763.$$

Thus, there is slightly less than a one in six chance you have the disease even though the test says you have it! The reason for this surprising result is that the overwhelming majority of the population is disease-free, which means that the second term in the denominator is quite a bit larger than the first term, which is also the term in the numerator. Improving the accuracy of the test (by reducing the chances for both false positive and false negative results) would only slightly improve this result (Exercise 4).

What does th is mean in terms of medical testing? The first conclusion to draw is that you should not assume a medical test is correct when it comes back positive. This is important in today's world because so much of the world is becoming dependent on your medical data, such as getting approved for insurance or maybe even getting a job. Care must be taken to protect the privacy of your medical data and to make sure it is correct.

Secondly, if you do receive a positive test, simply take the test again. Assuming the tests are independent, there is a much smaller chance of getting two positive results if you really do not have the disease. Or take a better test. This is done, for example, with the tine test for tuberculosis. If you get a positive result from the tine test, it is a red flag that you should look more closely for tuberculosis, such as by getting a chest X-ray. By itself, it does not mean you have the disease for certain. On the other hand, the tine test saves a lot of money by making it unnecessary to give everyone a chest X-ray (and probably safer, too, since a chest X-ray is more invasive.)

Exercises

1. In Example A.49, Costello's Private Eyes, write out the events symbolically and their probabilities for the calculation done earlier.
2. a. In Example A.49, Costello's Private Eyes, find the probability that a detective is based in St. Louis, given that she is a missing persons specialist.
   b. Find the probability that a detective is based in Boston, given that she is a missing persons specialist.
   c. Mr. Hall is a divorce specialist who works in the firm. What is the probability that he is based in Los Angeles?
3. In Example A.50, the medical test problem. Suppose you take the test and it comes back negative. What is the probability that you really do not have the disease?
4. In the medical test of Example A.50, suppose that the test was improved so that the probability of a false negative was 0.05 and a false positive was 0.02. Now compute $p(D|P)$.
5. An urn contains four black balls and two red balls. Two balls are drawn in sequence without replacement. If the second ball is red, what is the probability that the first ball was black?
6. Leitch Vintage Instruments, Ltd., is a company in Scotland that manufactures classic musical instruments such as bagpipes and hurdy-gurdies. Leitch has three factories – one in Aberdeen, one in Glasgow, and one in Perth. While most of the workmanship is high quality, sometimes a defective instrument is produced due to a malfunctioning valve or broken drone string. The probability of a defective instrument from Aberdeen is 0.01, the probability of a defective instrument from Glasgow is 0.02, and the probability of a defective instrument from Perth is 0.025. The plant in Aberdeen produces 45% of the company's output, the plant in Glasgow produces 20%, and the plant in Perth produces 35%. If a Leitch hurdy-gurdy chosen at random is found to be defective, what is the probability that it came from each of the three cities?

References

Bland, R. G. New finite pivoting rules for the simplex method. *Mathematics of Operations Research*, 2:103–110, 1977.

Brams, Steven J. *Rational Politics: Decisions, Games, and Strategy*. Academic Press, 1985a.

Brams, Steven J. *Superpower Games: Applying Game Theory to Superpower Conflict*. Yale University Press, 1985b.

Brams, Steven J. *Theory of Moves*. Cambridge University Press, 1994.

Brickman, Louis. *Mathematical Introduction to Linear Programming and Game Theory*. Springer, 1989.

Calvert, James E., and Voxman, William L. *Linear Programming*. Harcourt Brace Jovanovich, 1989.

Casti, John L. *Five Golden Rules: Great Theories of 20th Century Mathematics and Why They Matter*. Wiley, 1996.

Clark, John, and Holton, Derek Allan. *A First Look at Graph Theory*. World Scientific, 1991.

Dixit, Avinash, and Skeath, Susan. *Games of Strategy*. Norton, 1999.

Hodge, Jonathan K., and Klima, Richard E. *The Mathematics of Voting and Elections: A Hands-On Approach*. American Mathematical Society, 2005.

Karloff, Howard. *Linear Programming*. Birkhauser, 1991.

Kaufmann, Jerome. *Precalculus*, 3rd ed. PWS, 1995.

Lang, Serge. *Linear Algebra*, 2nd ed. Addison Wesley, 1971.

Lay, David C. *Linear Algebra and Its Applications*, 4th ed. Addison Wesley, 2012.

Loomba, Narendra Paul. *Linear Programming: A Managerial Perspective*, 2nd ed. MacMillan, 1976.

Luce, R. Duncan, and Raiffa, Howard. *Games and Decisions*. Wiley, 1957.

Milnor, John. Games against nature. In Thrall et al., editors, *Decision Processes*, pages 49–59. Wiley, 1954.

Nash, Stephen G., and Sofer, Ariela. *Linear and Nonlinear Programming*. McGraw-Hill, 1996.

Robinson, Jr. E. Arthur, and Ullman, Daniel H. *A Mathematical Look at Politics*. CRC Press, 2011.

Shapley, Lloyd, and Shubik, Martin. A method for evaluating the distribution of power in a committee system. *American Political Science Review*, 48:787–792, 1954.

Snyder, Glenn H. Prisoners dilemma and chicken models in international politics. *International Studies Quarterly*, 15:66–103, 1971.

Straffin, Philip D. *Game Theory and Strategy*. Mathematical Association of America, 1993.

Taylor, Alan D., and Pacelli, Allison M. *Mathematics and Politics: Strategy, Voting, Power and Proof*, 2nd ed. Springer, 2008.

Taylor, Alan D., and Zwicker, William S. A characterization of weighted voting. *Proceedings of the American Mathematical Society*, 115:1089–1094, 1992.

Taylor, Alan D., and Zwicker, William S. *Simple Games*. Princeton University Press, 1999.

Tomastik, Edmond C. *Applied Finite Mathematics*. Saunders College Publishing, 1994.

von Neumann, John, and Morgenstern, Oskar. *Theory of Games and Economic Behavior*. Princeton University Press, 1944.

Williams, J. D. *The Compleat Strategyst*. McGraw-Hill, 1954.

Index

abstraction
  in modeling process, 2
  linear equations and, 4–6
addition
  linear combination equations, 45–46
  matrices, 66–67
  vectors, 33–34
Addition Principle, 489–490
additive identity, vector addition and scalar multiplication, 34
adjoint of matrix, 79–81
artificial variables, phase I pivoting, 424–433
associative laws
  inverse matrix, 84–88
  matrix multiplication, 71
  set algebra, 486–487
  vector addition and scalar multiplication, 34
augmented coefficient matrix, 16
  duality construction, 174–177
  elimination operations, 19–20
  Gauss Jordan elimination algorithm, 25–28
  initial tableau set-up, 194
  linear combinations of vectors, 45–46
  non-degenerate system, 17–20
  reduced echelon form, 22–25
  simplex algorithm, 189
Axelrod, Robert, 379–384
axiomatic vectors, 34
axioms, negotiations and, 364–365

back substitution, Gauss Jordan elimination algorithm, 30
backwards induction, 359–361
Banhzhaf Power Index, 417–419
basic solutions, constraint space graphing, 137–139
basic variables
  constraint space graphing, 137–139
  phase I pivoting, 424–433
  Phase II simplex algorithm, tableaux maximization, 196–197
  simplex algorithm, 187–189
  tableaux, 189–194
basis vectors
  constraint space graphing, 139–140
  simplex algorithm, 187–189
Battle of the Bismarck Sea, 252–261

Battle of the Sexes, 378
Bayes Theorem, 512–517
binary operations
  matrix combination, 66–67
  set algebra, 483–487
binary tree, branching problems, integer programming, 472–474
binding constraints, 156–159
binomial coefficient, optimal points, constraint space, 142–144
Bland's Rule, 204–207, 431–433
bounding problems, integer programming, 474–477
branch and bound method, integer programming, 468–477
branching problems, integer programming, 470
brinksmanship, 378–384
butterfly effect, 148–150

Caesar cipher, 90–93
capacity changes
  sensitivity analysis, 148–150, 441–450
  simplex algorithm, 441–450
capacity coefficients
  objective coefficients and, 168
  sensitivity analysis, 148–150
  stable range of capacity numbers, 166–171
  stable ranges, 461–466
cardinality
  Addition Principle, 489–490
  Multiplication Principle, 490–491
Cartesian product, 486
  Multiplication Principle, 490–491
  sample space, 497–498
chaos theory, 148–150
characteristic function games, 405–406
  monotone voting systems, 412–413
  political power and, 413–419
chicken dilemma, 378, 382–384
  applications, 394–397
  non-myopic equilibrium, 390–393
  theory of moves and, 388–397
cipher, 90–93
circular paraboloid, 58
closed Leontief Model, 94–95

coalition formation
  cooperative games with side payments, 402–405
  critical defections from, 417–419
  disjoint coalitions, 405–406
  minimal winning coalitions, 417–419
  winning coalitions, 409–419
coefficient matrices, 16
  capacity changes, 441–450
  corner point theorem, 117
  Mathematica software applications, 220–225
  non-degenerate system, 17–20
  properties of, 72–73
coefficients
  elementary operations, 14–16
  objective coefficients, changes in, 452–460
column matrix, 63–64
combinations, enumeration and, 493–494
combinatorics, 488–497
communication
  Prisoner's dilemma, 374–384
  sensitivity analysis, 362–365
commutative laws
  matrix multiplication, 71
  set algebra, 486–487
  vector addition and scalar multiplication, 34,
    40–44
complementary slackness, 156–159, 216–219
  duality and, 181–182
  objective coefficient changes, 455–460
  square subgame theorem, 324
completing the square, 53, 55
computer-based solutions
  constant sum games, 326–333
  Excel software, 225–239
  Mathematica software, 220–225
  to linear programming, 220–239
condensed tableau
  constant sum games, linear programming solutions,
    316–320
  duality construction, 175–177, 214–219
  $m \times n$ game, minimax theorem, 321–324
  $m \times n$ problems, 321–324
  simplex algorithm, 189–194
  standard form minimization problem, 214–219
conditional probability, 273
  independent events and, 505–510
conditions of fairness, negotiations and, 364–365
conflicts, cooperative games and, 379–384
constant-sum games, 247–251
  Battle of the Bismarck Sea example, 252–261
  communication, 362–365
  computer-based solutions, 326–333
  dominant strategy, 256–257, 301–312
  expected value principle, 287–294
  linear programming solutions, 313–325
  minimax approach, 257–258
  mixed strategy games, 269–278
  $m \times n$ game, minimax theorem, 321–324

saddle point of payoff matrix, 262–266
  strategy choices, 353–354
constrained optimization, 56
constraint space
  $2 \times n$ problems, 132–144
  Algorithm, 144
  graphing in, 115–121, 133–142
  marginal values of constraint, 156–163
  mixed constraints, 160–163
  multiple decision variables, 140–142
  optimal points in, 142–144
  sensitivity analysis, 148–150
  simplex algorithm and, 186–187
consumer demand, open Leontief Model, 94–95
continuous game, 248
convex combinations of corner points, 124
convex sets
  corner point theorem, 121–124
  variable-sum games, payoff polygon, 339–342
cooperative games with side payments
  n-player games, 402–405
  politics and international diplomacy and, 362–365
cooperative game theory, 362–365
  n-player games, 401–406
  ordinal games, 378
  Tit for Tat algorithm, 379–384
corner point theorem
  constraint space graphing, 137–139
  convex sets, 121–124
  decision space graphing, 132–133
  $m \times 2$ problems, 115–121
  objective coefficient changes, 455–460
  optimal points in, 142–144
counter-prudential strategy, variable-sum games, 341–342
cross sections, quadric surfaces, 57–61
crossed type, 59–61
Cuban Missile Crisis, 382–384, 394–397
cycling
  Bland's Rule, 204–207
  simplex algorithm, 202–204

Dantzig, George, 185–186, 199–200
decimals, simplex algorithm, 202–204
decision space
  constraint space graphing, 137–139
  graphing in, 132–133
  $m \times 2$ problems, 115–121
  no-solution and infinite-solution linear programming
    problems, 124–127
decision variables, 102
  Bland's Rule, 204–207
  duality problems, 179–181
  initial tableau set-up, 194
  linear equation, 106–107
  non-negativity constraints, 207
  objective coefficient changes, 455–460
  Phase II simplex algorithm, tableaux maximization,
    197–199

decision vectors, constraint space graphing, 140–142, 144

Deegan-Packel Power Index, 418

degenerate problems
 Bland's Rule, 204–207
 constraint space graphing, 137–139
 simplex algorithm, 202–204
 square subgame theorem, 324

degenerate system, 12–14
 Gauss Jordan elimination algorithm, 25–28
 graphic representation of, 14–16
 reduced row echelon, 20–28

De Morgan's Laws, 486

De Morgan's Augustus, 486

derived game
 expected value principle, 290–294
 mixed strategies, 281–287
 mixed strategy variable sum games, 337–342

determinant of matrix, 79–81

diagonal matrix, 63–64

dictators, winning coalitions, 418

disjoint coalitions
 characteristic function games, 405–406
 monotone voting systems, 412–413

distributive laws
 conditional probability, 507–509
 matrix multiplication, 71
 set algebra, 486–487
 vector addition and scalar multiplication, 34

Divide the Dollar game, 405–406, 410–412

dominant strategy games
 constant-sum games, 301–312
 game theory, 256–261
 n-player games, 398–401
 Nash equilibrium and, 335–336
 ordinal game, 373–374
 square subgames, 308–309

dot product
 Excel software solutions for, 230–239
 matrix multiplication, 66–73

duality principle
 basic characteristics, 173–174
 complementary slackness, 181–182
 construction of, 174–177
 economic interpretation of, 177–181
 Mathematica software applications, 224–225
 $m \times n$ game, minimax theorem, 321–324
 sensitivity analysis, 173–182
 standard form minimization problem, 214–219

dual pivoting, 431–433

dummy voter, 414–419

economics
 duality interpretation, 177–181
 input-output models, 93–96

elementary operations
 defined, 12–14
 graphic representation, 14–16

elementary row operations
 defined, 16
 Gauss Jordan elimination algorithm, 25–28
 linear combinations of vectors, 45–46
 reduced echelon form, 22–25
 simplex algorithm, 192–194

elimination
 inverse matrices, 88
 matrices, 10–20
 non-degenerate system, 19–20
 solution, 4–6

empty set, linear programming problems, 125, 482–483

encryption, matrix cipher, 90–93

enumeration, 488–497
 combinations, 493–494
 permutation, 492–493

equalizing expectation
 reduced game matrix, 303–309
 3×2 game, 305–308
 two-person constant-sum games, 287–294

equalizing strategy, variable-sum games, 341–342

equivalent system
 elementary operations, 12–14
 graphic representation, 14–16
 linear equations, 4–6
 voting systems, 410–412

Euclidean space
 three-dimensional graphic solution, 11–12
 two-dimensional graphic solution, 10–11
 vectors in, 30–32

event, probability theory, 272–275

Excel software
 constant sum game solutions, 329–333
 linear programming applications, 225–239

Expected Value of E experiment, 69

expected value principle
 equalized expectations, 287–294
 games against nature, 256–257, 365–367
 probability theory, 274–276
 two-person constant-sum game, 287–294

experiments, 497–498

extensive form game, 249–251

extreme value theorem, 124

feasible points/feasible region
 constraint space graphing, 137–139
 in decision space, 132–133
 integer programming, branch and bound methods, 468–477
 linear equation, 105–106
 marginal value of constraints and, 156–159
 optimal points in, 142–144
 phase I pivoting, 433
 simplex algorithm, 193

final tableau
 capacity changes in, 441–450
 marginal values, 199–200
 objective coefficient changes, 452–460, 460

final tableau (cont)
  phase I pivoting, 425–433
  Phase II simplex algorithm, 198–199
5 × 2 game, 306–308
Follow the Leader dilemma, 378
fractional index, 416–419
free variable
  Gauss Jordan elimination algorithm, 25–28
  reduced echelon form, 22–25
fundamental theorem of linear programming
  convex sets, 121–124
  $m \times 2$ problems, 115–121

game theory. *See also* utility theory
  classification, 247–249
  computer solutions, constant-sum games, 326–333
  constant-sum games, 301–312
  defined, 243
  derived game, 281–287
  dominance and Nash equilibrium points, 335–336
  dominant strategies, 252–266
  dominated strategies, 301–312
  equalizing expectation, 287–294
  examples, 246–247
  expected value principle, 287–294
  formats, 249–251
  legislative voting, 409–413
  linear programming solutions, constant sum games, 313–325
  minimax approach, 257–258
  mixed strategy games, 281–294
  $m \times n$ case, minimax theorem, 321–324
  n-player games, 398–409
  Nash equilibria and, 258–261
  ordinal games, 352, 373–384
  political power and, 413–419
  rational behavior and, 245
  sensitivity analysis, 352–367
  sequential games, 388–397
  square subgame theorem, 310–312, 324
  variable-sum games, 335–347
Game Theory, matrix multiplication, 69
game tree, 249–251
  Battle of the Bismarck Sea example, 254–255
  reverse induction, 359–361
  sensitivity analysis, 356–361
games against nature, 256–257, 365–367
games of perfect information, 359–361
Gauss Jordan elimination algorithm, 25–28
  back substitution, 30
  linear combinations of vectors, 45–46
  redundant systems, 47
Gaussian elimination, inverse matrices, 79
general form equation, saddle surface, 59–61
general sum game, 247–249
global maximum point, optimization problems, 53, 55
grand coalition, 404–405

graphical solutions
  capacity changes, 441–450
  decision space, 132–133
  drawbacks of, 166–172
  elementary operations, 14–16
  linear equations, 10–11
  linear programming, 102–109
  $m \times 2$ problems, decision space, 115–121
  quadratic curves, 50–54
  simplex algorithm and, 186–187
greedy algorithm, 190–194

half-plane, linear inequality, 105–106
half-space, linear inequality, 105–106
Harsanyi, John C., 243
hidden variables, corner point theorem, 118
higher order dominant strategies, 258–261
Hill cipher, 90–93
Hill, Lester, 90–93
homogeneous system, linear equations, 43–44
Hurwicz method, cooperative games, 370
hyperbolic paraboloid, 59–61

identity matrix, 19–20, 63–64
  objective coefficient changes, 452–460
  properties of, 72–73
  reduced row-echelon form, 20–28
impossible events, 498
inactive strategies, 303, 306–308
inconsistent system, 12–14, 84–88
independent events
  conditional probability and, 505–510
  sample space, 509–510
inequality
  constant sum games, linear programming solutions, 316–320
  less-than inequalities, 423–433
  linear equations, 105
  phase I pivoting, 433
  reversal of, 212–219
infinite game, 248
infinite solutions, linear programming problems, 125–127
  constraint space graphing, 137–139
information sets, game trees, 359–361
initial tableau
  objective coefficient changes, 452–460
  phase I pivoting, 425–433
  set-up rules, 194
  simplex algorithm, 189–194
input-output matrix, 95–96
input-output models, 93–96
integer programming problems, 109
  branch and bound method, 468–477
  Mathematica software applications, 223
integers, matrix cipher, 90–93
interior point methods, 185–186
intermediate value theorem, 105–106

international diplomacy
  chicken dilemma and, 382–384
  game theory and, 362–365
interpretation, of mathematical model, 3
inverse matrix, 78–81
  capacity changes, 441–450
  input-output models, 93–96
  linear equations, 103
  linear system solutions, 84–88
iso-objective line, corner point theorem, 116

Johnston Power Index, 418

LaPlace theorem, 366–367
Lazy approach, cooperative games, 366–367
Least Cost Method, 440
legislative voting, game theory and, 409–413
Leontief, Wassily, 93–96
Less Naive Power Index, 416–419
less-than inequalities, phase I pivoting, 423–433
level objective line, corner point theorem, 116
linear dependence
  constraint space graphing, 137
  Span(S) of vectors, 38–39
linear equations
  constraint space graphing, 136–137
  inverse matrices, 79, 84–88
  matrix operations and, 72
  reduced echelon form, 22–25
  systems, 3–8
  vectors, 45–46
linear independence
  nontrivial linear combinations, 43–44
  Span(S) of vectors, 38–39
linear programming
  complementary slackness, 181–182
  computer solutions for, 220–239
  constant-sum games, 313–325
  defined, 107
  Excel software applications, 225–239
  graphical solutions, 102–109
  integer programming, 468–477
  marginal value of constraints and, 160–163
  Mathematica software applications, 220–225
  minimization problems, 435–439
  $m \times 2$ problems, 115–121
  $m \times n$ game, 321–324
  no-solution and infinite-solution problems, 124–127
  phase I pivoting, 423–433
  simplex algorithm and, 185–186
  square subgame theorem, 324
  stable ranges, 461–466
linear transformations, 46
  constraint space graphing, 140–142
linear vector combinations, 35–38
local maximum point, optimization problems, 53, 55
local properties, marginal value of constraints and, 159

marginal values
  constant sum games, linear programming solutions, 319–320
  drawbacks of, 166–172
  duality problems and, 179–181, 216–219
  Excel software solutions for, 233–239
  final tableau construction, 199–200
  linear programming, 160–163
  objective coefficient changes, 455–460
  resource values, 156–160
  sensitivity analysis, 148–150, 156–163
  stable range of capacity numbers, 166–171
mathematical model
  defined, 1
  game theory and, 243
  modeling process, 1–3
Mathematica software
  constant sum game solutions, 326–329
  linear programming applications, 220–225
matrices
  addition, 66–67
  coefficient matrix, 16
  defined, 16
  elimination and, 10–20
  Excel software applications, 225–239
  input-output matrix, 95–96
  input-output models, 93–96
  inversion, 78–81
  Mathematica software applications, 220–225
  multiplication, 68–73
  operations, 65
  properties of, 63–64
  sensitivity analysis and simplex algorithm, 441–450
  transpose of, 65–66
matrix cipher, 90–93
matrix multiplication, 68–73
  capacity changes, 441–450
  expected value principle, 287–294
  marginal value of constraints, 160–163
  mixed strategy zero-sum games, 276–277
maximization problem, 102
  phase I pivoting, 431–433
maximum point, optimization problems, 53, 55
method of oddments, 294
  dominated strategies, 301–312
Milnor, John, 366–367
minimal winning coalition, 417–419
minimax theorem
  constant sum games, linear programming solutions, 313–325
  dominant strategy games, 301–312
  game theory, 257–258
  mixed strategy games, 281–294
  mixed strategy variable sum games, 337–342
  $m \times n$ game, 321–324
  rationality assumption, 365–367
  square subgame theorem, 310–312

minimization problem, 102
  standard form, 108–109
minimum point, optimization problems, 53, 55
mixed constraints
  branching problems, integer programming, 470
  capacity changes and, 448–450
  duality construction, 177
  Excel software solutions for, 229–239
  linear programming, 109
  marginal value problems, 160–163
  Mathematica software applications, 223
  phase I pivoting, 428–433
mixed strategies
  constant-sum games, 269–278, 353–354
  derived game, 281–287
  expected value principle, 287–294
  game theory, 261–262
  minimax theorem, 2 × 2 case, 281–294
  probabilities and payoffs, 270–278
  variable-sum games, 337–342
  zero-sum games, 276–277
modular arithmetic, 92–93
monotone voting systems, 412–413
movement diagram
  n-player games, 399–401
  Nash equilibria, 258–261
  variable-sum games, 335–336
multi-move games, game tree and reverse induction,
    356–361
multiplication
  matrix multiplication, 68–73
  scalar, 38–39, 45–46, 66–67
  vectors, 32–34
Multiplication Principle, 490–491
multiplication rule
  Bayes Theorem, 512–517
  conditional probability and independent events,
    507–509
multiplicative identity
  inverse matrix, 84–88
  $n \times n$ matrices, 72–73
multiplicative inverse, 78–81
multisets, 482–483
$m \times 2$ problems
  decision space, 115–121
  marginal value of constraints, 160–163
  simplex algorithm, 202–204
  stable range of capacity numbers, 166–171
  zero-sum example, 303–309
$m \times n$ problems
  Bland's Rule, 204–207
  constraint space graphing, 137–139
  matrix properties, 63–64, 72–73
  minimax theorem and, 321–324
  payoff matrix, 248

n-dimensionality, linear programming problems, 119–121
n-player games, 398–409

characteristic function games, 405–406
cooperative games, 401–406
dominance and Nash equilibrium points, 398–401
n-tuples, 30–32
  corner point theorem, 121–124
  vector space, 34
Nash Arbitration Scheme, 364–365
Nash equilibria, 243, 258–261
  constant-sum games, 262–266
  cooperative games, 401–406
  derived game, 283
  expected value principle, 290–294
  goals of variable-sum games, 344–346
  interchangeability, 263–266
  mixed strategy constant-sum games, 269–278
  mixed strategy variable sum games, 337–342
  n-player games, 398–401
  ordinal game, 373–374
  Pareto efficiency, 340–342, 346–347
  Prisoner's dilemma, 374–384
  sensitivity analysis, 352
  two-person constant-sum games, 291–294
  variable-sum games, 261–262, 335–336, 355–356
Nash, John, 243
negative entries
  Phase II simplex algorithm, tableaux maximization,
    196–197
  simplex algorithm, 193
negotiation set, 363–364
negotiation, cooperative games, 362–365
no-solution linear programming problems, 125
non-basic variables, Phase II simplex algorithm, tableaux
    maximization, 197–199
non-binding constraints, 156–159
non-cooperative play
  ordinal games, 378
  variable-sum games, 362
non-degenerate system, 12–14
  examples, 17–20
  inverse matrix, 84–88
non-myopic equilibrium, Theory of Moves, 390–393
non-negative linear combinations
  phase I pivoting, 428–433
  Span of vectors, 39
  stable ranges, 461–466
non-negativity constraints
  constraint space graphing, 137–139
  Excel software solutions for, 233–239
  linear equation, 106–107
  marginal value of constraint, 160–163
  simplex algorithm, 207
non-positive solutions, marginal value of constraint,
    160–163
nonbasic variables, constraint space graphing,
    137–139
nontrivial linear combinations, homogeneous system,
    43–44
normal form game, 249–251

Normalized Pivot Row (NPR)
  example of, 200–204
  final tableau construction, 198–199
  $m \times n$ game, minimax theorem, 322–324
  Phase II simplex algorithm, tableaux maximization, 194–207
normalized pivot row, simplex algorithm, 192–194
Northwest Corner Method, 440
null set, 125, 482–483
Numerical Methods/Numerical Analysis, 202–204

objective coefficients
  capacity coefficients *vs.*, 168
  changes in, 150–154, 452–460
  marginal value constraints, 161–163, 170–171
  sensitivity analysis and, 148–150
  stable ranges, 461–466
objective function, 102
  corner point theorem, 119–121
  initial tableau set-up, 194
  linear equations, 106–107
  marginal value of constraints and, 156–159
  pivot placement and, 190–194
  sensitivity analysis, 148–150, 444–450
  simplex algorithm, 189–194, 202–204
  stable range of capacity numbers, 166–171
  standard form minimization problem, 212–219
open Leontief Model, 94–95
optimal points
  constraint space graphing, 142–144
  cooperative games, 363–364
  marginal value of constraints and, 156–159
  objective function coefficients, 150–154
  Phase II simplex algorithm, tableaux maximization, 194–207
  simplex algorithm, 186–187
optimization, 55–57. *See also* constrained optimization; linear programming
ordinal games, 352
  applications, 378–384
  dominance and Nash equilibrium points, 373–374
  Prisoner's and other dilemmas, 374–378
ordinal power, 419
outgoing row, simplex algorithm, 190–194

parabola
  local and global points, 53, 55
  optimization problems, 57
  quadratic curves, 50–54
  vertical and horizontal translations, 51–53
  vertical stretches and compressions, 50–51
parallel vectors, 33–34
parallelogram
  quadrilaterals and, 40–44
  span of vectors and, 35–38
Parallelogram Law, 34, 133–135
  constraint space graphing, 139–140
  nonnegative linear combinations, 140–142

parametric form, elementary row operations, 22–25
Pareto efficiency
  cooperative games, 363–364
  goals of variable sum game and, 343–347
  n-player games, 399–401
  Nash equilibrium, 346–347
  ordinal game, 373–374
  Prisoner's dilemma, 374–378
  sensitivity analysis, 352
  variable-sum games, 339–342, 355–356
Pareto, Vilfredo, 339–342
partial conflict, games of, 248
partial cooperation, variable-sum games, 362–365
partial order, political power, 419
payoff matrices
  Battle of the Bismarck Sea example, 252–261
  characteristic function games, 405–406
  derived game, 281–287
  dominant strategy and, 256–257
  Mathematica constant sum game solutions, 326–329
  mixed strategy constant-sum games, 269–278
  mixed strategy zero-sum games, 276–277
  multi-move games, 356–361
  n-player games, 398–401
  Nash equilibria, 258–261
  normal form games, 249–251
  partial conflict games, 248
  Prisoner's dilemma, 374–378
  saddle point of, 262–266
  strictly determined games, 263–266
  Theory of Moves, 389
  variable-sum games, 261–262
  zero-sum game, 276–277
payoff polygon
  cooperative games, 364–365
  variable-sum games, 339–342, 344–346
permutations, enumeration and, 492–493
Phase I pivoting
  linear programming, 423–433
  mixed constraints, 428–433
  sensitivity analysis and, 441–450
  simplex algorithm, 430–433
  standard form minimization problem, 435–439
Phase II pivoting
  duality construction, 214–219
  maximization problem tableaux, 194–207
  simplex algorithm, 194–207
  standard form minimization problem, 435–439
  termination in, 202–204
pivot number, simplex algorithm, 190–194
pivot operations, 19–20
  Bland's Rule, 204–207
  capacity changes, 441–450
  degenerate problems, 202–204
  phase I pivoting, 423–433
  Phase II simplex algorithm tableaux, 200–204
  reduced echelon form, 22–25

pivot operations (cont)
    simplex algorithm, 185–186
    standard form minimization problem, 212–219
pivot position, tableau set-up, 19–20, 190–194
pivot row
    Phase II simplex algorithm, tableaux maximization, 194–207
    simplex algorithm, 190–194
pivotal voter, 419
planar diagram, n-player games, 399–401
politics, game theory and, 362–365, 413–419
polytops, convex sets, 121–124
positive entries, simplex algorithm, 193
Power Set Principle, 490–491
Prisoner's dilemma, 374–378
    applications, 378–384
    theory of moves and, 396–397
probability matrix, 69
probability theory, 272–275
    Bayes Theorem, 512–517
    expected value, 274–276
    mixed strategy constant-sum games, 270–278
    mixed strategy zero-sum game, 276–277
    models, 498–501
    sample space, 272–275, 497–498
    uniform sample space, 502–504
promises, variable-sum games, 362, 368–371
prudential strategy
    goals of variable sum game and, 343–347
    variable-sum games, 341–342
pure strategies, constant-sum games, 270–278
    expected value principle, 287–294
    reduced game matrix, 303–309

quadratic curves
    completing the square, 53, 55
    constrained optimization, 56
    general form, 53, 55
    local and global points, 53, 55
    standard form, 50–54
    vertical and horizontal translations, 51–53
    vertical stretches and compressions, 50–51
quadric surfaces, 57–61
quadrilaterals, parallelograms and, 40–44

Rapoport, Anatol, 379–384
rational numbers, simplex algorithm, 202–204
rationality assumption
    dominant strategy and, 256–257
    game theory and, 245
    games against nature and, 365–367
    higher order dominant strategies, 258–261
    n-player games, 401–406
real numbers, defined, 30
reduced game matrix, 303–309
    $2 \times n$ and $m \times 2$ case, 303–309
reduced row-echelon form
    Gauss Jordan elimination algorithm, 25–28
    identity matrix, 20–28

redundant systems, 84–88
    vector equations, 47
relative zero-sum game, 344–346
resources, marginal values of, 156–160
reverse induction
    sensitivity analysis, 359–361
    Theory of Moves, 391–393
revised optimal data
    capacity changes and, 168, 444–450
    marginal value of constraint, 169–171
roster set description, 481–482
roundoff error, simplex algorithm, 202–204
row matrix, 63–64
row operations
    expected value principle, 287–294
    mixed strategy constant-sum games, 270–278
    non-degenerate systems, 19–20

saddle point, payoff matrix
    constant-sum games, 262–266, 353–354
    derived game, 283–285
    $m \times n$ game, minimax theorem, 322–324
    non-simultaneity in zero-sum games, 353–354
    strictly determined games, 263–266
    variable-sum games, 341–342
saddle surface, 59–61
    mixed strategy constant-sum games, 269–278
sample space
    defined, 497–498
    independent events, 509–510
    probability models, 498–501
    probability theory, 272–275
scalar matrix, 63–64
scalar multiplication
    constraint space graphing, 133–135
    linear combination equations, 45–46
    linearly independent Span of vectors, 38–39
    matrix multiplication, 71
    matrix operations, 66–67
    vectors, 32–34
second degree functions, quadratic curves, 50–54
secrecy, game theory and, 270–278
security level
    coalition formation and, 404–405
    variable-sum games, 341–342
Selten, Reinhard, 243
semi-standard form minimization problem, 108–109
    phase I pivoting, 423–433
sensitivity analysis
    basic principles, 148–150
    capacity changes, 441–450
    communication, 362–365
    drawbacks of graphical methods, 166–172
    duality, 173–182
    game theory and, 352–367
    game trees and reverse induction, 356–361
    integer programming, 468–477
    marginal value of constraint, 156–163
    non-simultaneous play, 353–356

sensitivity analysis (cont)
    objective coefficients, 150–154
    of mathematical model, 3
    ordinal games, 374–384
    sequential games and theory of moves, 388–397
    simplex algorithm and, 441–450
sequential games
    constant-sum games, 353–354
    sensitivity analysis, 353–356
    theory of moves, 388–397
    variable-sum games, 355–356
set-builder notation, 481–482
sets
    algebra of, 483–487
    enumeration, 488–497
    review of, 480–487
shadow pricing, 179–181
Shapley value, game theory, 413–419
Shapley-Shubik Power Index, 413–419
simple events, sample space, 497–498
simple figures, convex sets, 121–124
simplex algorithm, 108–109, 202–204
    Bland's Rule, 204–207
    capacity changes, 441–450
    characteristics of, 185–186
    constant sum games, linear programming solutions,
        319–320
    constraint space graphing, 137
    degenerate problems, 202–204
    disadvantages of, 199–200
    duality construction, 214–219
    marginal value of constraint and, 199–200
    non-negativity constraints, 207
    phase I pivoting, 430–433
    phase II pivoting, 194–207
    rational numbers *vs.* decimals and, 202–204
    roundoff error, 202–204
    sensitivity analysis and, 441–450
    square subgame theorem, 324
    standard form maximization problem, 186–194
    standard form minimization problem, 212–219
    termination in, 202–204
simplex polytope, 121–124
simultaneity, removal of
    constant-sum games, 353–354
    sensitivity analysis, 353–356
    variable-sum games, 355–356
single-move games, non-simultaneity and sequential play
        in, 353–356
singleton sets, 482–483
    sample space, 497–498
skew-symmetric matrix, 310–312
slack variables, 186–187
    Bland's Rule, 204–207
    duality construction, 214–219
    initial tableau set-up, 194
    phase I pivoting, 424–433
    Phase II simplex algorithm, tableaux maximization,
        197–204

social choice voting, 409–413
solution
    in mathematical model, 2
    linear equations, 4–6
Solver Add-in (Excel), 229–239
    constant sum game solutions, 329–333
Span(S) of vectors, 35–38
    linear independence of, 38–39
square matrix, 19–20, 63–64
    determinant or adjoint of, 79–81
    input-output models, 95–96
    inverse of, 79
square subgame theorem
    Nash equilibrium, 351
    proof, 324
    3 × 5 payoff matrix, 308–309
    3 × 3 example, 310–312
stable range of capacity numbers
    linear programming, 461–466
    marginal value of constraint, 166–171
stable range of coefficients, 148–150
    objective function, 150–154
standard basis, linearly independent Span of vectors,
        38–39
standard form maximization problem
    condensed tableau, 175–177
    constant sum games, linear programming solutions,
        316–320
    constraint space, 132–133, 144
    corner point theorem, 119–121
    Excel software solutions for, 229–239
    marginal value of constraints and, 159–160
    mixed constraints, 160–163
    non-negativity constraints, 207
    phase I pivoting, 430–433
    Phase II simplex algorithm, 194–207
    simplex algorithm, 186–194, 200–204
standard form minimization problem, 108–109
    corner point theorem, 119–121
    duality construction, 174–177, 214–219
    duality economics, 179–181
    Excel software solutions for, 229–239
    linear programming, 435–439
    marginal value of constraints and, 159–160
    Mathematica constant sum game solutions, 326–329
    mixed constraints, 160–163
    non-negativity constraints, 207
    simplex algorithm solutions, 212–219
story problems, 3–8
strategic behavior, analysis of. *See* game theory
strategy choices
    constant-sum games, 353–354
    games against nature, 256–257, 365–367
    multi-move games, 356–361
strictly determined games, 263–266
    constant-sum games, 353–354
    n-player games, 399–401
    ordinal games, 378
strong duality principle, 177

structural constraints
  constraint space graphing, 133–135
  Excel software solutions for, 233–239
  graphical interpretation of, 170–171
  linear equations, 106–107
  sensitivity analysis, 148–150
  standard form minimization problem, 108–109
subsets, defined, 482–483
substitution solution, 4–6
subtraction, matrices, 66–67
superadditive systems
  characteristic function games, 405–406
  monotone voting systems, 412–413
surplus variables, standard form minimization problem,
  216–219
symmetric matrix, 65–66

tableaus
  Bland's Rule, 204–207
  branching problems, integer programming, 470
  objective coefficient changes, 455–460
  phase I pivoting, 425–433
  Phase II pivoting algorithm, 194–207
  simplex algorithm, 189–194
Taylor, Alan, 421
terminal nodes
  bounding problems, integer programming, 474–477
  non-myopic equilibrium, 390–393
  Theory of Moves, 389
test point, linear equation, 105–106
Theory of Moves, 388–397
  applications, 394–397
  non-myopic equilibrium, 390–393
  tree diagram, 389
threats, variable-sum games, 362, 368–371
three-dimensional graphic solution, 11–12
three-dimensional space, span of vectors and, 35–38
3 × 2 game, equalizing expectations, 305–308
threshold value, theory of moves and, 396–397
time, functions of, Theory of Moves and, 396–397
Tit for Tat algorithm, 379–384
trade robustness, 421
transpose of matrix, 65–67
  condensed tableau, 175–177
transposition, matrix multiplication, 71
tree diagrams, 249–251
  Bayes Theorem, 512–517
  conditional probability, 507–509
  cooperative games and, 379–384
  non-myopic equilibrium, 390–393
  Theory of Moves, 389
2 × 2 constant-sum games, 281–294
  dominated strategies, 301–312
  expected value principle, 290–294
  reduced game matrix, 303–309
2 × n problems
  constraint space, 132–144
  zero-sum example, 303–309

two-dimensional graphic solution, 10–11
two-person constant-sum games
  mixed strategies, 269–278
  Nash equlibrium, 291–294
  payoff matrix, 277

unary operations, set algebra, 483–487
unbounded feasible set, linear programming problems,
  125
uniform sample space, 498, 502–504
  conditional probability and independent events,
  505–510
unit column, non-degenerate matrices, 19–20
utility theory, 244

value function
  characteristic function games, 405–406
  cooperative games with side payments, 404–405
  derived game, 283–285
  equivalent voting systems, 410–412
  game theory, 261–262
variable-sum games, 247–249, 335–347
  communication, 362–365
  dominance and Nash equilibrium points, 335–336
  goals of play, 343–347
  mixed strategies, 337–342
  Nash equilibrium and dominant strategy in,
  261–262
  non-cooperative play, 362
  non-rational assumptions and, 365–367
  non-simultaneity in, 355–356
  Pareto efficiency, 339–342
  partial cooperation, 362–365
  payoff polygon, 339–342
  Prisoner's dilemma, 374–378
  sensitivity analysis, 352
vectors
  addition of, 33–34
  constraint space graphing, 133–135, 137
  initial and terminal points, 30–34
  linear combinations as, 34, 45–46
  Mathematica software applications, 220–225
  non-negative linear combinations, 39
  properties of, 30–34
  scalar multiplication of, 32–34
Venn diagrams
  Addition Principle, 489–490
  Bayes Theorem, 512–517
  conditional probability and independent events,
  505–510
  defined, 482–483
  set algebra, 483–487
verbal set description, 481
Very Naive Power Index, 416–419
veto power
  equivalent voting systems, 410–412
  winning coalitions, 418
Vogel's Approximation Method, 440

Von Neumann, John, 281
voting systems
    legislative voting, 409–413
    political power and, 413–419
    Power Set Principle, 490–491
    trade robustness, 421

Wald approach, cooperative games,
    366–367
weak duality principle, 177
weighted voting system
    legislative voting, 410–412
    political power, 413–419
well-defined sets, 480–487

winning coalitions, 409–419
word problems, 3–8

yes-no voting, 409–413

zero-sum games, 247–251
    Battle of the Bismarck Sea example, 252–261
    Minimax Theorem, 281–294
    n-player games, 399–401
    non-simultaneity in, 353–354
    payoff matrix, 276–277, 322–324
    square subgame theorem, 310–312
    variable-sum games, 343–347
Zwicker, William, 421

Printed in the United States
by Baker & Taylor Publisher Services